Smart Innovation, Systems and Technologies

Volume 26

For further volumes:
http://www.springer.com/series/8767

About this Series

The Smart Innovation, Systems and Technologies book series encompasses the topics of knowledge, intelligence, innovation and sustainability. The aim of the series is to make available a platform for the publication of books on all aspects of single and multi-disciplinary research on these themes in order to make the latest results available in a readily-accessible form. Volumes on interdisciplinary research combining two or more of these areas is particularly sought.

The series covers systems and paradigms that employ knowledge and intelligence in a broad sense. Its scope is systems having embedded knowledge and intelligence, which may be applied to the solution of world problems in industry, the environment and the community. It also focusses on the knowledge-transfer methodologies and innovation strategies employed to make this happen effectively. The combination of intelligent systems tools and a broad range of applications introduces a need for a synergy of disciplines from science, technology, business and the humanities. The series will include conference proceedings, edited collections, monographs, handbooks, reference books, and other relevant types of book in areas of science and technology where smart systems and technologies can offer innovative solutions.

High quality content is an essential feature for all book proposals accepted for the series. It is expected that editors of all accepted volumes will ensure that contributions are subjected to an appropriate level of reviewing process and adhere to KES quality principles.

Simone Bassis · Anna Esposito
Francesco Carlo Morabito
Editors

Recent Advances of Neural Network Models and Applications

Proceedings of the 23rd Workshop of the Italian Neural Networks Society (SIREN), May 23–25, Vietri sul Mare, Salerno, Italy

Editors
Simone Bassis
Department of Computer Science
University of Milano
Milano
Italy

Anna Esposito
Department of Psychology
Second University of Naples Caserta,
and
Institute for Advanced Scientific Studies
(IIASS)
Vietri sul Mare (Salerno)
Italy

Francesco Carlo Morabito
DICEAM
Department of Civil, Energy,
Environmental, and Materials
Engineering
University Mediterranea of Reggio Calabria
Reggio Calabria
Italy

ISSN 2190-3018 ISSN 2190-3026 (electronic)
ISBN 978-3-319-37520-5 ISBN 978-3-319-04129-2 (eBook)
DOI 10.1007/978-3-319-04129-2
Springer Cham Heidelberg New York Dordrecht London

Softcover reprint of the hardcover 1st edition 2014

Printed on acid-free paper

Springer is part of Springer Science+Business Media (www.springer.com)

Preface

This volume collects a selection of contributions which have been presented at the 23rd Italian Workshop on Neural Networks (WIRN 2013), the yearly meeting of the Italian Society for Neural Networks (SIREN). The conference was held in Italy, Vietri sul Mare (Salerno), during May 23–24, 2013. The annual meeting of SIREN is sponsored by International Neural Network Society (INNS), European Neural Network Society (ENNS) and IEEE Computational Intelligence Society (CIS). The workshop, and thus this book, is organized in two main components, a special session and a group of regular sessions featuring different aspects and point of views of artificial neural networks, artificial and natural intelligence, as well as psychological and cognitive theories for modeling human behaviours and human machine interactions, including Information Communication applications of compelling interest.

More than 50 papers were presented at the Workshop, and most of them are reported here. The review process has been carried out in two steps, one before and one after the workshop in order to meet the Publisher requirements. The selection of the papers was made through a peer-review process, where each submission was evaluated by at least two reviewers. The submitted papers were authored by peer scholars from different countries (the Italian component was anyway preponderant). The acceptance rate was high because authors got the chance to review in two steps their work and also because they are experts in the field, being most of them involved in the organization of SIREN research activities for more than 20 years. In addition to regular papers, the technical program featured keynote plenary lectures by worldwide renowned scientists (Sankar Kumar Pal, India; Sara Rosenblum, Israel; Laurence Likforman, France; Virginio Cantoni, Italy).

The special session on Social and Emotional Networks for Interactional Exchanges was organized by Gennaro Cordasco, Anna Esposito and Maria Teresa Riviello (Department of Psychology, Second University of Naples, Italy). The Session explored new ideas and methods for developing automatic systems capable to detect and support users psychological wellbeing gathering information and meanings from the behavioral analysis of individual interactions both at the micro (dyadic and small groups) and macro level (information and opinion transfer over a large population). Of particular interest was the analysis of sequences of group actions explicated through a series of visual, written and audio signals and the corresponding computational efforts to detect and interpret their semantic and pragmatic contents. Social networking and connectedness as the ability to spread around thinking and related effects on social network behaviors were also considered, as well as ICT applications detecting the health status and affective states of their users. The special session's invited lectures were given in honour of Professors Maria Marinaro and Luigi Maria Ricciardi which directed the activities of the hosting Institute, the International Institute for Advanced Scientific Studies (IIASS), for more than a decade, sustaining the Workshop and sponsoring SIREN's activities.

The organization of an International Conference gathers the efforts of several people. We would like to express our gratitude to everyone that has cooperate to its organization, by offering their commitment, energy and spare time to make this event a successful one. Finally, we are grateful to the contributors of this volume for their cooperation, interest, enthusiasm and lively interactions, making it not only a scientifically stimulating gathering but also a memorable personal experience.

May 2013

Simone Bassis
Anna Esposito
Francesco Carlo Morabito

Organization

WIRN 2013 is organized by the Italian Society of Neural Networks (SIREN) in co-operation with the International Institute for Advanced Scientific Studies (IIASS) of Vietri S/M (Italy).

Executive Committee

Bruno Apolloni	University of Milano, Italy
Simone Bassis	University of Milano, Italy
Anna Esposito	Second University of Naples and IIASS, Italy
Francesco Masulli	University of Genova, Italy
Francesco Carlo Morabito	University Mediterranea of Reggio Calabria, Italy
Francesco Palmieri	Second University of Napoli, Italy
Eros Pasero	Polytechnic of Torino, Italy
Stefano Squartini	Polytechnic University of Marche, Italy
Roberto Tagliaferri	University of Salerno, Italy
Aurelio Uncini	University "La Sapienza" of Roma, Italy
Salvatore Vitabile	University of Palermo, Italy

Program Committee

Conference Chair:	Francesco Carlo Morabito (University Mediterranea of Reggio Calabria, Italy)
Conference Co-Chair:	Simone Bassis (University of Milan, Italy)
Program Chair:	Bruno Apolloni (University of Milan, Italy)
Organizing Chair:	Anna Esposito (Second University of Napoli and IIASS, Italy)
Special Tracks:	Gennaro Cordasco, Anna Esposito and Maria Teresa Riviello (Second University of Napoli and IIASS, Italy)

Referees

V. Arnaboldi
S. Bassis
G. Boccignone
A. Borghese
F. Camastra
P. Campadelli
R. Carbone
C. Ceruti
A. Ciaramella
M. Corazza
G. Cordasco
V. D'Amato
A. De Candia
R. De Rosa
F. Epifania
A. Esposito
A. Esposito
M. Fiasch
R. Folgieri
M. Frasca
C. Gallicchio
S. Giove
G. Grossi
F. La Foresta
D. Malchiodi
U. Maniscalco
F. Masulli
M. Mesiti
A. Micheli
P. Mottoros
C. Orovas
F. Palmieri
E. Pasero
F. Piazza
G. Piscopo
M. Re
M.T. Riviello
A. Roli
S. Rovetta
A. Rozza
M. Russolillo
G. Sarn
S. Scardapane
M. Scarpiniti
R. Serra
A. Sperduti
S. Squartini
A. Staiano
R. Tagliaferri
A. Uncini
G. Valentini
L. Valerio
S. Valtolina
M. Villani
S. Vitabile
J. Vitale
A. Zanaboni
A. Zippo
C. Zizzo
I. Zoppis

Sponsoring Institutions

International Institute for Advanced Scientific Studies (IIASS) of Vietri S/M (Italy)
Department of Psychology, Second University of Napoli (Italy)
Provincia di Salerno (Italy)
Comune di Vietri sul Mare, Salerno (Italy)

Contents

WIRN 2013

Part I: Algorithms

Part II: Signal Processing

Part III: Applications

Part IV: Special Session on "Social and Emotional Networks for Interactional Exchanges"

Part I
Algorithms

Identifying Emergent Dynamical Structures in Network Models

Marco Villani[1,2], Stefano Benedettini[1], Andrea Roli[1,3], David Lane[1,4], Irene Poli[1,5], and Roberto Serra[1,2]

[1] European Centre for Living Technology, Ca' Minich, S. Marco 2940, 30124 Venezia, Italy
[2] Dept. of Physics, Informatics and Mathematics, University of Modena e Reggio Emilia, v. Campi 213b, 41125 Modena, Italy
{marco.villani,roberto.serra}@unimore.it
[3] DISI Alma Mater Studiorum University of Bologna Campus of Cesena, via Venezia 52, I-47521 Cesena, Italy
andrea.roli@unibo.it
[4] Dept. of Communication and Economics, University of Modena e Reggio Emilia, v. Allegri 9, 41121 Reggio emilia, Italy
lane@unimore.it
[5] Department of Environmental Sciences, Informatics and Statistics, University Ca'Foscari, Venice, Italy
irene.poli@unive.it

Abstract. The identification of emergent structures in dynamical systems is a major challenge in complex systems science. In particular, the formation of intermediate-level dynamical structures is of particular interest for what concerns biological as well as artificial network models. In this work, we present a new technique aimed at identifying clusters of nodes in a network that behave in a coherent and coordinated way and that loosely interact with the remainder of the system. This method is based on an extension of a measure introduced for detecting clusters in biological neural networks. Even if our results are still preliminary, we have evidence for showing that our approach is able to identify these "emerging things" in some artificial network models and that it is way more powerful than usual measures based on statistical correlation. This method will make it possible to identify mesolevel dynamical structures in network models in general, from biological to social networks.

Keywords: Dynamical systems, emergent dynamical structures, cluster index, boolean networks, emergent properties.

1 Introduction

Emergent phenomena are among the most intriguing ones in natural as well as in artificial systems. Indeed, it can be argued [1] that neural networks represent an attempt at shaping the emergent properties of a set of models in order to perform some required tasks. An intriguing aspect is the "sandwiched" nature of most emergent

S. Bassis et al. (eds.), *Recent Advances of Neural Network Models and Applications*, Smart Innovation, Systems and Technologies 26,
DOI: 10.1007/978-3-319-04129-2_1, © Springer International Publishing Switzerland 2014

phenomena: while past researches were almost exclusively focused on bottom-up emergence in two-level systems (like e.g. Benard-Marangoni convection cells emerging from the interaction of the water molecules [2]) it is becoming increasingly clear that in the most interesting cases the new entities and levels do emerge between pre-existing ones. The paradigmatic example may be that of organs and tissues in multicellular organisms: both the lower (cellular) level and the upper one (organism) predate the appearance of the intermediate structures. Other examples come from the physical world (e.g. mesolevel structures in climate) and social systems (e.g. various factions within political parties). What is more interesting in the present case is that also in artificial systems, like neural networks, one observes the formation of intermediate-level circuits between the single neurons and the global properties. It goes without saying that some neural architectures have been devised precisely to stimulate the formation of these mesolevel structures, but here we are concerned with structures that come into being by spontaneous processes without being explicitly designed from the outside (although a certain type of design may ease or prevent the formation of these spontaneous structures).

A central question is then that of identifying the emerging "things": these may be either static entities or dynamical patterns, or some mixture of the two. In dynamical networks, static emergent structures take the form of topological features, like e.g. motifs in genetic networks or communities in a broader context. There is an extensive literature on community detection, so we will concentrate here on a different type of mesolevel structures, namely those that are created by the dynamical interactions in the network. Nodes may work together although they are not directly linked, since the dynamical laws may give rise to different parts working together. If the topology were regular, these nodes might be identified by visual inspection, but in the case of irregular topologies this approach seems hopeless.

In this paper we present a first step towards the development of formalized methods to identify these mesolevel "things": since they may have a topological as well as a dynamical nature, we refer to them as mesolevel dynamical structures (MDS). The task of identifying MDSs is a formidable one, so we will show here the outline of a promising approach and some preliminary results, while remarking that there are still more open questions than answers. However, the interest of the task motivates in our opinion the opportunity to report our preliminary results.

In order to escape "bird's eye" detection methods, we will consider different subsets of the network, looking for those whose nodes appear to be well coordinated among themselves and have a weaker interaction with the rest of the nodes. For each subset of nodes we will measure its so-called cluster index, a measure based on information theory that had been proposed by Tononi and Edelman [3]. After a suitable normalization procedure (see the following for the details) we rank the various subsets in order to identify those that are good candidates for the role of partially independent "organs" (note that they not necessarily exist in any network).

The cluster index has been defined so far for quasi-static systems, and we will discuss its extension to nonlinear dynamical systems. We will also show the result of the application of this ranking method to some model systems, including some synthetic dynamical networks and some genetic regulatory networks proposed by examining the biological literature. The method draws our attention on subsets that are functionally correlated and that represent an interesting hypothesis about possible MDSs. In the end we will also comment on the fact that our method, although not yet fully developed, already outperforms usual correlation techniques.

2 Some Useful Definitions

For the sake of definiteness, let us consider a system U, our "universe" that is a network of N nodes that can change in discrete time, taking one of a finite number l of discrete values (in the examples we will choose l=2 for simplicity). The value of node i at time t+1, $x_i(t+1)$, will depend upon the values of a fixed set of input nodes at time t, possibly including the i-th (self-loops are not prohibited). In several cases, networks start with a random state and change according to the evolution rules so the initial state may bear no relationship to the system itself. Since we are interested in finding out some properties of the networks themselves, we will consider their behaviors after an adequate relaxation time. For the time being we will also ignore external influences on some nodes, although these might be easily included.

The entropy of a single node is estimated from a long time series by taking frequencies f_v of observed values in time as proxies for probabilities, and is defined as

$$H_i = -\sum_{v=1}^{m} f_v \log f_v \tag{1}$$

where the sum is taken over all the possible values a node can take.

If the system is deterministic and is found in a fixed point attractor, H_i=0 for every node, since each node takes its value with frequency one. In order to apply entropy-based methods, Edelman and Tononi considered a system subject to gaussian noise around an equilibrium point. In our case it is however appropriate to deal with a richer time behavior since nonlinear networks can have several different attractors, each attractor contributing to the behavior of the system (though in different times). So our "long data series" will be composed by several repetitions of a single attractor, followed by repetitions of another one, etc. (ignoring the short transients between two attractors) . The number of times a single attractor is represented in the data series should be weighted in some way: there are possible several different strategies, depending on the nature of the system we are analyzing. In case of noisy systems a possibility is that of estimating the weights of the attractors by measuring the persistence time of the systems in each of them [4]; deterministic systems might be analyzed by weighting attractors with their basins of attraction. For simplicity in the following we opt for this second choice.

Now let us look for interesting sets of nodes (clusters, from now on). A good cluster should be composed by nodes (i) that possess high integration among themselves and (ii) that are more loosely coupled to other nodes of the system. The measure will define, called the cluster index, is not a Boolean one, but it provides a measure of "clusterness" that can be used to rank various candidate clusters (i.e., emergent intermediate-level sets of coordinated nodes).

3 Measuring the Cluster Index

Following Edelman and Tononi [3], we define the cluster index *C(S)* of a set of *k* nodes *S*, as the ratio of a measure of their integration *I(S)* to a measure of the mutual information *M(S\U-S)* of that cluster with the rest of the system.

The integration is defined as follows: let *H(S)* be the entropy (computed with time averages) of the elements of *S*. This means that each element is a vector of *k* nodes, and that the entropies are computed by counting the frequencies of the k-dimensional vectors. Then:

$$I(S)=\sum_{j\in S}H(x_j)-H(S) \tag{2}$$

The first term is the sum of the single-node entropies, the last one is computed using vectors of length *k*, so *I* measures the deviation from statistical independence of the *k* elements in *S*[1]. The mutual information of *S* to the rest of the world *U-S* is also defined by generalizing the usual notion of mutual information between nodes to *k* dimensional vectors

$$M(S;U-S)\equiv H(S)-H(S\mid U-S)=H(S)+H(U-S)-H(S,U-S) \tag{3}$$

where, as usual, *H(A|B)* is the conditional entropy and *H(A,B)* the joint entropy.

Finally, the cluster index *C(S)* is defined by

$$C(S)=\frac{I(S)}{M(S;U-S)} \tag{4}$$

The cluster index vanishes if *I*=0, *M*≠0, and is not defined whenever *M*=0. For this reason, the approach based upon cluster indices does not work properly when the mutual information of *S* with the rest of the system vanishes; these cases, in which *S* is statistically independent from the rest of the system – a significant property because they signal particularly strong structures - can be diagnosed in advance.

In this way, for every subsystem *S* we will get a measure of its quality as a cluster. In order to identify potential MDSs it is necessary to compare the indices of various candidate clusters. It is straightforward to compare clusters of the same size using *C(S)*, but unfortunately *C* scales with the size of the subsystem, so that a loosely connected subsystem may have a larger index than a more coherent, smaller one. In order to deal with these cases we need to normalize the clusters with respect to their size. The analysis may turn out quite cumbersome, but in most cases we found it sufficient to use a simple prescription, used by Tononi and Edelman in their original paper, which results in the calculation process outlined in the following.

The first step is to define a "null system", i.e., a non-clustered homogeneous system, from which we sample a series. This system provide us with a null hypothesis

---

[1] H(S) is estimated from the same time series used to calculate the frequencies f_v of eq. (1). So, to compute H(S) we calculate the frequencies $f^S{}_v$ of the observed values of S seen as a whole.

and allows us to calculate a set of normalization constants, one for each subsystem size. For each subsystem size, we compute average integration $<I_h>$ and mutual information $<M_h>$ (subscript h stands for "homogeneous"); we can then normalize the cluster index value of any subsystem *S* in it universe *U* using the appropriate normalization constants dependent on the size of *S*:

$$C'(S) = \frac{I(S)}{\langle I_h \rangle} \Big/ \frac{M(S;U-S)}{\langle M_h \rangle} \tag{5}$$

We apply this normalization to both the cluster indices in the analyzed system and in the null system.

The definition of "null system" is critical: it could be problem-specific, but we prefer a simple solution which is fairly general: given a series of Boolean vectors, we compute the frequency of ones *b* and generate a new random Boolean series where each bit has the same probability *b* of being one. This random null hypothesis is easy to calculate, related to the original data and parameter-free; moreover we believe it satisfies the requirements set by Tononi of homogeneity and cluster-freeness.

The second step involves the computation of a statistical significance index, called T_c, that is used to rank the clusters in the analyzed system. The T_c of a cluster *S* is:

$$T_c(S) = \frac{C'(S) - \langle C'_h \rangle}{\sigma(C'_h)} \tag{6}$$

where $<C'h>$ and $\sigma(C'h)$ are respectively the average and the standard deviation of the population of normalized cluster indices with the same size of S from the null system [5].

4 A Controlled Case Study

As a first step, we show the results of the application of our method on simple cases in which the systems analyzed have clusters by construction. These experiments make it possible to assess the effectiveness of the approach on controlled case studies, in which the expected outcome of the analysis is known a priori. Since our method aims at finding clusters of nodes which work together --- independently of their connections --- on the basis of sample trajectories of the system, we directly generated trajectories in which some groups of values behave coherently, i.e., they are the clusters to be detected. The trajectories are sequences of binary vectors of length *n*, $[x_1(t),x_2(t),...,x_n(t)]$. At each time step *t*, the values of the first *c* vector positions are generated according to the following procedure: $x_1(t)$, the leader, is a randomly chosen value in {0,1}; the values from position 2 to *c*, the followers, are a noisy copy of the leader, i.e., $x_i(t)=x_1(t)$ with probability *1-p* and $x_i(t)=\sim x_1(t)$ otherwise, being p the noise rate. Values $x_{c+1}(t),...,x_n(t)$ are randomly chosen in {0,1}. This way, the first block of the vector is composed of strongly correlated values and it should be clearly distinguished from the rest of the positions. Besides series with one cluster, with the same procedure we also generated trajectories with two independent clusters of size c_1 and c_2, respectively. In this case, the clusters can be found in positions $1,..,c_1$ and $c_{1+1},...,c_1+c_2$, where leaders are x_1 and x_{c1+1}. The trajectories were generated with p in $\{0, 0.01, 0.1\}$.

We applied our method based on the calculation of the normalized cluster index and we sorted the clusters as a function of the significance index T_c. In all the cases, the score based on T_c returns correctly the clusters in the first positions of the ranking. As an example, in Figure 1a we show the results of a representative case with two clusters with c_1=c_2=4 in a trajectory with 15 vector positions and p=0.01. The figure shows the ten T_c highest values and the corresponding cluster size. The bars are sorted in not increasing order of T_c. The highest peaks correspond indeed to the two clusters created in the trajectories. Each row of the matrix represents a cluster: white cells are the vector positions included in the cluster and they are ranked, from the top, by not increasing values of T_c. We can see that the first two clusters detected are indeed the ones corresponding to the positions 5,...,8 and 1,...,4 (small differences in T_c between the two clusters are due to noise).

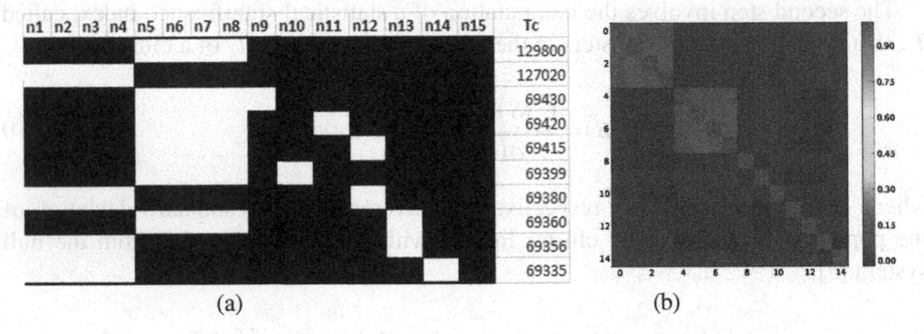

Fig. 1. (a) Matrix illustrating the elements of the clusters and the corresponding T_c values. The first two clusters are the ones introduced in the trajectory. (b) The heatmap shows the correlation values between pairs of vector positions in the trajectory.

The detected clusters are composed of correlated values, therefore we expect to find them also by simply calculating the correlation between every pair of positions. The correlation is computed by taking the value of the Pearson correlation coefficient for each pair of variables; in the case of binary variables, this quantity is usually called the phi coefficient. Since we are interested indifferently in both positive and negative correlations, we take the absolute value of the phi coefficients. Results can be plotted as heatmaps, with correlation values associated to colors from blue (lowest) to red (highest). An example is given in Figure 1b. As we can observe, the blocks composing the clusters are clearly detected and this result holds for all the trajectories we analyzed. This result is not surprising, as the vector positions composing a cluster are indeed strongly correlated (the only variance is introduced by noise). One might then object that the correlation measure is sufficient to detect clusters. In fact, this argument is only valid in some simple cases and does not extend to the general case. The reason is that correlation is a pairwise measure, while the cluster index accounts for multiple relations. These first tests enable us to state that our method based on the cluster index can be effectively used to capture multiple correlations among variables. In the next section, we will show that this approach can be particularly powerful in detecting clusters of nodes in networks.

5 Cluster Indices Applied to Network Models

The case study we are going to examine consists of three synchronous deterministic Boolean Networks (BNs) – the BN being a very interesting case of complex systems [6] [7], also applied to relevant biological data [8] [9] [10] and processes [11] [12]. The aim of this case study is to check whether CI analysis is capable of recognizing special topological cases, such as causally (in)dependent subnetworks and oscillators, where the causal relationships are more than binary. Note that in all the following cases the phi analysis is ineffective (doesn't relate any variable, having values different from zero only on the diagonal of the matrix).

The first example is a BN made of two independent sub-networks (RBN1 - Figure 2a); in this case we expect the analysis to identify the two subsystems. The second example (RBN2 - Figure 2b) is a BN made of an oscillator (nodes 0 and 1) and one of the subnetworks form the previous example, node 2 has no feedback connections. In the last example we simply merge the networks form the previous examples (RBN3 system). Figures 3 show the top 5 most relevant clusters according to T_c. CI analysis is able to correctly identify the two subnetworks in the first example (first and second rows). The analysis clusters together 5 of 6 nodes of RBN2: those already clustered in RBN1, plus nodes 1 and 2 (which negates each other - figure 2b) and the node that compute the XOR of the signal coming from the two just mentioned groups. Indeed, all these nodes are needed in order to correctly reconstruct the RBN2 series.

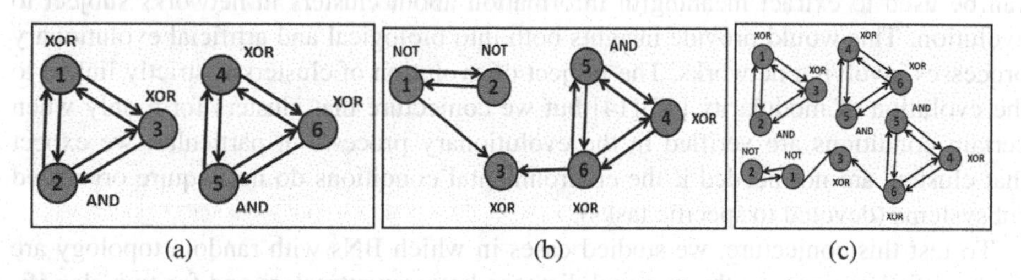

Fig. 2. (a) independent Boolean networks (RBN1); (b) interdependent networks (RBN2); (c) A system composed by both the previous networks (RBN3). Beside each boolean node there is the boolean function the node is realizing.

In the third example the top two clusters correspond respectively to the 5 nodes already recognized in RBN2 and to the whole RBN2 system, while the third and fourth rows correspond to the independent subgraphs of RBN1: all MDSs are therefore correctly identified.

We would like to point out that CI analysis does not require any knowledge about system topology or dynamics. This information is normally unavailable in real cases; on the other hand, our methodology just needs a data series.

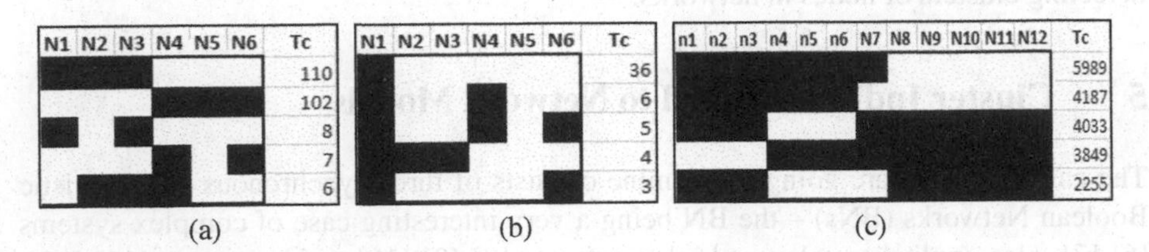

Fig. 3. Matrix illustrating the elements of the clusters and the corresponding T_c values, for (a) RBN1, (b) RBN2 and (c) RBN3 systems

As a final note, it is important to point out that covariance analysis is inadequate in this scenario as it is not able to identify any cluster. We took the same series we applied CI analysis upon and computed the correlation matrix between the node variables; the correlation indices between nodes are uniformly low in magnitude. The inadequacy of this method can be explained by the fact that correlation only takes into account binary linear interactions between variables as opposed to *CI*, which does not necessitate these hypotheses. Experiments performed using asynchronous update yielded essentially the same results with respect to both *CI* and correlation analyses.

6 Evolved Network: Some Examples

We have shown that our method makes it possible to discover clusters of coordinated nodes in a network. We may then raise the question as to what extent this technique can be used to extract meaningful information about clusters in networks subject to evolution. This would provide insights both into biological and artificial evolutionary processes involving networks. The subject of evolution of clusters is strictly linked to the evolution of modularity [13][14] but we conjecture that clusters form only when certain conditions are verified in the evolutionary process: in particular, we expect that clusters are not needed if the environmental conditions do not require organized subsystems (devoted to specific tasks).

To test this conjecture, we studied cases in which BNs with random topology are evolved for maximizing the minimal distance between attractors and for two classification tasks [15]. These tasks are static and not intrinsically modular; therefore, we expect not to find clusters in these evolved networks. The outcome of our analysis is that all these tasks can be optimally solved by BNs possessing only two complementary attractors. It can be easily shown that in homogeneous cases (systems without clusters) the cluster index scales linearly with the number of nodes of the cluster. Take a subsystem S and compute $I(S)$; all $H(X_i)$ are equal to 1 (I observes exactly two equally probable symbols on every node); moreover, $H(S)=H(X)=H(X \backslash S)=1$ because on any subsystem I again observes only two symbols with equal probability. To sum it up:

$$C(S) = \frac{I(S)}{M(S; X - S)} = \frac{N-1}{1} = N - 1 \qquad [7]$$

where N is the number of nodes in S.

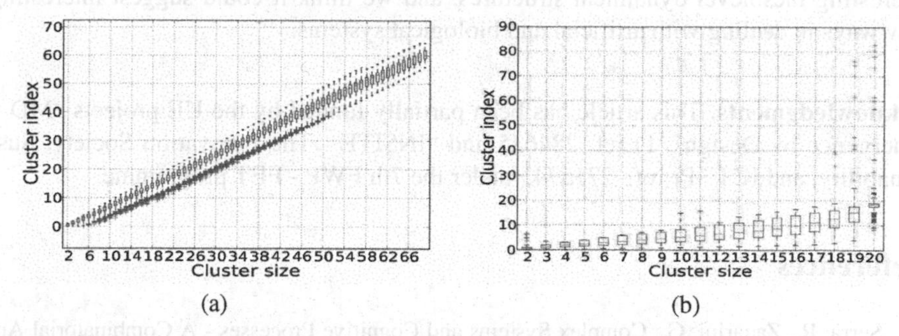

(a) (b)

Fig. 4. Distribution of maximum of CI for each cluster dimensions for evolved system, where (a) the task is the maximization of the minimal distance between attractors (systems with 70 nodes) and (b) the task is the Density Classification Problem (DCP), a simple counting problem [16] and a paradigmatic example of a problem hardly solvable for decentralized systems (the results regard networks with 21 nodes). Essentially, it requires that a binary dynamical system recognize whether an initial binary string contains more 0s or more 1s, by reaching a fixed attractor composed respectively by only 0s or 1s.

In figure 4 you can indeed observe this kind of behavior (note that only the averages have some meaning, because of no T_c has significant value – so, the few exceptions to the general behavior on the right side of figure 4b can be discarded. More details are available in [15]).

These are just preliminary experiments and we are currently studying cases in which the formation of clusters is indeed expected. Note however that there are data of evolved systems having well defined clusters: indeed, biological evolution is affecting living systems since 3.8 billion years.

In particular we are analyzing the gene regulatory network shaping the developmental process of Arabidopsis thaliana, a system composed by 15 genes and 10 different asymptotical behaviors [17]: our tool was able to group together the three genes core of the system (the first two clusters resulting from T_c ranking): in this case we are identifying clusters having T_c values very significant (see [18] for details).

7 Conclusions

A central question in distributed dynamical system is that of identifying the emerging "things": these may be either static entities or dynamical patterns, or some mixture of the two (neural networks representing an attempt at shaping the emergent properties of a set of models in order to perform some required tasks). In this paper we present a first step towards the development of formalized methods – a research initially started within studies on the brain activities [3] - to identify these mesolevel organizations

(MDSs in the work), which may have a topological as well as a dynamical nature. As examples of application we used time series of simple artificial systems and more complex data coming from Boolean Networks and biological gene regulatory systems (*A.thaliana*). So, the analysis performed by our system is able to identify several interesting mesolevel dynamical structures, and we think it could suggest interesting new ways in dealing with artificial and biological systems.

Acknowledgments. This article has been partially funded by the UE projects "MD – Emergence by Design", Pr.ref. 284625 and "INSITE - The Innovation Society, Sustainability, and ICT" Pr.ref. 271574, under the 7th FWP - FET programme.

References

1. Serra, R., Zanarini, G.: Complex Systems and Cognitive Processes - A Combinatorial Approach. Springer (1990)
2. Haken, H.: Synergetics. Springer, Heidelberg (2004)
3. Tononi, G., McIntosh, A.R., Russell, D.P., Edelman, G.M.: Functional Clustering: Identifying Strongly Interactive Brain Regions in Neuroimaging Data. Neuroimage 7 (1998)
4. Villani, M., Serra, R.: On the dynamical properties of a model of cell differentiation. EURASIP Journal on Bioinformatics and Systems Biology 2013, 4 (2013)
5. Benedettini, S.: Identifying mesolevel dynamical structures ECLT (European Center for Living Technologies) technical report, Venice (2013)
6. Kauffman, S.A.: The Origins of Order. Oxford University Press, Oxford (1993)
7. Kauffman, S.A.: At Home in the Universe. Oxford University Press, Oxford (1995)
8. Serra, R., Villani, M., Semeria, A.: Genetic network models and statistical properties of gene expression data in knock-out experiments. Journal of Theoretical Biology 227, 149–157 (2004)
9. Shmulevich, I., Kauffman, S.A., Aldana, M.: Eukaryotic cells are dynamically ordered or critical but not chaotic. Proc. Natl. Acad. Sci. 102, 13439–13444 (2005)
10. Villani, M., Serra, R., Graudenzi, A., Kauffman, S.A.: Why a simple model of genetic regulatory networks describes the distribution of avalanches in gene expression data. J. Theor. Biol. 249, 449–460 (2007)
11. Serra, R., Villani, M., Barbieri, B., Kauffman, S.A., Colacci, A.: On the dynamics of random boolean networks subject to noise: attractors, ergodic sets and cell types. Journal of Theoretical Biology 265, 185–193 (2010)
12. Villani, M., Barbieri, A., Serra, R.A.: Dynamical Model of Genetic Networks for Cell Differentiation. PLoS ONE 6(3), e17703 (2011), doi:10.1371/journal.pone.0017703
13. Espinosa-Soto, C., Wagner, A.: Specialization Can Drive the Evolution of Modularity. PLoS Comput. Biol. 6(3) (2010)
14. Clune, J., Mouret, J.-B., Lipson, H.: The evolutionary origins of modularity. Proceedings of the Royal Society B 280, 20122863 (2013)
15. Benedettini, S., Villani, M., Roli, A., Serra, R., Manfroni, M., Gagliardi, A., Pinciroli, C., Birattari, M.: Dynamical regimes and learning properties of evolved Boolean networks. Neurocomputing 99, 111–123 (2013)

16. Packard, N.: Adaptation toward the edge of chaos. In: Kelso, J., Mandell, A., Shlesinger, M. (eds.) Dynamic Patterns in Complex Systems. World Scientific, Singapore (1988)
17. Chaos, A., Aldana, M., Espinosa-Soto, C., Ponce de Leon, B.G., Garay Arroyo, A., Alvarez-Buylla, E.R.: From Genes to Flower Patterns and Evolution: Dynamic Models of Gene Regulatory Networks. J. Plant Growth Regul. 25, 278–289 (2006)
18. Villani, M., Filisetti, A., Benedettini, S., Roli, A., Lane, D., Serra, R.: The detection of intermediate-level emergent structures and patterns. In: Proceeding of ECAL 2013, the 12th European Conference on Artificial Life. MIT Press (2013) ISBN: 9780262317092

Experimental Guidelines for Semantic-Based Regularization

Claudio Saccà, Michelangelo Diligenti, and Marco Gori

Dipartimento di Ingegneria dell'Informazione,
Università di Siena, via Roma 54, Siena, Italy
claudiosacc@gmail.com, {diligmic,marco}@dii.unisi.it

Abstract. This paper presents a novel approach for learning with constraints called Semantic-Based Regularization. This paper shows how prior knowledge in form of First Order Logic (FOL) clauses, converted into a set of continuous constraints and integrated into a learning framework, allows to jointly learn from examples and semantic knowledge. A series of experiments on artificial learning tasks and application of text categorization in relational context will be presented to emphasize the benefits given by the introduction of logic rules into the learning process.

1 Introduction

Recent studies in machine learning enlightened the improvements given by the incorporation of a significant amount of prior knowledge into the learning process with a capable of bridging abstract descriptions of the environment with collections of supervised and unsupervised examples. In past few years remarkable approaches to provide a unified treatment of logic and learning were suggested by [3] in which the background knowledge on the problem at hand can be injected into the learning process mainly by encoding it into the kernel function. A related approach to combining first-order logic and probabilistic graphical models in a single representation are Markov Logic Networks [4]. In [2] and successively in [6],[7] it has been proposed a different approach to incorporate logic clauses, that are thought of as abstract and partial representations of the environment and are expected to dictate constraints on the development of an agent which also learns from examples. The approach is based on a framework that integrates kernel machines and logic to solve multi-task learning problems. The kernel machine mathematical apparatus allows casting the learning problem into a primal optimization of a function composed of the loss on the supervised examples, the regularization term, and a penalty term deriving from forcing the constraints converting the logic. This naturally allows to get advantage of unsupervised patterns in the learning task, as the degree of satisfaction of the constraints can be measured on unsupervised data. In particular, constraints are assumed to be expressed in First-Order Logic (FOL). The mathematical apparatus of Semantic Based Regularization (SBR) that converts the external knowledge into a set of real value constraints, which are enforced over the values assumed by

S. Bassis et al. (eds.), *Recent Advances of Neural Network Models and Applications*, Smart Innovation, Systems and Technologies 26,
DOI: 10.1007/978-3-319-04129-2_2, © Springer International Publishing Switzerland 2014

the learned classification functions, has been introduced in [2]. We introduced the new concept of binary predicates and given relations to exploit better the high expressivity of FOL rule. Furthermore, we developed a new software implementing SBR. Hence in this paper, we give some guidelines to the use of this software and present some representative benchmark experiments and a text-categorization task to show how we can take advantage of the integration of logic knowledge into the learning process to improve classification performace respect to a plain SVM classifier.

The paper is organized as follows: the next section introduces learning from constraints with kernel machines. The translation of any FOL knowledge into real-valued constraints is described in section 3, but can be examined in depth in [7] and some experimental results are reported in section 5 providing some guidelines in section 4 on how to execute the experiments through the related software for Semantic-Based Regularization (SBRS)[1].

2 Learning with Constraints

Let us consider a multitask learning problem as formulated in [7], where each task works on an input domain where labeled and unlabeled examples are sampled from. Each input pattern is described by a vector of features that are relevant to solve the tasks at hand. Let $\mathcal{D}_k$ and $f_k : \mathcal{D}_k \to I\!R$ be the input domain and the function implementing task k, respectively. We indicate as $\mathbf{x}_k \in \mathcal{D}_k$ a generic input vector for the k-th task. Task k is implemented by a function f_k, which may be known may be known a priori (*GIVEN* task) or it must be inferred (*LEARN* task). In this latter case it is assumed that each task function lives in an appropriate Reproducing Kernel Hilbert Space $\mathcal{H}_k$. Let us indicate with $\mathcal{T}$ the total number of tasks, of which the first T are assumed to be the learn tasks. For remaining evidence tasks, it will hold that all sampled data is supervised as the output of the task is known in all data points: $\mathcal{S}_k = \mathcal{L}_k, \mathcal{U}_k = \emptyset$ (*close-world assumption*).

The learning procedure can be cast as an optimization problem that aims at computing the optimal *LEARN* functions $f_k \in \mathcal{H}_k$, $k = 1, \ldots, T$ where $f_k : \mathcal{D}_k \to I\!R$. The optimization problem consists of three terms: a data fitting term, penalizing solutions that do not fit the example data, a regularization term, penalizing solutions that are too complex and a constraint term, penalizing solutions that do not respect the constraints:

$$E[f_1, \ldots, f_{\mathcal{T}}] = \sum_{k=1}^{T} \lambda_k^{\tau} \cdot \frac{1}{|\mathcal{L}_k|} \sum_{(\mathbf{x}_k, \mathbf{y}_k) \in \mathcal{L}_k} L_k^e(f_k(\mathbf{x}_k), \mathbf{y}_k) + \sum_{k=1}^{T} \lambda_k^r \cdot ||f_k||^2_{\mathcal{H}_k} + \sum_{h=1}^{H} \lambda_h^v \cdot L_h^c(\phi_h(f_1, \ldots, f_{\mathcal{T}})) \,. \tag{1}$$

[1] `https://sites.google.com/site/semanticbasedregularization/home/software`

where L_k^e and L_h^c are a loss function that measures the fitting quality respect to the target $\mathbf{y}_k$ for the data fitting term and the constraint degree of satisfaction, respectively. Clearly, if the tasks are uncorrelated, the optimization of the objective function is equivalent to T stand-alone optimization problems for each function.

The optimization of the overall error function is performed in the primal space using gradient descent [1]. The objective function is non-convex due to the constraint term. Hence, in order to face the problems connected with the presence of sub-optimal solutions, the optimization problem was split in two stages. In a first phase, as commonly done by kernel machines it is performed regularized fitting of the supervised examples. Only in a second phase, the constraints are enforced since requiring a higher abstraction level [2],[7]. The constraints can also be gradually introduced. As common practice in constraint satisfaction tasks, more restrictive constraints should be enforced earlier.

3 Translation of First-Order Logic (FOL) Clauses into Real-Valued Constraints

We have focused the attention on knowledge-based descriptions given by first-order logic. Let's consider as example that our knowledge-base (KB) is composed by generic FOL clauses in the following format $\forall v_i \; E(v_i, \mathcal{P})$, where $v_i \in \mathcal{D}_i$ is a variable belonging to the set of the variables $\mathcal{V} = \{v_1, \ldots, v_N\}$ used in the KB, $\mathcal{P}$ is the set of predicates used in the KB, and $E(v_i, \mathcal{P})$ represents the generic propositional (quantifier-free) part of the FOL formula. Without loss of generality, we focused our attention to FOL clauses in the Prenex-Conjunction Normal form [2]. A FOL rule is translated in a continuous form where a predicate $P(x)$ is approximated via a function $f_P(x)$ implemented by a Kernel Machine. The conversion process of a clause into a constraint functional consists of the following three steps:

I. PREDICATE SUBSTITUTION: substitution of the predicates with their continuous implementation realized by the functions f composed with a squash function, mapping the output values into the interval $[0,1]$.
II. CONVERSION OF THE PROPOSITIONAL EXPRESSION: conversion of the quantifier-free expression $E(v_i, \mathcal{P})$ using T-norms, where all atoms are grounded as detailed in [5],[7]
III. QUANTIFIER CONVERSION: conversion of the universal and existential quantifiers as explained in [2],[5],[7].

4 Semantic-Based Regularization Simulator Guidelines

Semantic-Based Regularization Simulator (SBRS) is the software implementing SBR that have used for the experiments presented in the next section. SBRS is able to use some of the most used kernels. Regarding to constraints evaluation, it is possible to use FOL clause with n-ary predicates and to learn or verify the satisfaction of logic constraints. The input is split in four separate ASCII files:

- data definition file: this file contains the patterns available in the dataset and each line represents a pattern. Each line is composed by fields which are delimited by the (;) symbol

```
patternID0;domainXY;X:0.2,Y:0.3
```

- predicate definition file: this file contains the definitions of predicates. As explained before, predicates are approximated with a function. They can be defined as *learn* if the correspondent function must be inferred from examples and constraints or *given* if it is known a priori. For *given* predicates can be specified a default value (T,F) for the correspondent function when the example is not provided.

```
DEF A(domainXY);LEARN
DEF R(domainXY,domainXY);GIVEN;F
```

- examples files: this files contains the desired output for specific groundings of the predicates (i.e. examples in machine learning context).

```
A(patternID0)=1
A(patternID1)=0
```

- FOL rules definition: this file containts the rules that have to be integrated in learning process. Rules are expressed in CNF and they are defined using a specific syntax. For each each rule, it must be specified the norms used to convert the propositional part (*product_tnorm*, *minimum_tnorm*) and quantifiers (*L1*,*LINF*) of the logic formula. Rules can be defined as *learn* if they have to be used in the learning process, or *verify* if we want to verify their satisfaction on a different sample of data.

```
forall p [(NOT A(p) OR NOT B(p) OR C(p))];LEARN;PRODUCT_TNORM;L1
forall p [(NOT C(p) OR A(p))];VERIFY;MINIMUM_TNORM;LINF
```

Due to a lack of space, more details on SBRS can be found in the manual[2]. A simple tutorial[3] with a few examples is also provided.

5 Experimental Results

For the fisrt part of this section we designed 2 artificial benchmarks to show how it is possible to define logic rules and the benefits of their integration into the learning process. All datasets assume a uniform density distribution and some prior knowledge is available on the classification task, that is expressed by logic clauses. A two-stage learning algorithm as described in [2,7] is exploited in the

[2] https://sites.google.com/site/semanticbasedregularization/SBRS_manual.pdf

[3] https://sites.google.com/site/semanticbasedregularization/SBRS_tutorial.pdf

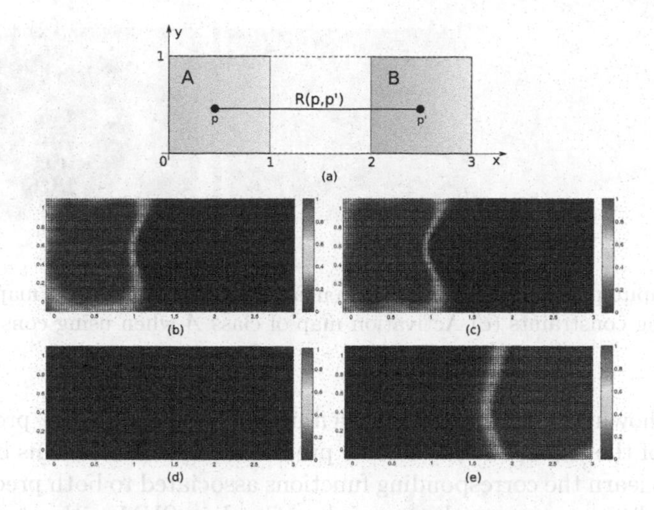

Fig. 1. (a) Input problem definition for benchmark 1 (b) Activation map of class A when not using constraints (c) Activation map of class A when using constraints (d) Activation map of class B when not using constraints (e) Activation map of class B when using constraints

experiments. All presented results are an average over multiple runs performed over different samples of the training and test sets. We considered learning tasks with examples drawn in $\mathbb{R}^2$ that allow us to plot the activation maps and assess the performance in a low dimensional space. Moreover, all the experiments have been designed so that we have small set of supervised examples and a great amount of unsupervised one since our approach it is expected to take advantage of unsupervised patterns to learn from the constraints. Therefore, we will present an application of text categorization on CiteSeer dataset where relational context permits to define different logic rules.

Benchmark 1: Universal and Existential Quantifiers. In this benchmark, it is supposed to check the effects of both universal and existential quantifier in a generic rule. In this benchmark it is also introduced the notion of *GIVEN* predicate and the difference between learn and given functions. The dataset is composed by patterns belonging to two classes: A and B. Dataset has been generated so that it is consistent with the following FOL rule:

$$\forall p\ A(p) \Rightarrow \exists p'\ B(p') \wedge R(p,p')$$

where $R(x,y)$ is *GIVEN* predicate. This means that, considering $p \equiv (x,y)$, its value is known for each groundings of its variables according to the following definition: $R(p,p') = 1$ if $\mid \|x' - x\| - 2 \mid \leq 0.01$, $\|y' - y\| \leq 0.01$, otherwise $R = 0$. Patterns are distributed uniformly over $\{(x,y) : x \in [0,3], y \in [0,1]\}$, but given a generic grounding for variable p, we have that $A(p) = 1$ iff $p \in \{p : 0 \leq x \leq 1\,, 0 \leq y \leq 1\}$, while $B(p) = 1$ iff $p \in \{p : 2 \leq x \leq 3\,, 2 \leq y \leq 3\}$. In figure

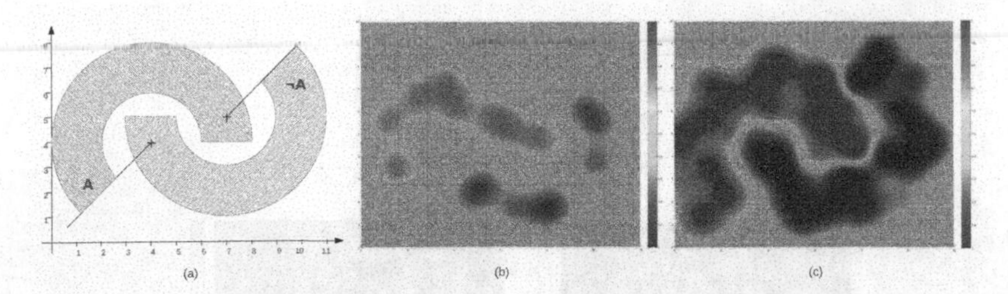

Fig. 2. (a) Input problem definition for benchmark 2 (b) Activation map of class A when not using constraints (c) Activation map of class A when using constraints

1 (a) it is shown the input problem definition. In addition, we provide some declaration of the examples only for the predicate $A(p)$. Also in this benchmark, the goal is to learn the corresponding functions associated to both predicate $A(p)$ and $B(p)$ and to compare with the results of a plain SVM without constraints, to show the benefits of integrating logic knowledge into the learning process. In particular, since we don't have examples for class B, we want to learn it only through the constraint. The parameters λ_l, λ_r and λ_c have been set to 1, 0.1 and 5, respectively. We exploited a Gaussian kernel with variance equal to 0.5. After a previous phase of cross-validation, we decided to use LINF/P-GAUSS norm to translate the FOL rule into a real-valued constraint.

As we expected, when no supervised examples for B are provided, in figure 1 (c) the activation area is all set to zero because without example it is not possible to infer a function. On the other hand, SBR can benefits of the logic knowledge to infer the activation map for class B with a good approximation.

Benchmark 2: Generalization of Manifold Regularization - The Two Moons. In this experiment, it will be shown how manifold regularization can be considered a special case of SBR. It will be assumed that for each couple of patterns in the dataset the information about their neighbourhood is provided through a *GIVEN* predicate $N(p,p')$ that is true if the patterns are near ($\|p - p'\| \leq 0.2$) to each other or false otherwise. The dataset is composed by patterns distributed uniformly over the two moons shown in figure 2 (a) and could belong or not to class A. The assumption that two points connected by an edge in the manifold must belong to the same class could be translated in logic by the following rule:

$$\forall p \; \forall p' \; N(p,p') \Rightarrow A(p) \Leftrightarrow A(p').$$

This logic formula, can be seen as the logic equivalent of manifold assumption, because predicate $N(p,p')$ holds true if and only if p, p' are connected on the manifold built using the relation of neighbourhood. Dataset has been generated so that, given a generic grounding for variable p, we have $A(p) = 1$ iff p belongs to the red moon, while $A(p) = 0$ iff p belongs to the blue moon. In addition, we provide very few supervised examples for the class A while a great amount of patterns remain unsupervised. The classification task consists to learn the function associated to predicate $A(p)$ when integrating the logic rule into the learning process. The parameters λ_l, λ_r and λ_c have been set to 1, 0.1 and 1, respectively. A Gaussian kernel, with variance equal to 0.8, has been exploited. Results in figures 2 (b)(c) show that, when adding the FOL rule, SBR can infer the activation map for class A with a better approximation respect to the case when no rules are used to train the classifier.

Using a test-set to evaluate the classification performance of the learned functions, we can se that our approach improves consistently the F1-score from 0.806 $\pm$ 0.009 when no constraints are used to 0.982 $\pm$ 0.014 when we force the satisfaction of the FOL rule, defined before, during the learning process.

CiteSeer: Text Categorization in Relational Context. The CiteSeer dataset[4] consists of 3312 scientific publications classified into at least one of six classes. The citation network consists of 4732 links. Each publication in the dataset is described by a 0/1-valued word vector indicating the absence/presence of the corresponding word from the dictionary. Five folds have been generated by selecting randomly 15% and 35% of the papers for validation and test set respectively. For the remaining 50% of the training set, n% (n=5,10,25,50) of the papers where selected randomly keeping the supervisions. The others remain unsupervised. The knowledge base collects different collateral information which is available on the dataset. CiteSeer makes available a list of citations for each papers. Our algorithm can exploit these relations assuming that a citation represents a common intent between the papers that are therefore suggested to belong to the same set of categories. This can be expressed via a set of 6 clauses (one per category) such that foreach $i = 1, \ldots, 6$:

$$\forall x \in \mathcal{P}\ \forall y \in \mathcal{P}\ \ Link(x, y) \Rightarrow (C_i(x) \wedge C_i(y)) \vee (\neg C_i(x) \wedge \neg C_i(y))$$

where $\mathcal{P}$ is the domain of all papers in the dataset and $Link(x, y)$ is a binary predicate which holds true iff paper x cites paper y. In the dataset we know that all papers belonging to ML class have been also tagged as AI. This information can be exploited through the following rule: $\forall\ x \in \mathcal{P}\ \ ML(x) \Rightarrow AI(x)$. Furthermore, the following rule defines a close-world assumption $\forall\ x \in \mathcal{P}\ C_1(x) \vee \ldots \vee C_6(x)$, where $C_1 \ldots C_6$ are the six categories of this problem. Finally using the supervised examples available in training, we add a prior to each class adding for each category this rule:

$$\exists_n x\ \ C_i(x) \wedge \exists_m\ x\ \ \neg C_i(x) \quad : \quad n + m = N$$

where n and consequently m are choosen basing on the number of supervised examples in training set for that class. For each subsample size of the training set, one classifier has been trained. As a comparison, we also trained for each set a standard SVM (using only the supervised labels), a Transductive SVM (implemented in the svmlight software package). The validation set has been used to select the best values for λ_r and λ_c. The F1-score has been compute as

[4] Available at: `http://linqs.cs.umd.edu/projects//projects/lbc/index.html`

Table 1. Micro F1 metrics averaged over 5 runs using SVM, Transductive SVM (TSVM) and Semantic Based Regularization (SBR)

	5%	10%	25%	50%
SVM	0.237 ±0.015	0.442 ±0.023	0.589 ±0.016	0.644 ±0.008
TSVM	0.604 ±0.023	0.623 ±0.021	0.631 ±0.02	0.655 ±0.007
SBR	0.637 ±0.019	0.656 ±0.02	0.661 ±0.022	0.679 ±0.013

an average over five fold. Table 1 summarizes the results for a different number of supervised data. SBR provides a statistically significant F1 gain with the respect to a standard SVM that do not exploit logic knowledge and it improves in average the classification performance of a trasductive SVM.

6 Conclusions

In this paper we give some insights on how to integrate prior-knowledge in form of logic clause into the general framework of regularization with kernel machines. This apparatus makes it possible to use a semi-supervised scheme in which the unsupervised examples, often abundant, play a crucial role in the approximation of the penalty term associated with the logic constraints. These preliminary experiments suggest the possibility to exploit a new class of semantic-based regularization machines in which the introduction of prior knowledge takes into account constraints on the tasks. The general principles at the base of this approach can be applied to several fields like bioinformatics for prediction of proteins interactions that we are going to explore.

Acknowledgements. This research was partially supported by the research grant PRIN2009 "Learning Techniques in Relational Domains and Their Applications" (2009LNP494) from the Italian MURST.

References

1. Chapelle, O.: Training a support vector machine in the primal. Neural Computation 19(5), 1155–1178 (2007)
2. Diligenti, M., Gori, M., Maggini, M., Rigutini, L.: Bridging logic and kernel machines. Machine Learning, 1–32 (2011)
3. Frasconi, P., Passerini, A.: Learning with kernels and logical representations. In: De Raedt, L., Frasconi, P., Kersting, K., Muggleton, S.H. (eds.) Probabilistic ILP 2007. LNCS (LNAI), vol. 4911, pp. 56–91. Springer, Heidelberg (2008)
4. Richardson, M., Domingos, P.: Markov logic networks. Machine Learning 62(1-2), 107–136 (2006)

5. Saccà, C., Diligenti, M., Maggini, M., Gori, M.: Integrating logic knowledge into graph regularization: an application to image tagging. In: Ninth Workshop on Mining and Learning with Graphs - MLG (KDD) (2011)
6. Saccà, C., Diligenti, M., Maggini, M., Gori, M.: Learning to tag from logic constraints in hyperlinked environments. In: ICMLA, pp. 251–256 (2011)
7. Saccà, C., Frandina, S., Diligenti, M., Gori, M.: Constrained-based learning for text categorization. In: Workshop on COmbining COnstraint solving with MIning and LEarning - CoCoMiLe (ECAI) (2012)

A Preliminary Study on Transductive Extreme Learning Machines

Simone Scardapane, Danilo Comminiello, Michele Scarpiniti, and Aurelio Uncini

Department of Information Engineering, Electronics and Telecommunications (DIET), "Sapienza" University of Rome, via Eudossiana 18, 00184, Rome
{simone.scardapane,danilo.comminiello, michele.scarpiniti}@uniroma1.it, aurel@ieee.org

Abstract. Transductive learning is the problem of designing learning machines that succesfully generalize only on a given set of input patterns. In this paper we begin the study towards the extension of Extreme Learning Machine (ELM) theory to the transductive setting, focusing on the binary classification case. To this end, we analyze previous work on Transductive Support Vector Machines (TSVM) learning, and introduce the Transductive ELM (TELM) model. Contrary to TSVM, we show that the optimization of TELM results in a purely combinatorial search over the unknown labels. Some preliminary results on an artifical dataset show substained improvements with respect to a standard ELM model.

Keywords: Transductive learning, extreme learning machine, semi-supervised learning.

1 Introduction

In the classical Machine Learning setting [1], starting from a limited set of data sampled from an unknown stochastic process, the goal is to infer a general predictive rule for the overall system. Vapnik [2] was the first to argue that in some situations, this target may be unnecessarily complex with respect to the actual requirements. In particular, if we are interested on predictions limited to a given set of input patterns, then a learning system tuned to this specific set should outperform a general predictive one. In Vapnik words, the advice is that, *"when solving a problem of interest, do not solve a more general problem as an intermediate step"* [2]. Vapnik also coined a term for this setting, which he called *Transductive Learning* (TL).

In [2] he studied extensively the theoretical properties of TL, and his insights led him to propose an extension to the standard Support Vector Machine (SVM) algorithm, namely the Tranductive SVM (TSVM). While SVM learning results in a quadratic optimization problem, TSVM learning is partly combinatorial, making it a difficult non-convex optimization procedure. However, a number of interesting algorithms have been proposed for its efficient solution. The interested reader can find a comprehensive review of them in Chapelle et al. [3].

By drawing theoretical and practical ideas from TSVMs, in this paper we extend *Extreme Learning Machine* (ELM) theory [4] to the transductive setting. ELM models

S. Bassis et al. (eds.), *Recent Advances of Neural Network Models and Applications*,
Smart Innovation, Systems and Technologies 26,
DOI: 10.1007/978-3-319-04129-2_3, © Springer International Publishing Switzerland 2014

have gained some attention as a conceptual unifying framework for several families of learning algorithms, and possess interesting properties of speed and efficiency. An ELM is a two-layer feed-forward network, where the input is initially projected to an highly dimensional feature space, on which a linear model is subsequently applied. Differently from other algorithms, the feature space is fully fixed before observing the data, thus learning is equivalent to finding the optimal output weights for our data. We show that, in the binary classification case, Transductive ELM (TELM) learning results in a purely combinatorial search over a set of binary variables, thus it can be solved more efficiently with respect to TSVM. In this preliminary work we use a simple Genetic Algorithm (GA) [5] as a global optimizer and test the resulting algorithm on an artificial dataset. Results show promising increase in performance for different sizes of the datasets.

Transductive learning has been throughly studied lately due to the interest in Semi-Supervised Learning (SSL) [6]. In SSL, additional unlabelled data is provided to the algorithm (as in TL), but the goal is to infer a general predictive rule as in classical inductive learning. In this respect, unlabelled data is seen as additional information that the algorithm can use to deduce general properties about the geometry of input patterns. Despite TL and SSL have different objectives, their inner workings are in some respects similar, and many TL and SSL algorithms can be used interchangeably in the two situations. In particular, TSVMs are known as Semi-Supervised SVM (S3VM) [3] in the SSL community. Hence, our work on TELM may be of interest as a first step towards the use of ELM models in a SSL setting.

The rest of this paper is organized as follows: in Section 2 we introduce some basic concepts on TL, and detail the TSVM optimization procedure. Section 3 summarizes the main theory of ELM. Section 4, the main contribution of this work, extends ELM theory using concepts from Section 2. Section 5 shows some preliminary results on an artificial dataset. Although we provide a working algorithm, two fundamental questions remain open, and we confront with them in Section 6. Finally, we make some final remarks in Section 7.

2 Transductive Learning

2.1 Inductive Learning and Support Vector Machines

Consider an unknown stochastic process described by the joint probability function $p(\mathbf{x},y) = p(\mathbf{x})p(y|\mathbf{x}), \mathbf{x} \in X, y \in Y$, where X and Y are known as the *input* and *output* spaces respectively. In this work we restrict ourselves to the binary classification case, i.e., $Y = \{-1,+1\}$. Given a loss function $L(\mathbf{x},y,\hat{y}) : X \times Y \times Y \to \mathbb{R}$ that measures the loss we incur by estimating $\hat{y} = f(\mathbf{x})$ instead of the true y, and a set of possible models H, the goal of inductive learning is to find a function that minimizes the *expected risk*:

$$I[f] = \int_{X \times Y} L(\mathbf{x},y,f(\mathbf{x}))p(\mathbf{x},y)d\mathbf{x}dy \tag{1}$$

We are given only a limited dataset of N samplings from the process $S = (\mathbf{x}_i, y_i)_{i=1}^N$, that we call the *training set*. The *empirical risk* is defined as:

$$I_{emp}[f;S] = \frac{1}{N}\sum_{i=1}^{N} L(\mathbf{x}_i, y_i, f(\mathbf{x}_i)) \tag{2}$$

Vapnik [2] derived several bounds, known as *VC bounds*, on the relation between (1) and (2) for limited datasets. All bounds are in the following form and are valid with probability $1-\eta$:

$$I[f] \leq I_{emp}[f;S] + \Phi(h, N, \eta) \tag{3}$$

where h is the *VC-dimension* of the set H, and $\Phi(h, N, \eta)$ is known as a *capacity* term. In general, such term is directly proportional to h. Thus, for two functions $f_1, f_2 \in H$ with the same error on the dataset, the one with lower VC-dimension is preferable. Practically, this observation can be implemented in the Support Vector Machine (SVM) algorithm, as we describe below.

Consider a generic *Reproducing Kernel Hilbert Space H* as set of models. There is a direct relationship [7] between h and the inverse of $\|f\|_H$, $f \in H$, where $\|f\|_H$ is the norm of f in H. Thus, the optimal function is the one minimizing the error on the dataset and of minimum norm. When using the *hinge loss* $L(\mathbf{x}_i, y_i, f(\mathbf{x})) = \max\{0, 1 - y_i f(\mathbf{x}_i)\}$ as loss function, we obtain the SVM for classification [8]. It can be shown that learning corresponds to a quadratic optimization problem:

$$\begin{aligned} \underset{f}{\text{minimize}} \quad & \frac{1}{2}\|f\|_H^2 + C_s \sum_{i=1}^{N} \zeta_i \\ \text{subject to} \quad & y_i f(\mathbf{x}_i) \geq 1 - \zeta_i,\ \zeta_i \geq 0,\ i = 1, \ldots, N. \end{aligned} \tag{4}$$

where ζ_i are a set of *slack variables* that measures the error between predicted and desired output and C_s is a regularization parameter set by the user. Solution to (4) is of the form $f(\mathbf{x}) = \sum_{i=1}^N a_i k(\mathbf{x}, \mathbf{x}_i)$, where $k(\cdot,\cdot)$ is the *reproducing kernel* associated to H.

2.2 Transductive Learning and Transductive SVM

In *Transductive learning* (TL) we are given an additional set[1] $U = (\mathbf{x}_i)_{i=N+1}^{N+M}$, called the *testing set*, and we aim at minimizing $I_{emp}[f;U]$. An extension of the theory described above [2] leads to minimizing the error on both S and U.

By denoting with $\mathbf{y}^* = [y_{N+1}^*, \ldots, y_{N+M}^*]^T$ a possible labelling of the elements in U, this results in the following (partly combinatorial) optimization problem, known as the *Transductive SVM* (TSVM):

$$\begin{aligned} \underset{f, \mathbf{y}^*}{\text{minimize}} \quad & \frac{1}{2}\|f\|_H^2 + C_s \sum_{i=1}^{N} \zeta_i + C_u \sum_{i=N+1}^{N+M} \zeta_i \\ \text{subject to} \quad & y_i f(\mathbf{x}_i) \geq 1 - \zeta_i,\ \zeta_i \geq 0,\ i = 1, \ldots, N. \\ & y_i^* f(\mathbf{x}_i) \geq 1 - \zeta_i,\ \zeta_i \geq 0,\ i = N+1, \ldots, N+M \end{aligned} \tag{5}$$

[1] Note the peculiar numbering on the dataset.

where we introduce an additional regularization term C_u. In particular, equation (5) is combinatorial over $\mathbf{y}^*$, since each label is constrained to be binary. This makes the overall problem highly non-convex and difficult to optimize in general. Some of the algorithms designed to efficiently solve it are presented in [3].

Typically, we also try to enforce an additional constraint on the proportion of labellings over $\mathbf{U}$, of the form:

$$\rho = \frac{1}{M}\sum_{i=1}^{M} y_i^*$$

where ρ is set *a-priori* by the user. This avoids unbalanced solutions in which all patterns are assigned to the same class.

3 Extreme Learning Machine

An Extreme Learning Machine (ELM) [9,4] is a linear combination of an L-dimensional feature mapping of the original input:

$$f(\mathbf{x}) = \sum_{i=1}^{L} h_i(\mathbf{x})\beta_i = \mathbf{h}(\mathbf{x})^T\beta \tag{6}$$

where $\mathbf{h}(\mathbf{x}) = [h_1(\mathbf{x}), \ldots, h_L(\mathbf{x})]^T$ is called the *ELM feature vector* and β is the vector of expansion coefficients. The feature mapping is considered fixed, so the problem is that of estimating the optimal β. Starting from a known function $g(\mathbf{x}, \theta)$, where θ is a vector of parameters, it is possible to obtain an ELM feature mapping by drawing parameters θ at random from an uniform probability distribution, and repeating the operation L times. Huang et al. [4] showed that almost any non-linear function can be used in this way, and the resulting network will continue to be an universal approximator. Moreover, they proposed the following regularized optimization problem, where we aim at finding the weight vector that minimizes the error on S and is of minimum norm:

$$\begin{aligned} \underset{\beta}{\text{minimize}} \quad & \frac{1}{2}\|\beta\|_2^2 + \frac{C_s}{2}\sum_{i=1}^{N}\zeta_i^2 \\ \text{subject to} \quad & \mathbf{h}^T(\mathbf{x}_i)\beta = y_i - \zeta_i,\ i = 1, \ldots, N. \end{aligned} \tag{7}$$

As for SVM, C_s is a regularization parameter that can be adjusted by the user, and $\zeta_i, i = 1, \ldots, N$ measure the error between desired and predicted output. The problem is similar to (4), but has a solution in closed form. In particular, a possible solution to (7) is given by [4]:

$$\beta = \mathbf{H}^T\left(\frac{1}{C_s}\mathbf{I}_{N\times N} + \mathbf{H}\mathbf{H}^T\right)^{-1}\mathbf{y} \tag{8}$$

where $\mathbf{I}_{N\times N}$ is the $N \times N$ identity matrix, and we defined the hidden matrix $\mathbf{H} = [\mathbf{h}(\mathbf{x}_1), \ldots, \mathbf{h}(\mathbf{x}_N)]$ and the output vector $\mathbf{y} = [y_1, \ldots, y_N]^T$. When using ELM for classification, a decision function can be easily computed as:

$$f'(\mathbf{x}) = sign(f(\mathbf{x}))$$

4 Transductive ELM

Remember that in the TL setting we are given an additional dataset $U = (\mathbf{x}_i)_{i=N+1}^{N+M}$ over which we desire to minimize the error. To this end, similarly to the case of TSVM, we consider the following modified optimization problem:

$$\begin{aligned} \underset{\beta, y^*}{\text{minimize}} \quad & \frac{1}{2}\|\beta\|_2^2 + \frac{C_s}{2}\sum_{i=1}^{N}\zeta_i^2 + \frac{C_u}{2}\sum_{i=N+1}^{N+M}\zeta_i^2 \\ \text{subject to} \quad & \mathbf{h}^T(x_i)\beta = y_i - \zeta_i,\ i = 1,\dots,N. \\ & \mathbf{h}^T(x_i)\beta = y_i^* - \zeta_i,\ i = N+1,\dots,N+M. \end{aligned} \tag{9}$$

We call (9) the *Transductive ELM* (TELM). At first sight, this may seems partly combinatorial as in the case of TSVM. However, for any possible choice of the labelling $\mathbf{y}^*$, the optimal β is given by (8), or more precisely, by a slightly modified version to take into account different parameters for C_s and C_u:

$$\beta = \mathbf{H}^T(\mathbf{C}^{-1}\mathbf{I} + \mathbf{H}\mathbf{H}^T)^{-1}\begin{bmatrix}\mathbf{y} \\ \mathbf{y}^*\end{bmatrix} \tag{10}$$

Where $\mathbf{C}$ is a diagonal matrix with the first N elements equal to C_s and the last M elements equal to C_u, and the hidden matrix is computed over all $N+M$ input patterns:

$$\mathbf{H} = [\mathbf{h}(\mathbf{x}_1),\dots,\mathbf{h}(\mathbf{x}_N),\mathbf{h}(\mathbf{x}_{N+1}),\dots,\mathbf{h}(\mathbf{x}_{N+M})]$$

Back-substituting (10) into (9), we obtain a fully combinatorial search problem over $\mathbf{y}^*$. This can be further simplified by considering:

$$\hat{\mathbf{H}} = \mathbf{H}^T(\mathbf{C}^{-1}\mathbf{I} + \mathbf{H}\mathbf{H}^T)^{-1} = \begin{bmatrix}\hat{\mathbf{H}}_1 \ \hat{\mathbf{H}}_2\end{bmatrix} \tag{11}$$

Where $\hat{\mathbf{H}}_1$ is the submatrix containing the first N columns of $\hat{\mathbf{H}}$, and the other block follow. Equation (10) can be rewritten as:

$$\beta = \hat{\mathbf{H}}_1\mathbf{y} + \hat{\mathbf{H}}_2\mathbf{y}^* \tag{12}$$

Where the vector $\hat{\mathbf{H}}_1\mathbf{y}$ and the matrix $\hat{\mathbf{H}}_2$ are fixed for any choice of the labeling of U. Any known algorithm for combinatorial optimization [5] can be used to train a TELM model, and form (12) is particularly convenient for computations. We do not try to enforce a specific proportion of positive labels (although this would be relatively easy) since in our experiments the additional constraint never improved performance.

5 Results

The TELM algorithm was tested on an artificial dataset known in literature as *the two moons*, a sample of which is shown in Fig. 1. Two points, one for each class, are shown in red and blue respectively. All simulations were performed by MATLAB 2012a, on an Intel i3 3.07 GHz processor at 64 bit, with 4 GB of RAM available, and each result is averaged over 100 runs. The TELM is solved using a standard *Genetic*

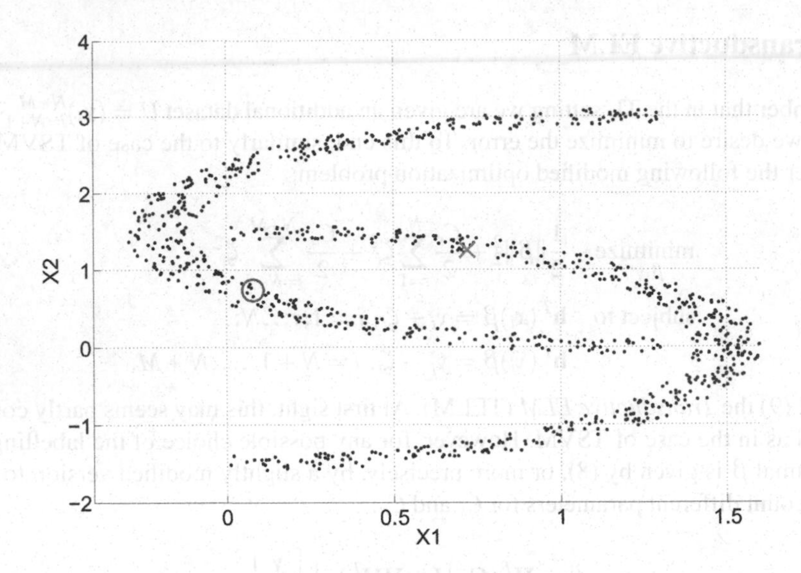

Fig. 1. Sample of the dataset

Algorithm [5]. For comparison, we implemented as baseline a standard ELM model and a binary SVM.

Sigmoid additive activation functions are used to construct the ELM feature space:

$$g(\mathbf{x}) = \frac{1}{1 + e^{-(\mathbf{a}\mathbf{x}+b)}} \tag{13}$$

Using standard default choices for the parameters, we consider 40 hidden nodes, and set $C = 1$. Parameters $\mathbf{a}$ and b of equation (13) were generated according to an uniform probability distribution. The SVM uses the Gaussian kernel:

$$k(\mathbf{x},\mathbf{y}) = \exp\{-\gamma\|\mathbf{x} - \mathbf{y}\|_2^2\} \tag{14}$$

Parameter γ in (14) was also set to 1 in all the experiments. Algorithms were tested using five different sizes of the datasets. For the first four experiments, a total of 100 samples was considered, and the training size was gradually increased. In the last experiment, instead, we considered two datasets of 100 elements each. For each method we present the classification accuracy in Table 1, where the highest accuracy in each row is highlighted in boldface.

As can be seen, TELM outperforms both methods for every combination we considered. In particular, it gives a small improvement when trained using very small training datasets (first two rows), very large increments with datasets of medium size (third and fourth row), and is able to reach 100% classification accuracy with sufficient samples (fifth row).

Table 1. Experimental results: classification accuracy

	SVM	ELM	TELM
$N = 4, M = 98$	0.77	0.75	**0.79**
$N = 10, M = 90$	0.81	0.75	**0.86**
$N = 40, M = 60$	0.85	0.80	**0.93**
$N = 60, M = 40$	0.85	0.81	**0.97**
$N = 100, M = 100$	0.93	0.95	**1**

6 Open Questions

Two main questions remain to be answered for an effective implementation of the TELM algorithm. We detail them briefly in this Section.

1. Our formulation suffers from a major drawback which is encountered also on TSVMs. In particular, it cannot be easily extended to the regression case. It is easy to show that any minimizer β of the first two terms of equation (10) automatically minimizes the third with the trivial choice $y_i^* = h(\mathbf{x}_i)^T \beta$. Thus, some modifications are needed, for example following [10].
2. The genetic algorithm imposes a strong computational effort in minimizing (10). This can be addressed by developing specialized solvers able to take into consideration the specific nature of the problem. To this end, we imagine that many of the algorithms used for TSVMs can be readily extended to our context.

7 Conclusions

In this work we presented an initial study for the extension of ELM theory to the transductive learning framework. We showed that this results in a fully combinatorial optimization problem. In our experiments, we solved it using a standard GA. Results are highly promising in the dataset we considered. However, there is the need of further optimizing the learning algorithm before a successful real-world application.

References

1. Cherkassky, V., Mulier, F.: Learning from data: concepts, theory, and methods (2007)
2. Vapnik, V.: The nature of statistical learning theory, 2nd edn., vol. 8. Springer (January 1999)
3. Chapelle, O., Sindhwani, V., Keerthi, S.: Optimization techniques for semi-supervised support vector machines. Journal of Machine Learning Research 9, 203–233 (2008)
4. Huang, G.B., Zhou, H., Ding, X., Zhang, R.: Extreme learning machine for regression and multiclass classification. IEEE Transactions on Systems, Man, and Cybernetics 42(2), 513–529 (2012)
5. Luke, S.: Essentials of metaheuristics (2009)

6. Chapelle, O., Schölkopf, B., Zien, A.: Semi-supervised learning (2006)
7. Evgeniou, T., Pontil, M., Poggio, T.: Regularization networks and support vector machines. Advances in Computational Mathematics 13, 1–50 (2000)
8. Steinwart, I., Christmann, A.: Support vector machines, 1st edn. (2008)
9. Huang, G.B., Zhu, Q.Y., Siew, C.K.: Extreme learning machine: Theory and applications. Neurocomputing 70(1-3), 489–501 (2006)
10. Cortes, C., Mohri, M.: On transductive regression. In: Advances in Neural Information Processing Systems (2007)

Avoiding the Cluster Hypothesis in SV Classification of Partially Labeled Data

Dario Malchiodi[1] and Tommaso Legnani[2]

[1] Dipartimento di Informatica, Università degli Studi di Milano
malchiodi@di.unimi.it

[2] Dipartimento di Matematica "F. Enriques", Università degli Studi di Milano

Abstract. We propose a Support Vector-based methodology for learning classifiers from partially labeled data. Its novelty stands in a formulation not based on the *cluster hypothesis*, stating that learning algorithms should search among classifiers whose decision surface is far from the unlabeled points. On the contrary, we assume such points as specimens of uncertain labels which should lay in a region containing the decision surface. The proposed approach is tested against synthetic data sets and subsequently applied to well-known benchmarks, attaining better or at least comparable performance w.r.t. methods described in the literature.

1 Introduction

The problem of classification consists in assigning objects in a given domain to one among a prefixed set of classes. In its simplest version, this problem is solved in the machine learning context through *learning* a classifier (that is, a mapping from objects to classes) on the basis of a *labeled sample* consisting of pairs (point, class). Such a problem admits several variations, and this paper focuses on a special setting characterized by the presence in the sample of points not associated to any specific class. This happens for instance in many real world situations where collecting objects is extremely easy (e.g., when mining the Internet) but labeling them is expensive (tipycally because some sort of human intervention is required). Such cases are dealt within the field of semi supervised learning [1] taking into account the so-called *cluster hypothesis*, stating that unlabeled points should not be close to the decision surfaces of the learnt classifiers. In this paper, instead, we require that unlabeled points be confined in a region of the space containing the decision function of the learnt classifier. Indeed, in some situations unlabeled points are characterized by some inherent form of uncertainty, rather than on the difficulty of labeling them. Web spam detection is a typical example of such a situation: in many cases, even humans reading the text contained in a web page are not able to definitely classify it as spam or non-spam, or experts produce different classifications on a same page [2].

The paper is organized as follows: Sect. 2 describes the proposed method, while Sect. 3 applies it to artificial and real-world data sets. Finally, Sect. 4 is devoted to outlooks and concluding remarks.

S. Bassis et al. (eds.), *Recent Advances of Neural Network Models and Applications*, Smart Innovation, Systems and Technologies 26,
DOI: 10.1007/978-3-319-04129-2_4,

2 The Learning Algorithm

The standard algorithm for learning SV classifiers is centered around the minimization of $\frac{1}{2}w\cdot w+C\sum_{i=1}^{m}\xi_i$, constrained to $y_i(w\cdot\Phi(x_i)+b)\geq 1-\xi_i$ and $\xi_i\geq 0$ for $i=1,\ldots,m$. In this formulation, x_i denotes a point in a given space X, while y_i is a label in $\{-1,1\}$[1] whose value ties x_i to a class. Variables w and b refer to a hyperplane in the image H of a mapping Φ having equation $w\cdot\Phi(x)+b=0$. Constraints aim at dividing H in two half-spaces, each containing only points associated to a given class: when this is possible, the constraints will be satisfied with $\xi_i\leq 1$, otherwise the value for ξ_i represents the amount of *error* for x_i. The objective function rewards hyperplanes inducing a small cumulative error, meanwhile having a small slope (in order to enhance the classifier performance). The parameter $C>0$ specifies the weight to be given to these two factors. This problem is generally solved through the analysis of its Wolfe Dual:

$$\max\sum_{i=1}^{m}\alpha_i-\frac{1}{2}\sum_{i,j=1}^{m}\alpha_i\alpha_j y_i y_j k(x_i,x_j)\ ,$$

$$\sum_{i=1}^{m}\alpha_i y_i=0,\quad 0\leq\alpha_i\leq C,\quad i=1,\ldots,m\ ,$$

where k denotes a kernel function mapping its arguments through Φ and computing the dot product of their images. The optimal solution $\alpha_1^*,\ldots,\alpha_m^*$ is used in order to infer the label of a generic point $x^{\mathrm{N}}\in X$ as the signum of $f(x^{\mathrm{N}})=\sum_{i=1}^{m}\alpha_i^* y_i k(x_i,x^{\mathrm{N}})+b$, being $b=y_i-\sum_{j=1}^{m}\alpha_j^* y_j k(x_j,x_i)$ for any i such that $0<\alpha_i^*<C$.

When adding to the available data a set of unlabeled points $U=\{x_1^{\mathrm{u}},\ldots,x_n^{\mathrm{u}}\}$, the assumption that the latter should not lay afar from the separating surface of the learnt classifier allows for easily modifying the above formulation. It will suffice to find an hyperplane acting in X both as a classifier for the labeled points and a regressor for the unlabeled ones. The latter goal can be achieved through the standard SV technique for learning regressors [3], confining the images of points in U within an ϵ-tube centered around the above mentioned hyperplane. Aiming at minimizing ϵ leads to:

$$\min\frac{1}{2}w\cdot w+C\sum_{i=1}^{m}\xi_i+\epsilon\ , \tag{1}$$

$$y_i(w\cdot\Phi(x_i)+b)\geq 1-\xi_i\quad i=1,\ldots,m\ ,$$
$$-\epsilon\leq w\cdot\Phi(x_s^{\mathrm{u}})+b\leq\epsilon\quad s=1,\ldots,n\ ,$$
$$\epsilon\geq 0,\quad \xi_i\geq 0\quad i=1,\ldots,m\ .$$

[1] In this section we restrict to the problem of binary classification, where only two classes are involved. See Sect. 3 in order to deal with multi-class classification.

The dual formulation will maximize the objective function:

$$\sum_{i=1}^{m} \alpha_i - \sum_{i,t} \alpha_i y_i (\gamma_t - \delta_t) k(x_i, x_t^{\mathrm{u}})$$
$$-\frac{1}{2}\sum_{i,j=1}^{m} \alpha_i \alpha_j y_i y_j k(x_i, x_j) - \frac{1}{2}\sum_{s,t=1}^{n} (\gamma_s - \delta_s)(\gamma_t - \delta_t) k(x_s^{\mathrm{u}}, x_t^{\mathrm{u}})$$

subject to contstraints

$$\sum_{i=1}^{m} \alpha_i y_i + \sum_{s=1}^{n} (\gamma_s - \delta_s) = 0, \quad \sum_{s=1}^{n} (\gamma_s + \delta_s) \le 1 \ ,$$
$$0 \le \alpha_i \le C \quad i = 1, \ldots, m, \quad \gamma_s \ge 0, \quad \delta_s \ge 0 \quad s = 1, \ldots, n \ .$$

The analysis of duality conditions shows that once the solution $\langle(\alpha_1^*, \ldots, \alpha_m^*), (\gamma_1^*, \ldots, \gamma_n^*), (\delta_1^*, \ldots, \delta_n^*)\rangle$ has been found, the optimal value for b is

$$b^* = y_i - \sum_{j=1}^{m} \alpha_j^* y_j k(x_j, x_i) - \sum_{s=1}^{n} (\gamma_s^* - \delta_s^*) k(x_s^{\mathrm{u}}, x_i)$$

for any i such that $0 < \alpha_i^* < C$, while ϵ^* can be obtained either as:

$$\epsilon^* = -\sum_{i=1}^{m} \alpha_i^* y_i k(x_i, x_s^{\mathrm{u}}) - \sum_{t=1}^{n} (\gamma_t^* - \delta_t^*) k(x_t^{\mathrm{u}}, x_s^{\mathrm{u}}) - b^* \ \text{, or} \tag{2}$$

$$\epsilon^* = \sum_{i=1}^{m} \alpha_i^* y_i k(x_i, x_s^{\mathrm{u}}) + \sum_{t=1}^{n} (\gamma_t^* - \delta_t^*) k(x_t^{\mathrm{u}}, x_s^{\mathrm{u}}) + b^* \ , \tag{3}$$

where the index s is chosen such that $\gamma_s^* > 0$ in (2) and such that $\delta_s^* > 0$ in (3). A new point x^{N} is classified computing the signum of

$$\sum_{i=1}^{m} \alpha_i^* y_i k(x_i, x^{\mathrm{N}}) + \sum_{s=1}^{n} (\gamma_s^* - \delta_s^*) k(x_s^{\mathrm{u}}, x^{\mathrm{N}}) + b^* \ .$$

3 Experiments

We applied the proposed approach to the different contexts of learning shadowed sets, semi supervised classification, and learning fuzzy regions.

3.1 Learning Shadowed Sets

A straightforward application of the proposed method is that concerning how to learn a shadowed set [4]. Such sets represent an abstraction of standard sets whose membership function assumes the values 1 (for elements belonging to the set), 0 (for elements excluded from the set), or [0, 1] (for elements whose membership to the set is completely uncertain, belonging to the so-called *shadow* of the set). A shadowed set A can be learnt through the proposed procedure using a sample made up by elements in A (coupled with the label 1), elements excluded by A (coupled with the label -1), and elements in the shadow of A (as unlabeled points). The result of the learning procedure induces on X a tripartition $\langle X^-, X^{\mathrm{s}}, X^+\rangle$ such that:

$$X^- = \{x \in X \text{ such that } w^* \cdot \Phi(x) + b^* < -\epsilon\} \ ,$$
$$X^{\mathrm{s}} = \{x \in X \text{ such that } -\epsilon \leq w^* \cdot \Phi(x) + b^* \leq \epsilon\} \ ,$$
$$X^+ = \{x \in X \text{ such that } w^* \cdot \Phi(x) + b^* > \epsilon\} \ ,$$

where X^-, X^+, and X^{s} represent the inferred forms of $X \backslash A$, A, and of the shadow of A.

Figure 1 shows the results of two simple experiments on synthetic data sets learnt using the polynomial kernel $k(x_1, x_2) = (1 + x_1 \cdot x_2)^p$, with $p = 2$. It can be noted how ϵ^* changes in order to accommodate for a bigger shadowed region (delimited by dashed curves). Moreover, the graphs show how the inferred classifier differs from that obtained through a standard SV approach.

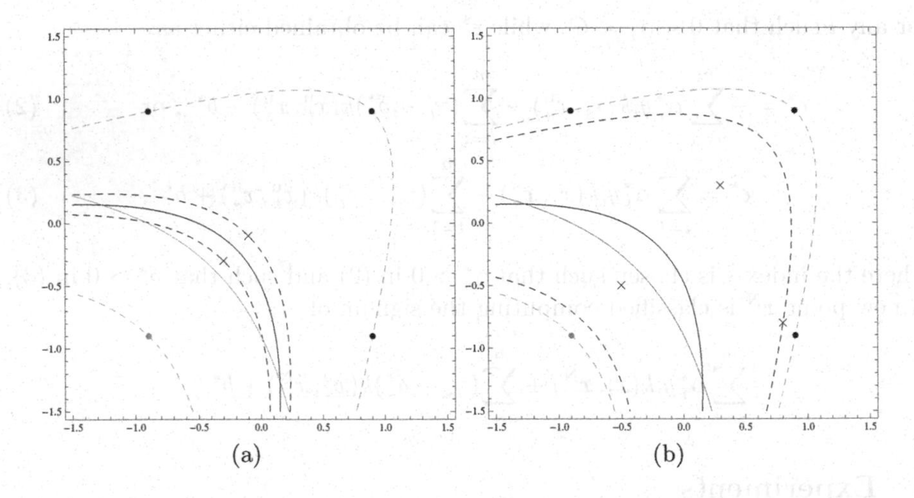

Fig. 1. Learning a shadowed set A through quadratic SV classifiers. Black and grey bullets respectively points belonging to A and excluded by A, while crosses denote points in the shadow of A. The black curves denote the learnt frontier of A (plain curve) and the ϵ-tube containing unlabeled points (dashed curve). The gray curves show the SV classifier learnt using exclusively the labeled points (plain curve) and the SVs (sample points at unit normalized distance from the separating surface, dashed curve). (a) and (b) refer to two different sets of unlabeled points in input to the learning algorithm.

3.2 Semi Supervised Learning

We also tested the proposed method on the problem of learning a binary classifier from semi supervised data, in order to compare performances with methods based on the cluster hypothesis. We worked on the eight benchmarks selected from the UCI repository [5] listed in Table 1, along with their size and dimensionality, intended respectively as the number of examples and the dimension of points (shown in columns size and $\dim(X)$). After standard pre-processing techniques (namely, normalization and shifting), we followed on each data set the same approach described in [6]: we shuffled the available examples and subsequently divided them into two groups containing respectively 90% and 10% of the available data. We trained a classifier using all examples in the first group as labeled data, together with the points in the second group as unlabeled data. Finally we tested the obtained classifier on the unlabeled points, whose labels were available although they were not used during the training process. Table 1 shows the fraction of misclassified examples, averaged over 10 experiments, each starting after data reshuffling. Initially we used linear kernels and set $C = \frac{1-\lambda}{\lambda(m+n)}$, where $\lambda = 0.001$ while m and n respectively denote the number of labeled and unlabeled points, exactly as in [6], obtaining results in line with the cited paper. Subsequently we repeated all experiments using a gaussian kernel

$$k(x_1, x_2) = \mathrm{e}^{-\frac{||x_1 - x_2||^2}{2\sigma^2}},$$

coupled with a cross-validation procedure in order to select the optimal values for C and σ within a grid of candidate values. Table 1 shows that the results of the proposed method (column $\mathrm{TS}^3\mathrm{VM}$) always outperform those obtained in the above mentioned paper (column $\mathrm{S}^3\mathrm{VM}$). Moreover, last column in Table 1 highlights that in 6 out of 8 benchmarks the proposed methodology scored a better error than the best results we could find in the literature, not necessarily using a SV-based approach, listed on the UCI web site with reference to the single data sets.

3.3 Learning Fuzzy Sets

The proposed method can be applied also to the problem of learning fuzzy sets when the only information available consists in elements having unit or null membership to the set (that is, elements definitely contained in the set or excluded by it), together with elements whose membership to the set is unknown. Indeed, it is possible to associate all elements in the first group to a label encoding their membership value, meanwhile considering the remaining elements as unlabeled. Such approach has been applied to the widely known Iris data set, consisting in measurements for 150 different samples from species *Iris virginica*, *Iris versicolor*, and *Iris setosa*.

Focusing, for the sake of visualization, only on the projection of the sample onto its first two principal components [14], a fuzzy membership function μ for any of the species can be inferred through the following procedure:

Table 1. Performance of the proposed method (column TS^3VM) against competitors in literature on eight UCI benchmarks, compared with the corresponding results in [6] (column S^3VM) and with the best result shown in the UCI Web site (column Literature). The best result on each row is highlighted in bold. Dimension of points nd benchmark size are reported, respectively, in columns dim(X) and size.

Data set	dim(X)	size	S^3VM	TS^3VM	Literature
Breast Cancer Wisconsin (Original)	9	699	0.034	0.025	**0.019** [7]
Breast Cancer Wisconsin (Diagnostic)	30	569	0.033	**0.007**	0.027 [8]
Statlog (Heart)	13	297	0.160	0.156	**0.155** [9]
Housing	13	506	0.151	**0.057**	0.129 [10]
Ionosphere	34	351	0.106	**0.043**	0.053 [11]
Musk (Version 1)	166	476	0.173	**0.054**	0.076 [12]
Pima Indians Diabetes	8	769	0.222	**0.204**	0.222 [6]
Connectionist Bench (Sonar, Mines vs. Rocks)	60	208	0.219	**0.062**	0.096 [13]

1. select among all measurements those laying in the space region between a class and the remaining ones, and consider them as unlabeled points;
2. consider the remaining measurements along with the species they refer to as labeled examples;
3. run the TS^3VM algorithm and obtain: i) a classification function g mapping measurements x into real values whose sign determines a crisp membership to the selected species, and ii) the optimal value ϵ^* for the width of the tube enclosing uncertain measurements, so that it is possible to define μ as

$$\mu(x) := \begin{cases} 1 & \text{if } g(x) \geq \varepsilon^* \ , \\ 0 & \text{if } g(x) \leq -\varepsilon^* \ , \\ \frac{g(x)+\varepsilon^*}{2\varepsilon^*} & \text{otherwise} \ . \end{cases}$$

The above procedure specializes to the various classes only in the step selecting the uncertain measurements. In particular:

- the *Iris setosa* species is separable w.r.t. the remaining two species through a hyperplane, so there are no "uncertain" measurements. However, the TS^3VM method requires the presence of at least two unlabeled points in order to be executed. Thus we drew two points on the smallest segment joining measurement from the setosa to other species;
- the *Iris virginica* and *Iris versicolor* classes slightly overlap, thus some of their measurements can be interpreted as having an intermediate membership to both species. In order to find them we merged all measurements from the two species and applied the fuzzy c-means algorithm [15], with the aim of finding two clusters (ideally, one for each species). The output of this algorithm is a couple of fuzzy membership values to the two clusters for each measurement. When a measurement does not definitely belong to any of the two species, it will have a low membership values to both clusters. Thus, denoted u_i^1 and u_i^2 the membership values of i-th measurement to the

two found clusters, we selected as uncertain all measurements x_i such that $\max(u_i^1, u_i^2) < \tau$. We fixed $\tau = 0.6$ after some experiments.

As a result, we were able to run three learning processes setting $C = 100$ and using a gaussian kernel with parameter $\sigma = 0.7$, obtaining the three membership functions shown in Fig. 2. In order to evaluate the performance of the inferred membership function we used a winner-takes-all defuzzification procedure: each measurement in the data set was assigned to the species scoring the highest membership value, getting only two misclassifications.

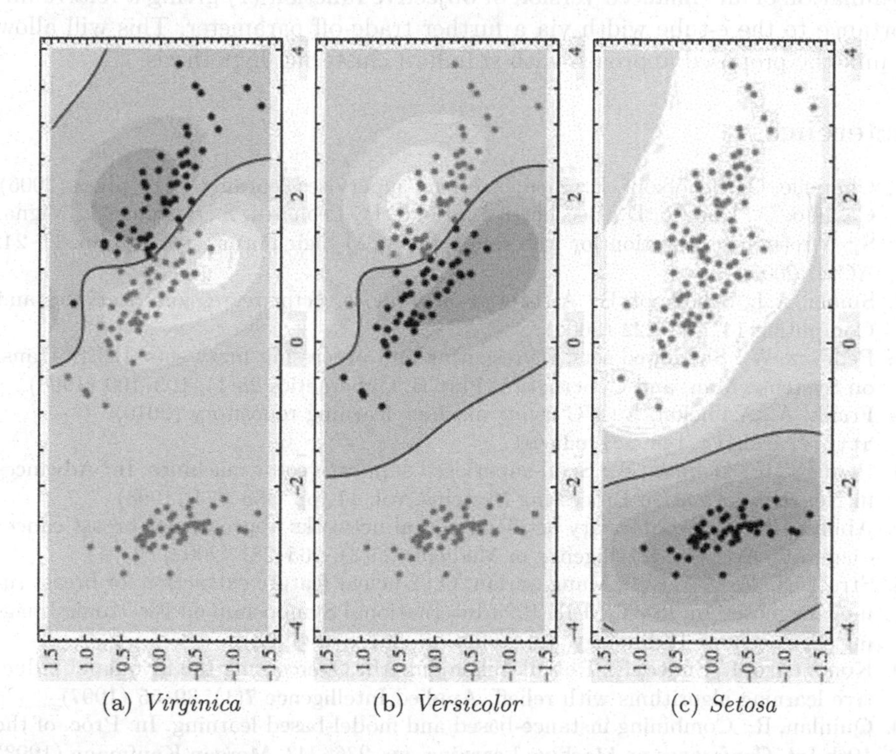

(a) *Virginica* (b) *Versicolor* (c) *Setosa*

Fig. 2. Results for the Iris data set. Each figure shows the inferred fuzzy membership function for a specific species (with a darker shade denoting a higher value), the learning set (black and gray bullets denote respectively positive and negative points) and the crisp classifier obtained through a winner-takes-all defuzzification procedure (black curve).

A more extensive experiment, selecting C and σ with a multiple hold-out scheme consisting in randomly dividing 100 times the data set into a training and a testing set containing respectively 70% and 30% of the examples, shows that the proposed method attains a result in line with the best performance described in the literature (see for instance [16]), both in terms of misclassification rate and number of fuzzy rules.

4 Conclusions

We proposed a SV-based algorithm for learning binary classifiers from partially labeled data explicitly avoiding the cluster hypothesis. The algorithm was tested on different data sets and in various fields, exhibiting performances either in line with or better w.r.t. comparable methodologies. In the future we plan to work on bigger data sets, also characterized by inherent uncertainty in the classification process, such as in the fields of Web spam detection and in written digit recognition. We also plan to further develop the learning algorithm trough the formulation of an enhanced version of objective function (1) giving a relaive importance to the ϵ-tube width via a further trade-off parameter. This will allow to mix the proposed approach with standard clustering hypothesis.

References

1. Chapelle, O., Schölkopf, B., Zien, A.: Semi-supervised learning. MIT press (2006)
2. Castillo, C., Donato, D., Becchetti, L., Boldi, P., Leonardi, S., Santini, M., Vigna, S.: A reference collection for web spam. In: ACM Sigir Forum, vol. 40, pp. 11–24. ACM (2006)
3. Smola, A.J., Schölkopf, B.: A tutorial on support vector regression. Statistics and Computing 14, 199–222 (2004)
4. Pedrycz, W.: Shadowed sets: representing and processing fuzzy sets. IEEE Trans. on Systems, Man, and Cybernetics, Part B: Cybernetics 28(1), 103–109 (1998)
5. Frank, A., Asuncion, A.: UC irvine machine learning repository (2010), `http://archive.ics.uci.edu/ml`
6. Bennett, K., Demiriz, A.: Semi-supervised support vector machines. In: Advances in Neural Information Processing Systems, vol. 11, pp. 368–374 (1998)
7. Abbass, H.: An evolutionary artificial neural networks approach for breast cancer diagnosis. Artificial Intelligence in Medicine 25(3), 265–281 (2002)
8. Street, N., Wolberg, W., Mangasarian, O.: Nuclear feature extraction for breast tumor diagnosis. In: IS&T/SPIE 1993 International Symposium on Electronic Imaging: Science and Technology, vol. 1905, pp. 861–870 (1993)
9. Kononenko, I., Šimec, E., Robnik-Šikonjam, M.: Overcoming the myopia of inductive learning algorithms with relieff. Applied Intelligence 7(1), 39–55 (1997)
10. Quinlan, R.: Combining instance-based and model-based learning. In: Proc. of the 10th Int. Conference on Machine Learning, pp. 236–243. Morgan Kaufmann (1993)
11. Fung, G., Dundar, M., Bi, J., Rao, B.: A fast iterative algorithm for fisher discriminant using heterogeneous kernels. In: Proceedings of the 21st International Conference on Machine Learning, pp. 40–47. ACM Press (2004)
12. Dietterich, T., Lathrop, R., Lozano-Pérez, T.: Solving the multiple instance problem with axis-parallel rectangles. Artificial Intelligence 89(1), 31–71 (1997)
13. Gorman, P., Sejnowski, T.: Analysis of hidden units in a layered network trained to classify sonar targets. Neural Networks 1(1), 75–89 (1988)
14. Abdi, H., Williams, L.J.: Principal component analysis. Wiley Interdisciplinary Reviews: Computational Statistics 2, 433–459 (2010)
15. Bezdek, J.: Pattern Recognition with Fuzzy Objective Function Algorithms. Plenum Press, New York (1981)
16. Chen, Y., Wang, J.: Support vector learning for fuzzy rule-based classification systems. IEEE Trans. on Fuzzy Systems 11(6), 716–728 (2003)

Learning Capabilities of ELM-Trained Time-Varying Neural Networks

Stefano Squartini, Yibin Ye, and Francesco Piazza

Università Politecnica delle Marche
{s.squartini,f.piazza}@univpm.it, yeyibin@gmail.com

Abstract. System identification in nonstationary environments surely represents a challenging problem. The authors have recently proposed an innovative neural architecture, namely Time-Varying Neural Network (TV-NN), which has shown remarkable identification capabilities in this kind of scenarios. It is characterized by time-varying weights, each being a linear combination of a certain set of basis functions. This inevitably increases the network complexity with respect to the stationary NN counterpart and in order to keep the training time low, an Extreme Learning Machine (ELM) approach has been proposed by the same authors for TV-NN learning, instead of Back-Propagation based techniques. However the learning capabilities of TV-NN trained by means of ELM have not been investigated in the literature and in this contribution such a lack is faced: the theoretical foundations of ELM usage for TV-NN are analytically discussed, by extending the corresponding results obtained in the stationary case study.

Keywords: Extreme Learning Machine, Time-Varying Neural Networks, Nonstationary System Identification.

1 Introduction

Extreme Learning Machine (ELM) is a fast learning algorithm designed for single hidden layer feedforward neural networks (SLFNs) [5,6]. In ELM, the input weights of SLFNs do not need to be tuned and can be randomly generated, whereas the output weights are analytically determined using the least-square method, thus allowing a significant training time reduction. In recent years, ELM has been raising much attention and interest among scientific community, in both theoretical studies and applications [7,2]. Among the various extensions, an ELM application for time-varying neural networks has been proposed in [1], taking full advantage of speedy training procedure with certain matrix transformations.

Time-Varying Neural Networks (TV-NN) represent a relevant example in neural architectures able to operate in nonstationary environments. Such networks implement time-varying weights, each being a linear combination of a certain set of basis functions, whose independent variable is time. The candidate orthogonal basis function types employed in time-varying networks include Legendre polynomials, Chebyshev polynomials, Fourier sinusoidal functions, Prolate

S. Bassis et al. (eds.), *Recent Advances of Neural Network Models and Applications*, Smart Innovation, Systems and Technologies 26,
DOI: 10.1007/978-3-319-04129-2_5,

spheroidal functions, and more. An extended Back-Propagation algorithm has been first advanced to train these networks (BP-TV) [13]. Then, an Extreme Learning Machine approach has been developed for TV-NN, accelerating the training procedure significantly (ELM-TV) [1]. Recently, several ELM-TV variants and related applications have been proposed. A group selection evolutionary approach applies differential evolutionary with group selection to set the type of basis function and its input weight parameters [15], while two other algorithms referred as EM-ELM-TV [14] and EM-OB [16] are able to automatically determine the number of hidden neurons or output basis functions, respectively. An online extreme learning machine is extended to deal with time-varying neural networks and referred as OS-ELM-TV [17,18], which can learn the training data one-by-one or chunk-by-chunk. The data which have already been used in the training can be discarded, so as to save more memory and computational load to process the new coming data.

As addressed in [10], research on the approximation capabilities of feedforward neural networks has focused on two aspects: (i) universal approximation on compact input sets, and (ii) approximation in a finite set of training samples. For the first aspect, it is proved in [9] that given any bounded nonconstant piecewise continuous activation function for additive nodes, SLFNs with random hidden nodes can approximate any continuous functions on any compact sets. Nevertheless, in real applications, the neural networks are trained over a finite set of input-output patterns. For function approximation over a finite training set, Extreme Learning Machine (ELM) for standard single-hidden layer feedforward neural network (SLFN) with at most N hidden nodes and with almost any nonlinear activation function can exactly learn N distinct observations [8,10]. This is addressed as the network learning capability. In this contribution, we aim at providing a theoretical analysis of ELM-TV learning capabilities over a finite training set. It will be shown that, on the basis of rank properties of Khatri-Rao product, ELM trained time-varying neural networks, with at most N hidden nodes or N orders of basis functions, are able to approximate N distinct training samples.

The details of theoretical analysis are arranged in Section 5. Prior to this, some preliminary knowledge and notations about Extreme Learning Machine, Time-Varying Neural Networks and ELM-TV algorithm are described in Section 2, 3 and 4 respectively. Section 6 deals with the computer simulations carried out in support of what shown in Section 5. Finally, conclusions are drawn in Section 7.

2 Extreme Learning Machine

Let's assume that an SLFN with I input neurons, K hidden neurons, L output neurons and activation function $g(\cdot)$ is trained to learn N distinct samples $(\mathbf{x}[n], \mathbf{t}[n])$, where $\mathbf{x}[n] \in \mathbb{R}^I$ and $\mathbf{t}[n] \in \mathbb{R}^L$. In ELM, the input weights $\{w_{ik}\}$ and hidden biases $\{b_k\}$ are randomly generated, where w_{ik} is the weight connecting i input neuron to k-th hidden neuron, and b_k is the bias of k-th hidden neuron. Further let $w_{0k} = b_k$ and $x_0[n] = 1$, define extended input matrix

$\mathbf{X} = \{x_i[n]\} \in \mathbb{R}^{N\times(I+1)}$ and extended input weights matrix $\mathbf{W} = \{w_{ik}\} \in \mathbb{R}^{(I+1)\times K}$ $(i = 0, 1, \ldots, I)$. Hence, the hidden-layer output matrix $\mathbf{H} = \{h_k[n]\} \in \mathbb{R}^{N\times K}$ can be obtained by:

$$h_k[n] = g\left(\sum_{i=0}^{I} x_i[n] \cdot w_{ik}\right) \tag{1}$$

$$\text{or} \quad \mathbf{H} = \mathbf{g}(\mathbf{X} \cdot \mathbf{W}) \tag{2}$$

where $\mathbf{g} : \mathbb{R}^{N\times K} \to \mathbb{R}^{N\times K}$, mapping each element of matrix variable to the element of corresponding position in $\mathbf{H}$ with activation function $g(\cdot)$.

Let $\beta = \{\beta_{kl}\} \in \mathbb{R}^{K\times L}$ be the matrix of output weights, where β_{kl} denotes the weight connection between k-th hidden neuron and l-th output neuron; and $\mathbf{Y} = \{y_l[n]\} \in \mathbb{R}^{N\times L}$ be the matrix of network output data, with $y_l[n]$ the output data in l-th output neuron at n-th time instant. Therefore, this equation can be obtained for the linear output neurons:

$$y_l[n] = \sum_{k=1}^{K} h_k[n] \cdot \beta_{kl} \tag{3}$$

$$\text{or} \quad \mathbf{Y} = \mathbf{H} \cdot \beta \tag{4}$$

Thus, given the hidden-layer output matrix $\mathbf{H}$ and the target matrix $\mathbf{T} = \{t_l[n]\} \in \mathbb{R}^{N\times L}$, with $t_l[n]$ the desired output in l-th output neuron at n-th time instant, to minimize $\|\mathbf{Y} - \mathbf{T}\|_2$, the output weights can be calculated as the minimum norm least-square(LS) solution of the linear system:

$$\hat{\beta} = \mathbf{H}^{\dagger} \cdot \mathbf{T}, \tag{5}$$

where $\mathbf{H}^{\dagger}$ is the Moore-Penrose generalized inverse of matrix $\mathbf{H}$. By computing output weights analytically, ELM allows achieving good generalization performance with speedy training phase.

3 Time-Varying Neural Networks

The time-varying version of Extreme Learning Machine has been studied in [1]. In a time-varying neural network as shown in Fig. 1, the input weights, or output weights, or both are changing with time (both in training and testing phases). The generic weight $w[n]$ can be expressed as a linear combination of a certain set of basis functions [3,13]:

$$w[n] = \sum_{b=1}^{B} f_b[n] w_b \tag{6}$$

in which $f_b[n]$ is the known orthogonal function at n-th time instant of b-th order, w_b is the b-th order coefficient of the basis function to construct time-varying weight $w[n]$, while B (preset by the user) is the total number of bases.

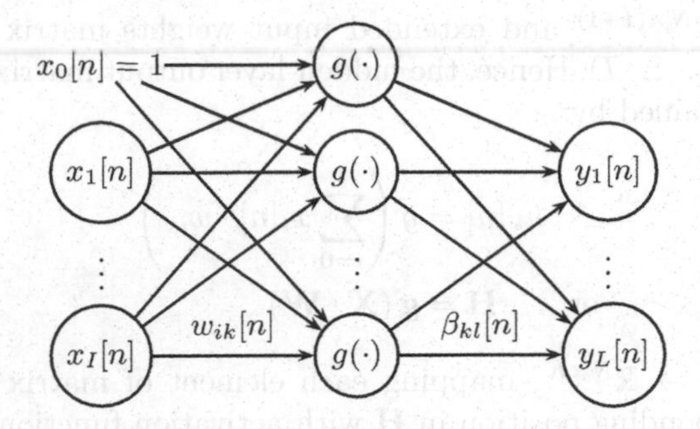

Fig. 1. Architecture of Time-Varying Neural Networks

If time-varying input weights are introduced in a SLFN, the hidden neuron output function can be written as:

$$h_k[n] = g\left(\sum_{i=0}^{I} x_i[n] \cdot w_{ik}[n]\right) \tag{7}$$

where $w_{ik}[n] = \sum_{b=1}^{B} f_b[n] w_{b,ik}$. Similarly, if time-varying output weights are introduced, the standard output equation can be written as:

$$y_l[n] = \sum_{k=1}^{K} h_k[n]\beta_{kl}[n] \tag{8}$$

where $\beta_{kl}[n] = \sum_{b=1}^{B} f_b[n]\beta_{b,kl}$. Note that we require that the $B = 1$ case study means that the TV-NN become a standard NN: this means that $f_1[n] = 1, \forall n$.

It will be shown in the next sections that, assuming that the TV-NN has an infinitely differentiable activation function g, with K hidden nodes and B basis functions, it is thus able to approximate N samples with zero error.

It must be underlined that the number of input and output basis functions can differ in general. However, in this paper, to improve the readability of the manuscript and without affecting the generality of the proofs given in the following, it is assumed to be the same.

4 ELM-TV Algorithm

To train a TV-NN, input weights $\{w_{b,ik}\}$ are randomly generated and hidden-layer output matrix $\mathbf{H}$ is computed according to equation (7). However, the values of a set of output weight parameters $\{\beta_{b,kl}\}$ can not be so straightforward calculated. Some transformations are needed. Expanding the time-varying output weights $\beta_{kl}[n]$ in (8) and assuming:

$$\mathbf{f}[n] = [f_1[n], f_2[n], \dots, f_B[n]]^T \in \mathbb{R}^B,$$
$$\mathbf{h}[n] = [h_1[n], h_2[n], \dots, h_K[n]]^T \in \mathbb{R}^K,$$
$$\boldsymbol{\beta}_{kl} = [\beta_{1,kl}, \beta_{2,kl}, \dots, \beta_{B,kl}]^T \in \mathbb{R}^B,$$

$$\boldsymbol{\beta}_{(l)} = [\boldsymbol{\beta}_{1l}, \boldsymbol{\beta}_{2l}, \ldots, \boldsymbol{\beta}_{Kl}] \in \mathbb{R}^{B \times K},$$
$$\boldsymbol{\omega}_{(l)} = [\boldsymbol{\beta}_{1l}^T, \boldsymbol{\beta}_{2l}^T, \ldots, \boldsymbol{\beta}_{Kl}^T] \in \mathbb{R}^{B \cdot K \times 1},$$

the following hold:

$$\begin{aligned} y_l[n] &= \sum_{k=1}^{K} h_k[n] \cdot \left(\sum_{b=1}^{B} f_b[n] \cdot \beta_{b,kl} \right) \\ &= \sum_{k=1}^{K} \mathbf{f}[n]^T \cdot \boldsymbol{\beta}_{kl} \cdot h_k[n] \\ &= \mathbf{f}[n]^T \cdot \boldsymbol{\beta}_{(l)} \cdot \mathbf{h}[n] \\ &= (\mathbf{h}[n]^T \otimes \mathbf{f}[n]^T) \cdot \boldsymbol{\omega}_{(l)} \end{aligned} \tag{9}$$

where $\otimes$ denotes the *Kronecker product* of $\mathbf{h}[n]^T$ and $\mathbf{f}[n]^T$. The last step consists in: $vec(\mathbf{AXB}) = (\mathbf{B}^T \otimes \mathbf{A})vec(\mathbf{X})$ [4], (note that $vec(y_l[n]) = y_l[n]$) and $\boldsymbol{\omega}_{(l)} = vec(\boldsymbol{\beta}_{(l)})$ is the vectorization of the matrix $\boldsymbol{\beta}_{(l)}$ formed by stacking the columns of $\boldsymbol{\beta}_{(l)}$ into a single column vector. Moreover, let's define:

$$\mathbf{G} = \mathbf{H} * \mathbf{F} = \begin{bmatrix} \mathbf{h}[1]^T \otimes \mathbf{f}[1]^T \\ \vdots \\ \mathbf{h}[N]^T \otimes \mathbf{f}[N]^T \end{bmatrix}_{N \times B \cdot K} \tag{10}$$

where $\mathbf{H} = [\mathbf{h}[1], \mathbf{h}[2], \ldots, \mathbf{h}[N]]^T \in \mathbb{R}^{N \times K}$, $\mathbf{F} = [\mathbf{f}[1], \mathbf{f}[2], \ldots, \mathbf{f}[N]]^T \in \mathbb{R}^{N \times B}$, $*$ denotes the *Khatri-Rao product* of matrices $\mathbf{H}$ and $\mathbf{F}$, with $\mathbf{h}[n]^T$ and $\mathbf{f}[n]^T$ as their submatrices, respectively. Further assuming that $\mathbf{Y} = \{y_l[n]\} \in \mathbb{R}^{N \times L}$, $\mathbf{T} = \{t_l[n]\} \in \mathbb{R}^{N \times L}$, $\boldsymbol{\Omega} = [\boldsymbol{\omega}_{(1)}, \boldsymbol{\omega}_{(2)}, \ldots, \boldsymbol{\omega}_{(L)}] \in \mathbb{R}^{B \cdot K \times L}$, the following holds:

$$\mathbf{G} \cdot \boldsymbol{\Omega} = \mathbf{Y} \tag{11}$$

Since $\mathbf{F}$ is obtained by the type of the basis function predetermined by the user and $\mathbf{H}$ can be calculated by (7) once input weight parameters are randomly generated, hence $\mathbf{G}$ can be computed. Similarly to the ELM algorithm described in previous section, the time-varying output weight matrix $\boldsymbol{\Omega}$ can be computed by:

$$\hat{\boldsymbol{\Omega}} = \mathbf{G}^{\dagger} \cdot \mathbf{T} \tag{12}$$

where $\mathbf{G}^{\dagger}$ is the Moore-Penrose inverse of matrix $\mathbf{G}$, and consequently, $\hat{\boldsymbol{\Omega}}$ is a set of optimal output weight parameters minimizing the training error.

Similarly, in the input layer, if we define

$$\mathbf{S} = \mathbf{X} * \mathbf{F} \tag{13}$$
$$= \begin{bmatrix} x_0[1]f_1[1] & \dots & x_I[1]f_B[1] \\ \vdots & \ddots & \vdots \\ x_0[N]f_1[N] & \dots & x_I[N]f_B[N] \end{bmatrix}_{N\times B(I+1)}$$

$$\mathbf{U} = \begin{bmatrix} w_{1,01} & \dots & w_{1,0K} \\ \vdots & \ddots & \vdots \\ w_{B,01} & \dots & w_{B,0K} \\ w_{1,11} & \dots & w_{1,1K} \\ \vdots & \ddots & \vdots \\ w_{B,I1} & \dots & w_{B,IK} \end{bmatrix}_{B(I+1)\times K} \tag{14}$$

we can express the hidden layer output matrix as follows

$$\mathbf{H} = \mathbf{g}(\mathbf{S} \cdot \mathbf{U}) \tag{15}$$

Finally, we summarize ELM-TV in Algorithm 1.

Algorithm 1. ELM-TV

1: Randomly generate the input weights set $\{w_{b,ik}\}$.
2: Calculate matrix $\mathbf{H}$ by Eq. (15) or Eq. (7).
3: Calculate matrix $\mathbf{G}$ according to Eq. (10).
4: Calculate the output weight matrix $\mathbf{\Omega}$ by Eq. (12).

5 Learning Capabilities of the ELM-TV Algorithm

In this section, we present some theoretical results of ELM-TV based on rank properties of Khatri-Rao product. A little more details about Khatri-Rao product is described below:

Given two matrices $\mathbf{A} = [\mathbf{a}_1, \dots, \mathbf{a}_N]^T \in \mathbb{R}^{N\times K}$ and $\mathbf{B} = [\mathbf{b}_1, \dots, \mathbf{b}_N]^T \in \mathbb{R}^{N\times B}$ (with $\mathbf{a} \in \mathbb{R}^K$ and $\mathbf{b} \in \mathbb{R}^B$) of identical number of rows, their *Khatri-Rao product* is denoted by

$$\mathbf{A} * \mathbf{B} = \begin{bmatrix} \mathbf{a}_1^T \otimes \mathbf{b}_1^T \\ \vdots \\ \mathbf{a}_N^T \otimes \mathbf{b}_N^T \end{bmatrix}_{N\times B\cdot K} \tag{16}$$

where $\otimes$ denotes the *Kronecker product*. For two row vectors $\mathbf{a}^T$ and $\mathbf{b}^T$, the Kronecker product is given by

$$\mathbf{a}^T \otimes \mathbf{b}^T = [a_1\mathbf{b}^T, \dots, a_K\mathbf{b}^T] \tag{17}$$

The rank properties of Khatri-Rao product have some relationships with *Kruskal rank* [12,11]. Here, we define kruskal rank of a matrix $\mathbf{A}$, denoted by $krank(\mathbf{A})$, equal to r when every collection of r rows of $\mathbf{A}$ is linearly independent but there exists a collection of $r+1$ linearly dependent rows. Therefore, we always have $krank(\mathbf{A}) \leqslant rank(\mathbf{A})$. Note that when $\mathbf{A}$ has full rows rank N, we also have $krank(\mathbf{A}) = rank(\mathbf{A}) = N$.

Lemma 1. *For two matrices* $\mathbf{A} \in \mathbb{R}^{N\times K}$ *and* $\mathbf{B} \in \mathbb{R}^{N\times B}$*, with* $krank(\mathbf{A}) \geqslant 1$ *and* $krank(\mathbf{B}) \geqslant 1$*, it holds true that*

$$min\{N, krank(\mathbf{A}) + krank(\mathbf{B}) - 1\} \leqslant krank(\mathbf{A} * \mathbf{B}) \tag{18}$$

Proof. Detailed proof can be found in Lemma 1 (K-rank of Khatri-Rao product) of [12].

Lemma 2. *(i) For one full rank matrix* $\mathbf{A} \in \mathbb{R}^{N\times N}$ *and* $\mathbf{B} \in \mathbb{R}^{N\times B}$ *with* $krank(\mathbf{B}) \geqslant 1$*, or (ii) one full rank matrix* $\mathbf{B} \in \mathbb{R}^{N\times N}$ *and* $\mathbf{A} \in \mathbb{R}^{N\times K}$ *with* $krank(\mathbf{A}) \geqslant 1$*, it holds true that*

$$rank(\mathbf{A} * \mathbf{B}) = N \tag{19}$$

Proof. Consider condition (i), Since $\mathbf{A}$ is full rank matrix, we have $krank(\mathbf{A}) = rank(\mathbf{A}) = N$, and $\mathbf{A}*\mathbf{B} \in \mathbb{R}^{N\times NB}$, according to Lemma 1, $N \leqslant krank(\mathbf{A}*\mathbf{B}) \leqslant N$, hence $rank(\mathbf{A} * \mathbf{B}) = krank(\mathbf{A} * \mathbf{B}) = N$. The same proof can also apply for condition (ii).

The following Lemma is an extension to the time-varying case study of what already provided in [10], which solely applies to stationary SLFNs.

Lemma 3. *Given a time-varying SLFN with* K *hidden nodes,* B *basis function and activation function* $g : \mathbb{R} \to \mathbb{R}$ *which is infinitely differentiable in any interval, for* N *arbitrary distinct input samples* $\mathbf{x}[n]$*, where* $\mathbf{x}[n] \in \mathbb{R}^I$*, for any* $w_{b,ik}$ *randomly chosen from any interval of* $\mathbb{R}$*, according to any continuous probability distribution, when* $K = N$*, with probability one, the hidden layer output matrix* $\mathbf{H} = \mathbf{g}(\mathbf{SU})$ *of the SLFN is full rank.*

Proof. Once observed that $x_0[n] = f_1[n] = 1$, we can formulate the hidden neuron outputs in (7) as follows:

$$h_k[n] = g\left(\sum_{i=0}^{I}\sum_{b=1}^{B} x_i[n]f_b[n]w_{b,ik}\right) = g\left(w_{1,0k} + d_k[n]\right) \tag{20}$$

where $d_k[n] = \sum_{i=1}^{I}\sum_{b=1}^{B} x_i[n]f_b[n]w_{b,ik} + \sum_{b=2}^{B} f_b[n]w_{b,0k}$ Therefore, the k-th column of $\mathbf{H}$ can be written as $\mathbf{c}(w_{1,0k}) = [g(w_{1,0k} + d_k[1]), \dots, g(w_{1,0k} + d_k[N])]^T \in \mathbb{R}^N$. Since $w_{b,ik}$ are randomly generated based on a continuous probability distribution, it follows that that $d_k[n] \neq d_{k'}[n]$ for all $k \neq k'$. It can be proved by contradiction that vector $\mathbf{c}(w_{1,0k})$ does not belong to any subspace whose dimension is less than N.

Suppose that $\mathbf{c}(w_{1,0k})$ belongs to a subspace of dimension $N-1$, there exists a non-zero vector α which is orthogonal to this subspace

$$(\alpha, \mathbf{c}(w_{1,0k}) - \mathbf{c}(\tilde{w})) = \alpha_1 g(w_{1,0k} + d_k[1]) + \cdots + \alpha_N g(w_{1,0k} + d_k[N]) - (\alpha, \mathbf{c}(\tilde{w})) = 0 \quad (21)$$

Moreover, assuming without loss of generality that $\alpha_N \neq 0$, Eq.(21) can be written as

$$g(w_{1,0k} + d_k[N]) = -\sum_{n=1}^{N-1} \gamma_n g(w_{1,0k} + d_k[n]) + \frac{(\alpha, \mathbf{c}(\tilde{w}))}{\alpha_N} \quad (22)$$

where $\gamma_n = \alpha_n/\alpha_N$. Since $g(x)$ is infinitely differentiable in any interval of $\mathbb{R}$, we have

$$g^{(l)}(w_{1,0k} + d_k[N]) = -\sum_{n=1}^{N-1} \gamma_n g^{(l)}(w_{1,0k} + d_k[n]) \quad (23)$$

where $g^{(l)}$ is the l-th derivative of function g of $w_{1,0k}$, and $l = 1, 2, \ldots, N, N+1, \ldots$. However, there are only $N-1$ free coefficients $\gamma_1, \ldots, \gamma_{N-1}$ for more than $N-1$ derived linear equations, which is contradictory. Therefore, vector $\mathbf{c}(w_{1,0k})$ does not belong to any subspace whose dimension is less than N. Concluding, for any weights $w_{b,ik}$, $(b = 1, 2, \ldots, B; i = 0, 1, \ldots, I; k = 1, \ldots, K)$ chosen from any interval of $\mathbb{R}$, when there exist $K = N$ hidden nodes, according to any continuous probability distribution, with probability one, the matrix $\mathbf{H}$ is full rank. Indeed we can always select suitable values for $w_{1,01}, \ldots, w_{1,0N}$ so that the corresponding column vectors in $\mathbf{H}$, i.e. $\mathbf{c}(w_{1,01}), \ldots \ldots, \mathbf{c}(w_{1,0N})$ are linearly independent.

Remark 1. It has to be underlined that, under the assumptions in Lemma 3, the rank of matrix $\mathbf{H}$ is always equal to K: this easily follows from the discussion made in the proof above.

Theorem 1. *Given a time-varying SLFN with K hidden nodes, a selected basis function matrix $\mathbf{F}$ and activation function $g : \mathbb{R} \to \mathbb{R}$ which is infinitely differentiable in any interval, for N arbitrary distinct samples $(\mathbf{x}[n], \mathbf{t}[n])$, where $\mathbf{x}[n] \in \mathbb{R}^I$ and $\mathbf{t}[n] \in \mathbb{R}^L$, for any $w_{b,ik}$ randomly chosen from any interval of $\mathbb{R}$, according to any continuous probability distribution, when $B = N$ or $K = N$, with probability one, matrix $\mathbf{G}$ is full row rank and there exists $\mathbf{\Omega}$ so that $\|\mathbf{G\Omega} - \mathbf{T}\| = 0$.*

Proof. In time-varying case, the hidden layer output matrix is $\mathbf{H} = \mathbf{g}(\mathbf{SU})$. If $\mathbf{x}[n], (n = 1, \ldots, N)$ are arbitrary distinct samples, so would also be $\mathbf{s}[n] = \mathbf{x}[n] \otimes \mathbf{f}[n], (n = 1, \ldots, N)$, fulfilling every equivalent condition in Lemma 3. Hence, $\mathbf{H}$ will be full rank when $K = N$. On the other hand, since $\mathbf{F}$ is a basis function matrix, $\mathbf{F}$ is also full rank when $B = N$.

With $\mathbf{G} = \mathbf{H} * \mathbf{F}$, according to Lemma 2, $rank(\mathbf{G}) = N$, i.e. $\mathbf{G}$ is a full row rank matrix. Therefore, there exists $\mathbf{\Omega}$ so that $\mathbf{G\Omega} = \mathbf{T}$. i.e. $\|\mathbf{G\Omega} - \mathbf{T}\| = 0$.

Remark 2. It must be noted that, in contrast to the time-invariant case study, in this proof the value of K can be smaller than N and zero-error system invertibility can be achieved due to the full-rankness of the **F** matrix.

Actually, if we have $krank(\mathbf{H}) + krank(\mathbf{F}) \geqslant N+1$, even though K and B may $< N$, we can still have a full row rank matrix according to Lemma 1, therefore making $\|\mathbf{G\Omega} - \mathbf{T}\| = 0$. However this is not guaranteed. For example let us assume to have $BK = N$, with B, K both bigger than 1. According to Lemma 1 and taking into account that $rank(\mathbf{H}) = K$ and $rank(\mathbf{F}) = B$, we are able to get a lower bound for the rank of **G**, such as:

$$rank(\mathbf{G}) \geqslant min(N, B+K-1) = B+K-1 \tag{24}$$

For instance, let us assume that:

$$\mathbf{H}^T = \begin{pmatrix} 1 & 1 & 1 & 1 & 1 & 1 \\ 1 & -1 & 1 & -1 & 1 & -1 \end{pmatrix} \tag{25}$$

with $rank(\mathbf{H}) = 2$ and

$$\mathbf{F}^T = \begin{pmatrix} 1 & 1 & 1 & 1 & 1 & 1 \\ 1 & 1 & 1 & -1 & -1 & -1 \\ 1 & -1 & 1 & -1 & 1 & -1 \end{pmatrix} \tag{26}$$

with $rank(\mathbf{F}) = 3$, it can be easily proved that $rank(\mathbf{G}) = 4$, thus **G** is not full-rank.

Theorem 2. *Given any small positive value $\epsilon > 0$ and activation function g: $\mathbb{R} \to \mathbb{R}$ which is infinitely differentiable in any interval, there exists $K \leqslant N$ and $B \leqslant N$, such that for N arbitrary distinct samples $(\mathbf{x}[n], \mathbf{t}[n])$, where $\mathbf{x}[n] \in \mathbb{R}^I$ and $\mathbf{t}[n] \in \mathbb{R}^L$, for any $w_{b,ik}$ randomly chosen from any interval of $\mathbb{R}$, according to any continuous probability distribution, with probability one, $\|\mathbf{G\Omega} - \mathbf{T}\| < \epsilon$.*

Proof. The statement is valid, otherwise K or B can be simply set equal to N so that $\|\mathbf{G\Omega} - \mathbf{T}\| = 0 < \epsilon$, according to what proved in Theorem 1.

Remark 3. When $K \ll N$ and $B \ll N$ with $KB < N$, the rows number of $\mathbf{\Omega}$ ($= KB$) is less than that of $\mathbf{G}$ ($= N$). Considering equation $\|\mathbf{G\Omega} = \mathbf{T}\|$, the free parameters in $\mathbf{\Omega}$ are less than the equations set to be satisfied. Usually this leads to $\|\mathbf{G\Omega} - \mathbf{T}\| > 0$.

As K or B increases, more free parameters in $\mathbf{\Omega}$ can be used to deal with N equations set, thus usually making $\|\mathbf{G\Omega} - \mathbf{T}\|$ smaller.

6 Computer Simulations

In this section some computer simulations are described to experimentally prove what stated above from a theoretically point of view. They have been performed in a MATLAB 7.8.0 environment running on an Intel Core2 Duo CPU P8400 2.26GHz, with Windows Vista as operating system.

Table 1. RMSE training results (averaged over 10 runs) of SLFNs with different values of K and B. Bold values are relative to the $BK = N$ case study.

	K						
	350	400	450	**500**	550	600	650
$B = 1$	-24.38	-25.48	-27.82	**-117.23**	-131.93	-131.98	-134.43

	K						
	175	200	225	**250**	275	300	325
$B = 2$	-29.24	-30.32	-32.76	**-123.89**	-134.50	-135.70	-136.74

	K						
	70	80	90	**100**	110	120	130
$B = 5$	-30.41	-32.73	-35.86	**-127.48**	-136.99	-138.50	-140.16

	K						
	35	40	45	**50**	55	60	65
$B = 10$	-26.01	-26.75	-28.76	**-117.56**	-135.52	-140.83	-142.69

The system to be identified here is the same considered in [1] and [13]: a time-variant IIR (Infinite Impulse Response)-buffered MLP (Multilayer Perceptron) with 11 input lines, one 5 neurons hidden layer, one single output neuron. The input weights and output weights are combination of 3 Chebyshev basis functions; the lengths of the input and output time delay lines are equal to 6 and 5 respectively. Both the hidden and output neurons use tangent sigmoid activation function. The input signal is white noise with normal distribution.

The SLFNs used to identify such a system have K hidden neurons, with tangent sigmoid activation function, Legendre-type basis function and one single linear output neuron, whereas the number of output basis function is equal to B and N to 500. The number of input basis functions is equal to 1. The training results, expressed in terms of root mean-square error (RMSE), have been averaged over 10 runs and then reported in Table 1. The input signal is again white noise with normal distribution.

It can be easily observed that the RMSE values decrease as K increases with a remarkable decrement attained in correspondence of the $BK = N$ case study (represented by the central column in Table 1). The rank of matrix $\mathbf{G}$ is always equal to BK: It must be noted that this is not guaranteed by the theoretical analysis carried out in previous Section (see Remark 2 in particular), but it is what practically occurs in most cases like in this task. Nevertheless, as expected by such analysis, when $\mathbf{G}$ is full-rank we are able to get the lowest RMSEs (less than -100dB) as result of the ELM-based training phase: A clear difference with respect to the values attained in correspondence of those cases where $\mathbf{G}$ is not full-rank is registered. This confirms that, in contrast to what pointed out in [10], the TV-NN architecture gives us more flexilibity in selecting the number of hidden neurons to get the highest training accuracies, due to the presence of the output basis functions.

It must be observed that the best result for $BK = N$ is obtained for $B = 5$, corresponding to a better matching with the architectural characteristics of the nonstationary system generating the training data. The generalization performance of the SLFNs involved in these simulations has not been addressed since the focus here is on the learning capabilities of the ELM-TV algorithm, at the light of the theoretical results shown in this paper. Typically best identification accuracies in the testing phase are achieved for low K values: Please refer to [1] and [13] for an extensive evaluation from such a perspective.

Finally, it has to be underlined that similar results have been obtained by varying the number of input basis functions and with other system identification tasks, as in [1] and [13]: related results have not been reported for the sake of conciseness.

7 Conclusions

The Extreme Learning Machine paradigm has been recently extended and developed to speed up the training procedure for Time-Varying Neural Networks, keeping unchanged the network generalization performances obtained by other common training algorithms, like Back-Propagation. The effectiveness of the approach has been proved by several experiments, as shown in [1,13] and also in the several ELM-TV variants already proposed by the authors in the recent past.

In this paper, moving from the analysis already provided in the stationary case study [10], the applicability of the ELM training algorithm to TV-NN has been theoretically discussed and proved in this paper by introducing suitable algebraic tools to deal with the more complex neural architecture due to the presence of time-varying weights. Some computer simulations have been also provided in support of this.

Future work will be conducted to extend these results to other ELM-TV algorithmic variants and explore the universal approximation capability of ELM-TV on compact input sets.

References

1. Cingolani, C., Squartini, S., Piazza, F.: An extreme learning machine approach for training Time Variant Neural Networks. In: Proc. IEEE Asia Pacific Conference on Circuits and Systems, APCCAS 2008, pp. 384–387 (2008)
2. Ding, S., Zhao, H., Zhang, Y., Xu, X., Nie, R.: Extreme learning machine: algorithm, theory and applications. Artificial Intelligence Review, 1–13 (2013)
3. Grenier, Y.: Time-dependent ARMA modeling of nonstationary signals. IEEE Transactions on Acoustics, Speech and Signal Processing 31(4), 899–911 (1983)
4. Horn, R., Johnson, C.: Topics in matrix analysis. Cambridge University Press (1994)
5. Huang, G., Ding, X., Zhou, H.: Optimization method based extreme learning machine for classification. Neurocomputing 74(1), 155–163 (2010)
6. Huang, G., Wang, D., Lan, Y.: Extreme learning machines: a survey. International Journal of Machine Learning and Cybernetics, 1–16 (2011)

7. Huang, G., Zhou, H., Ding, X., Zhang, R.: Extreme learning machine for regression and multiclass classification. IEEE Transactions on Systems, Man, and Cybernetics, Part B: Cybernetics 42(2), 513–529 (2012)
8. Huang, G.B.: Learning capability and storage capacity of two-hidden-layer feedforward networks. IEEE Transactions on Neural Networks 14(2), 274–281 (2003), doi:10.1109/TNN.2003.809401
9. Huang, G.B., Chen, L., Siew, C.K.: Universal approximation using incremental constructive feedforward networks with random hidden nodes. IEEE Transactions on Neural Networks 17(4), 879–892 (2006)
10. Huang, G.B., Zhu, Q.Y., Siew, C.K.: Extreme learning machine: Theory and applications. Neurocomputing 70(1-3), 489–501 (2006)
11. Ma, W., Hsieh, T., Chi, C.: Doa estimation of quasi-stationary signals via khatri-rao subspace. In: IEEE International Conference on Acoustics, Speech and Signal Processing, ICASSP 2009, pp. 2165–2168 (2009)
12. Sidiropoulos, N., Bro, R.: On the uniqueness of multilinear decomposition of n-way arrays. Journal of Chemometrics 14(3), 229–239 (2000)
13. Titti, A., Squartini, S., Piazza, F.: A new time-variant neural based approach for nonstationary and non-linear system identification. In: Proc. IEEE International Symposium on Circuits and Systems, ISCAS 2005, pp. 5134–5137 (2005)
14. Ye, Y., Squartini, S., Piazza, F.: Incremental-Based Extreme Learning Machine Algorithms for Time-Variant Neural Networks. In: Huang, D.-S., Zhao, Z., Bevilacqua, V., Figueroa, J.C. (eds.) ICIC 2010. LNCS, vol. 6215, pp. 9–16. Springer, Heidelberg (2010)
15. Ye, Y., Squartini, S., Piazza, F.: A Group Selection Evolutionary Extreme Learning Machine Approach for Time-Variant Neural Networks. In: Neural Nets WIRN10: Proceedings of the 20th Italian Workshop on Neural Nets, pp. 22–33. IOS Press (2011)
16. Ye, Y., Squartini, S., Piazza, F.: ELM-Based Time-Variant Neural Networks with Incremental Number of Output Basis Functions. In: Liu, D., Zhang, H., Polycarpou, M., Alippi, C., He, H. (eds.) ISNN 2011, Part I. LNCS, vol. 6675, pp. 403–410. Springer, Heidelberg (2011)
17. Ye, Y., Squartini, S., Piazza, F.: On-Line Extreme Learning Machine for Training Time-Varying Neural Networks. In: Huang, D.-S., Gan, Y., Premaratne, P., Han, K. (eds.) ICIC 2011. LNCS, vol. 6840, pp. 49–54. Springer, Heidelberg (2012)
18. Ye, Y., Squartini, S., Piazza, F.: Online sequential extreme learning machine in nonstationary environments. Neurocomputing 116, 94–101 (2013)

A Quality-Driven Ensemble Approach to Automatic Model Selection in Clustering

Raffaella Rosasco[1], Hassan Mahmoud[1], Stefano Rovetta[1], and Francesco Masulli[1,2]

[1] DIBRIS, University of Genoa, Italy
[2] Center for Biotechnology, Temple University, Philadelphia, USA
{raffaella.rosasco,hassan.mahmoud,stefano.rovetta,
francesco.masulli}@unige.it

Abstract. A fundamental limitation of the data clustering task is that it has an inherent, ill-defined model selection problem: the choice of a clustering technique also implies some a-priori decision on cluster geometry. In this work we explore the combined use of two different clustering paradigms and their combination by means of an ensemble technique. Mixing coefficients are computed on the basis of partition quality, so that the ensemble is automatically tuned so as to give more weight to the best-performing (in terms of the selected quality indices) clustering method.

Keywords: Central clustering, Fuzzy clustering, Possibilistic c-Means, Spectral clustering, Clustering quality, Ensemble clustering.

1 Introduction

Clustering techniques differ in several respects, but one especially crucial area of diversity is the underlying hypothesis on what kind of aggregation should constitute a cluster. This a-priori model selection is in most cases implied in how the clustering procedure is specified.

Centroid-based clustering techniques [14,3], for instance, use a prototypical point in the data space, usually, but not exclusively, defined as the barycenter of points weighted by memberships, to represent each cluster; membership of a point in a cluster is decided on a distance-based criterion. This results in clusters which, in crisp cases, are the tiles of a Voronoi tessellation, and are therefore convex; in fuzzy clustering, tile boundaries are fuzzy, but their shape is still convex.

A completely different road is taken by affinity-based techniques, where clustering is performed on the pairwise distance or similarity (affinity) matrix. Here the implied cluster model is that of a component whose points feature a mutual intracluster similarity that is larger than the similarity to points in other clusters. The use of local connectivity may yield very complex overall shapes.

Instances of the latter family range from the single-linkage or MST clustering [13] to density-based approaches [15]. However in the last decade spectral clustering [23], borrowing from spectral graph partitioning [6], has been undisputably the most prominent example, enjoying a wealth of publications [20,21,10].

S. Bassis et al. (eds.), *Recent Advances of Neural Network Models and Applications*, Smart Innovation, Systems and Technologies 26,
DOI: 10.1007/978-3-319-04129-2_6, © Springer International Publishing Switzerland 2014

These two clustering paradigms have their own strength and weaknesses. Unfortunately, the choice is an inherent, ill-defined model selection problem. This is a fundamental limitation of clustering, for which no theoretical approach can provide a thorough answer because of the amount of subjectivity involved.

In this work we explore the combined use of two different clustering paradigms and their combination by means of an ensemble technique. Mixing coefficients are computed on the basis of partition quality, so that the ensemble is automatically tuned so as to give more weight to the best-performing clustering method. To estimate the quality of partitions, we employ well-established cluster validity indices appropriate to each individual clustering method.

2 Cluster Models

Central clustering works with distances from a reference point acting as a cluster prototype. For metric data, this results in a Voronoi tessellation of the data space, so that each cluster is a Voronoi region (convex, possibly open - although not in the unnormalized "possibilistic" case –, boundary half-way between two centroids). Spectral clustering, on the other hand, can theoretically represent any cluster shape that is an arbitrary manifold in the data space, but is very sensitive to cluster separation.

In practice, when applied to metric data, as opposed to graph data, there are limitations due to non-sparse affinity matrices, especially when working with Euclidean data. The eigensystem is usually not partitionable in an unequivocal way whenpoint density and cluster size differ across the data space by more than some threshold [19]. As a consequence, the method fails to differentiate clusters if they are joined by points with a density that is not sufficiently lower than the inner cluster density. Several ways to cope with this problem have been proposed in the literature, from completely heuristic to deeply grounded in theory [2,11,5].

In this study we include one fuzzy and two crisp centroid-based methods (k-means, fuzzy c-means, and asymmetric graded possibilistic c-means), and 3 variants of spectral clustering. All the centroid-based methods considered compute centroids $\mathbf{y}_j$ as the weighted means of points $\mathbf{x}_l$ in each cluster, $\mathbf{y}_j = \sum_l U_{lj}\mathbf{x}_l$, for a suitably defined membership matrix U whose rows correspond to points and columns to clusters.

2.1 k-Means

We use the standard k-means algorithm as the crisp centroid-based method. k-means optimizes the empiric distortion $\sum_l \sum_j U_{lj}\|\mathbf{x}_l - \mathbf{y}_j\|^2$, where U_{lj} is integer and $\sum_j U_{lj} = 1$ $\forall l$. We refer to this method as KM.

2.2 Fuzzy c-Means

The first fuzzy method is Dunn's and Bezdek's fuzzy c-means [8,4], here FCM. Here the distortion is defined as $\sum_l \sum_j U_{lj}^m\|\mathbf{x}_l - \mathbf{y}_j\|^2$, with m a fuzziness parameter, $m > 1$, again subject to $\sum_j U_{lj} = 1$ $\forall l$.

2.3 Asymmetric Graded Possibilistic *c*-Means

The *Asymmetric Graded Possibilistic c-Means* [18] is a variation over the possibilistic *c*-means by Krishnapuram and Keller [16]. The possibilistic approach consists in removing the above mentioned constraint on the rows of U but, differently from the original version of [16], we replace it with a constraint that forces rows to sum up to *at most* 1; there is also a constraint on the lower value that depends on a parameter α. We label this method as AGPCM.

2.4 Spectral Methods

As representatives of manifold-based cluster representation, we employ the Ng-Jordan-Weiss algorithm [20] in three different variations. Here the normalized Laplacian is defined as follows:

$$L_{sym} := D^{-1/2} L D^{-1/2} = I - D^{-1/2} W D^{-1/2}, \tag{1}$$

where $L = D - W$, $W_{ij} = e^{-\frac{\|\mathbf{x}_i - \mathbf{x}_j\|^2}{\sigma_i \sigma_j}}$ for $i \neq j$, $W_{ii} = 0$, $D_{ii} = \sum_j W_{ij}$, and $D_{ij} = 0$ for $i \neq j$.

The variations are related to three different ways to overcome the problem of selecting the kernel parameter, σ_i. The first way, labeled SC(fix), simply uses a fixed value computed by trial and error, i.e., $\sigma_i = \sigma \; \forall i$. Adaptive σ_i is instead computed locally for each point $\mathbf{x}_i$ in two ways. The computation is based on the ranked list of neighbors. The distances of each point from its top K neighbors (in the experiment we selected $K = 10$) are weighted with a coefficient that depends on the neighbor order, so that the nearest neighbor has the maximum weight, the second nearest a lower weight, and so on until the last neighbor considered. These weights are used to compute an aggregate weighted distance representative of the distribution, that is assumed as the value of σ_i.

The two techniques differ in the decay of the weights as a function of the neighbor's position in the ranked list. The first one, SC(var1), is based on a linear decay: the weight corresponding to any given rank ρ has weight $1 - \frac{\rho - 1}{K - 1}$. In the second variant, SC(var2), the decay is exponential, so the weight for the same neighbor rank ρ is $e^{\left(\frac{\rho-1}{K-1} - 1\right)}$.

3 Clustering Quality

A general goal of data clustering is to have low inter-cluster similarity (separation) among members of different clusters, in addition to high intra-cluster similarity (cohesion) among members of different clusters. Quality, or validity, can be estimated by exploiting different aspects of the data distribution among clusters, such as well-separateness, total number of clusters, stability, significance, and reproducibility of results even in case of perturbations. Validity indices are statistical measures used to judge the quality of a clustering by evaluating some of these criteria. There is a huge literature devoted to the subject. Some indices are specific of a given clustering method, while others are general, and work on the membership matrix only. The quality of a clustering may be measured on the basis of point distribution only, as in the case of *internal* quality measures, or with the aid of external knowledge attached to the available data, such as class labels, as in the case of *external* measures. Finally, we can mention stability-based approaches, that relate cluster quality to how much they change as data or algorithm parameters are perturbed.

Here we are mainly interested in well-established measures that can be applied to each of the cases of our interest, i.e., crisp centroid-based clusters, fuzzy centroid-based clusters, and clusters obtained by spectral partitioning.

3.1 Davies-Bouldin Index

The Davies-Bouldin index [7] is a popular measure of cluster quality for standard (crisp) clustering. It is defined as $R_{\mathrm{DB}} = \frac{1}{c}\sum_{i=1}^{c} R_i$, with $R_i = \max_{j=1,\ldots,c,\, i\neq j} \frac{s_i+s_j}{\|\mathbf{y}_i-\mathbf{y}_j\|}$ with $s_i = \frac{1}{\sum_l U_{li}} \sum_l U_{li} \|\mathbf{x}_l - \mathbf{y}_i\|$. Since its optimal value is 0, we will turn it into a quality score $q_{\mathrm{DB}} = \kappa_1 e^{-R_{\mathrm{DB}}}$ that takes on values from 0 (worst) to 1 (best), with a suitable κ_1.

3.2 Beni-Xie Index

The Beni-Xie index [24] was explicitly proposed for fuzzy clustering. It is defined as $S_{\mathrm{BX}} = \frac{\sum_l \sum_j U_{lj}^2 \|\mathbf{x}_l-\mathbf{y}_j\|^2}{(\sum_l \sum_j U_{lj})\, \min_{i,j}\|\mathbf{y}_i-\mathbf{y}_j\|^2}$. This index, like R_{DB}, can be turned into a quality score $q_{\mathrm{BX}} = \kappa_2 e^{-S_{\mathrm{BX}}}$ that takes on values from 0 (worst) to 1 (best), again with a suitable κ_2.

3.3 Eigengap Index

For spectral clustering, an usual criterion for binary partitioning quality is algebraic connectivity [9] (the second eigenvalue). Since this is not a valid criterion for completely connected, weighted graphs as those obtained from Euclidean data, we use the following related index: $q_{\mathrm{GAP}} = 1 - \lambda_{k-1}/\lambda_k$, valid for multi-way clustering as well, where λ_k is the k-th eigenvalue of the graph Laplacian. This index is positively related to quality and has 1 as its maximum value.

4 A Quality-Driven Ensemble Clustering Approach

4.1 Ensemble Clustering

Ensemble methods were originally devised and studied for classification [17], but subsequently applied to clustering. Various approaches may be used to produce diverse ensembles: using different initialization parameters, using resampling methods like bootstrapping or boosting, representing the data set from different views (e.g., subspaces). It should be noted, however, that ensemble clustering is not the same problem as ensemble classification; on the contrary, in some respects it can be stated as a dual problem. Therefore, specific techniques have been developed.

A popular approach is based on combining coassociation matrices. For each pair of data points, a coassociation matrix indicates whether (or, if fuzzy, how much) they belong to the same cluster. This is a powerful approach, adopted for instance in Fred and Jain's "evidence accumulation" method [12] or in Strehl and Ghosh' method [22].

In this case, the consensus partition is simply obtained by averaging the coassociation matrices: this works even for partitions with different cluster numbers. However, the resulting consensus matrix has a size that grows quadratically with the number of data items. Moreover, once obtained, the consensus matrix must be re-clustered, since it contains only implicit partition information.

Directly combining partition matrices, on the other hand, is possible only in restrictive hypotheses, namely, when the correspondence problem admits a satisfactory solution. This can obviously happen only if the number of clusters is the same for both partitions, or if it is possible to match several clusters from one partition on only one cluster of the other, at least to a reasonable degree.

Solving the general correspondence problem also implies finding the best-matching partition among several, as opposed to two, partitions. This is a problem of minimizing some average measure of match of the resulting partition with respect to all other partitions in the ensemble, a possibly complex optimization task.

If the above issues can be satisfactorily solved, combining partition matrices involves working with matrices whose size is linear in the number of data items and in the number of clusters.

4.2 The Proposed Approach

In the applications that motivate this study the number of data items is possibly very large; on the other hand, the number of base clustering methods is not high, with one partition per method, so the search for the best match is not computationally demanding. Therefore we opt for partition matrix combination.

Let U_k^F be the membership matrix obtained from the k-th clustering from family F, where $F = \{C, S\}$ (C for Centroid-based, S for Spectral). The consensus clustering is given by the aggregate membership matrix

$$U = \sum_k \mu_F w_k U_k^F \tag{2}$$

where μ_F is the mixing coefficient for family F, $\mu_C = 1 - \mu_S$, and w_k is the weight computed from the quality index obtained for the k-th clustering, $\sum_k w_k = 1$.

The mixing coefficients μ_F are computed from the quality indices $q(U)$ of all the clusterings in family F (U_k^F for one of the possible values of F), normalized to sum up to 1:

$$\mu'_C = \sum_k q(U_k^C) , \quad \mu'_S = \sum_k q(U_k^S) \tag{3}$$

and

$$\mu_C = \frac{\mu'_C}{\mu'_C + \mu'_S} , \quad \mu_S = \frac{\mu'_S}{\mu'_C + \mu'_S} \tag{4}$$

where $q(U)$ is the quality index computed on a generic partition matrix U, in our specific case chosen among those described above.

Of course this formulation, here presented for just two families of clustering paradigms, is readily generalizable to more than two.

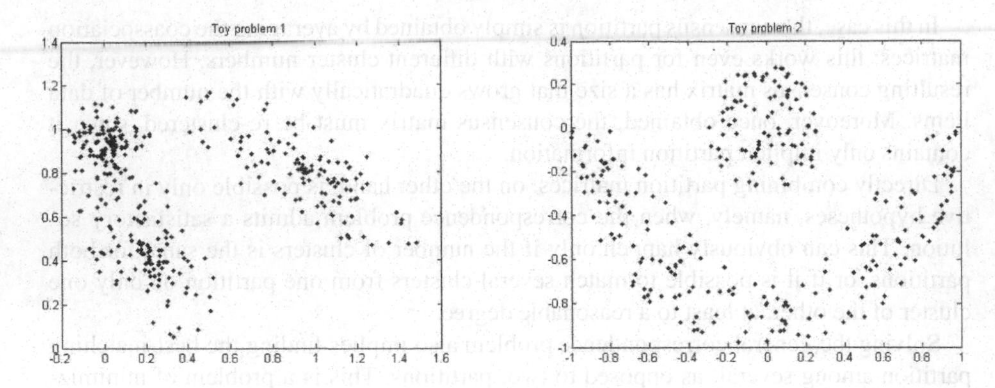

Fig. 1. Synthetic data sets 1 and 2

5 Experiments

5.1 Experimental Setup

For the experimental verification we first applied the method to two synthetic data sets for which different cluster paradigms are clearly required. They are composed of two clusters in the plane, for a total of 300 points. The first data set is composed of three elongated Gaussian blobs of 100 points each. Two blobs are partially overlapping, so that they configure one elongated cluster with two density peaks. The second data set has no linearly separable clusters, so we expect that it will be more challenging for centroid-based methods. Its two clusters are composed of 100 and 200 points respectively; the smaller one is an isotropic Gaussian cluster, while the larger one is a half-moon shaped distribution. Both data sets are shown in Figure 1.

5.2 Experimental Results

The quality index values for the individual clustering methods, computed for several choices of the number of clusters c, is shown in Table 1 for the two problems. The optimal value is highlighted in bold. It is interesting to note that AGPCM, being a variation over a *mode seeking* technique rather than a partitional one, refuses to place excess centroids in unsuitable positions, making them overlap instead. This is an inherent cluster validation measure, which not always agrees with other measures, as seen in the table, but tends to reflect the actual cluster structure better than the indices themselves. Stars (*) indicate situations where the values of validity indices are not available because solutions with that particular number of (non-coincident) centroids were not obtained. This automatic validity criterion is useful because the possibilistic clustering model is not a fuzzy generalization of crisp partitions, so the usual hypotheses underlying standard quality indices may not apply.

Table 2 indicates the mixing weights obtained for the two problems. In the first case the values are approximately the same; this indicates that the two paradigms reach about the same level of clustering quality.

Table 1. Clustering quality for the two data sets

Problem 1 (blobs)

	KM		FCM		AGPCM		SC(fix)	SC(var1)	SC(var2)
c	R_{DB}	S_{BX}	R_{DB}	S_{BX}	R_{DB}	S_{BX}	G	G	G
2	**0,290**	**0,119**	**0,363**	**0,096**	**1,044**	**0,871**	**0,0000**	0,0000	0,0000
3	0,319	0,138	0,424	0,099	1,491	1,828	0,0053	**0,0000**	**0,0000**
4	0,502	0,408	0,707	0,142	(*)	(*)	0,0721	0,0002	0,0006
5	0,540	0,270	1,737	0,429	(*)	(*)	0,2710	0,0006	0,0017
6	0,506	0,292	2,286	1,940	(*)	(*)	0,3976	0,0007	0,0020
7	0,603	0,520	3,795	6,003	(*)	(*)	0,4829	0,0011	0,0033
8	0,693	0,669	2,211	0,585	(*)	(*)	0,5208	0,0013	0,0034

Problem 2 (moon)

	KM		FCM		AGPCM		SC(fix)	SC(var1)	SC(var2)
c	R_{DB}	S_{BX}	R_{DB}	S_{BX}	R_{DB}	S_{BX}	G	G	G
2	0,455	0,273	0,5897	0,208	**0,5205**	**0,148**	**0,0000**	0,0000	0,0000
3	0,463	0,196	0,6437	0,131	1,6058	0,989	0,0036	**0,0000**	**0,0000**
4	**0,416**	**0,092**	**0,5131**	**0,068**	1,5820	0,624	0,0172	0,0001	0,0003
5	0,470	0,150	0,8019	0,121	2,5449	1,491	0,0642	0,0003	0,0009
6	0,537	0,590	1,4408	0,535	(*)	(*)	0,1687	0,0008	0,0020
7	0,539	0,431	3,0569	5,567	(*)	(*)	0,2312	0,0017	0,0041
8	0,505	0,647	3,1629	1,839	(*)	(*)	0,3239	0,0022	0,0048

(*) Values not available for AGPCM. See text.

Table 2. Mixing weights obtained for the two datasets

	μ_C	μ_S
Problem 1	0.44	0.56
Problem 2	0.23	0.77

In the second case, however, the preference for the spectral methods (manifold-based clusters) is clear, with a weight more than triple with respect to the centroidal representation.

To measure the ensemble quality, for lack of a suitable internal validation index, we resorted to the external index of cluster purity, the percentage of points in a cluster that belong to the majority class; classes here correspond to the individual distribution components used to generate the data.

The purity of clusters defined by the ensemble method is compared in Table 3 with that of individual clusterings. Although the overall purity of the ensembles is slightly smaller than that of the best performing clusterings, the method clearly points out the most suitable paradigm in each case.

Table 3. Quality, as measured by cluster purity, for the two datasets

Problem 1 (blobs)	KM	FCM	AGPCM	SC(fix)	SC(var1)	SC(var2)	Ensemble
Cluster 1	93%	93%	93%	100%	100%	100%	100%
Cluster 2	85%	85%	85%	85%	85%	85%	85%
Cluster 3	93%	93%	93%	100%	100%	100%	93%
Problem 2 (moon)	KM	FCM	AGPCM	SC(fix)	SC(var1)	SC(var2)	Ensemble
Cluster 1	90%	43%	45%	100%	100%	100%	100%
Cluster 2	57%	54%	53%	100%	100%	100%	99%

6 Conclusions

Acknowledging that no clustering paradigm is adequate for all data sets, the method presented here helps in deciding the most suitable representation scheme on an empirical basis, by evaluating the quality of clusterings and by providing an ensemble method that balances its components according to their clustering quality, as measured by quality indices.

The procedure as presented leaves room for several developments. Mixing coefficients are computed as (normalized) average qualities; other aggregation schemes may be devised, where different computations of the μ_F may be more appropriate.

Working with membership matrices is clearly a suboptimal choice with respect to coassociation methods, although with real data it is often a necessity (the method has also been applied to a database of 4 000 000 observations, for which each coassociation matrix has 16×10^{12} entries, requiring 64 TB of RAM in single precision floating point). On-line clustering procedures may be devised allowing to compute clusters without storing the whole dataset; this is an interesting avenue of research especially in the field of spectral methods.

Acknowledgments. Work partially funded by a grant of the University of Genova. Hassan Mahmoud is a PhD student in Computer Science at DIBRIS, University of Genova.

References

1. Asuncion, A., Newman, D.J.: UCI machine learning repository (2007)
2. Bach, F.R., Jordan, M.I.: Learning spectral clustering. Tech. Rep. UCB/CSD-03-1249, EECS Department, University of California, Berkeley (2003)
3. Baraldi, A., Blonda, P.: A survey of fuzzy clustering algorithms for pattern recognition. I. IEEE Transactions on Systems, Man and Cybernetics, Part B (Cybernetics) 29, 778–785 (1999)
4. Bezdek, J.C.: Pattern Recognition with Fuzzy Objective Function Algorithms. Kluwer Academic Publishers, Norwell (1981)

5. Chaudhuri, K., Chung, F., Tsiatas, A.: Spectral clustering of graphs with general degrees in the extended planted partition model. Journal of Machine Learning Research 2012, 1–23 (2012)
6. Chung, F.R.K.: Spectral Graph Theory. CBMS Regional Conference Series in Mathematics, vol. 92. American Mathematical Society (February 1997)
7. Davies, D.L., Bouldin, D.W.: A cluster separation measure. IEEE Transactions on Pattern Analysis and Machine Intelligence, PAMI 1(2), 224–227 (1979)
8. Dunn, J.C.: A fuzzy relative of the ISODATA process and its use in detecting compact well-separated clusters. Journal of Cybernetics 3, 32–57 (1974)
9. Fiedler, M.: Algebraic connectivity of graphs. Czechoslovak Mathematical Journal 23(2), 298–305 (1973)
10. Filippone, M., Camastra, F., Masulli, F., Rovetta, S.: A survey of kernel and spectral methods for clustering. Pattern Recognition 40(1), 176–190 (2008)
11. Fischer, I., Poland, J.: New methods for spectral clustering. Tech. rep., IDSIA/USI-SUPSI (2004)
12. Fred, A.L.N., Jain, A.K.: Data clustering using evidence accumulation. In: International Conference on Pattern Recognition, vol. 4 (2002)
13. Gower, J.C., Ross, G.J.S.: Minimum spanning trees and single linkage cluster analysis. Journal of the Royal Statistical Society 18(1), 54–64 (1969)
14. Jain, A.K., Dubes, R.C.: Algorithms for Clustering Data. Prentice Hall, Englewood Cliffs (1988)
15. Kriegel, H.P., Kröger, P., Sander, J., Zimek, A.: Density-based clustering. Wiley Interdisciplinary Reviews: Data Mining and Knowledge Discovery 1(3), 231–240 (2011)
16. Krishnapuram, R., Keller, J.M.: A possibilistic approach to clustering. IEEE Transactions on Fuzzy Systems 1(2), 98–110 (1993)
17. Kuncheva, L.: Combining pattern classifiers. Methods and Algorithms. Wiley, Chichester (2004)
18. Masulli, F., Rovetta, S.: Soft transition from probabilistic to possibilistic fuzzy clustering. IEEE Transactions on Fuzzy Systems 14(4), 516–527 (2006)
19. Nadler, B., Galun, M.: Fundamental limitations of spectral clustering. In: Advances in Neural Information Processing Systems, vol. 19, p. 1017 (2007)
20. Ng, A.Y., Jordan, M.I., Weiss, Y.: On spectral clustering: Analysis and an algorithm. In: Dietterich, T.G., Becker, S., Ghahramani, Z. (eds.) Advances in Neural Information Processing Systems, vol. 14. MIT Press, Cambridge (2002)
21. Shi, J., Malik, J.: Normalized cuts and image segmentation. IEEE Transactions on Pattern Analysis and Machine Intelligence 22(8), 888–905 (2000)
22. Strehl, A., Ghosh, J.: Cluster ensembles — a knowledge reuse framework for combining multiple partitions. J. Mach. Learn. Res. 3, 583–617 (2003)
23. Von Luxburg, U.: A tutorial on spectral clustering. Statistics and Computing 17(4), 395–416 (2007)
24. Xie, X.L., Beni, G.: A validity measure for fuzzy clustering. IEEE Transactions on Pattern Analysis and Machine Intelligence 13(8), 841–847 (1991)

An Adaptive Reference Point Approach to Efficiently Search Large Chemical Databases

Francesco Napolitano[1,2], Roberto Tagliaferri[1], and Pierre Baldi[2]

[1] Department of Informatics,
University of Salerno, Fisciano (SA), Italy
[2] Institute for Genomics and Bioinformatics,
School of Information and Computer Sciences,
University of California-Irvine, Irvine, CA 92697-3435, USA
{f.napolitano,r.tagliaferri}@unisa.it, pfbaldi@uci.edu

Abstract. The ability to rapidly search large repositories of molecules is a crucial task in chemoinformatics. In this work we propose AOR, an approach based on adaptive reference points to improve state of the art performances in querying large repositories of binary fingerprints basing on the Tanimoto distance. We propose a unifying view between the context of reference points and the previously proposed hashing techniques. We also provide a mathematical model to forecast and generalize the results, that is validated by simulating queries over an excerpt of the ChemDB. Clustering techniques are finally introduced to improve the performances. For typical situations the proposed algorithm is shown to resolve queries up to 4 times faster than compared methods.

Keywords: molecular fingerprits, chemical database, binary vector search.

1 Introduction

Repositories of molecules are of huge importance in chemoinformatics [1–4]. They provide the researchers with the ability of exploring the space of molecules, starting from a given compound of interest and looking for similar molecules with desired additional features. Even though there exist complex and accurate molecular descriptors, like 3D atom to atom distances, simple binary vectors (molecular fingerprints) are commonly used to allow for faster search. The actual meaning of the figerprints bits is irrelevant for the purpose of this paper, thus our work is valid for any database of binary vectors, as long as the Tanimoto Similarity (also knowns as Jaccard Similarity) is used to compare them. The performance analysis, on the other hand, is bound to the probability distribution of the 1-bits across the vectors, which is expected to be strongly non linear [5].

Let $\boldsymbol{A} = (A_i)$ be a binary molecular fingerprint and $A = \sum_i A_i$ the number of 1s in the fingerprint. A is also called *length of* $\boldsymbol{A}$. If a set S contains only molecules of the same length, we call *length of* S the length of the molecules it contains. Let $\boldsymbol{B}$ also be a binary fingerprint. We define $A \cap B = |\boldsymbol{A} \cap \boldsymbol{B}|$ and

S. Bassis et al. (eds.), *Recent Advances of Neural Network Models and Applications*, Smart Innovation, Systems and Technologies 26,
DOI: 10.1007/978-3-319-04129-2_7, © Springer International Publishing Switzerland 2014

$A \cup B = |\boldsymbol{A} \cup \boldsymbol{B}|$[1]. According to this notation, the Tanimoto Similarity between two molecules is given by:

$$T(\boldsymbol{A}, \boldsymbol{B}) = \frac{A \cap B}{A \cup B} \tag{1}$$

This measure is often used to search for similar molecules in a database given a query molecule (for example in the ChemDB [13]). Since current repositories contain millions of molecules, efficient techniques are necessary to perform such searches. [10] showed a simple bound (we will refer to it as the *Bit Bound*), based on the number of bits that are equal to 1 in the fingerprints being compared:

$$T(\boldsymbol{A}, \boldsymbol{B}) \leq \frac{min(A, B)}{max(A, B)} \tag{2}$$

which is very efficient and can be applied in constant time if the fingerprints are sorted basing on their 1-bit count. In fact, requiring a minimum similarity t between a query fingerprint $\boldsymbol{A}$ and any target fingerprint $\boldsymbol{B}$, we can discard $\boldsymbol{B}$ if $B < tA$ or $B > A/t$.

In the past [12] showed how the triangular inequality can be exploited to prune data from a generic metric space, organizing it into clusters whose distance from an assigned reference vector is similar. The authors state that this very general approach is not efficient in high-dimensional spaces. On the other hand, [11] recently introduced the Intersection Inequality and proved that it is sharper than the triangular inequality for the Tanimoto Similarity. Such inequality was exploited in [8] in the form of a hashing approach, where the bits of each fingerprint are partitioned into subsets and the number of 1-bits in each subset is stored in order to prune the search (see next Section). In particular, in [8] subsets formed according to bit positions $k \, mod \, M$ were studied, with $k = 1, \ldots, M$ and M as a parameter to be chosen. A pruning approach composed of a first cut with $M = 2$ and a second cut with $M = 256$ was shown to be optimal according to simulated search running time and mathematical predictions. In the next Section we will show that the hashing approach with $M = 2$ can be equivalently seen as a single reference point technique based on the Intersection Inequality.

In this paper we propose an adaptive criterion for the choice of a reference point based on the set of molecules to be searched. We also provide a mathematical model to predict the performances of the algorithm and validate it through simulated queries against an excerpt of the ChemDB [13]. We show how the actual performances of the adaptive scheme are typically 2 to 4 times better than the $k \, mod \, M, M = 2$ fixed scheme. This is shown indirectly through comparison with the performance of the Bit Bound, which is in practice not worse than the $k \, mod \, M, M = 2$ hashing approach. As a last improvement we show how clustering techniques can be used to produce more efficient partitions of the database allowing better pruning.

[1] With abuse of notation, we use $\boldsymbol{A}$ both as an ordered vector of binary values and as the corresponding set of positions i such that $A_i = 1$.

In Section 2 we estabilish a link beetween the concepts of Hashing Sets and Reference Points, which, to our knowledge, has not been pointed out before. In Section 3 we introduce the concept of adaptive reference point. Section 4 shows an algorithm exploiting the concept. Section 5 provides a mathematical model for predicting and generalizing the performance of the algorithm. Section 6 shows experimental results that are compatible with the model and comparisons with the predicted performance. Section 7 introduces a strategy to improve the algorithm through the use of clustering. Overall performance is reported in Section 8. Conclusions and ideas for future work are reported in section 9.

2 Hashing Sets and Reference Points

Reference [12] proposed a general framework for pruning databases during searches that can be applied to any metric space. Let X be such a space, $\boldsymbol{A}, \boldsymbol{B}, \boldsymbol{C} \in X$ and d any metric on X. By definition, the triangular inequality holds for d, that is $d(\boldsymbol{A}, \boldsymbol{B}) \leq d(\boldsymbol{B}, \boldsymbol{C}) + d(\boldsymbol{A}, \boldsymbol{C})$. Consider $\boldsymbol{A}$ as a *reference point*, $\boldsymbol{B}$ as an object in a database to be searched and $\boldsymbol{C}$ as a query object. When searching for all the objects $\boldsymbol{B} \in X$ whose distance from $\boldsymbol{C}$ is smaller than a threshold t, all objects $\boldsymbol{B}$ such that $|d(\boldsymbol{A}, \boldsymbol{C}) - d(\boldsymbol{A}, \boldsymbol{B})| \geq t$ can be discarded. If the database is structured in bins such that all the objects belonging to the same bin are at a constant distance from the reference point $\boldsymbol{A}$, that is $\forall \boldsymbol{B} \in X^1\, d(\boldsymbol{A}, \boldsymbol{B}) = k$, $X^1 \subseteq X$, than the whole cluster can be bound at once and eventually discarded.

If $T(\boldsymbol{A}, \boldsymbol{B})$ is the Tanimoto Similarity for the two binary vectors $\boldsymbol{A}$ and $\boldsymbol{B}$, than $1 - T(\boldsymbol{A}, \boldsymbol{B})$ is known to be a metric (the Tanimoto distance). Thus, the reference points approach can be applied to any database of molecular binary fingerprints based on the Tanimoto distance. Unfortunately this is not efficient due to typical high-dimensionality of the fingerprints. On the other hand reference [11] shows an inequality for the Tanimoto distance that can be used in place of the triangular inequality and is proven to be sharper. It is the Intersection Inequality:

$$T(\boldsymbol{B}, \boldsymbol{C}) \leq \frac{min(\beta, \gamma) + min(B - \beta, C - \gamma)}{B + C - min(\beta, \gamma) - min(B - \beta, C - \gamma)} \tag{3}$$

where $\beta = A \cap B$ and $\gamma = A \cap C$. The Bit Bound [10] algorithm can be seen as a reference point technique where the reference point is chosen to be the fingerprint made of all 1-bits and the Intersection Inequality is used in place of the triangular inequality. In such case inequality (3) reduces to inequality (2) since: $\beta = B$, $\gamma = C$ and $B + C - \min(\beta, \gamma) = \max(B, C)$. Moreover fingerprints with the same number of 1-bits are at a constant Tanimoto distance from such reference point since in this case $A \cap B = A \cap C = A \cup B = A \cup C = B = C$.

Reference [8] showed a generalization of the Bit Bound called Modulo M Hashing. Let a_i^M be the number of 1-bits in a fingerprint $\boldsymbol{A}$ falling in positions that are congruent to i modulo M[2]. Than the following inequality holds:

[2] The bound actually holds for any partition of the bit positions, not necessarily obtained with the modulo M scheme.

$$T(\boldsymbol{B},\boldsymbol{C}) \leq \frac{\sum_{i=1}^{M} \min(b_i^M, c_i^M)}{\sum_{i=1}^{M} \max(b_i^M, c_i^M)} \tag{4}$$

Let $\boldsymbol{A}_i^M$ be a reference point having 1s in positions that are congruent to i modulo M. Then $A_i^M \cap B = \beta_i^M = b_i^M$, $A_i^M \cap C = \gamma_i^M = c_i^M$. Thus we conclude that the hashing scheme can be seen as a particular set of reference points.

Let us consider the case $M = 2$. In this case $b_2^2 = B - b_1^2$ and $c_2^2 = C - c_1^2$. Renaming $b_1^2 = \beta$ and $c_1^2 = \gamma$, inequality (4) for $M = 2$ can be rewritten as:

$$T(\boldsymbol{B},\boldsymbol{C}) \leq \frac{\min(\beta,\gamma) + \min(B-\beta, C-\gamma)}{\max(\beta,\gamma) + \max(B-\beta, C-\gamma)} \tag{5}$$

which reduces to inequality (3), that is a single reference point bound. In fact, choosing the reference point $\overline{\boldsymbol{A}_i^2}$, i.e. the complement of $\boldsymbol{A}_i^2$, is the same as switching the values of b_2^2 with b_1^2 and c_1^2 with c_2^2 and has no effect on the bound. When reference points are used explicitly, as in the adaptive algorithm we present, this property can be exploited to store the shorter of any complementary reference points couple, thus gaining improvements in both memory consumption and, due to faster intersection computations, overall algorithm speed.

In the rest of the paper we will use the reference point perspective for its wider generality.

3 An Adaptive Reference Point Scheme

From inequality (3), applying a similarity threshold t, the following cut-off criterion can be derived:

$$\beta > \gamma + \frac{B - tC}{t+1}, \; \beta < \gamma + \frac{tB - C}{t+1} \tag{6}$$

This was shown in [8] as a bound on the $M = 2$ hashing approach and is reported here in a slightly different form with notational adaptation. This criterion can be used to prune the search in a purposely structured database (see Fig. 1). The main problem for the application of the criterion is of course the choice of the reference point. When $\boldsymbol{A} = [1, 1, \ldots, 1]$ and no recursion is performed the algorithm reduces to the Bit Bound. In this context we define an algorithm *adaptive* if the choice of $\boldsymbol{A}$ depends on S_i (see Algorithm 1). We introduce an adaptive scheme in the next section.

Now let us consider the particular choice $\boldsymbol{A} = \bigcup_{\boldsymbol{B} \in S} \boldsymbol{B}$ for a reference point $\boldsymbol{A}$ in a bin S. In this case $\forall \boldsymbol{B} \in S$, $A \cap B = \beta = B$ and the inequalities 6 reduces to:

$$B > \frac{\gamma(t+1)}{t} - C, \; B < \gamma(t+1) - C \tag{7}$$

The first inequality is verified when γ is large enough, while the second one is verified when γ is small enough. Notice that when $\gamma \to C$ the adaptive OR-based criterion (AOR) reduces to the Bit Bound. Since this method will be

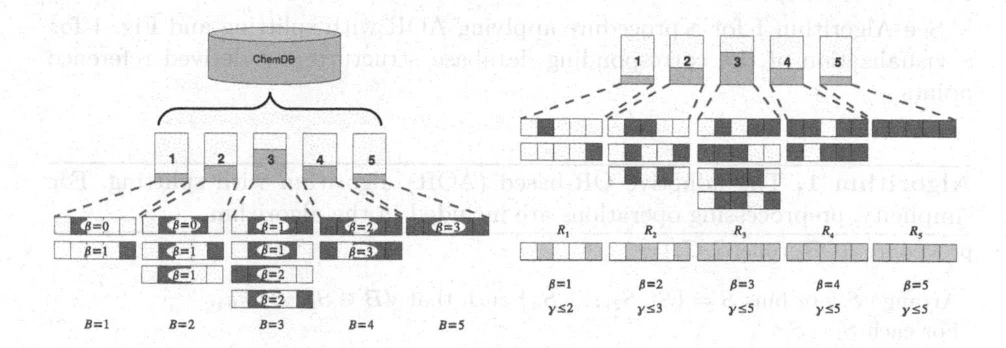

Fig. 1. Organization of a molecular database for the application of a generic reference point scheme (left) and the AOR scheme (right). For the former, molecules of equal length B_i are stored in the i-th bin. Inside each bin molecules are sorted according to their β values, that are computed from the intersection with the reference point assigned to the bin. In the example the reference point is the fingerprint having all ones in odd positions. The bound on the values of B is first applied. Then, inside each bin i, with B_i holding constant, the bound on β is applied. For AOR (right) reference points are chosen as the binary OR of all the molecules in a bin. As a consequence, molecules inside a bin have the same value of β and it is equal to B. Corresponding acceptable values of γ are shown.

applied after the Bit Bound (see next Section), $\gamma \to 0$ is desired. Moreover, this choice of the reference point implicitly maximizes β. We also experimentally verified that the contribution of the second inequality to the amount of pruning after the application of the Bit Bound is negligible. For these reasons we focus on the first inequality.

4 The AOR Algorithm with Splitting

The AOR scheme can not be used recursively inside each bin, since the same reference point would be chosen at each level. To allow for further pruning, *sub-bin splitting* is introduced, enabling the use of different reference points per bin while keeping the property $\beta = B$.

Each S_i is split into k sub-bins S_{ij} producing A_{ij}-s with $\beta_{ij} = B_j$, $i = 1, \ldots, k$, but smaller γ_{ij}-s, which is desirable for the efficacy of the criterion. Note that when $k \to |S_i|$, $|S_{ij}| \to 1$ and the AOR $\to$ Tanimoto similarity. In fact, since $B = \beta$ and $\gamma = B \cap C$,

$$B + \gamma - \beta = B \cap C, \quad C + B - \gamma = B \cup C \tag{8}$$

and since $B \cap C \leq B \cup C$, inequality (3) becomes:

$$\frac{min[(B + \gamma - \beta),\, (C + \beta - \gamma)]}{max[(B + \gamma - \beta),\, (C + \beta - \gamma)]} = \frac{B \cap C}{C \cup B} = T(\boldsymbol{B}, \boldsymbol{C})$$

See Algorithm 1 for a procedure applying AOR with splitting and Fig. 1 for a visualization of the corresponding database structure and derived reference points.

Algorithm 1. The adaptive OR-based (AOR) algorithm with splitting. For simplicity, preprocessing operations are included in the algorithm.

---

procedure AORprune($\mathcal{S}, \boldsymbol{C}, t$)

Arrange $\mathcal{S}$ into bins $S = \{S_1, S_2, \ldots, S_n\}$ such that $\forall \boldsymbol{B} \in S_i,\ B = k_i$.
For each $S_i \in S$:
  $T' \leftarrow \emptyset$, $\beta \leftarrow B \leftarrow k_i$.
  Split S_i into sub-bins $T = \{T_1, T_2, \ldots, T_s\}$.
  For each $T_i \in T$:
    Choose the reference point $\boldsymbol{R} = \bigcup_{\boldsymbol{F} \in T_i}$.
    Compute $\gamma = R \cap C$.
    If $B > \frac{\gamma(t+1)}{t} - C$ or $B < \gamma(t+1) - C$, prune T_i.

---

5 Mathematical Model

In order to assess the probability of pruning a bin S_i, our mathematical model has the purpose of assessing how likely is the following to happen:

$$\gamma_i > \lambda_i = (B_i + C)\frac{t}{t+1}, \quad \gamma_i < \mu_i = \frac{B_i + C}{t+1} \tag{9}$$

inside each bin S_i, Inequality (9) is derived from the AOR criterion, inequality (7), and expresses the possibility of pruning S_i as a function of γ_i.

5.1 Probability Profiles

We define probability profiles to characterize basic statistics for a set of fingerprints. The probability profile for a set composed of m-bits fingerprints is a stochastic process $\{F_t : t \in [1, \ldots, n]\}$, where each random variable F_t models the probability that the t-th bit of a fingerprint randomly chosen from the bin is equal to 1. The idea is detailed in the following.

Let $S = \{\boldsymbol{B}_1, \boldsymbol{B}_2, \ldots, \boldsymbol{B}_n\}$ a bin of binary fingerprints and $\boldsymbol{B}_j(i)$ the value of the i-th bit of $\boldsymbol{B}_j$. We approximate the probability $p(i)$ that the $\underline{i}$-th bit of a randomly chosen fingerprint $\boldsymbol{B} \in S$ as its frequency, and define it as $p_S(i) = \frac{1}{n}\sum \boldsymbol{B}_j(i)$, $B_j \in S$, $i = 1, \ldots, m$. This way statistical independence between $p_S(i)$-s is implicitly assumed, thus inducing the model to slightly overestimates the value of γ (see next subsections and 3). The results predicted by the model can thus be assumed to be slightly pessimistic.

We define *probability profile of a set of fingerprints* S the ordered set of $p_S(i)$-s, $i = 1, \ldots, m$, such that $p_S(1) \leq p_S(2), \ldots, \leq p_S(m)$.

If S_i is a binary fingerprint conversion (with $m = 1024$) of the i-th bin of molecules from the ChemDB (in which bin S_i contains molecules of length i), than Fig. 2 shows the profile of the S_i-s of length between 50 and 450 (we call them *reliable*).

For each one of such bins S_i we built a composition of two linear models, the first one modeling the probabilities $p_{S_i}(1), \ldots, p_{S_i}(i)$ and the second one modeling the probabilities $p_{S_i}(i+1), \ldots, p_{S_i}(m)$. The corresponding fitting is shown in Fig. 2.

To obtain a profile for the remaining bins, which contain a small number of fingerprints, we in turn extrapolate a model of the parameters defining the models of the reliable bins.

Let M_i be the model of S_i, such that:

$$p_{S_i}(j) \simeq M_i(j|a_i, b_i, c_i, d_i) = \begin{cases} a_i j + c_i & i = 1, \ldots, j \\ b_i j + d_i & i = j+1, \ldots, n \end{cases} \tag{10}$$

Then, in order to model the parameters a, b, c, and d, we use the following four models:

$$a_i \simeq M^1(i|p_1, p_2) = p_1 i + p_2 \tag{11}$$

$$b_i \simeq M^2(i|p_3, p_4) = p_3 i + p_4 \tag{12}$$

$$c_i \simeq M^3(i|p_5, p_6, p_7,) = p_5 i^{p_6} + p_7 \tag{13}$$

$$d_i \simeq M^4(i|p_8, p_9, p_{10}) = p_8 i^{p_9} + p_{10} \tag{14}$$

Once the parameters $p_1, \ldots, p_{10}$ are estimated, the models M_i can be built for $i = 1, \ldots, 1024$.

Consider now the profile of the reference point $\boldsymbol{A}_i$ of the bin S_i, that is p_{A_i}. Since $\boldsymbol{A}_i(j) = \bigcup_{\boldsymbol{B} \in S_i} \boldsymbol{B}(i)$, under assumption of independence, the probability of the j-th bit of $\boldsymbol{A}_i$ of being 1 is given by:

$$p_{A_i}(j) = 1 - [1 - p_{S_i}(j)]^{|S_i|} \tag{15}$$

Let $\boldsymbol{C}$ be a query fingerprint. We take $p_{S_i}, i = C$ as its profile. We can then easily derive the profile of $\boldsymbol{\gamma}_i$ for each bin i, as $p_{\gamma_i} = p_{A_i} \cdot p_{S_i}$.

5.2 Expected Results

To study the likely outcome of the application of the AOR criterion, we need to forecast the value of γ_i and compare it with λ_i and μ_i. Such bounds can be easily computed for each bin S_i from ineq. 9, letting $B_i = \beta_i = i$ and assuming, for example, $t = 0.9$.

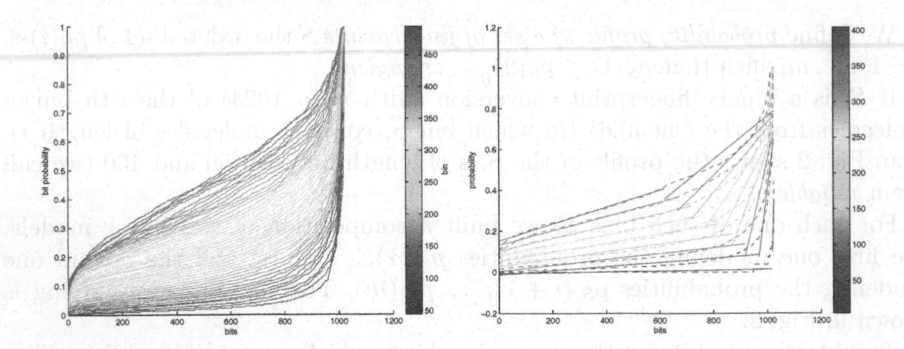

Fig. 2. Left: profiles of the ChemDB reliable (see main text) bins. Different bin lengths are represented by colors, while bits positions are on the x axis and probabilities of having a 1, estimated by frequency in the bin, on the y axis. Right: linear fitting of a subset of the same profiles.

The value of γ_i can be modeled as the outcome of a Bernoulli experiment made of n coin-flips, each one having probability $\hat{p}_i = \frac{1}{n}\sum_{i=1}^{n} p_{\gamma_i}$. Thus we have:

$$P(\gamma_i = k) = \binom{n}{k} \cdot \hat{p}_i^k (1 - \hat{p}_i)^{n-k} \tag{16}$$

and finally:

$$P(\gamma_i < \lambda) = \sum_{k=1}^{\lfloor \lambda \rfloor - 1} P(\gamma_i = k) \tag{17}$$

$$P(\gamma_i > \mu) = \sum_{k=\lfloor \mu \rfloor + 1}^{n} P(\gamma_i = k) \tag{18}$$

The two probabilities can be computed faster exploiting the incomplete regularized beta function [6]:

$$P(\gamma_i \leq \lambda_i) = I_{1-\hat{p}_i}(n - \lambda_i, \lambda_i + 1) = (n - \lambda_i)\binom{n}{\lambda_i} \int_0^{1-\hat{p}_i} t^{n-\lambda_i-1}(1-t)^{\lambda_i} dt \tag{19}$$

$$P(\gamma_i \geq \mu_i) = 1 - I_{1-\hat{p}_i}(n - \mu_i, \mu_i + 1) = 1 - (n - \mu_i)\binom{n}{\mu_i} \int_0^{1-\hat{p}_i} t^{n-\mu_i-1}(1-t)^{\mu_i} dt \tag{20}$$

Fig. 3 shows a comparison between the values of γ_i for the reliable bins computed on the sample DB (see next section) and the values of γ_i for the same bins as forecasted by the model, together with the prediction on all the bins, using the extrapolated profiles. Fig. 4 shows the probability of pruning a sub-bin as a function of its length for queries of 5 different sizes when using $t = 0.9$, both as simulated and predicted, together with the same predictions for all the query sizes.

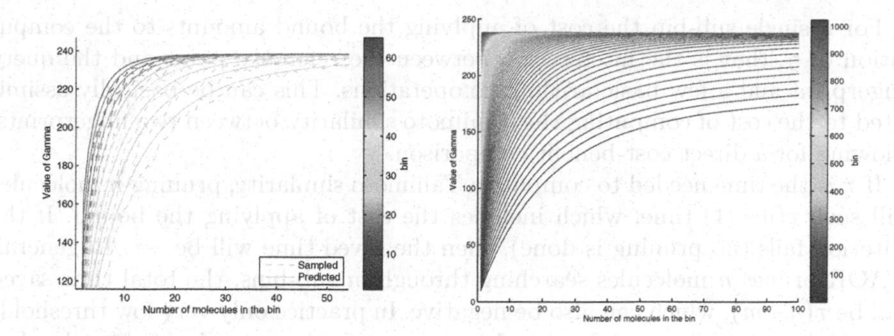

Fig. 3. Left: accuracy of the prediction of the values of γ_i for the reliable bins. Number of molecules in the sub-bin is reported on the x axis. The dashed lines represent values of γ_i directly computed on sample data and averaged over 100 trials. The continuous lines show the corresponding predictions. Different colors represent different bin lengths. Right: prediction of the values of γ_i over all the possible bins for molecules of length 1024.

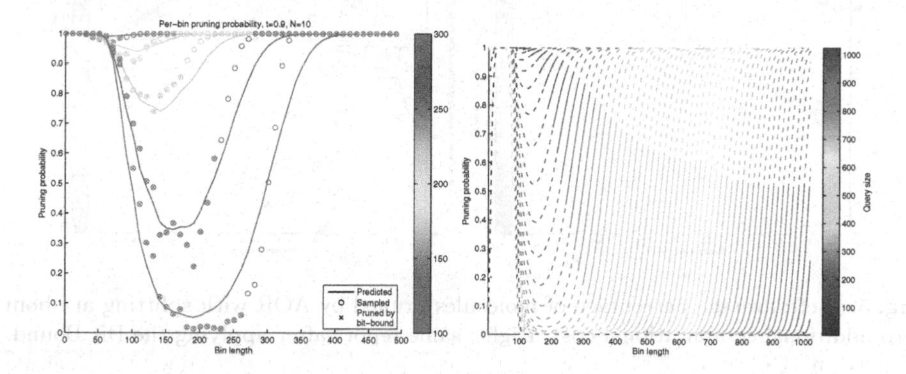

Fig. 4. Left: pruning probability as estimated by the mathematical model (lines) and obtained by simulation (circles). Bin length is on the x axis, probability of pruning each sub-bin is shown on the y axis and color represent different query sizes. The prediction assumes $t = 0.9$ and sub-bins of size 10. Circles marked with an 'x' represent bins that would be pruned by the Bit Bound. Right: Pruning probability as estimated by the mathematical model for all bin sizes up to 1024. Dashed parts represent cases that would be pruned by the bit bound.

6 Experiments

In the following, experimental results are shown, using AOR with splitting and an increasing number of sub-bins. Since the bin size varies, we split any bin of n molecules into $max(\lfloor \frac{n}{k} \rfloor, 1)$ sub-bins, where N is the whole database size and k is the parameter varying across experiments. The database used is an excerpt of 100.000 molecules evenly extracted from the ChemDB plus another 100 fingerprints used as queries. We tried 30 different threshold values, linearly spaced between 0.1 and 0.9 and the value of k between 1 and 40.

For a single sub-bin the cost of applying the bound amounts to the computation of γ, that is the intersection between the reference point and the query fingerprint and a few basic arithmetic operations. This can be basically assimilated to the cost of computing the Tanimoto similarity between two fingerprints, allowing for a direct cost-benefit comparison.

If τ is the time needed to compute a Tanimoto similarity, pruning n molecules will save $\tau(n-1)$ time, which includes the cost of applying the bound. If the criterion fails (no pruning is done), then the saved time will be $-\tau$. In general, if AOR prunes n molecules searching through m sub-bins, the total time saved will be $\tau(n-m)$, which can also be negative. In practice only very low thresholds can produce negative speed-up and in such cases the second recursion level of the algorithm could be skipped. Fig. 5 show the cost-corrected performances of the algorithm.

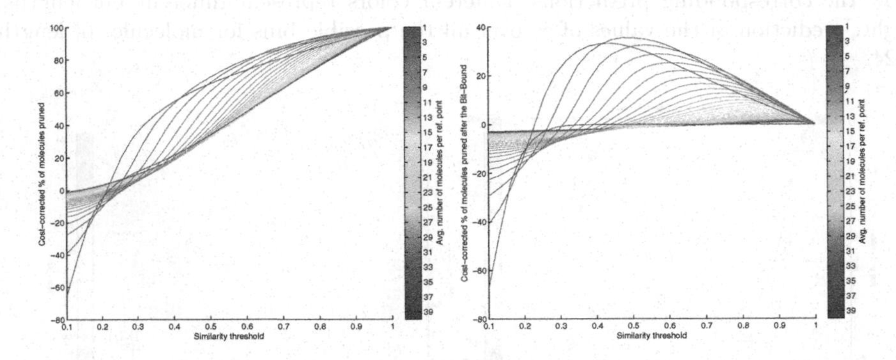

Fig. 5. Left: average percentage of molecules pruned by AOR with splitting at about zero additional computational cost. Right: same report after applying the Bit Bound.

7 Further Improvement with Clustering

A significant improvement in performance can be obtained optimizing the split process, which was considered to be performed randomly up to this point. The aim of splitting the bins into sub-clusters is to reduce the expected value of γ. Note that, for a query fingerprint $\boldsymbol{C}$ and a reference point A_i, γ_i is bounded by:

$$\gamma_i = A_i \cap C \leq min(A_i, C) \tag{21}$$

thus γ_i can be made smaller on average by reducing the length of the reference points $\boldsymbol{A}_i = \bigcup_{\boldsymbol{B} \in S_i}$. This happens if molecules with high intersection are in the same sub-bin. Clustering techniques can be applied to the aim. We replaced the random splitting step in the AOR algorithm with Hierarchical Clustering through Hamming distance. We chose the cutting threshold according to the number of clusters to form (related with the value of k from the previous section). Ward's linkage method was chosen for its tendency to build balanced clusters,

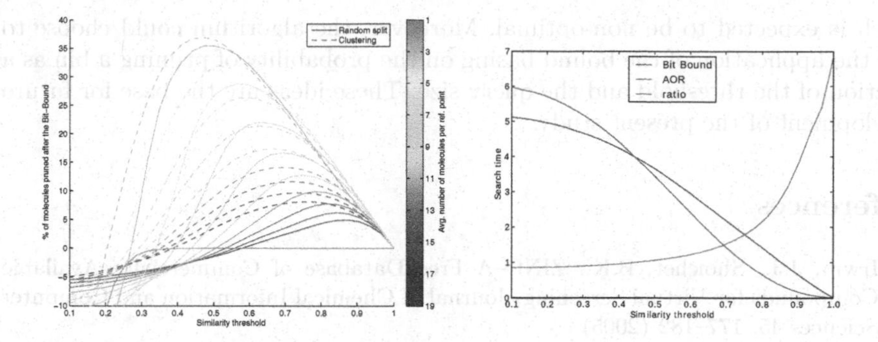

Fig. 6. Left: average percentage of molecules pruned by AOR with splitting (continuous lines) and clustering (dashed lines), excluding the pruning made by the Bit Bound and subtracting the cost of applying the bound, over 100 queries, as a function of the similarity threshold and the number of clusters. Right: actual timing performance for AOR (green), the Bit Bound (blue) and their ratio (red).

which is useful to avoid high dynamics in the cost-benefit ratio that would be tricky to control. This attempt brings significant improvements, as shown in Fig. 6.

8 Actual Search Time Test

As a final experiment, we measured the average search time using the same data as in previous sections. The modulo M approach for $M = 2$ is known not to be more efficient than the Bit Bound, which we compare AOR with. Fig. 6 shows the results of the test. AOR is faster than the Bit Bound for treshold values larger than about 0.44: in the other cases the second phase of the algorithm can be simply skipped. For thresholds between 0.8 and 0.9, which is a usual choice, AOR is 2-folds to 4-folds faster than the Bit Bound. Other bounds, like the modulo M with $M = 256$ can be applied on the remaining fingerprints.

9 Conclusions and Future Work

In this paper we showed how the Hashing approach can be seen as a specialized Intersection Inequality based istance of the earlier reference points framework. Under this framework, we derived an adaptive approach, an algorithm to apply it efficiently and a mathematical model to predict and generalize its results. By direct experiments we showed that with a 0.9 threshold AOR resolves an average query in almost one fourth of the time needed by the Bit Bound. We finally showed that clustering techniques can be applied to further improve the results.

There is still room for improvement with the AOR algorithm. For simplicity, in our simulations we used the same sub-bin (or sub-clusters) size for each bin, which is expected to be non-optimal. Moreover, the algorithm could choose to skip the application of the bound basing on the probability of pruning a bin as a function of the threshold and the query size. These ideas are the base for future development of the present study.

References

1. Irwin, J.J., Shoichet, B.K.: ZINC–A Free Database of Commercially Available Compounds for Virtual Screening. Journal of Chemical Information and Computer Sciences 45, 177–182 (2005)
2. Chen, J., Linstead, E., Swamidass, S.J., Wang, D., Baldi, P.: ChemDB Update–Full Text Search and Virtual Chemical Space. Bioinformatics 23, 2348–2351 (2007)
3. Wang, Y., Xiao, J., Suzek, T., Zhang, J., Wang, J., Bryant, S.: PubChem: a public information system for analyzing bioactivities of small molecules. Nucleic Acids Research 37, W623–W633 (2009)
4. Sayers, E., Barrett, T., Benson, D., Bolton, E., Bryant, S., Canese, K., Chetvernin, V., Church, D., DiCuccio, M., Federhen, S., et al.: Database resources of the National Center for Biotechnology Information. Nucleic Acids Research 38, D5–D16 (2010)
5. Benz, R.W., Swamidass, S.J., Baldi, P.: Discovery of Power-Laws in Chemical Space. Journal of Chemical Information and Modeling 48, 1138–1151 (2008)
6. Paris, R.B.: Incomplete beta functions. In: Olver, F.W.J., Lozier, D.M., Boisvert, R.F., et al. (eds.) NIST Handbook of Mathematical Functions. Cambridge University Press (2010) ISBN 978-0521192255
7. Shapiro, M.: The choice of reference points in best-match file searching. Communications of the ACM 20, 339–343 (1977)
8. Nasr, R., Hirschberg, D.S., Baldi, P.: Hashing Algorithms and Data Structures for Rapid Searches of Fingerprint Vectors. J. Chem. Inf. Model. 50(8), 1358–1368 (2010), doi:10.1021/ci100132g
9. Bohacek, R.S., McMartin, C., Guida, W.C.: The art and practice of structure-based drug design: A molecular modelling perspective. Medicinal Research Reviews 16(1), 3–50 (1996)
10. Swamidass, S.J., Baldi, P.: Bounds and Algorithms for Fast Exact Searches of Chemical Fingerprints in Linear and Sublinear Time. Journal of Chemical Information and Modeling 47(2), 302–317 (2007)
11. Baldi, P., Hirschberg, D.S.: An Intersection Inequality Sharper than the Tanimoto Triangle Inequality for Efficiently Searching Large Databases. J. Chem. Inf. Model. 49(8), 1866–1870 (2009), doi:10.1021/ci900133j
12. Burkhard, W.A., Keller, R.M.: Some approaches to best-match file searching. Communications of the ACM Archive 16(4), 230–236 (1973)
13. Chen, J.H., Linstead, E., Swamidass, S.J., Wang, D., Baldi, P.: ChemDB update–full-text search and virtual chemical space. Bioinformatics 23(17), 2348–2351 (2007)

A Methodological Proposal for an Evolutionary Approach to Parameter Inference in MURAME-Based Problems

Marco Corazza[1,3], Stefania Funari[2], and Riccardo Gusso[1]

[1] Department of Economics, Ca' Foscari University of Venice
[2] Department of Management, Ca' Foscari University of Venice
[3] Advanced School of Economics of Venice
Sestiere Cannaregio n. 873, 30121 Venice, Italy
{corazza,funari,rgusso}@unive.it

Abstract. In this paper we propose an evolutionary approach in order to infer the values of the parameters for applying the MURAME, a multicriteria method which allows to score/rank a set of alternatives according to a set of evaluation criteria. This problem, known as preference disaggregation, consists in finding the MURAME parameter values that minimize the inconsistency between the model obtained with those parameters and the true preference model on the basis of a reference set of decisions of the Decision Maker. In order to represent a measure of inconsistency of the MURAME model compared to the true preference one, we consider a fitness function which puts emphasis on the distance between the scoring of the alternatives given by the Decision Maker and the one determined by the MURAME. The problem of finding a numerical solution of the involved mathematical programming problem is tackled by using an evolutionary solution algorithm based on the Particle Swarm Optimization. An application is finally provided in order to give an initial assessment of the proposed approach.

Keywords: Preference disaggregation, MURAME, Particle Swarm Optimization.

1 Introduction

In this paper we propose an evolutionary approach in order to infer the values of the parameters (weights of criteria, preference, indifference and veto thresholds) for applying the MURAME, a multicriteria method which allows to score/rank a set of alternatives, from the best ones to the worst ones, according to a set of evaluation criteria.

This problem is known in multicriteria analysis as preference disaggregation. It consists in determining the true preference model of the Decision Maker (DM) from a given reference set of decisions, so that the MURAME model is as consistent as possible with the decisions made by the DM (see [12] for a review on preference disaggregation in multicriteria decision making methods).

As a matter of fact, in multicriteria outranking methods (such as the ones belonging to the well known ELECTRE and PROMETHEE families, and also the MURAME

S. Bassis et al. (eds.), *Recent Advances of Neural Network Models and Applications*,
Smart Innovation, Systems and Technologies 26,
DOI: 10.1007/978-3-319-04129-2_8, © Springer International Publishing Switzerland 2014

which is a particular combination of them), the considered model is generally characterized by several parameters which are the weights and the preference, the indifference and the veto thresholds associated to the evaluation criteria. However, the explicit direct determination of these parameters by the DM cannot be considered realistic for several applications, so the use of preference disaggregation is often desirable. Nevertheless, the estimation of the parameters of these models is not easy because of the size and the complexity of the involved optimization problems.

Recently, some evolutionary algorithms have been used to deal with the problem of the preference disaggregation. For example, [1] focuses on the multiple criteria classification method PROAFTN and uses an approach based on Variable Neighborhood Search metaheuristic in order to disaggregate preferences. Also [9] handles classification problems, but it undertake the analysis in the ELECTRE TRI context. In particular, for determining the parameter values, the use of a procedure based on an evolutionary methodology is proposed, namely the Differential Evolution algorithm, that allows to obtain a simultaneous estimation of all the parameters of the considered multicriteria model.

In this paper we advance a methodological proposal based on an evolutionary approach in order to infer the values of the above-mentioned parameters. Nevertheless, unlike the cited contributions, we consider sorting/ranking problems and focus on preference disaggregation in the context of MURAME, a multicriteria method developed in [11]. Moreover, compared to [9], we employ a different evolutionary approach based on swarm intelligence, namely the Particle Swarm Optimization (PSO) ([2]). Finally, in order to provide an initial evaluation of the training and of the predictive performances of the proposed methodology we consider an application to a problem of sorting a set of Italian small and medium-sized firms according to a set of balance-sheet indicators.

The remainder of the paper is organized as follows. Section 2 focuses on the MURAME and summarizes the two phases according to which it can be implemented. Section 3 formulates the optimization problem that has to be solved to disaggregate the preference structure in a MURAME framework. Section 4 illustrates the implementation of the evolutionary solution approach based on the PSO. Section 5 presents the numerical application to the above sorting problem. Finally, Section 6 reports some closing remarks.

2 MURAME

The MUlticriteria RAnking MEthod (MURAME) is a methodology proposed in [11] which allows to sort/rank a set of alternatives $A = \{a_1, ..., a_i, ..., a_m\}$ according to a set of criteria $\{crit_1, ..., crit_j, ..., crit_n\}$. In the following we give a brief description about it.

Let us denote by g_{ij} the score of alternative a_i in relation to $crit_j$, and let us assume below that $crit_j$ is maximized. Moreover, let us denote by:

$\mathbf{q} = (q_1, \dots, q_n)$ the vector of the indifference thresholds;
$\mathbf{p} = (p_1, \dots, p_n)$ the vector of the preference thresholds;

$\mathbf{v} = (v_1, \ldots, v_n)$ the vector of the veto thresholds;
$\mathbf{w} = (w_1, \ldots, w_n)$ the vector of the normalized weights.

Notice that $0 \leq q_j \leq p_j \leq v_j$ and that $w_j > 0$ $(j = 1, \ldots, n)$ and $\sum_j w_j = 1$.

The MURAME is implemented in two phases which take inspiration from two well known multicriteria methods: the ELECTRE III ([16]) and the PROMETHEE II ([4]).

Table 1 summarizes this implementation. In the first phase, an outranking index $O(a_i, a_k)$ is constructed in order to evaluate for each pair of alternatives (a_i, a_k) the strength of the assertion "alternative a_i is at least as good as alternative a_k". The outranking index is obtained by means of the calculation of proper concordance and discordance indexes which delineate the strength and the discordance of the hypothesis that alternative a_i dominates a_k according to any criterion $crit_j$. In the second phase, starting from the previously calculated outranking indexes, a final net score $\varphi(a_i)$ is computed for each alternative a_i, according to which a final sorting/ranking of the alternatives can be obtained.

It is to note that, unlike the approaches based on classical preference structures, MURAME makes use of the concepts of indifference, preference and veto thresholds. The introduction of such thresholds allows to take into consideration not only the case in which the DM is perfectly sure to prefer a given alternative to another one and the case in which she/he is indifferent between the two alternatives, but it permits also to consider an hesitation area in which she/he is not completely sure to prefer the alternative a_i to the alternative a_j, with $i \neq j$ (this concept is known as "weak preference")[1].

The region of weak preference is represented in Figure 1, which shows the local concordance index (formula (1)). Figure 2 illustrates the discordance index (formula (2)), in which the role played by the veto threshold v_j to reject the assertion "alternative a_i is at least as good as the alternative a_k" is put in evidence.

An appropriate setting of the MURAME parameters (the weights $\mathbf{w}$ and the thresholds $\mathbf{q}, \mathbf{p}, \mathbf{v}$) entails that this methodology is able to obtain a sorting/ranking of the alternatives that reflects the DM's preferences. Actually, the values of the net flows (5) depend on all the parameters values, that is $\varphi(a_i) \equiv \varphi(a_i; \mathbf{w}, \mathbf{q}, \mathbf{p}, \mathbf{v})$. However, in case where the DM has not fixed the value of such parameters in advance, one has to deal with the problem of their determination. In this paper we propose to infer the values of the MURAME parameters through the preferences disaggregation. In next section we illustrate the formulation of the involved optimization problem.

3 Mathematical Formulation of the Preference Disaggregation Method

In order to infer values of the MURAME parameters $\mathbf{w}$, $\mathbf{q}$, $\mathbf{p}$ and $\mathbf{v}$ which are coherent with the DM's preference model, we start from a reference set of decisions of the DM

[1] In effect, the preference between two alternatives is not often clearly defined also in real life situations, as could be for the example proposed in [5]: a person who likes sweet beverages has to choose between two cups of tea, the first one containing 10 mg of sugar and the second one containg 11 mg of sugar. According to the traditional preference models, the person would prefer the second cup. But can a normal person perceive such a little difference?

Table 1. The two-phase implementation of MURAME

[Phase I] **I.1** Computation of a local concordance index $C_j(a_i, a_k)$ and a local discordance index $D_j(a_i, a_k)$ for each pair of alternatives (a_i, a_k) and for each criterion $crit_j$:

$$C_j(a_i, a_k) = \begin{cases} 1 & \text{if } g_{kj} \leq g_{ij} + q_j \\ 0 & \text{if } g_{kj} \geq g_{ij} + p_j, \\ \frac{g_{ij} - g_{kj} + p_j}{p_j - q_j} & \text{otherwise} \end{cases} \tag{1}$$

$$D_j(a_i, a_k) = \begin{cases} 0 & \text{if } g_{kj} \leq g_{ij} + p_j \\ 1 & \text{if } g_{kj} \geq g_{ij} + v_j . \\ \frac{g_{kj} - g_{ij} - p_j}{v_j - p_j} & \text{otherwise} \end{cases} \tag{2}$$

I.2 Computation of a global concordance index $C(a_i, a_k)$ for each pair of alternatives (a_i, a_k):

$$C(a_i, a_k) = \sum_{j=1}^{n} w_j C_j(a_i, a_k). \tag{3}$$

I.3 Computation of an outranking index $O(a_i, a_k)$ for each pair of alternatives (a_i, a_k):

$$O(a_i, a_k) = \begin{cases} C(a_i, a_k) & \text{if } D_j(a_i, a_k) \leq C(a_i, a_k) \quad \forall j \\ C(a_i, a_k) \prod_{j \in T} \frac{1 - D_j(a_i, a_k)}{1 - C(a_i, a_k)} & \text{otherwise} \end{cases}, \tag{4}$$

where $T \subseteq \{1, \ldots, n\}$ is the subset of the index of the criteria for which $D_j(a_i, a_k) > C(a_i, a_k)$.

[Phase II] Computation of a final net flow, that is a final score, for each alternative a_i:

$$\varphi(a_i) = \varphi^+(a_i) - \varphi^-(a_i), \tag{5}$$

where $\varphi^+(a_i) = \sum_{k \neq i} O(a_i, a_k)$, and $\varphi^-(a_i) = \sum_{k \neq i} O(a_k, a_i)$.

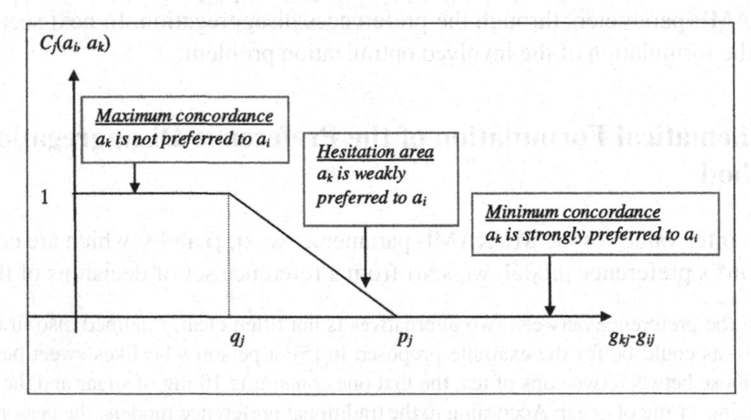

Fig. 1. The local concordance index $C_j(a_i, a_k)$

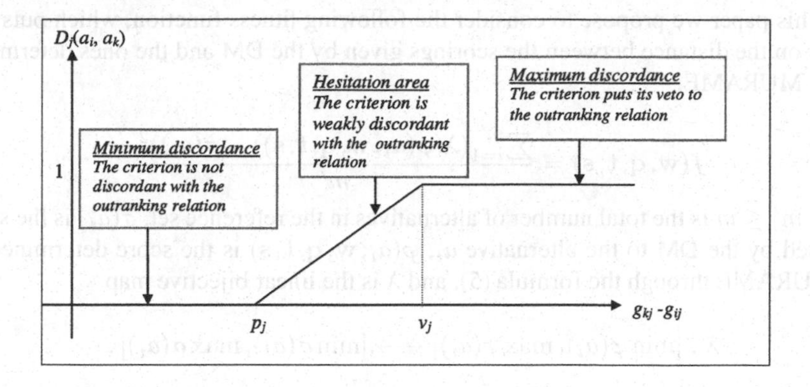

Fig. 2. The local discordance index $D_j(a_i, a_k)$.

herself/himself, that is a set of past decisions regarding the considered alternatives or regarding a subset $A' \subseteq A$ of the whole set of the alternatives. The problem consists in determining the value of such parameters that minimizes a measure of inconsistency $f(\mathbf{w}, \mathbf{q}, \mathbf{p}, \mathbf{v})$ between the scoring/ranking produced by the MURAME with that set of parameters and the scoring/ranking provided by the reference set. The problem can therefore be formalized as follows:

$$\begin{array}{l} \min\limits_{\mathbf{w},\mathbf{q},\mathbf{p},\mathbf{v}} \ f(\mathbf{w}, \mathbf{q}, \mathbf{p}, \mathbf{v}) \\ \text{s.t. } \mathbf{w} \geq \mathbf{0} \\ \sum\limits_{j=1}^{n} w_j = 1 \\ \mathbf{q} \geq \mathbf{0} \\ \mathbf{p} \geq \mathbf{q} \\ \mathbf{v} \geq \mathbf{p} \end{array}, \qquad (6)$$

where $\mathbf{0}$ is an M-vector of zeros.

By introducing the auxiliary variables $\mathbf{t} = \mathbf{p} - \mathbf{q}$ and $\mathbf{s} = \mathbf{v} - \mathbf{p}$, problem (6) can be reformulated as the following one:

$$\begin{array}{l} \min\limits_{\mathbf{w},\mathbf{q},\mathbf{t},\mathbf{s}} \ f(\mathbf{w}, \mathbf{q}, \mathbf{t}, \mathbf{s}) \\ \text{s.t. } \mathbf{w}, \mathbf{q}, \mathbf{t}, \mathbf{s} \geq \mathbf{0} \\ \sum\limits_{j=1}^{n} w_j = 1 \end{array}. \qquad (7)$$

This apparently simple mathematical programming problems hides its complexity in the objective function $f(\mathbf{w}, \mathbf{q}, \mathbf{t}, \mathbf{s})$. Indeed, every choice of f requires that it produces an order of the alternatives and then that a measure of the consistency is calculated. It is then hard to write an exact analytical expression for f in term of its variables $\mathbf{w}$, $\mathbf{q}$, $\mathbf{t}$ and $\mathbf{s}$, so that the use of gradient methods for the optimization task is discouraged, and an evolutionary approach seems more appropriate.

In this paper we propose to consider the following fitness function, which puts emphasis on the distance between the scorings given by the DM and the ones determined by the MURAME:

$$f(\mathbf{w},\mathbf{q},\mathbf{t},\mathbf{s}) = \frac{\sum_{i=1}^{m'}(\lambda(\varphi(a_i;\mathbf{w},\mathbf{q},\mathbf{t},\mathbf{s})) - \sigma(a_i))^2}{m'}, \tag{8}$$

where $m' \leq m$ is the total number of alternatives in the reference set, $\sigma(a_i)$ is the score assigned by the DM to the alternative a_i, $\varphi(a_i;\mathbf{w},\mathbf{q},\mathbf{t},\mathbf{s})$ is the score determined by the MURAME through the formula (5), and λ is the linear bijective map

$$\lambda : [\min_i \varphi(a_i), \max_i \varphi(a_i)] \longleftrightarrow [\min_i \sigma(a_i), \max_i \sigma(a_i)] .$$

4 The PSO Evolutionary Approach to Preference Disaggregation

In this paper we suggest to adopt an evolutionary methodology, the PSO ([2], [13]), in order to deal with the problem of preference disaggregation in a MURAME context, that is in order to find a numerical solution of the mathematical programming problem (7), with fitness function given by (8).

The basic idea of the PSO in solving optimization problems is to model the so called “swarm intelligence” ([3]) that drives groups of individuals belonging to the same species when they move all together looking for food. On this purpose, every member of the swarm explores the search area keeping memory of its best position reached so far, and it exchanges this information with the neighbors in the swarm. Thus, the whole swarm is supposed to eventually converge to the best global position reached by the swarm members.

In order to give a synthetic quantitative description of the PSO, let us consider the global optimization problem

$$\min_{\mathbf{x}\in\mathbb{R}^d} \mathcal{F}(\mathbf{x}),$$

where $\mathcal{F} : \mathbb{R}^d \mapsto \mathbb{R}$ is the objective function in the minimization problem.

The PSO algorithm, in the version with inertia weight ([17]) which we consider in this paper, is summarized in table 2. Every member of the swarm (namely a particle) represents a possible solution of the considered optimization problem, and it is initially randomly positioned in the feasible set of the problem. The algorithm consists of some steps that are repeated until a stopping condition is verified: computation of the objective function value associated to each particle; updating of the best position for each particle and of the best position in a neighborhood of each particle; updating of the velocity and of the position of each particle.

For more details about PSO, the specification of its parameters and of the neighborhood topology, we refer the reader to [2].

In our proposal, in order to specify the neighborhood of each particle, we suggest to considered the so called *gbest* topology, that is $g(l) = g$ for every $l = 1,\ldots,M$, where g indicates the index of the best particle in the whole swarm, that is $g =$ $= \arg\min_{l=1,\ldots,M} \mathcal{F}(\mathbf{p}_l)$. This choice implies that the whole swarm is used as the neighborhood of each particle.

Table 2. The PSO algorithm

Notation:

$\mathbf{x}_l^k \in \mathbb{R}^d$ is the position at step k of particle l ($l = 1, \dots, M$);
$\mathbf{v}_l^k \in \mathbb{R}^d$ is the velocity at step k of particle l ($l = 1, \dots, M$);
$\mathbf{p}_l \in \mathbb{R}^d$ is the best position visited so far by particle l ($l = 1, \dots, M$);
$pbest_l = \mathcal{F}(\mathbf{p}_l)$ is the value of objective function in position $\mathbf{p}_l$ ($l = 1, \dots, M$);
$\mathbf{p}_{g(l)}$ is the best position in a neighborhood of particle l ($l = 1, \dots, M$).

Algorithm:

1. Set $k = 1$ and evaluate $\mathcal{F}(\mathbf{x}_l^k)$ for $l = 1, \dots, M$. Set $pbest_l = +\infty$ for $l = 1, \dots, M$.
2. If $\mathcal{F}(\mathbf{x}_l^k) < pbest_l$ then set $\mathbf{p}_l = \mathbf{x}_l^k$ and $pbest_l = \mathcal{F}(\mathbf{x}_l^k)$ for $l = 1, \dots, M$.
3. Update velocity and position of the particle l for $l = 1, \dots, M$:

$$\mathbf{v}_l^{k+1} = w^{k+1}\mathbf{v}_l^k + \mathbf{U}_{\phi_1} \otimes (\mathbf{p}_l - \mathbf{x}_l^k) + \mathbf{U}_{\phi_2} \otimes (\mathbf{p}_{g(l)} - \mathbf{x}_l^k), \quad (9)$$

$$\mathbf{x}_l^{k+1} = \mathbf{x}_l^k + \mathbf{v}_l^{k+1}, \quad (10)$$

where $\mathbf{U}_{\phi_1}, \mathbf{U}_{\phi_2} \in \mathbb{R}^d$ are vectors whose components are uniformly randomly distributed in $[0, \phi_1]$ and $[0, \phi_2]$, ϕ_1 and ϕ_2 are the so called acceleration coefficients, the symbol $\otimes$ denotes component-wise product, and w^k is the inertia weight.

4. If a convergence test is not satisfied then set $k = k + 1$ and go to 2.

Moreover, as stopping criterion, we propose that the PSO algorithm is terminated when the objective function do not have a decrease of at least 10^{-4} in a prefixed number of steps (which depends on the nature of the investigated problem).

Finally, the inertia weight w^k islinearly decreasing with the number of steps:

$$w^k = w_{max} + \frac{w_{min} - w_{max}}{K} k,$$

where K is the maximum number of steps allowed.

Since PSO was conceived for unconstrained problems, the algorithm above cannot prevent from generating infeasible positions of the particles, that is infeasible solutions, when constraints are considered. To avoid this problem, different strategies have been proposed in the literature, and most of them involve the repositioning of the particles ([20]) or the introduction of some external criteria to rearrange the components of the particles ([8,18]). In this paper we suggest to follow the same approach adopted in [6], which consists in keeping PSO as in its original formulation and in reformulating the optimization problem into the following unconstrained one:

$$\min_{\mathbf{w},\mathbf{q},\mathbf{t},\mathbf{s}} P(\mathbf{w}, \mathbf{q}, \mathbf{t}, \mathbf{s}; \varepsilon), \quad (11)$$

where the objective function $P(\mathbf{w}, \mathbf{q}, \mathbf{t}, \mathbf{s}; \varepsilon)$ is defined as follows:

$$\begin{aligned} P(\mathbf{w}, \mathbf{q}, \mathbf{t}, \mathbf{s}; \varepsilon) = f(\mathbf{w}, \mathbf{q}, \mathbf{t}, \mathbf{s}) + \frac{1}{\varepsilon} \Bigg[\left| \sum_{j=1}^{n} w_j - 1 \right| + \sum_{j=1}^{n} \max\{0, -w_j\} + \\ + \sum_{j=1}^{n} \max\{0, -q_j\} + \sum_{j=1}^{n} \max\{0, -t_j\} + \\ + \sum_{j=1}^{n} \max\{0, -s_j\} \Bigg] \end{aligned}, \quad (12)$$

with ε being a penalty parameter. The adopted approach is called ℓ_1 penalty function method. For more details about it and about the relationships between the solutions of the constrained problem (7) and those of the unconstrained problem (11) see [19,10,15,6].

We may observe that the penalty function $P(\mathbf{w}, \mathbf{q}, \mathbf{t}, \mathbf{s}; \varepsilon)$ is clearly nondifferentiable because of the ℓ_1-norm in the objective function (12). This feature contributes to motivate the choice of using the PSO, since it does not require the derivatives of $P(\mathbf{w}, \mathbf{q}, \mathbf{t}, \mathbf{s}; \varepsilon)$. However, as PSO is a metaheuristics, the minimization of the penalty function $P(\mathbf{w}, \mathbf{q}, \mathbf{t}, \mathbf{s}; \varepsilon)$ does not theoretically ensure that a global minimum of problem (7) is detected. Nevertheless, PSO often provides a suitable compromise between the performance of the approach (i.e. a satisfactory estimate of the global minimum solution for problem (7)) and its computational cost.

Finally, with regard to the initialization procedure of the particles, in order to obtain the initial weights we suggest to generate $d_1 < \ldots < d_{n-1}$ random numbers uniformly distributed in $[0, 1]$, and then to set:

$$w_1^0 = d_1, w_2^0 = d_2 - d_1, \ldots, w_n^0 = 1 - d_{n-1}. \quad (13)$$

Then, to obtain the initial values of the variables q_j, t_j and s_j, three random numbers $a_j^1 < a_j^2 < a_j^3$ $(j = 1, \ldots, n)$ uniformly distributed in $[0, 2]$ are generated, through which we set:

$$q_j^0 = (\overline{g}_j - \underline{g}_j)\frac{a_j^1}{10}, \; t_j^0 = (\overline{g}_j - \underline{g}_j)\frac{a_j^2}{10}, \; s_j^0 = (\overline{g}_j - \underline{g}_j)\frac{a_j^3}{10}, \quad (14)$$

where $\overline{g}_j = \max_{1 \leq i \leq m} g_{ij}$, and $\underline{g}_j = \min_{1 \leq i \leq m} g_{ij}$.
Moreover, for the particle $\mathbf{x}_l^0 = (\mathbf{w}_l^0, \mathbf{q}_l^0, \mathbf{t}_l^0, \mathbf{s}_l^0)$ $(l = 1, \ldots, m)$, the components of the initial velocity $\mathbf{v}_l^0$ are generated as random numbers uniformly distributed in $[-x_h^0, x_h^0]$ for every $h = 1, \ldots, 4n$.

5 A Numerical Application

In this section we give an initial assessment of the performance of the proposed methodology by applying it to a problem of sorting a given set of firms according to a given

Table 3. Balance-sheet indicators used as evaluation criteria

Indicator	Definition
$crit_1$	$\frac{Net\ assets}{Debts}$
$crit_2$	$\frac{Net\ assets}{Total\ assets}$
$crit_3$	$\frac{Current\ assets}{Current\ liabilities}$
$crit_4$	$\frac{Current\ assets-Inventory}{Current\ liabilities}$
$crit_5$	$\frac{Assets+Liabilities}{Total\ assets}$

set of evaluation criteria. As input data we consider around 3,000 Italian small and medium-sized firms as defined in EU recommendation 2003/361. The evaluation criteria are represented by five balance-sheet indicators which have been collected for year 2008 in the AIDA database[2]. Their definitions are reported in table 3.

We carried out an experimental analysis which has a twofold aim: first, to evaluate the training performance of the proposed methodology, that is its capability to find the values of the MURAME parameters such that the scores obtained on a given reference set of firms is the same as the scores initially associated on the same reference set; second, to evaluate the predictive performance of the proposed methodology, that is its capability to produce the same scores as the ones associated on a sample of firms outside of the reference set.

To pursue this double aim, we employed a bootstrap analysis which consists in repeating $N = 500$ times the procedure which follows.

First we determine as reference set a sample of $H = 100$ firms, which is randomly selected without replacement from the 3,000 initial firms. By using the MURAME with the initial values of parameters showed in table 4, we calculate for each firm in the reference set the final net score computed according to formula (5). The initial values for the MURAME parameters $(\mathbf{w}, \mathbf{q}, \mathbf{p}, \mathbf{v})$ illustrated in table 4 have been computed for the whole sample of 3,000 firms indicator data, according to the following rules:

$$w_j = \frac{1}{n},\ q_j = \frac{1}{6}(\overline{g}_j - \underline{g}_j),\ p_j = 4q_j,\ v_j = 5q_j,$$

where $\overline{g}_j = \max_{1 \leq i \leq m} g_{ij}$, and $\underline{g}_j = \min_{1 \leq i \leq m} g_{ij}$.

Notice that such initial values are consistent with the initialization rules (13) and (14).

[2] AIDA is a database containing information about around 1,000,000 Italian enterprises. The balance-sheet indexes selected for this application are a subset of those used in [7] for evaluating the creditworthiness of a given set of firms.

Table 4. Initial values of the MURAME parameters

	$crit_1$	$crit_2$	$crit_3$	$crit_4$	$crit_5$
w	0.200	0.200	0.200	0.200	0.200
q	3.523	15.808	1.637	1.563	1018.220
p	14.093	63.233	6.547	6.253	4072.880
v	17.616	79.042	8.183	7.816	5091.110

The coefficients of the PSO algorithm have been set, after some preliminary experiments[3], as follows:

$$\phi_1 = \phi_2 = 1.75, w_{max} = 0.9, w_{min} = 0.4, K = 200, \text{ and } \varepsilon = 0.0001.$$

Afterwards, in the *training step* we use the PSO-based solution algorithm in order to determine new (generally good sub-optimal) values for the MURAME parameters by solving the optimization problem (11) with fitness function f determined as in the formula (8).

Finally, in the *predictive step* another sample of firms of the same size $H = 100$ is randomly selected from the 2,900 ones which are remained after the selection of those belonging to the sample used in the training step. In this step, first we apply the MURAME with the initial values of the parameters in order to obtain final scores for these H firms. Then we apply again the MURAME, but now with the values of the parameters as determined in the previous step, i.e. in the training step, so obtaining other final scores for the same H firms. The values of the fitness function f is finally computed in order to respectively evaluate the distance between the two obtained scorings.

In table 5 we present the mean and the standard deviation of the fitness function values obtained after the 500 iterations for both the training step and the predictive one, with a population size of $M = 100$ and $M = 200$ particles.

Table 5. Mean and standard deviation of f from the bootstrap analysis

	$M = 100$		$M = 200$	
	Training	Predictive	Training	Predictive
Mean	0.032	0.480	0.020	0.074
Standard deviation	0.017	0.640	0.015	0.038

[3] The experiments aimed at determining the best values for: the acceleration coefficients ϕ_1 and ϕ_2; the maximum number of steps K; the value of the penalty parameter ε. In order to do so, we have undertaken the bootstrap procedure first by testing different values for each parameter (while keeping fixed the values of the remaining ones), then by choosing that value which allowed to obtain on average the best results for the objective function. Instead, for the initial and the final values of the inertia weights, w_{min} and w_{max}, we considered the ones most used in the literature.

It may be seen that a population of 100 particles ensures to obtain a good consistency between the MURAME model produced by the PSO-based solution algorithm and the true preference one for the training step since the average value of the fitness function is close to zero, but 200 particles are needed to get analogous results for the predictive step.

As for the obtained values of the MURAME parameters, which are reported in table 6, it may be observed that the values of the weights are quite similar to the initial ones, whereas there are notable differences regarding the values of the thresholds. However, since the scoring performance of the MURAME model is quite satisfactory, this seems to suggest that there are some flexibility degrees in the specification of such parameters, thus opening the possibility to a more precise calibration of the MURAME model according to other inputs provided by the DM.

Table 6. Values of the MURAME parameters determined by the PSO-based solution algorithm in the bootstrap analysis

	$crit_1$	$crit_2$	$crit_3$	$crit_4$	$crit_5$
w	0.210	0.266	0.184	0.154	0.185
q	1.596	13.569	1.155	1.867	226.000
p	8.675	75.818	5.669	4.797	753.840
v	39.342	398.830	29.952	25.039	10303.400

6 Some Final Remarks

The novelty of our methodological proposal consists both in dealing with the problem of preference disaggregation in a MURAME context and in tackling this problem by means of a solution algorithm based on the PSO.

Since the MURAME determines an outranking index for each pair of considered alternatives in the reference set, the computational effort needed to determine the parameter values is notably higher than in other multicriteria models. For instance, in the ELECTRE TRI-based models the comparison of each alternative is performed only with some reference profiles ([14]).

In the numerical application in which we have considered the problem of sorting around 3,000 Italian small and medium-sized firms, we find that our methodological proposal leads to a good consistency. Actually, the results show that the inconsistency between the MURAME model produced by the PSO-based solution algorithm and true preference model is very close to zero, even if an higher number of particles is needed to obtain consistency in the predictive step.

References

1. Belacel, N., Bhasker Raval, H., Punnen, A.P.: Learning multicriteria fuzzy classification method. PROAFTN from data. Computers & Operations Research 34(7), 1885–1898 (2007)
2. Blackwell, T., Kennedy, J., Poli, R.: Particle swarm optimization – An overview. Swarm Intelligence 1(1), 33–57 (2007)
3. Bonabeau, E., Dorigo, M., Theraulaz, G.: From Natural to Artificial Swarm Intelligence. Oxford University Press (1999)
4. Brans, J.P., Vincke, P.: A preference ranking organisation method (The PROMETHEE method for multiple criteria decision-making). Management Science 31(6), 647–656 (1985)
5. Buchanan, J., Sheppard, P., Vanderpooten, D.: Project ranking using ELECTRE III. Research report 99-01, Department of Management Systems, University of Waikato, New-Zealand (1999)
6. Corazza, M., Fasano, G., Gusso, R.: Portfolio selection with an alternative measure of risk: Computational performances of particle swarm optimization and genetic algorithms. In: Perna, C., Sibillo, M. (eds.) Mathematical and Statistical Methods for Actuarial Sciences and Finance, pp. 123–130. Springer (2012)
7. Corazza, M., Funari, S., Gusso, R.: Il merito creditizio delle Pmi italiane durante la crisi finanziaria: l'utilizzo di più fonti informative per l'analisi e lo scoring. Bancaria 1(1), 47–63 (2012) (in Italian)
8. Cura, T.: Particle swarm optimization approach to portfolio optimization. Nonlinear Analysis: Real World Applications 10(4), 2396–2406 (2009)
9. Doumpos, M., Marinakis, Y., Marinaki, M., Zopounidis, C.: An evolutionary approach to construction of outranking models for multicriteria classification: The case of the ELECTRE TRI method. European Journal of Operational Research 199(2), 496–505 (2009)
10. Fletcher, R.: Practical Methods of Optimization. John Wiley & Sons (1991)
11. Goletsis, Y., Askounis, D.T., Psarras, J.: Multicriteria judgments for project ranking: An integrated methodology. Economic Financial Modelling 8(3), 127–148 (2001)
12. Jacquet-Lagrèze, E., Siskos, Y.: Preference disaggregation: 20 years of MCDA experience. European Journal of Operational Research 130(2), 233–245 (2001)
13. Kennedy, J., Eberhart, R.C.: Particle Swarm Optimization. Proceedings of the IEEE International Conference on Neural Networks 4, 1942–1948 (1995)
14. Mousseau, V., Slowinski, R., Zielniewicz, P.: ELECTRE TRI 2.0a: Methodological guide and user's documentation. Université de Paris-Dauphine (1999)
15. Di Pillo, G., Grippo, L.: Exact penalty functions in constrained optimization. SIAM Journal on Control and Optimization 27(6), 1333–1360 (1989)
16. Roy, B.: ELECTRE III: Un algorithme de classements fondé sur une representation floue des préférences en présence de critères multiples. Cahiers du CERO 20(1), 3–24 (1978)
17. Shi, Y., Eberhart, R.: A modified particle swarm optimizer. In: The 1998 IEEE International Conference on Evolutionary Computation Proceedings, pp. 69–73 (1998)
18. Thomaidis, N., Angelidis, T., Vassiliadis, V., Dounias, G.: Active portfolio management with cardinality constraints: An application of Particle Swarm Optimization. New Mathematics and Natural Computation 5(3), 535–555 (2009)
19. Zangwill, W.I.: Non-linear programming via penalty functions. Management Science 13(5), 344–358 (1967)
20. Zhang, W.J., Xie, X.F., Bi, D.C.: Handling boundary constraints for numerical optimization by particle swarm flying in periodic search space. In: Proceedings of the 2004 Congress on Evolutionary Computation IEEE, pp. 2307–2311 (2005)

Genetic Art in Perspective

Rachele Bellini and N. Alberto Borghese

Department of Computer Science
Università degli Studi di Milano
borghese@di.unimi.it

Abstract. Since the pioneer observations of Alan Turing, emotional and aesthetical capabilities have been considered as one of the fundamental element of a genuinely intelligent machine. Among the proposed approaches, genetic algorithms try to combine intuitively a generative impulse with a critical capacity that steers the production towards a valuable goal. The approach here presented is based on Karl Sim's approach in which a set of possible primitives is defined and it represent the genotype of the system. Such expressions are combined using genetic algorithms rules to obtain more complex functions that describe new images. At each step, images are evaluated by the user and this implicitly drives the evolution process. Results can be impressive, however a clear understanding of the determinants of our aesthetic evaluation is presently beyond reach.

1 Introduction

When Alan Turing in the forties of last century proposed the reasons for which a machine could never become intelligent, the lack of creativity and of aesthetic sense was one of the main arguments [1]. The introduction of genetic art in the eighties sounds like a counter-example of this, but, although results have been impressive the understanding of underlying mechanisms is still far beyond reach.

Genetic (or evolutionary) art is a type of digital art in which the artwork is created by a genetic algorithm [2, 3]. Goal is to achieve an aesthetically valuable digital artwork. The final result is usually mathematically described by an ensemble of functions, but the process of its creation involves the human judgment and taste.

We review here the pipeline of creating artistic images through genetic algorithms and highlight some critical points.

2 Algorithm Description

The algorithm starts from a set of images usually randomly generated from the ensemble of available functions. These images constitute what are named parents of the first generation [4, 5]. A new set of images, called son images, are generated from each parent image by transforming each pixel of each image through a function obtained combining simple functions with genetic operators.

S. Bassis et al. (eds.), *Recent Advances of Neural Network Models and Applications*,
Smart Innovation, Systems and Technologies 26,
DOI: 10.1007/978-3-319-04129-2_9, © Springer International Publishing Switzerland 2014

The resulting images are judged by a human user who implicitly assigns a fitness value to each of them that represents the personal degree of preference according to the image aesthetic value or interest. Images are ordered with respect to their fitness and the best images are chosen as parents for the next generation.

The functions that define the child-images are composed starting from the functions that created the parent image: they inherit all of the sub-functions and the operators of the parent-image, except for one of them (a function or operator), extracted randomly, which is replaced from a new one (function or operator) also randomly extracted from the database. In such way, an evolution of the images can be envisaged. The process can be iterated ad libitum, evolving the images from one generation to the other, until a satisfactory image is reached or the user decides to interrupt the evolutionary process. Some examples of this evolutionary process are provided in the Results section.

2.1 Images Definition

Each image is digitized into a N x N pixels. If color images are considered, the value of each pixel will be its RGB value, where each channel is coded over 8 bits for a total of 24 bits per pixel. If gray level images are considered, the value will be the pixel gray value discretized over 8 bits: from 0 to 255. The considered images are N x N, with N = 300 pixels. Each image is described by a 2-dimensional mathematical function, which is sampled in the center of the pixels of the image.

2.2 Sampling Modalities

Several functions and sampling strategies can be used. In general, the selected functions are those trigonometric combined with power functions (cf. Table 1). For sake of simplicity, trigonometric functions are defined with unitary frequency. Here we show three different sampling approaches, each of them having different aesthetic features. A first approach consists in choosing a unitary sampling step, s = 1. In this case, the trigonometric components will have a period of 2 π and this approach leads to images characterized by a high-frequency texture, as shown in Figure 1.

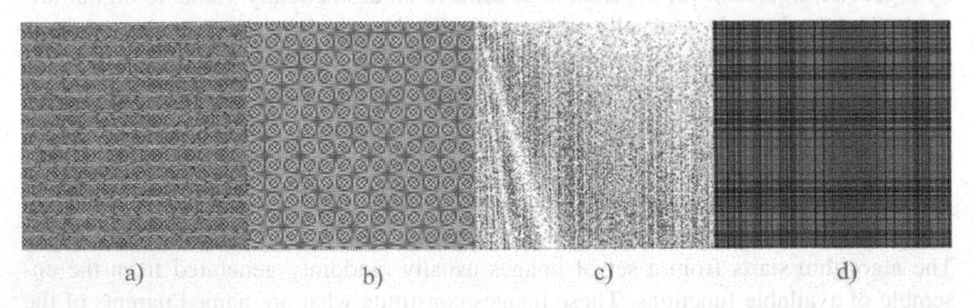

Fig. 1. An example of a set of four images generated with functions with sampling step size s = 1: a) sin(x*y)*cos(x*y)/cos(y)+sin(y)+cos(x)+sin(x); b) tan(cos(sqrt(x*y*x*y))); c) sin(x*y*x)+cos(y*x*y)/sin(x*x)/sin(y/x); d) sin(x*x)*cos(y*y)

A lower frequency content can be obtained if the functions are sampled with a less dense granularity. We take a sampling step, s = 1 / N. This approach leads to much more smoothed images, sometimes just gradients, aesthetically very different from the ones produced with the former approach (see Figure 2).

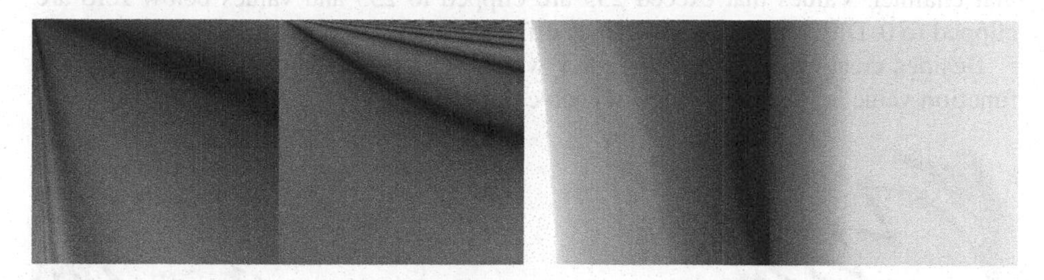

Fig. 2. An example of a set of four images generated sampling the same functions of Fig. 1, with a step size of 1 / N: a) cos(sqrt(x+y))*y+sqrt(cos(y)*sin(x))*sin(y/x)+y-sin(x); b) sin(x*y*x)+cos(y*x*y)*cos(x/y); c) sin(x*y*x)+cos(y*x*y)-sqrt(abs(x)+abs(y))/abs(x); d) abs(x*3)

In a half-way approach we evaluate the function with a period of the trigonometric component approximately equal to the image width, i.e. s = 2 π / N. We have chosen this step size because of its more pleasant aesthetic results. In this paper, all the following images are generated sampling the functions with a step size s = 0.02, that approximates this condition (Figure 3 shows an example of this).

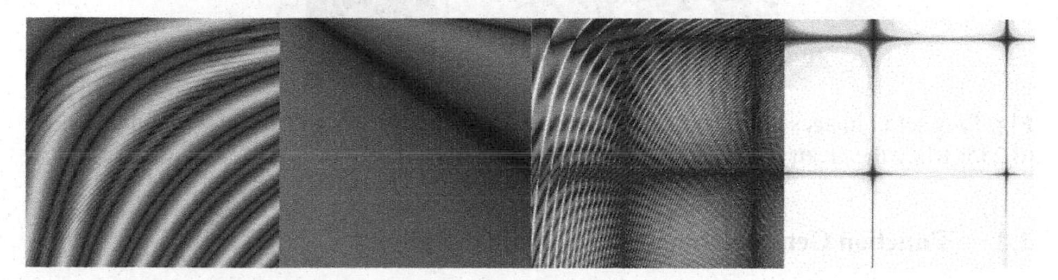

Fig. 3. An example of a set of four images generated evaluating the same functions of Fig. 1 with a sampling step, s = 0.02: a) sin(abs(cos(x+y))+abs(cos(y*x*y)))+tan(cos(sqrt(x*y*x*y))); b) cos(x/y); c) sin(x)*cos(y)/sin(x*y*x)+cos(y*x*y)/abs(x); d) ((x+y)*y*x*sin(x)*cos(y))

2.3 Intensity Adaptation

The computed pixel values have to be adapted to the image range. No constraint is made about the values assumed by the generating functions: trigonometric functions can assume both positive and negative values and polynomial and absolute value functions can have a maximum value that may either not span the entire range of 255 values or exceed this range.

Moreover, at each iteration, only one function is selected to produce each image. Different scaling values, that we will name "color weight", will be applied to the R, G and B channels separately to obtain a color image. When the color weight is equal on the three channels, the image will appear in gray levels (Fig. 4), otherwise a color image is produced, whose tone changes at each generation (Fig. 5). To obtain this, the color weight of each channel is randomly generated inside the interval between 0 and 255. The value of that channel for all the pixels is multiplied by the color weight of that channel. Values that exceed 255 are clipped to 255 and values below zero are clipped to 0. Different images will have different triplets of color values.

Besides creating color images, color weights have also the aim of spreading the function value in the useful range of each color channel.

Fig. 4. A set of gray-scale images generated weighting the three color channels equally

Fig. 5. A set of images generated with the same color-weights: the red and blue weights are set to 255, while the green one is set to 100

2.4 Function Generation and the Genetic Operators

The function applied to the pixels of the parent images is obtained as a combination of a variable number of sub-functions (from a minimum of one to a maximum of five sub-functions), chosen among those defined in Table 1. Such functions are mainly trigonometric and polynomial functions of x and y and were defined experimentally by trial and error [6].

Once the sub-functions are extracted, these are combined through a set of simple operators, restricted here to the basic four operations: multiplication, '*', division, '/', sum, '+' and subtraction, '-', also extracted randomly. Therefore, the final function that is applied to each parent image is created joining the chosen sub-functions with the chosen operators.

A simple example of the obtained functions is shown in Fig. 7. Here the sub-functions randomly extracted are: `sin(x/y)`, `cos(x/y)` and `abs(y*3)` and the extracted operators are '*' and '-'. Therefore the final composed function is the following: `sin(x/y)*cos(x/y)-abs(y*3)`.

Table 1. The sub-functions used to create the function applied to the pixels of a parent image at each generation

sin(x*x + y*y)	cos(x*x/y)+sin(y*y/x)
sin(x*x)*cos(y*y)	sin(x)+sin(x)+cos(y)+cos(y)
sin(x/y)*cos(x/y)	cos(x)+cos(x)+sin(y)+sin(y)
cos(x/y)	sin(x)+cos(x)+sin(y)+cos(y)
sin(y/x)	cos(y)+sin(y)+cos(x)+sin(x)
abs(y)-x	tan(cos(sqrt(x*y*x*y)))
x+abs(y)	sqrt(abs(x)+abs(y))
abs(x)	sin(x*y*x)+cos(y*x*y)
abs(y)	sin(sqrt(abs(x)))-cos(sqrt(abs(y)))
abs(x)*abs(y)	sqrt(cos(x)+sqrt(x)*sin(y)+sqrt(y))
sin(x)*cos(y)	cos(x)*sin(x*y)
sin(x*y)*cos(x*y)	cos(y)*sin(x*y)
sin(x*x-y*y)	sin(x+y*x*y+x*x)
sin(x*x)	sin(y+x*y*x+y*y)
y-abs(x)	abs(x*y+x*x+y*y)
y-sin(x)	((x+y)*y*x*sin(x)*cos(y))
x-cos(y)	((x+y*x)+sin(x*y)+cos(y/x))
abs(x)+y	sin(x*y+x)+cos(y*x+y)
sin(x*x*x-y*y*y)	cos(x+y)*sin(x+y)/2
sin(y*y*y)+sin(x*x*x)	cos(sqrt(x+y))*y+sqrt(cos(y)*sin(x))
cos(y*y*y+x*x*x)	sin(sqrt(y+x))*x+sqrt(sin(x)*cos(y))
cos(y*y*y)+cos(x*x*x)	cos(x)*sin(x)+cos(y)*sin(y)
abs(y*3)	sin(abs(cos(x+y))+abs(cos(y*x*y)))
abs(x*3)	sin(cos(x)*abs(y)*abs(y))
sin(x*x/y-y*y/x)	cos(x)*sin(y)*cos(x*y)

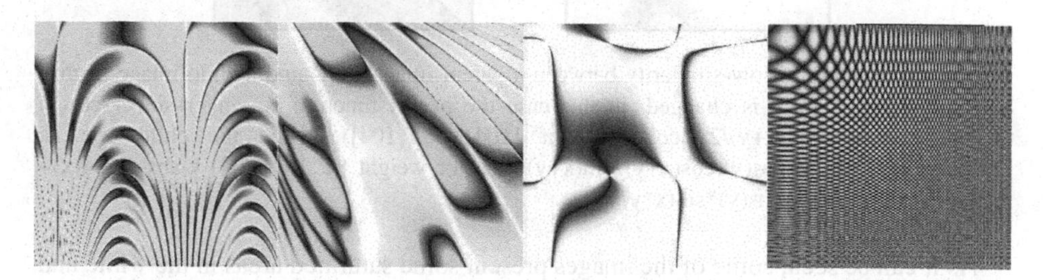

Fig. 6. Some examples of obtained functions. From left to right: a) sin(x)+sin(x)+cos(y)+cos(y)/sin(cos(x)*abs(y)*abs(y)) with the color weights [25, 218, 238]. b) cos(x)+cos(x)+sin(y)+sin(y)+cos(x/y)/cos(x*x/y)+sin(y*y/x) with the color weights [47, 252, 110]. c) y-sin(x)/cos(y)+sin(y)+cos(x)+sin(x)/cos(x)+cos(x)+sin(y)+sin(y) with the color weights [123, 144, 105]. d) cos(y*y*y)+cos(x*x*x) with the color weights [200, 151, 21].

3 Results

In nature there is no aesthetic rule that binds somatic aspect between parents and their children. Even if some basic characteristics are expected, children can be either extremely similar to their parents, or very different.

This phenomenon is reproduced with our genetic art algorithm: since any operator and any function can be replaced during the evolutionary process between a parent-image and a child-image, there is no control over the amount of aesthetic change. The amount of change can be sometimes even non-noticeable (Fig. 7), while in other situations the child-image can be aesthetically totally different from the parent one (Fig. 8).

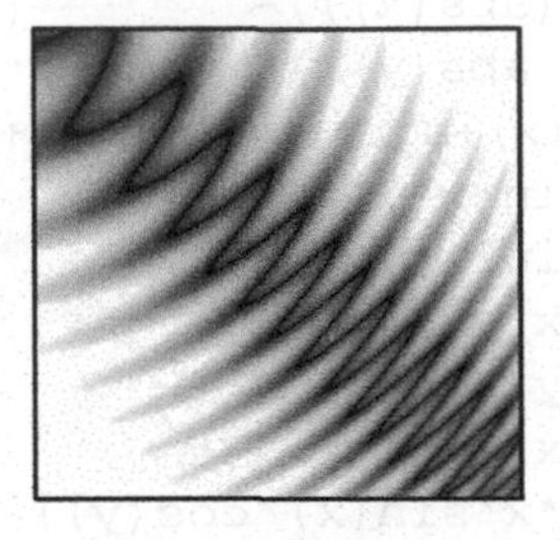
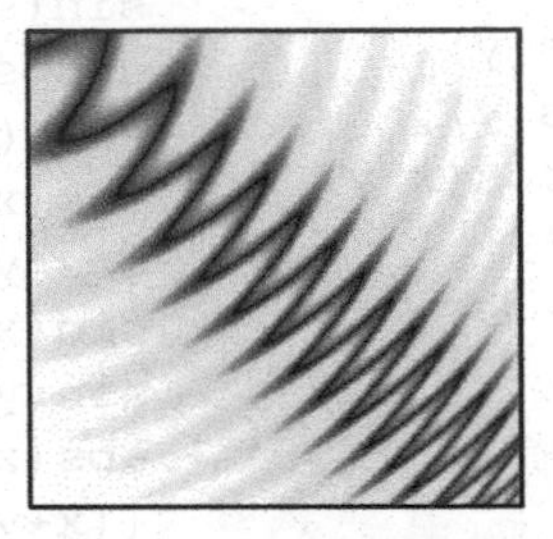

Fig. 7. An example of high-similarity between a parent image (left) and a child-image (right). In this case the function is exactly the same `(y-abs(x)-sin(x*x + y*y))`, while the color weight is different ([204, 82, 131] in the first image, [70, 222, 252] in the second).

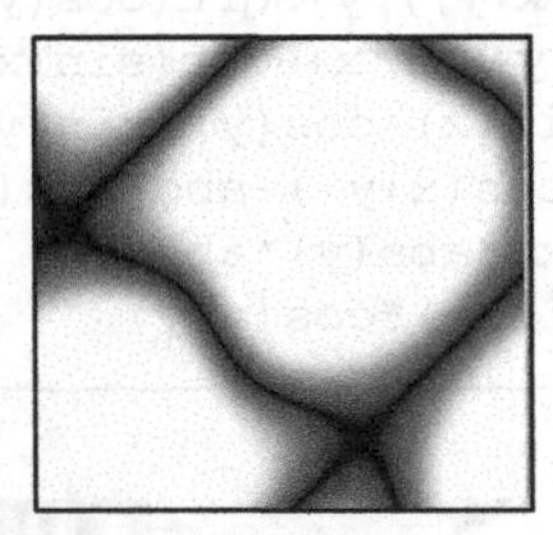
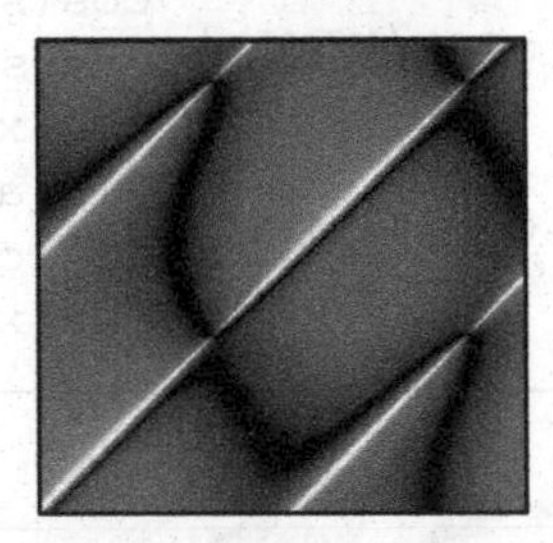

Fig. 8. An example of low-similarity between a parent image (left) and a child image (right), where just an operator is changed. In this case the parent function is: sin(x)+sin(x)+cos(y) +cos(y)+cos(x+y)*sin(x+y)/2 (color weight [104, 66, 104]), while the child one is: sin(x)+sin(x)+cos(y)+cos(y)/cos(x+y)*sin(x+y)/2 (color weight [4, 16, 163]) y-abs(x)/sin(x)+ cos(x)+sin(y)+cos(y)+cos(y)*sin(x*y).

As it can be seen, some of the images present some saturated areas in the white and black region. In fact there is no explicit control on saturation when applying the transforming function to the pixels of an image. For instance, we have analyzed the amount of saturated pixels over a set of 40 images and found out that 1/3 of the images has at least 90% of saturated pixels, 1/3 has less than 10% and 1/3 has a percentage between 10% and 90%.

An additional example of the evolution over four generations is shown in the following lines. In this case, only one parent is considered at each generation as the son with the best fitness value. Each parent generates four children images that are evaluated and ranked. The initial random functions extracted for the four images, a-d in the first generation, are the followings:

```
1a. tan(cos(sqrt(x*y*x*y)))+sin(x*x)
1b. y-abs(x)
1c. abs(x*3)-sin(x*x*x-y*y*y)-sin(y+x*y*x+y*y)
1d. y-abs(x)+sin(x)+cos(x)+sin(y)+cos(y)+cos(y)*sin(x*y)
```

The image with the highest fitness was the fourth one (1d) in Fig. 9, which becomes the parent-image for the new generation. The three following sets of functions represent the three generations descending from the former parent (cf. Fig. 9):

```
2a. y-abs(x)/sin(x)+cos(x)+sin(y)+cos(y)+cos(y)*sin(x*y)
2b. y-abs(x)+sin(x)+cos(x)+sin(y)+cos(y)+cos(y)*sin(x*y)
2c. y-abs(x)*sin(x)+cos(x)+sin(y)+cos(y)+cos(y)*sin(x*y)
2d. y-abs(x)*cos(x)*sin(x)+cos(y)*sin(y)+cos(y)*sin(x*y)

3a. y-abs(x)*tan(cos(sqrt(x*y*x*y)))+cos(y)*sin(x*y)
3b. y-abs(x)-tan(cos(sqrt(x*y*x*y)))+cos(y)*sin(x*y)
3c. y-abs(x)-cos(y*y*y)+cos(x*x*x)+cos(y)*sin(x*y)
3d. y-abs(x)-cos(y)+sin(y)+cos(x)+sin(x)+cos(y)*sin(x*y)

4a. y-abs(x)/cos(y)+sin(y)+cos(x)+sin(x)+cos(y)*sin(x*y)
4b. y-abs(x)/sin(x*y*x)+cos(y*x*y)+cos(y)*sin(x*y)
4c. y-abs(x)/sin(x)+cos(x)+sin(y)+cos(y)+cos(y)*sin(x*y)
4d. y-abs(x)/x+abs(y)+cos(y)*sin(x*y)
```

4 Discussion and Conclusion

Overall, the generated images exhibit a high variability and richness also within the same generation: a simple change in one of the function components that realize the transformation function may produce a very similar or different result, depending on the overall function shape and on the component. Such property, observed experimentally, can be object of further investigation to determine which can be the determinant characteristics that make an image more attractive than another. The correlation between the fitness value of an image and its aesthetic value can be a tool that may allow investigating the processes underlying our aesthetic evaluation [7, 8].

This analysis could also allow improving the genetic algorithm introducing some elitist mutation rule for which, as far as the generations progress, some of the subcomponents, those with the most impact, are kept fixed and only the other subcomponents are mutated [5, 9].

From first experimental observations, the heavy use of trigonometric functions has been introduced as these produce gentle oscillating variations that are often interpreted as "motion" and add aesthetic value to the images. The frequency of the oscillations introduced is also important as these should be related to the amplitude of the image: too high or too small frequencies tend to produce less pleasant images.

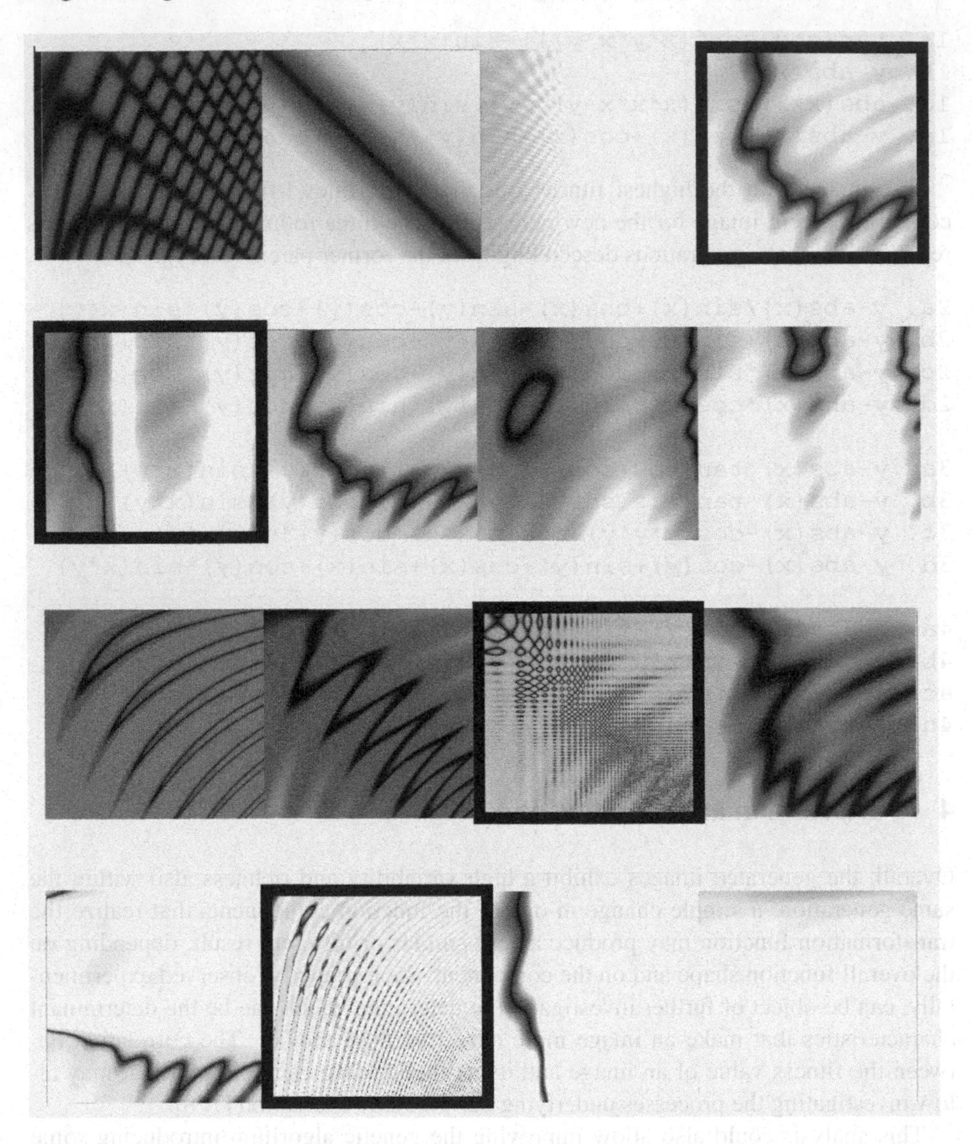

Fig. 9. The images of four different generations. The correspondent functions are presented above. The selected sub-image in each image is highlighted by a black frame. Notice how some parent-child relations are aesthetically very clear, while some others are not.

Genetic algorithms turn out to be a particularly effective tool to generate images with aesthetic value that do not contain structured scenes. The fitness value cannot be easily captured by analytical functions as usually done in the computer science field. This opens the challenge to identify the models, the features and the determinants for our aesthetic evaluation. Besides a comprehension of the mechanisms of aesthetic evaluation, this can add value in all the manufacts in which the external shape can be designed or colored arbitrarily.

References

1. Turing, A.: Computing Machinery and Intelligence. Mind LIX (236), 433–460 (1950)
2. Crow, F., Demos, G., Hardy, J., McLaughlin, J., Sims, K.: 3D Image Synthesis on the Connection Machine. International Journal of High Speed Computing, 329–347 (1989)
3. Romero, J., Machado, P. (eds.): The Art of Artificial Evolution: A Handbook on Evolutionary Art and Musi. Springer (2007)
4. Koza, J.: Genetic Programming: On the Programming of Computers by Means of Natural Selection. MIT Press, Cambridge (1992)
5. Schmitt, L.M.: Theory of Genetic Algorithms. Theoretical Computer Science 259, 1–61 (2001)
6. `http://softologyblog.wordpress.com/category/genetic-art/`
7. Galanter, P.: Computational Aesthetic Evaluation: Past and Future. In: McCormack, J., D'Inverno, M. (eds.) Computers and Creativity. Springer, Berlin (2012)
8. Lewis, M.: Evolutionary Visual Art and Design. In: Romero, J., Machado, P. (eds.) The Art of Artificial Evolution: a Handbook on Evolutionary Art and Music, pp. 3–37. Springer, Berlin (2008)
9. Cerveri, P., Pedotti, A., Borghese, N.A.: Combined evolution strategies for dynamic calibration of video based measurement systems. IEEE Trans. Evolutionary Computation 5(3), 271–282 (2001)

Part II
Signal Processing

Proportionate Algorithms for Blind Source Separation

Michele Scarpiniti, Danilo Comminiello, Simone Scardapane,
Raffaele Parisi, and Aurelio Uncini

Department of Information Engineering, Electronics and Telecommunications (DIET),
"Sapienza" University of Rome, via Eudossiana 18, 00184, Rome
{michele.scarpiniti,danilo.comminiello,simone.scardapane,
raffaele.parisi,aurelio.uncini}@uniroma1.it

Abstract. In this paper we propose an extension of time-domain Blind Source Separation algorithms by applying the well known proportionate and improved proportionate adaptive algorithms. These algorithms, known in the context of adaptive filtering, are able to use the sparseness of acoustic impulse responses of mixing environments and give better performances than standard algorithms. Some preliminary experimental results show the effectiveness of the proposed approach in terms of convergence speed.

Keywords: Blind Source Separation, Independent Component Analysis, Proportionate algorithms, Improved proportionate.

1 Introduction

Blind Source Separation (BSS) applied to speech and audio signals is an attractive research topic in the field of adaptive signal processing [1,2]. The problem is to recover original sources from a set of mixtures recorded in an unknown environment. While several well-performing approaches exist when the mixing environment is instantaneous, some problems arise in convolutive environments.

Several solutions were proposed to solve BSS in a convolutive environment [3,2]. Some of these solutions work in time domain, others in frequency domain. Each of them have some advantages and disadvantages, but there is not a unique winner approach [4].

In addition, when working with speech and audio signals, convergence speed is an important task to be performed. Since impulse responses of standard environments, e.g. office rooms, are quite sparse, some authors have proposed to incorporate sparseness in the learning algorithm [5,6]. The idea is to introduce a weighting matrix in the update equation, that can give more emphasis to the most important part of the impulse response. In particular a proportionate [5] and an improved proportionate [7] algorithms were proposed for supervised signal processing applications, like acoustic echo cancellation.

In this paper we aim to extend these proportionate algorithms to the BSS problem, in the hope that they can be effective also for unsupervised case. Hence a proportionate and an improved proportionate version of the well-known time-domain Torkkola's algorithm [8,9] will be proposed. Some preliminary results, that demonstrate the effectiveness of the proposed idea in terms of convergence speed, are also presented.

S. Bassis et al. (eds.), *Recent Advances of Neural Network Models and Applications*,
Smart Innovation, Systems and Technologies 26,
DOI: 10.1007/978-3-319-04129-2_10,

The rest of the paper is organized as follows: Section 2 introduces the BSS problem in convolutive environments. Then Section 3 describes the proposed algorithm, while Section 4 shows some experimental results. Finally Section 5 draws our conclusions.

2 Blind Source Separation for Convolutive Mixtures

Let us consider a set of N unknown and independent sources denoted as $\mathbf{s}[n] = [s_1[n], \ldots, s_N[n]]^T$, such that the components $s_i[n]$ are zero-mean and mutually independent. Signals received by an array of M sensors are denoted by $\mathbf{x}[n] = [x_1[n], \ldots, x_M[n]]^T$ and are called mixtures. For simplicity we consider the case of $N = M$.

The convolutive model introduces the following relation between the i-th mixed signal and the original source signals

$$x_i[n] = \sum_{j=1}^{N} \sum_{k=0}^{K-1} a_{ij}[k] s_j[n-k], \quad i = 1, \ldots, M \tag{1}$$

The mixed signal is a linear mixture of filtered versions of the source signals, $a_{ij}[k]$ represents the k-th mixing filter coefficient and K is the number of filter taps. The task is to estimate the independent components from the observations without resorting to *a priori* knowledge about the mixing system and obtaining an estimate $\mathbf{u}[n]$ of the original source vector $\mathbf{s}[n]$:

$$u_i[n] = \sum_{j=1}^{M} \sum_{l=0}^{L-1} w_{ij}[l] x_j[n-l], \quad i = 1, \ldots, N \tag{2}$$

where $w_{ij}[l]$ denotes the l-th mixing filter coefficient and L is the number of filter taps.

The weights $w_{ij}[l]$ can be adapted by minimizing some suitable cost function. A particular good choice is to maximize joint entropy or, equivalently, to minimize the mutual information [1,3]. Different approaches can be used, for example implementing the de-mixing algorithm in the time domain or in the frequency domain. In this paper we adopt the time-domain approach, using the algorithm proposed by Torkkola in [8] and based on the feedback network shown in Figure 1 (for the particular case of $M = N = 2$) and described mathematically by:

$$u_i[n] = \sum_{l=0}^{L-1} w_{ii}[l] x_i[n-l] + \sum_{j=1, j \neq i}^{N} \sum_{l=1}^{L-1} w_{ij}[l] u_j[n-l], \quad i = 1, \ldots, N \tag{3}$$

This latter will be used in the paper for achieving source separation in time domain by maximizing the joint entropy of a nonlinear transformation of network output: $y_i[n] = f(u_i[n])$, with $f(\cdot)$ a suitable nonlinear function, very close to the source cumulative density function [1,8]. In this work we use $f(\cdot) = \tanh(\cdot)$. The k-th weight of the de-mixing filter $w_{ij}[k]$ is adapted by using the general rule:

$$w_{ij}^{p+1}[k] = w_{ij}^{p}[k] + \mu \Delta w_{ij}^{p}[k], \tag{4}$$

where in particular the stochastic gradient method can be used, μ is the learning rate and p is the iteration index.

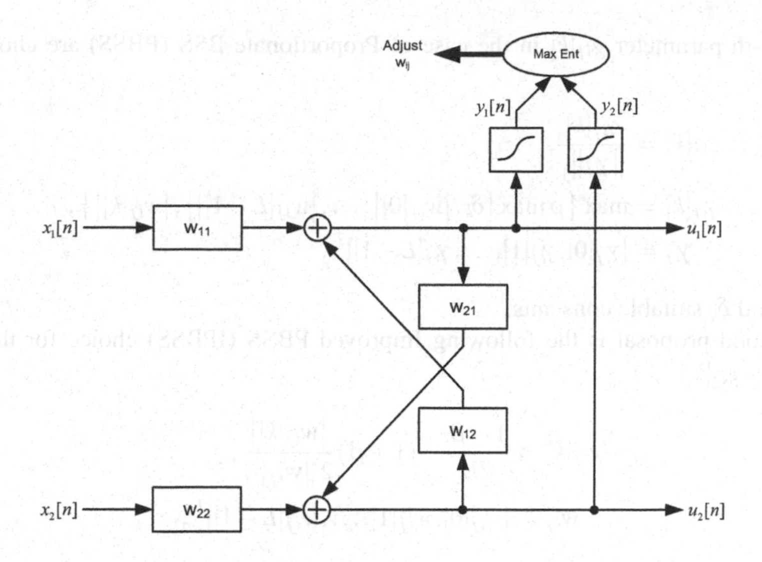

Fig. 1. The architecture for BSS in time domain in the particular case of $M = N = 2$

3 The Proposed Algorithm

Standard BSS algorithms were proposed so far in several works [3,9,10,8]. More recently, some authors have underlined the importance of sparseness of the impulse response $a_{ij}[n]$ [6]. In order to improve the performance of standard adaptive algorithms (LMS and NLMS [11]), that do not take into account sparseness, these authors proposed some modifications introducing the so-called Proportionate NLMS (PNLMS) [5] and Improved Proportionate NLMS (IPNLMS) [7]. The resulting algorithms derive from a more general class of regularized gradient adaptive algorithms [12,13].

In this kind of algorithms, the update term is simply multiplied with a matrix $\mathbf{G}[k]$, whose entries $g_{ij}[k]$ are chosen using different criteria (NLMS, PNLMS and IPNLMS) and take into account the sparseness of the impulse response: the parameters variation is proportional to the impulse response itself

$$\mathbf{w}^{p+1} = \mathbf{w}^p + \mu \Delta \mathbf{w}^p \quad \Rightarrow \quad \mathbf{w}^{p+1} = \mathbf{w}^p + \mu \mathbf{G}\{\mathbf{w}^p\} \Delta \mathbf{w}^p.$$

The aim of this section is to extend the previous ideas on proportionate adaptive algorithms to time-domain BSS algorithm proposed by Torkkola [8], deriving a new updating rule for the de-mixing filter matrix $\mathbf{W}[k]$.

Based on [8] and [6], the proposed modification to the Torkkola's algorithm results in the following algorithm:

$$\begin{aligned} \Delta w_{ii}^p[0] &\propto g_{ii}[0] \left(\frac{1}{w_{ii}[0]} - 2y_i[n]x_i[n] \right), \\ \Delta w_{ii}^p[k] &\propto -2g_{ii}[k]y_i[n]x_i[n-k], \quad \text{for } k \geq 1 \\ \Delta w_{ij}^p[k] &\propto -2g_{ij}[k]y_i[n]u_j[n-k], \quad \text{for } k \geq 1 \text{ and } i \neq j \end{aligned} \tag{5}$$

where $y_i[n] = f(u_i[n]) = \tanh(u_i[n])$.

The k-th parameter $g_{ij}[k]$ in the case of Proportionate BSS (PBSS) are chosen as follows

$$g_{ij}[k] = \frac{\gamma_{ij}[k]}{\left\|\gamma_{ij}\right\|_1}, \quad (6)$$

$$\gamma_{ij}[k] = \max\left\{\rho \max\left[\delta_k, \left|w_{ij}[0]\right|, \ldots, \left|w_{ij}[L-1]\right|\right], \left|w_{ij}[k]\right|\right\},$$
$$\gamma_{ij} = \left[\gamma_{ij}[0], \gamma_{ij}[1], \ldots, \gamma_{ij}[L-1]\right]^T, \quad (7)$$

with ρ and δ_k suitable constants.

A second proposal is the following Improved PBSS (IPBSS) choice for the k-th parameter $g_{ij}[k]$:

$$g_{ij}[k] = \frac{1-\beta}{2L} + (1+\beta)\frac{\left|w_{ij}[k]\right|}{2\left\|\mathbf{w}_{ij}\right\|_1}, \quad (8)$$

$$\mathbf{w}_{ij} = \left[w_{ij}[0], w_{ij}[1], \ldots, w_{ij}[L-1]\right]^T, \quad (9)$$

where $-1 \leq \beta \leq 1$ is a constant.

4 Experimental Results

Some experimental results are proposed using two speech signals sampled at 8 kHz. Three different synthetic mixing weight sets were proposed and a total of 5000 samples are used.

The first set of weights is a very simple and sparse set given by

$$\begin{array}{l} a_{11}[0] = 1, \quad a_{22}[0] = \frac{5}{6}, \; a_{12}[10] = \frac{4}{6}, \\ a_{21}[10] = \frac{3}{6}, \; a_{12}[40] = \frac{2}{6}, \; a_{21}[40] = \frac{1}{6}. \end{array} \quad (10)$$

A more dense set of weights is used in the second set, perhaps more realistically in a room that produces notable reverberation, where it was imagined that you have one microphone closer to the first source and another microphone close to the other, which results in:

$$\begin{array}{l} a_{11}[0] = a_{22}[0] = 1 \\ a_{ij}[5n-1] = \exp(-n), \quad \text{for } n = 1, \ldots, 120 \text{ and } i, j = 1, 2 \end{array} \quad (11)$$

that is, we have the strongest input respectively from the two sources in the input, with subsequent echoes every five taps of the filter that decay exponentially.

The third set has only non-zero elements for the length of the filter, but is otherwise similar to the previous weights, and is meant to simulate the same situation:

$$\begin{array}{l} a_{11}[0] = a_{22}[0] = 1 \\ a_{ij}[n] = \exp\left(-(1+0.4n)\right), \quad \text{for } n = 1, \ldots, 200 \text{ and } i, j = 1, 2 \end{array} \quad (12)$$

Furthermore, the de-mixing weights are not updated for each iteration of the learning rule, rather we sum the contributions from the learning rule over 150 iterations between each update of the filter for a robust estimation. When using the **G** matrix from the PBSS algorithm, we always use $\rho = 0.01$ and $\delta_k = 0.01$, as suggested as good values by [6]. When using IPBSS, we always use the parameter $\beta = 0.1$. In both cases we use a de-mixing filter length of $L = 200$ samples and the learning rate is set to $\mu = 1.5 \times 10^{-5}$. We will face the more complicated problem of separating real world mixtures in a future work.

Now, we ask ourself the natural question: Does the proposed method using a **G** matrix from either the PBSS method or the IPBSS method improve the results from the standard algorithm? The answer found, is that it depends on the filter length. Or likely, more generally, how sparse the filter is.

In order to evaluate the convergence of the algorithm, an estimate I of the mutual information is used as convergence and performance index [14,15].

4.1 Results of Proportionate Algorithm

As we will see in this section, there are improvements in the convergence speed over the standard algorithm. Figure 2 shows the convergence history using the estimated mutual information, and we can clearly see that there is a considerable speed-up in convergence: already at epoch number 10, we can see that using the PBSS has almost reached convergence, while the standard algorithm does not reach this level until around epoch number 30. At epoch number 20, using PBSS, the algorithm has practically reached convergence, while convergence is reached for the standard algorithm before around epoch number 80.

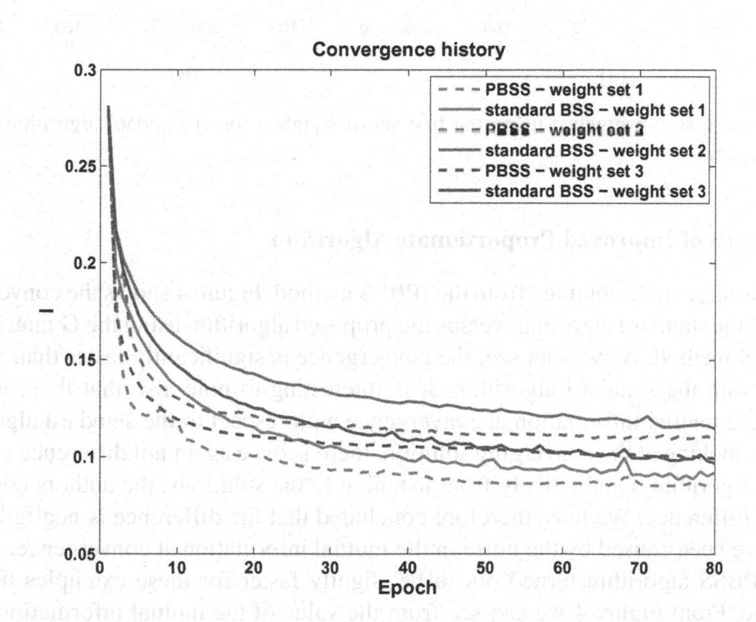

Fig. 2. Comparison of the standard algorithm to the algorithm with the **G** matrix from the PBSS method

This suggests that the algorithm converges between 3 and 4 times faster when using the **G** matrix from the PBSS method, which shows that by adaptively changing the learning rate for the different taps of the filter, we get a significantly better performance. As it can be seen in Figure 2, at convergence the solution of both algorithms reaches the same level of mutual information. It would be interesting to see if there are any differences, looking at the recovered de-mixing filters directly. Figure 3 shows the solution at convergence for both methods using the first weight set, and we easily see that the main features of both solutions are present in both filters and are approximately the same filter taps, albeit at slightly different scalings. Except for that, the are other small differences, but it seems to be mostly noise in the filter. Qualitatively, from hearing the de-mixing of the the mixed sound files, we could discern no differences between the two solutions at convergence. Thus it seems reasonable to conclude that both methods eventually converge to the same solution, but that using the **G** matrix from the PBSS method, we obtain convergence and a good solution significantly faster.

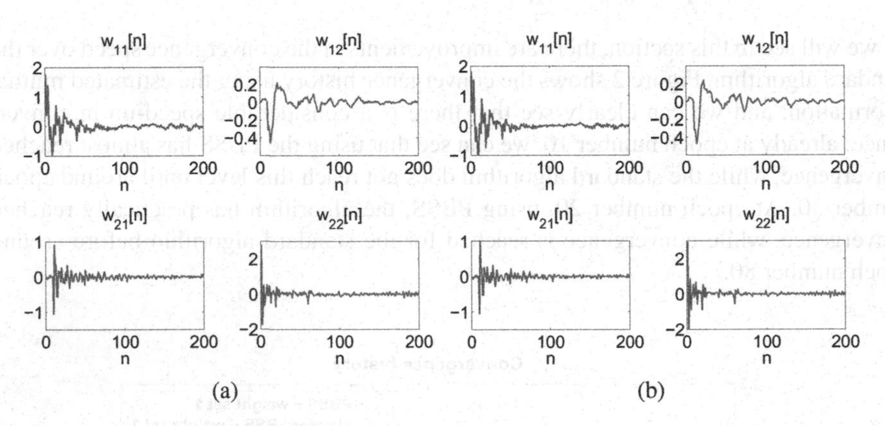

Fig. 3. Solution at convergence using the first set of weights, for a) standard algorithm, and b) PBSS algorithm

4.2 Results of Improved Proportionate Algorithm

Similar results can be obtained from the IPBSS method. Figure 4 shows the convergence history of the standard algorithm versus the proposed algorithm using the **G** matrix from the IPBSS method. As we can see, the convergence is significantly faster than what is possible with the standard algorithm. It is interesting to note also that there is more noise in the mutual information at convergence with respect to the standard algorithm. However, looking at the converging solution, there is once again not difference with the standard algorithm. Qualitatively from listening to the solutions, the authors could not hear any difference. We have therefore concluded that the difference is negligible and might have been caused by the noise in the mutual information at convergence.

The IPBSS algorithm turned out to be slightly faster for these examples than the PBSS one. From Figure 4 we can see from the value of the mutual information of the proposed algorithm using IPBSS from the first epoch, that the same value is reached for the standard algorithm between epoch 4 and 5, while for PBSS the value of the mutual

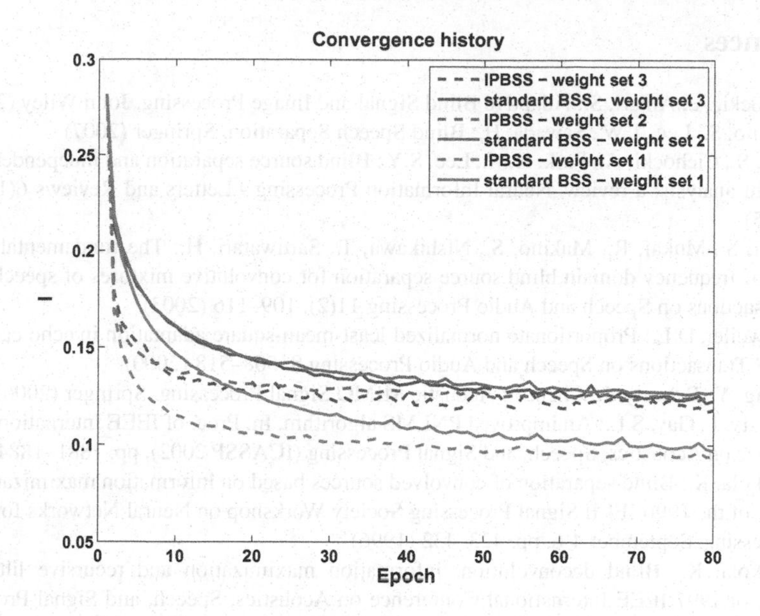

Fig. 4. Comparison of the standard algorithm to the algorithm with the **G** matrix from the IPBSS method

information at the first iteration is reached between epoch 2 and 3. A similar reasoning can be made for other epochs of iterations.

As concluded in the previous section, that indicates that the proposed algorithm using the PBSS method increases the speed of convergence of the first iterations of 3-4 times. From this, we can also infer that the IPBSS method further increases the speed of convergence, suggesting an improvement of 20-25% over the PBSS algorithm.

5 Conclusions

In this paper some preliminary results on a proportionate and improved proportionate version of the well-known time-domain Torkkola BSS algorithm have been proposed. The proposed idea is to use the sparseness of the acoustic impulse responses of the mixing environment, by applying some proportionate algorithm known in the context of speech adaptive filtering. Some experimental results have shown the effectiveness of the proposed approach in terms of convergence speed and encourage us for a more theoretical introduction of these novel classes of separation algorithms.

Acknowledgment. The authors would like to thank Mr. Gaute Halvorsen for his precious help in performing some of the presented experimental results.

References

1. Cichocki, A., Amari, S.: Adaptive Blind Signal and Image Processing. John Wiley (2002)
2. Makino, S., Lee, T.W., Sawada, H.: Blind Speech Separation. Springer (2007)
3. Choi, S., Cichocki, A., Park, H.M., Lee, S.Y.: Blind source separation and independent component analysis: a review. Neural Information Processing - Letters and Reviews 6(1), 1–57 (2005)
4. Araki, S., Mukai, R., Makino, S., Nishikawa, T., Saruwatari, H.: The fundamental limitation of frequency domain blind source separation for convolutive mixtures of speech. IEEE Transactions on Speech and Audio Processing 11(2), 109–116 (2003)
5. Duttweiler, D.L.: Proportionate normalized least-mean-square adaptation in echo cancelers. IEEE Transactions on Speech and Audio Processing 8, 508–518 (2000)
6. Huang, Y., Benesty, J., Chen, J.: Acoustic MIMO Signal Processing. Springer (2006)
7. Benesty, J., Gay, S.L.: An improved PNLMS algorithm. In: Proc. of IEEE International Conference on Acoustics, Speech, and Signal Processing (ICASSP 2002), pp. 1881–1884 (2002)
8. Torkkola, K.: Blind separation of convolved sources based on information maximization. In: Proc. of the 1996 IEEE Signal Processing Society Workshop on Neural Networks for Signal Processing, September 4-6, pp. 423–432 (1996)
9. Torkkola, K.: Blind deconvolution, information maximization and recursive filters. In: Proc. of 1997 IEEE International Conference on Acoustics, Speech, and Signal Processing (ICASSP 1997), April 21-24, pp. 3301–3304 (1997)
10. Torkkola, K.: Blind separation of delayed sources based on information maximization. In: Proc. of 1996 IEEE International Conference on Acoustics, Speech, and Signal Processing (ICASSP 1996), May 7-10, pp. 3509–3512 (1996)
11. Haykin, S.: Adaptive Filter Theory, 4th edn. Prentice-Hall (2001)
12. Ouedraogo, W.S.B., Jaidane, M., Souloumiac, A., Jutten, C.: Regularized gradient algorithm for non-negative independent component analysis. In: Proc. of IEEE International Conference on Acoustics, Speech and Signal Processing (ICASSP 2011), Prague, Czech Republic, May 22-27, pp. 2524–2527 (2011)
13. Boulmezaoud, T.Z., El Rhabi, M., Fenniri, H., Moreau, E.: On convolutive blind source separation in a noisy context and a total variation regularization. In: Proc. of IEEE Eleventh International Workshop on Signal Processing Advances in Wireless Communications (SPAWC 2010), Marrakech, June 20-23, pp. 1–5 (2010)
14. Masulli, F., Valentini, G.: Mutual information methods for evaluating dependence among outputs in learning machines. Technical Report TR-01-02, Dipartimento di Informatica e Scienze dell'Informazione, Università di Genova (2001)
15. Torkkola, K.: Learning feature transforms is an easier problem than feature selection. In: Proc. of 16th International Conference on Pattern Recognition, August 11-15, pp. 104–107 (2002)

Pupillometric Study of the Dysregulation of the Autonomous Nervous System by SVM Networks

Luca Mesin[1], Ruggero Cattaneo[2], Annalisa Monaco[2], and Eros Pasero[1]

[1] Dipartimento di Elettronica e Telecomunicazioni Politecnico di Torino, Turin, Italy
[2] Department of Life, Health and Environmental Sciences, Dental Unit, Universitá di L'Aquila, L'Aquila, Italy
{luca.mesin,eros.pasero}@polito.it,
catrug@libero.it, annalisamonaco@yahoo.it

Abstract. Pupil is controlled by the autonomous nervous system. Patients with temporomandibular disorders (TMD) and with obstructive sleep apnea syndrome (OSAS) are affected by a dysregulation of the autonomous system. Pupillometry is here used to investigate the state of the autonomous system in 3 groups: control, TMD and OSAS. Different indexes are extracted from the pupillogram to characterize pupil dynamics investigated in rest and under stationary stimulations. All possible sets of 3 and 4 indexes are used as features to train support vector machines (SVM) to identify the correct groups. The indexes providing optimal classification are identified.

Keywords: Support vector machine (SVM), Temporomandibular disorders (TMD), obstructive sleep apnea syndrome (OSAS), Pupillometry.

1 Introduction

The control of pupil dilation (mydriasis) and constriction (miosis) is due to the sympathetic nerve centers (Budges ciliospinalis centre) and parasympathetic centre (Edinger–Wesphal Nucleus), respectively. The investigation of pupil dynamics is a simple and non-invasive tool to assess the effects of drugs on the autonomous nervous system [1]. Pupil diameter significantly correlates with heart rate variability [2] and showed its usefulness for assessing the dysregulation in clinical conditions in which the autonomous nervous system is involved [3]. Moreover, some oscillation frequencies of different systems controlled by the autonomous nervous system (e.g., cardiovascular and respiratory systems) are coupled [4], and the pupil shares some of the common rhythms [5]. Pupil indexes in healthy normal subjects were also found to be positively correlated with the level of daytime alertness and were similar to daytime variations in the multiple sleep latency test (MSLT) [6].

The role of the autonomous nervous system was recently investigated in temporo-mandibular disorders (TMD) and in obstructive sleep apnea syndrome (OSAS). Some authors investigating patients with TMD [7] reported a dysregulation of the autonomous nervous system, in part genetics [8]. Patients show

S. Bassis et al. (eds.), *Recent Advances of Neural Network Models and Applications*,
Smart Innovation, Systems and Technologies 26,
DOI: 10.1007/978-3-319-04129-2_11,

an enhanced sympathetic drive, which inhibits normal catecholamine release resulting in significant effects on peripheral target organs and functions of the autonomous nervous system, which would become less efficient in adapting to the needs of environmental and physiologic demands [9]. Recently, it has been proposed that patients with OSAS could be characterized by an imbalance in the autonomous nervous system tone. In particular, some authors suggested that the sympathetic hypertonus is positively correlated with daytime sleepiness [10]. Sympathetic hypertonus could be related to an impaired reaction to several different physiological stimuli, which depend on the severity of OSAS [11]. On the other hand, sympathetic hyperactivity would largely be responsible for other diseases (e.g., cardiac and/or metabolic) that frequently affect these patients [12]. Also a parasympathetic system dysfunction may play a key role in the dysregulation of the autonomous nervous system in OSAS patients [13].

Despite the pupil is easily assessable, its study in diseases involving the autonomous system is still limited to disorders characterized by peripheral or central anatomic nerve lesion. In addition, much of the work regarding the dynamics of the pupil was performed using light stimulation or accommodation reflexes. On the other hand, this paper investigates pupil dynamics in stationary tasks. The considered stationary conditions induce weak activations of the autonomous system. Our study is devoted to assess if these weak responses can characterize disorders, like TMD and OSAS, that are believed to depend on an imbalance of the autonomous system rather than an anatomo-pathologic change of central or peripheral tissues. Different indexes (e.g., based on Fourier theory or on recurrence quantification analysis RQA, [14]) are extracted from the time series describing the size of pupil over time. Moreover, an average measure of the saccadic motion of pupil is used to characterize its dynamics. Different sets of indexes are used to distinguish the patients from a control group using support vector machines (SVM). The indexes providing minimal error can be assumed to better characterize pupil dynamics.

2 Methods

2.1 Data Acquisition

Images of the pupil were acquired by the Oculus system (Inventis srl, Padova, Italy), using infrared CCD cameras (resolution 720x576 pixels, 256 grey levels) mounted on a light helmet, with sampling frequency of 25 frames/s. Each eye was illuminated with three infrared diodes with 880 nm of wavelength. Experiments on pupil dynamics under constant light conditions were obtained providing additional illumination by a LED with wavelength 540 nm and intensity 1.5 mcd.

One minute long acquisitions were obtained from 8 healthy subjects (control group), 7 patients with TMD and 5 with OSAS, in different stationary conditions, which require a different involvement of the sympathetic and of the parasympathetic control. Specifically, the neutral position of the jaw (rest position: RP) in darkness was considered as the rest condition; habitual dental occlusion (HDO) in darkness and light condition with mandible in RP were considered as two different stimulations.

2.2 Signal Processing

The pupil was tracked identifying it with the region growing algorithm, estimating the borders and applying the analytical method proposed in [15] to estimate the radius and the centre of an interpolating circle. Pupil size was computed as the sum of pixels identified by the region growing algorithm. Pupil position was given by the centre of the interpolating circle.

The time series of the position of the pupil was considered as a bidimensional trajectory. The scalar time series describing pupil size was embedded in a high dimensional phase space, obtaining again a trajectory. It was considered as a time series extracted from a deterministic physiological system. The methods of time series embedding [16] were used to extract some information on the system from which the pupillogram was recorded. Given the single measured time series $s(t)$, a time delayed vector was considered with a time delay τ and a number m of elements of the vector called embedding dimension. The time delay corresponding to a 90% decrease of mutual information between the value at τ=0 and the average for 100< τ <200 (delay measured in samples) was considered as optimal. In order to choose the proper embedding dimension, Cao's method [17] was used.

2.3 Pupil Indexes

The average pupil size (A) was considered. Mean frequency (MNF) was computed from the Fourier spectrum of pupil size. The percentages of energy of the spectrum of pupil size in the ranges 0.04-0.15 Hz (F_{low}) and 0.15-0.5 Hz (F_{high}) were computed. Moreover, the amplitude of the movements was measured by the standard deviation of the displacement (STD).

Recurrence quantification analysis (RQA) was recently applied to process pupil dynamics under stationary conditions [18]. It quantifies the number and duration of recurrences of a dynamical system presented by its phase space trajectory. It is based on the recurrence plot, which is a binary map obtained by assigning value 1 to the entry (i,j) if the Euclidean distance between the ith and the jth point along the trajectory is smaller than a threshold, and value 0 otherwise [14]. A simple measure which can be extracted is the recurrence rate (RR), which is the density of recurrence points in a recurrence plot. Another interesting measure is the determinism (DET), which is the percentage of recurrence points which form diagonal lines in the recurrence plot of a given minimal length L_{min}. This index is related with the predictability of the dynamical system.

Here, we consider DET computed after fixing the value of RR to 0.1, by choosing properly the threshold distance between neighbors. Imposing a specific number of recurrences instead of fixing the threshold was suggested as a more objective way to compare the resulting topological properties of different recurrence plots [19]. Moreover, the value of DET depends on the length of the diagonal lines L_{min}. A robust estimation of the index was computed as the average of the values obtained for L_{min} varying in the range 2-10.

In summary, the following 6 indexes extracted from each experiment and from each pupil were considered: mean area (A), mean frequency (MNF), percentage energy of the low frequency (F_{low}) and high frequency (F_{high}) components of the spectrum, average movement (STD) and determinism (DET). The corresponding indexes extracted from the two pupils were averaged. Then, the resulting indexes extracted from each of the two stimulation conditions (light and HDO) were divided by the corresponding value estimated in rest condition (darkness and mandible in RP). Thus, we obtained two ratios for each index (12 indexes in total), measuring the variation of each index estimated during stimulation with respect to rest.

2.4 Classification by SVMs

Pupil dynamics was characterized by sets of indexes extracted from the pupillograms recorded in different conditions. Different sets were used to train a SVM [20] to perform a binary classification, discriminating the TMD or the OSAS patients from the control subjects. As the classes were not linearly separable, the input space was mapped into a feature space using a polynomial kernel, with order finely tuned in order to increase the classification performances (specifically, the order of the polynomial corresponding to the best classification was selected out of the range 2-5; the optimal order was always 3 for identifying TMD patients, whereas it was 3, 2 for identifying OSAS patients using 3, 4 inputs, respectively). All possible sets of 3 and 4 indexes (out of 12, as explained in the previous subsection) were considered as input data for the SVM. The classification performance of each case was assessed with a leave-one-out approach. Then, the sets of indexes providing the best classifications were indicated. Moreover, the distribution of the classification error of the SVMs including a specific index as an input was computed, to assess the importance of such an index in characterizing the pupil dynamics.

3 Results

Figure 1 shows all the indexes used within this study. Data extracted from a stimulated condition (considering light and HDO as stimuli) were normalized with respect to the rest condition. Data from the same subject in different conditions are linked by a segment.

Figure 2 shows the result of the classification. From the 12 different indexes extracted from each subject (the 6 indexes A, MNF, F_{low}, F_{high}, STD, DET in the two stimulated conditions normalized with respect to rest), all possible choices of 3 or 4 inputs were used to discriminate TMD or OSAS patients from control subjects. Each of the 12 indexes was used for a large number of SVMs (including also other different indexes as inputs). The distributions of the classification errors of all classifiers including each index were considered. The indexes

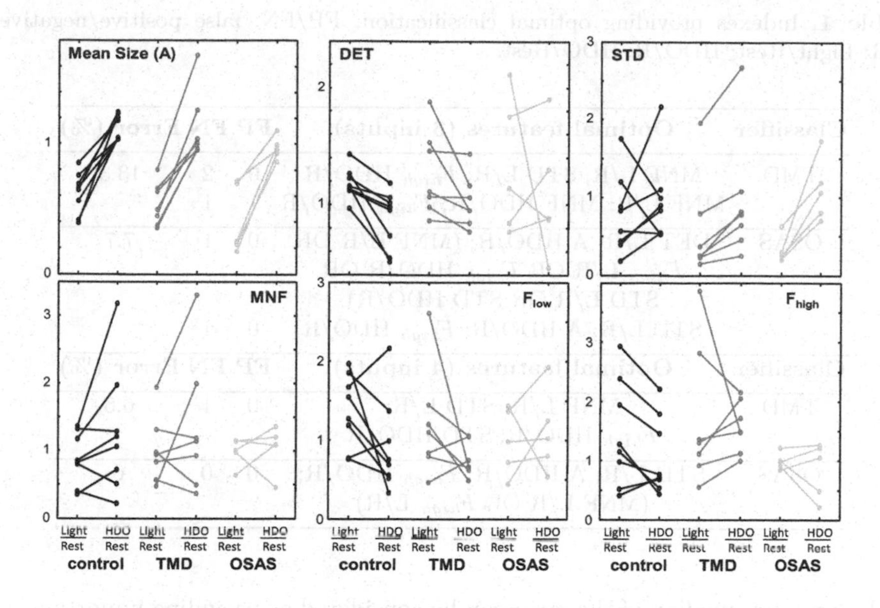

Fig. 1. Indexes considered to describe pupil dynamics

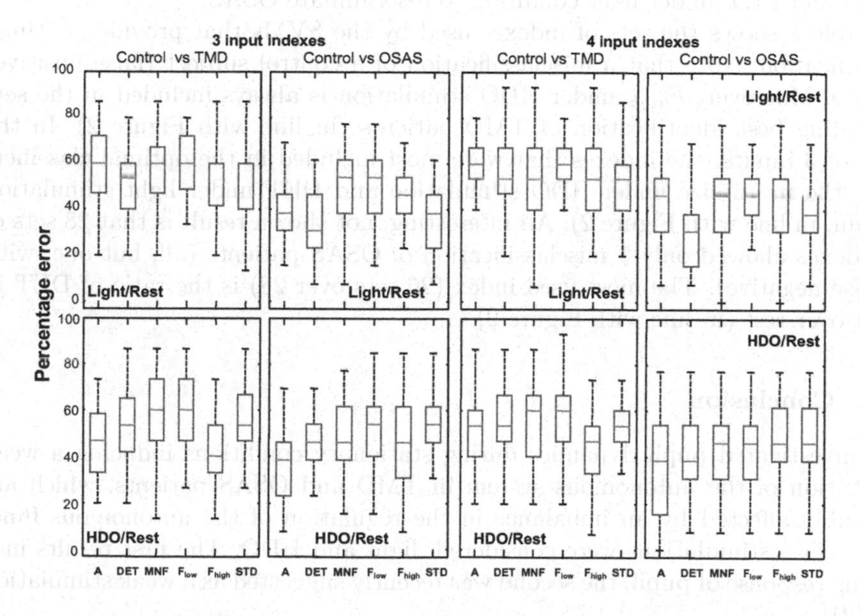

Fig. 2. Median, quartiles and range (excluding outliers) of the classification error for SVMs including specific indexes, considering all classifiers with 3 or 4 inputs

Table 1. Indexes providing optimal classification. FP/FN: false positive/negative; L/R: Light/Rest; HDO/R: HDO/Rest.

Classifier	Optimal features (3 inputs)	FP	FN	Error (%)
TMD	MNF L/R; STD L/R; F_{high} HDO/R	0	2	13.3
	MNF L/R; MNF HDO/R; F_{high} HDO/R	1	1	
OSAS	DET L/R; A HDO/R; (MNF L/R OR F_{high} L/R OR F_{high} HDO/R OR STD L/R OR STD HDO/R)	0	1	7.7
	STD L/R; A HDO/R; F_{high} HDO/R	0	1	
Classifier	**Optimal features (4 inputs)**	**FP**	**FN**	**Error (%)**
TMD	MNF L/R; STD L/R; F_{high} HDO/R; STD HDO/R	0	1	6.6
OSAS	STD L/R; A HDO/R; F_{high} HDO/R; (MNF L/R OR F_{high} L/R)	0	0	0

with minimum median of the error can be considered as providing important information to identify the specific pathology. From Figure 2, we deduce that such interesting indexes are the followings: F_{high} under HDO stimulation to identify TMD and DET under light condition to discriminate OSAS.

Table 1 shows the sets of indexes used by the SVMs that provided optimal classification. Note that a misclassification of a control subject (false positive) is rare. Moreover, F_{high} under HDO stimulation is always included in the sets providing best identification of TMD patients (in line with Figure 2). In the case of 3 inputs, the indexes that were most included in the optimal classifiers were the mean size under HDO stimulation and DET under light stimulation (again, in line with Figure 2). An interesting, not shown result is that 28 sets of 4 indexes allowed only 1 misclassification of OSAS patients (all, but one, with a false negative). The most used index (26 sets over 28) is the ratio of DET in light over rest (in line with Figure 2).

3.1 Conclusion

We investigated pupil dynamics during stationary conditions inducing a weak activation of the autonomous system in TMD and OSAS patients, which are probably affected by an imbalance in the regulation of the autonomous functions. Two stimulations were considered: light and HDO. The first results in a strong response of pupil, the second was recently suggested as a weak stimulation [18][21].

Different indexes were extracted to characterize pupil dynamics and were used to train SVMs to discriminate patients from a control group. The number of study cases was small, so that the problem of overfitting data exists. Indeed, if the number of parameters is the same as or greater than the number of observations, an optimal performance on the training data could be obtained, but with poor generalization. Therefore, we used only 3 or 4 indexes for classification. Nevertheless, some interesting indications were observed. Indeed, classification was possible, indicating that pupil dynamics reflect the considered diseases. Moreover, some pupil indexes, when included in the input set of the SVMs, allowed to improve the classification. Specifically, TMD patients can be identified including in the classifier some spectral indexes under light or HDO stimulation. OSAS patients can be discriminated from the control group considering DET under light and mean size under HDO stimulation.

In the following, we focus on 2 inputs, that we consider more indicative: DET under light and F_{high} under HDO. On the other hand, we don't discuss the extension of the eye movement (STD), which allows greater accuracy of some classification models, but is difficult to interpret, as the eye movement and the pupil size have two control systems which are coordinated, but distinct.

DET can be considered as a measure of the activation of the investigated physiological system, as it increases with stress, illness or aging [22]. OSAS patients show larger variations of DET compared to control subjects (Figure 1). Moreover, their response to different stimuli varies a lot between different subjects, whereas in the control group a higher DET resulted from the light stimulation. Using a dynamical system interpretation, possibly the control subjects could be considered as driven around a similar attractor (reflecting a healthy dynamics), whereas patients are unstable or their dynamics is evolving on different attractors, each characterizing the individual specific response to the pathology. DET was the only index sensitive to the nonlinear dynamical evolutions of the physiological system, so that, augmented with the information of other indexes, it can provide optimal identification of OSAS.

The energy of the high frequency components of the spectrum (F_{high}) under HDO shows an increase for TMD and a decrease for OSAS. This index is considered to represent the parasympathetic component of data reflecting the autonomic response (e.g., heart rate variability) [4]. An increase in F_{high} might indicate the prevalence of the parasympathetic activation. Thus, data could suggest that the increase of F_{high} in TMD during HDO is related to a predominance of the parasympathetic activity, probably due to the inability to activate the sympathetic component in response to the muscle load. The reduction of F_{high} in OSAS in HDO would show a deficit in the parasympathetic system, or its inability to exert an inhibitory activity on stress-induced muscle sympathetic activity. These data support the difficulty of adjustment of the autonomous system sympathetic (TMD) or parasympathetic (OSAS) branches [24].

In conclusion, the study shows the efficacy of the combined use of linear and nonlinear measurements of the pupil dynamics to classify subjects suffering from disorders clinically different (TMD and OSAS) and characterized, probably, by different types of dysfunction of the sympathetic and parasympathetic balance.

References

1. Murillo, R., Crucilla, C., Schmittner, J., Hotchkiss, E., Pickworth, W.B.: Pupillometry in the detection of concomitant drug use in opioid-maintained Patients. Methods Find Exp. Clin. Pharmacol. 26, 271–275 (2004)
2. Br, K.J., Schulz, S., Koschke, M., Harzendorf, C., Gayde, S., Berg, W., Voss, A., Yeragani, V.K., Boettger, M.K.: Correlations between the autonomic modulation of heart rate, blood pressure and the pupillary light reflex in healthy subjects. J. Neurol. Sci. 279(1-2), 9–13 (2009)
3. Keivanidou, A., Fotiou, D., Arnaoutoglou, C., Arnaoutoglou, M., Fotiou, F., Karlovasitou, A.: Evaluation of autonomic imbalance in patients with heart failure: a preliminary study of pupillomotor function. Cardiol. J. 17(1), 65–72 (2010)
4. Censi, F., Calcagnini, G., Cerutti, S.: Coupling patterns between spontaneous rhythms and respiration in cardiovascular variability signals. Comput. Methods Programs Biomed. 68(1), 37–47 (2002)
5. Borgdorff, P.: Respiratory fluctuations in pupil size. Am. J. Physiol. 228(4), 1094–1102 (1975)
6. Wilhelm, B., Giedke, H., Ludtke, H., Bittner, E., Hofmann, A., Wilhelm, H.: Daytime variations in central nervous system activation measured by a pupillographic sleepiness test. J. Sleep Res. 10(1), 1–7 (2001)
7. Eze-Nliam, C.M., Quartana, P.J., Quain, A.M., Smith, M.T.: Nocturnal heart rate variability is lower in temporomandibular disorder patients than in healthy, pain-free individuals. J. Orofac. Pain 25(3), 232–239 (2011)
8. Diatchenko, L., Slade, G.D., Nackley, A.G., Bhalang, K., Sigurdsson, A., Belfer, A., Goldman, D., Xu, K., Shabalina, S.A., Shagin, D., Max, M.B., Makarov, S.S., Maixner, W.: Genetic basis for individual variations in pain perception and the development of a chronic pain condition. Hum. Mol. Genet. 14, 135–143 (2005)
9. Light, K.C., Bragdon, E.E., Grewen, K.M., Brownley, K.A., Girdler, S.S., Maixner, W.: Adrenergic dysregulation and pain with and without acute beta-blockade in women with fibromyalgia and temporomandibular disorder. J. Pain 10(5), 542–552 (2009)
10. Donadio, V., Liguori, R., Vetrugno, R., Contin, M., Elam, M., Wallin, B.G., Karlsson, T., Bugiardini, E., Baruzzi, A., Montagna, P.: Daytime sympathetic hyperactivity in OSAS is related to excessive daytime sleepiness. J. Sleep Res. 16(3), 327–332 (2007)
11. Montesano, M., Miano, S., Paolino, M.C., Massolo, A.C., Ianniello, F., Forlani, M., Villa, M.P.: Autonomic cardiovascular tests in children with obstructive sleep apnea syndrome. Sleep 33(10), 1349–1355 (2010)
12. Yun, A.J., Lee, P.Y., Bazar, K.A.: Autonomic dysregulation as a basis of cardiovascular, endocrine, and inflammatory disturbances associated with obstructive sleep apnea and other conditions of chronic hypoxia, hypercapnia, and acidosis. Med. Hypotheses 62(6), 852–856 (2004)
13. Guilleminault, C., Poyares, D., Rosa, A., Huang, Y.S.: Heart rate variability, sympathetic and vagal balance and EEG arousals in upper airway resistance and mild obstructive sleep apnea syndromes. Sleep Med. 6(5), 451–457 (2005)
14. Marwan, N., Romano, M.C., Thiel, M., Kurths, J.: Recurrence Plots for the Analysis of Complex Systems. Physics Reports 438, 237–329 (2007)
15. Chaudhuri, B.B., Kundu, P.: Optimum circular fit to weighted data in multidimensional space. Pattern Recogn. Lett. 14, 1–6 (1993)

16. Kantz, H., Schreiber, T.: Nonlinear Time-series Analysis. Cambridge Univ. Press (1997)
17. Cao, L.Y.: Practical method for determining the minimum embedding dimension of a scalar time series. Physical D 110, 43–50 (1997)
18. Mesin, L., Monaco, A., Cattaneo, R.: Investigation of Nonlinear Pupil Dynamics by Recurrence Quantification Analysis. BioMed. Research International (in press, 2013)
19. Zou, Y., Donner, R.V., Donges, J.F., Marwan, N., Kurths, J.: Identifying complex periodic windows in continuous-time dynamical systems using recurrence-based methods. Chaos 20(4), 043130 (2010)
20. Cortes, C., Vapnik, V.N.: Support-Vector Networks. Machine Learning, 20 (1995)
21. Monaco, A., Cattaneo, R., Mesin, L., Ciarrocchi, I., Sgolastra, F., Pietropaoli, D.: Dysregulation of the Autonomous Nervous System in Patients with Temporomandibular Disorder: A Pupillometric Study. PLoS One 7(9), e45424 (2012)
22. Javorka, M., Turianikova, Z., Tonhajzerova, I., Javorka, K., Baumert, M.: The effect of orthostasis on recurrence quantification analysis of heart rate and blood pressure dynamics. Physiol. Meas. 30(1), 29–41 (2009)
23. Hayashi, N., Someya, N., Fukuba, Y.: Effect of intensity of dynamic exercise on pupil diameter in humans. J. Physiol. Anthropol. 29(3), 119–122 (2010)
24. Eckberg, D.L.: Sympathovagal Balance: A Critical Appraisal. Circulation 96, 3224–3232 (1997)

A Memristor Circuit Using Basic Elements with Memory Capability

Amedeo Troiano, Fernando Corinto, and Eros Pasero

Politecnico di Torino, Department of Electronic and Telecommunication (DET),
Corso Duca degli Abruzzi 24, 10129 Torino, Italy
{amedeo.troiano,fernando.corinto,eros.pasero}@polito.it

Abstract. After the introduction of the memory–resistor (i.e. memristor), a fundamental two-terminal circuit element defined as a nonlinear relationship between the integral of the voltage and the integral of the current, the class of memristor-based systems was extended by L.O. Chua and S. Kang in 1976. In the literature, the research interest devoted to the discover of novel physical systems with memristor behavior is growing. In 2012, an elementary electronic circuit discovered by F. Corinto and A. Ascoli, based on a Graetz bridge loaded with a RLC filter, was classified to be a memrisor-based system without memory capability. In this paper, the possibility of adding memory on the memristor-based system proposed in 2012 is exploited. In fact, the circuit proposed by F. Corinto and A. Ascoli does not have the memory capability, one of the more important characteristics of the memristor. The memory is added to the system using elementary components in a transfer charge circuit. Results show the memristor-based nature and the memory capability of the presented electronic system.

Keywords: Memristor, Memristor-based system, Transfer charge circuit, Non-linear circuit, Memory devices.

1 Introduction

In 1971, L. O. Chua predicted the existence of a component that works as a resistor with memory capability, named memory-resistor, or (ideal) memristor for short [1]. Five years later, L. O. Chua and S. Kang realized that the ideal memristor is just one element from a class of nonlinear dynamical systems, the memristor-based systems [2]. In 2008, at Hewlett-Packard (HP) Laboratories (Labs), Williams announced the first memristor nano-device [3]. Since then, the research interests devoted to the discovery of novel memristor-based systems have been growing. Recently, a great number of research papers are devoted to investigate the physical realizations of memristor-based systems [4], the modeling of memristor nanodevices [5,6,7,8,9] and its PSpice implementation [10,11]. In addition, several innovative memristor-based nonlinear circuits and neuromorphic systems have been considered (see for example [12,13,15,16,17]).

In 2012, the first electronic circuit with only passive components which represents a memristor-based system was presented in [18]. The electronic circuit

S. Bassis et al. (eds.), *Recent Advances of Neural Network Models and Applications*, Smart Innovation, Systems and Technologies 26,
DOI: 10.1007/978-3-319-04129-2_12, © Springer International Publishing Switzerland 2014

includes only a Graetz bridge with a RLC series filter, and it has the more important features of a memristor-based systems [20], i.e. a pinched hysteretic current-voltage loop for any state initial condition and for any nonzero amplitude and any nonzero and non-infinite frequency of any periodic sign-varying driving source [18]. Moreover, the circuit acts as a nonlinear resistor at direct current (DC) and at infinite frequency of any periodic sign-varying driving source with nonzero amplitude. However, this system has not memory capability, which is one of the more important innovations of the memristor-based system.

In this paper, the memristor-based system discovered by F. Corinto and A. Ascoli was revised, adding a memory capability. The idea is to add storage possibility using one of the elementary electronic components, in order to do not add complexity to the system. The chosen component is the capacitor, which has the characteristic to maintain a quantity of charge. The charge is transferred to the memory capacitor with the introduction of a transfer charge circuit [19] in the electronic scheme described in [18]. The electronic circuit comprises only passive and active already-existing components (four diodes, two transistors MOSFET, an inductor, two capacitors and a resistor).

2 Methods

2.1 Electronic Circuit

The electronic circuit of the innovative memristor system with memory capability is presented in Fig. 1. The heart of the circuit, with memristor-based characteristics, comprises the full-wave rectifier cascaded with the components L_1, R_1, and C_1, as described in [18]. Moreover, the capacitor C_2 is inserted in the circuit for adding memory to the system, and the NMOS transistors M_1 and M_2 to implement the transfer charge circuit. The transistors M_1 and M_2 act as switches, and their bulk terminal is connected to a negative potential (V_3) in order to allow the complete opening and closing of the channels of the transistors in case of negative sign of the input voltage V_1.

Transistors M_1 and M_2 are driven from the square signal V_2, which closes the transistors if its value is 5V, and opens them if it is -5V, to implement the transfer charge phenomena. When the transistors are closed, the circuit is acting as a memristor-based system, as descibed in [18], and capacitance value is C_1+C_2. Instead, when the transistors are opened, no signal is applied to the circuit and the voltage value in which the circuit acts before the opening of the transistor M_2 is stored in the capacitance C_2, adding memory capability to the system. Then, when the transistors closing again, the voltage stored in C_2 is transfered to the input, allowing to read the stored value.

2.2 Mathematical Validation

In [2] a memristor-based system is a nonlinear dynamical circuit element defined by the following differential-algebraic system of equations:

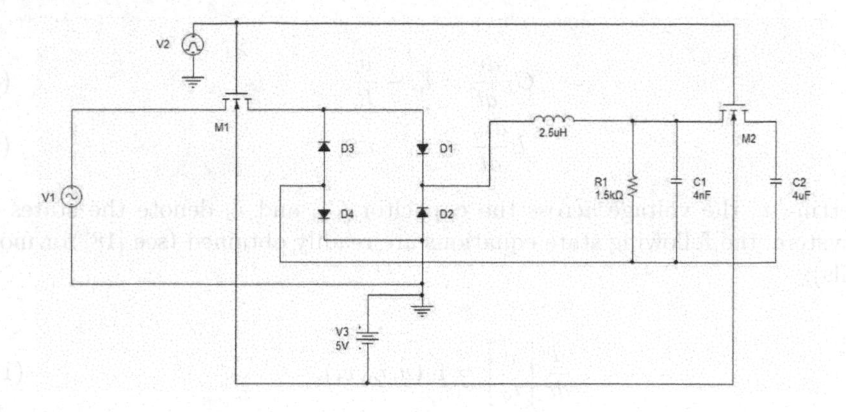

Fig. 1. Electronic circuit of the memristor-based system with memory capability

$$\frac{d\mathbf{x}}{dt} = \mathbf{f}(\mathbf{x}, u), \tag{1}$$

$$y = \mathbf{h}(\mathbf{x}, u), \tag{2}$$

where $\mathbf{x} \in \mathbb{R}^n$ is the state, $u \in \mathbb{R}$ refers to the input, $y \in \mathbb{R}$ describes the output, $\mathbf{f}(\mathbf{x}, u) : \mathbb{R}^n \times \mathbb{R} \to \mathbb{R}^n$ stands for the *state evolution function*, while $\mathbf{h}(\mathbf{x}, u) : \mathbb{R}^n \times \mathbb{R} \to \mathbb{R}$, denoting the *input-output function*, may be factorized as

$$\mathbf{h}(\mathbf{x}, u) = \mathbf{g}(\mathbf{x}, u)u, \tag{3}$$

where $\mathbf{g}(\mathbf{x}, u) : \mathbb{R}^n \times \mathbb{R} \to \mathbb{R}$ is the *memductance* (*memristance*) for a voltage (current)-controlled memristor-based element.

The circuits presented in Fig. 1 can be characterized by a system of differential-algebraic equations of the kind given in (1)-(2), with $\mathbf{h}(\cdot, \cdot)$ fulfilling the factorization property described in (3).

As shown in [18], current and voltage of the diode-bridge (i.e. (i_i, v_i) at the input and (i_o, v_o) at the output) composed of the four diodes satisfy the following constrains (where v_{D_k} and i_{D_k}, with $k = 1, \ldots, 4$ are voltage and current of the diode D_k)

$$i_i = i_{D_1} - i_{D_2}, \tag{4}$$

$$v_i = v_{D_1} - v_{D_2}, \tag{5}$$

$$i_o = i_{D_1} + i_{D_2}, \tag{6}$$

$$v_o = -v_{D_1} - v_{D_2}. \tag{7}$$

Let us assume $V_2 = 0$ (if $V_2 \neq 0$ then $i_i = 0$ and the charge is transferred from C_1 to C_2). Under this assumption, the output port of the diode-bridge is closed on a linear dynamical bipole with equations:

$$C_1 \frac{dv}{dt} = i_o - \frac{v}{R}, \tag{8}$$

$$L \frac{di_o}{dt} = v_o - v. \tag{9}$$

Letting v, the voltage across the capacitor C_1, and i_o denote the states of the system, the following state equations are readily obtained (see [18] for more details):

$$\frac{d}{dt} \begin{bmatrix} v \\ i_o \end{bmatrix} = \mathbf{f_2}(v, i_o, v_i), \tag{10}$$

where

$$\mathbf{f_2}(v, i_o, v_i) = \begin{bmatrix} \frac{1}{C} \left(i_o - \frac{v}{R} \right) \\ -\frac{1}{L} v + f_2(i_o, v_i) \end{bmatrix}, \tag{11}$$

with $f_2(\cdot, \cdot)$ described by

$$f_2(i_o, v_i) = -\frac{2V_T}{L} \ln \left(\frac{i_o + 2I_S}{2I_S \cosh\left(\frac{v_i}{2V_T}\right)} \right) - \frac{R}{L} i_L. \tag{12}$$

In conclusion, the defining equations of this second-order voltage-controlled current-state memristor-based circuit are the state equation (10) and $i_i = g_2(i_o, v_i) v_i$ with

$$g_2(i_o, v_i) = \frac{(i_o + 2I_S)}{2V_T} \left(\sum_{n=0}^{\infty} \frac{\left(\frac{v_i}{2V_T}\right)^{2n}}{(2n)!} \right)^{-1} \sum_{n=0}^{\infty} \frac{\left(\frac{v_i}{2V_T}\right)^{2n}}{(2n+1)!}, \tag{13}$$

As a consequence, the proposed circuit, the current–voltage i_i–v_i (with $v_1 = v_i$) falls into the class memristor-based system defined by (1)-(2).

Numerical simulation are reported in the next section.

3 Results

3.1 Memristive Behaviour

PSpice simulations of the circuit of Fig. 1 with both transistors closed are accomplished to verify the memristor-based behaviour of the system. Diodes 1N4148 are inserted in the Graetz bridge, and the MbreakN Psipce model is used for the simulation of NMOS transistors. The values of the passive components are those indicated in Fig. 1. The circuit is driven by a sine-wave voltage source (V_1) with amplitude of 2.5V and frequency of 500Hz. Result of the simulation is provided in Fig. 2, where the current of the voltage source is plotted as a function of the voltage V_1. The circuit manifests the typical pinched hysteretic current-voltage loop characterising memristor-based systems. As described in [18], for infinite frequency the circuit behaves as a nonlinear resistor.

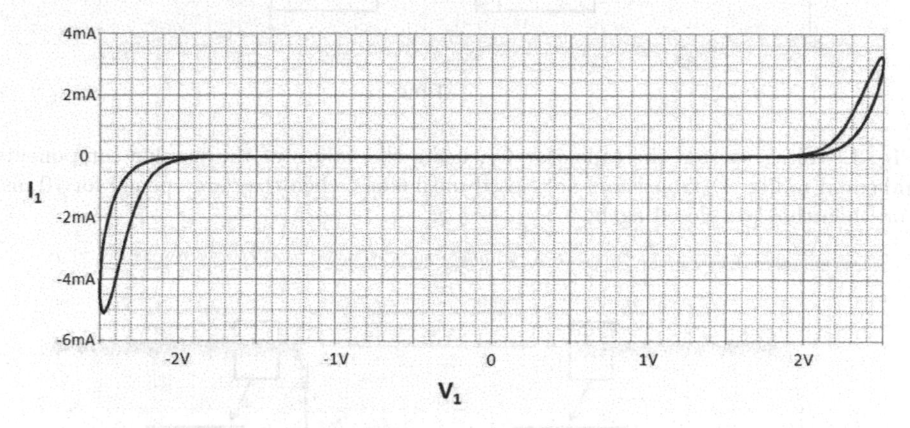

Fig. 2. Current-voltage characteristics obtained from the PSpice simulations of the circuit of Fig. 1 with both transistors closed. The input signal is sine-wave with amplitude of 2.5V and frequency equal to 500Hz.

3.2 Memory Capability

PSpice simulations of the voltage across the capacitor C_2 with transistors opened and closed are performed to verify the memory capability of the system. The values of the passive components are those indicated in Fig. 1. The capacitor C_1 was first set at 4nF, and then at 4μF (the same value of C_2) in order to see differences. In both cases, transistors are initially closed until 60ms, then they are opened for 70 ms, and then they are closed again.

Voltage across the capacitor C_2, with C_1 set at 4nF, is showed in Fig. 3. When the transistors are opened (from 60ms to 130ms) the voltage across C_2 is maintained, and this value is stable even when M_1 and M_2 are closed again, verifying the memory capability of the system.

On the other hand, if the capacitor C_1 is set at 4μF, the voltage across the capacitor C_2 is maintained when the transistors are opened but drops when M_1 and M_2 are closed again, as showed in Fig. 4. This effect is due since the capacitance values of C_1 and C_2 are the same. In fact, when transistor M_2 closed, half of the charge in C_2 is transferred in C_1, halving the voltage across C_2. Instead, if the value of C_2 is one thousand times that of C_1, as showed in Fig. 3, only one thousandth of the charge in C_2 is transferred in C_1, maintaining almost

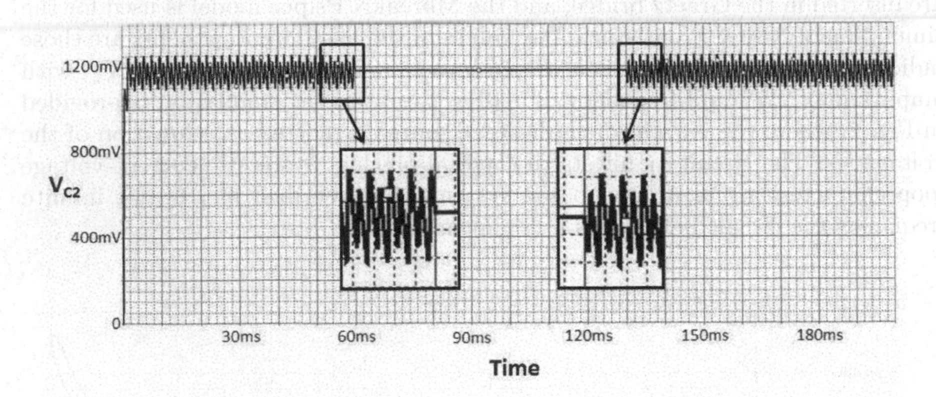

Fig. 3. Voltage across the capacitor C_2, using the values of the passive components indicated in Fig. 1. Transistors are closed until 60ms, then they are opened for 70 ms, and then they are closed again.

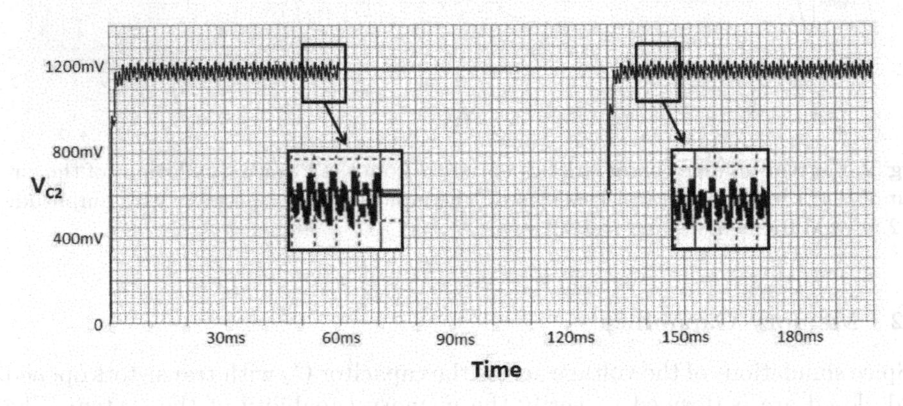

Fig. 4. Voltage across the capacitor C_2, using the values of the passive components indicated in Fig. 1 except for C_1 that is set at 4μF (the same value of C_2). Transistors are closed until 60ms, then they are opened for 70 ms, and then they are closed again.

constant the voltage across C_2. In order to garantee the memory capability of the system C_2 needs to be bigger than C_1, higher is the value of C_2 with respect that of C_1 and smaller will be the error on the maintenance of the voltage on C_2.

PSpice simulations of the circuit of Fig. 1 with C_1 set at 4μF and C_2 at 4nF, with transistors T_1 and T_2 closed until 60ms, then opened for 70 ms, and then closed again is provided in Fig. 5. The value of C_2 is three orders of magnitude higher than that of C_1, assuring the memory capability. In the figure, the current of the voltage source is plotted as a function of the voltage source V_1. The circuit manifests the typical pinched hysteretic current-voltage loop characterising memristor-based systems. The highlighted pick in the current value of the figure is due to the reading at the input of the system of the stored value. Moreover, it can be seen that the system reach the loop immediately after the reading of the stored value when the transistors closed, as highlighted in Fig. 5.

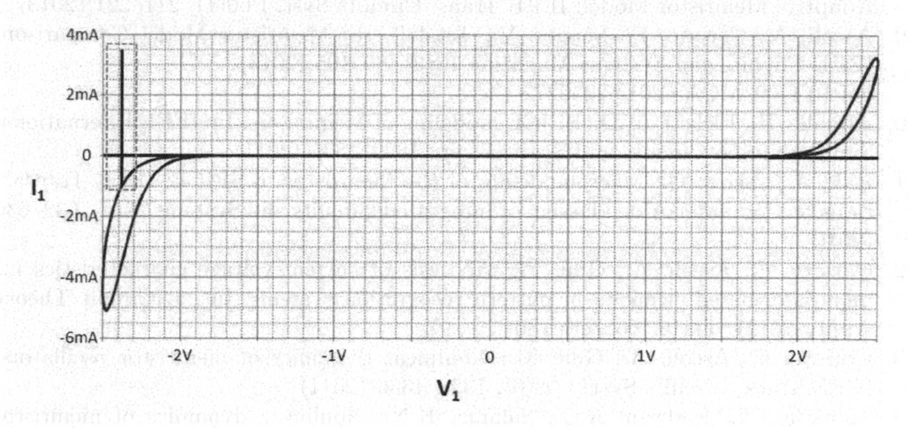

Fig. 5. Current-voltage characteristics obtained from the PSpice simulations of the circuit of Fig. 1, with transistors closed until 60ms, then opened for 70 ms, and then closed again

4 Conclusions

In this paper, a theoretical proof for the memristor-based nature and the memory capability of a elementary electronic system is presented. The electronic circuit is composed of passive and active already-existing components, introducing nonlinearity and dynamical behaviour into the system, and allowing to store data, in a cheap and simple way. Agreement between PSpice simulations of the electronic circuit and numerical solutions of the system model validates the theoretical proof.

References

1. Chua, L.O.: Memristor: the missing circuit element. IEEE Transactions on Circuit Theory 18(5), 507–519 (1971)
2. Chua, L.O., Kang, S.M.: Memristive devices and systems. Proc. IEEE 64(2), 209–223 (1976)
3. Strukov, D.B., Snider, G.S., Stewart, D.R., Williams, R.S.: The missing memristor found. Nature 14, 80–83 (2008)
4. Pickett, M.D., Strukov, D.B., Borghetti, J.L., Yang, J.J., Snider, G.S., Stewart, D.R., Williams, R.S.: Switching Dynamics in Titanium Dioxide Memristive Devices. J. Appl. Phys. 106, 74508 (2009)
5. Joglekar, Y.N., Wolf, S.T.: The elusive memristor: properties of basic electrical circuits. Eur. J. Phys. 30, 661–675 (2009)

6. Biolek, Z., Biolek, D., Biolková, B.: Spice model of memristor with nonlinear dopant drift. Radio. Eng. 18(2), 210–214 (2009)
7. Corinto, F., Ascoli, A.: A boundary condition-based approach to the modeling of memristor nanostructures. IEEE Trans. Circuits Syst. I 59, 2713–2726 (2012), doi:10.1109/TCSI.2012.2190563
8. Kvatinsky, S., Friedman, E.G., Kolodny, A., Weiser, U.C.: TEAM - ThrEshold Adaptive Memristor Model. IEEE Trans. Circuits Syst. I 60(1), 211–221 (2013)
9. Ascoli, A., Corinto, F., Senger, V., Tetzlaff, R.: Memristor Model Comparison. IEEE Circuits and Systems Magazine 13(2), 89–105 (2013), doi:10.1109/MCAS.2013.2256272
10. Abdalla, H., Pickett, M.D.: SPICE modeling of Memristors. In: IEEE International Symposium on Circuits and Systems, pp. 1832–1835 (2011)
11. Rak, A., Cserey, G.: Macromodeling of the Memristor in SPICE. IEEE Transactions on Computer-Aided Design of Integrated Circuits and Systems 29(4), 632–636 (2010)
12. Corinto, F., Ascoli, A., Gilli, M.: Analysis of current-voltage characteristics for memristor-based elements in pattern recognition systems. Int. J. Circuit Theory Appl. (2012), doi:10.1002/cta.1804
13. Corinto, F., Ascoli, A., Gilli, M.: Nonlinear dynamics of memristor oscillators. IEEE Trans. Circuits Syst. I 58(6), 1323–1336 (2011)
14. Talukdar, A., Radwan, A.G., Salama, K.N.: Nonlinear dynamics of memristor based 3rd order oscillatory system. Microelectronics Journal 43(3), 169–175 (2012)
15. Strukov, D.B., Stewart, D.R., Borghetti, J.L., Li, X., Pickett, M., Ribeiro, G.M., Robinett, W., Snider, G.S., Strachan, J.P., Wu, W., Xia, Q., Yang, J.J., Williams, R.S.: Hybrid CMOS/memristor circuits. In: Int. Symp. on Circuits and Syst., pp. 1967–1970 (2010)
16. Versace, M., Chandler, B.: MoNETA: A Mind Made from Memristors. IEEE Spectrum (2010)
17. Johnsen, G.K.: An introduction to the memristor: a valuable circuit element in bioelectricity and bioimpedance. J. Electr. Bioimp. 3, 20–28 (2012)
18. Corinto, F., Ascoli, A.: Memristive diode bridge with LCR filter. Electronics Letters 48(14), 824–825 (2012)
19. Troiano, A., Pasero, E., Mesin, L.: New System for Detecting Road Ice Formation. IEEE Transactions on Instrumentation and Measurement 60(3), 1091–1101 (2011)
20. Chua, L.O.: Resistance switching memories are memristors. Appl. Phys. A 102(4), 765–783 (2011)

Effects of Pruning on Phase-Coding and Storage Capacity of a Spiking Network

Silvia Scarpetta[1,3] and Antonio De Candia[2,3,4]

[1] Dipartimento di Fisica "E.R. Caianiello", Università di Salerno, Italy
[2] Dipartimento di Scienze Fisiche, Università di Napoli Federico II
[3] INFN, Sezione di Napoli e Gruppo Coll. di Salerno
[4] CNR-SPIN, Unità di Napoli
sscarpetta@unisa.it, decandia@na.infn.it

Abstract. Synaptic pruning is a crucial process during development. We study the imprinting and replay of spatiotemporal patterns in a spiking network, as a function of pruning degree. After a Spike Timing Dependent Plasticity-based learning of synaptic efficacies, the weak synapses are removed through a competitive pruning process. Surprisingly, after this pruning stage, the storage capacity for spatiotemporal patterns is relatively high also for very high diluition ratio.

Introduction

The capacity to code and memorize information is fundamental to normal cognition. Here we study the effects of Synaptic pruning — the elimination of synapses — on the information processing capacity. The study of the relation between synaptic efficacy and synaptic pruning suggests that the weak synapses may be modified and removed through competitive learning. In our strategy the strength of connections is determined by a learning rule based on spike-time-dependent plasticity (STDP), with an asymmetric time window depending on the relative timing between pre- and post-synaptic activity. After storage of multiple spatiotemporal patterns, we apply a synaptic pruning strategy, eliminating the weaker connections. We eliminate all the weaker connections, in such a way that at the end each neuron is connected only to a small number $z \ll N$ of other neurons. We find that the ability to replay the stored spatiotemporal patterns is relatively high also for very small value of z.

In this paper we analyze the role of synaptic pruning, eliminating the weaker connections after a learning rule based on STDP, in storing multiple phase-coded memories as attractor states of the neural dynamics. The framework of storing and retrieval of memories as attractors of the dynamics is widely accepted, but the effects of pruning on such dynamical periodic spatiotemporal patterns haven't been investigated before.

In Section 1 we present the model, the learning rule and pruning rule. In Section 2 we introduce some parameters which will be useful to study the capacity of the network. Lastly, in Section 3 we report a summary of the obtained results along with a short discussion.

S. Bassis et al. (eds.), *Recent Advances of Neural Network Models and Applications*, Smart Innovation, Systems and Technologies 26,
DOI: 10.1007/978-3-319-04129-2_13,

1 The Model

We consider a network of N spiking Leaky Integrate and Firing (LIF) neurons, with $N(N-1)$ possible directed connections J_{ij}. The fully-connected model has been studied in [16]. Here we study the effect of pruning. Briefly, the post-synaptic membrane potential of a neuron i is given by:

$$h_i(t) = \sum_j J_{ij} \sum_{t^*{}_j > t^*{}_i} \epsilon(t - t^*{}_j), \tag{1}$$

where $i = 1, \ldots, N$, J_{ij} are the synaptic connections, the sum over $t^*{}_j$ runs over all pre-synaptic firing times following the last spike of neuron i, and $\epsilon(t)$ describes the response kernel to incoming spikes on neuron i:

$$\epsilon(t - t^*{}_j) = K \left[\exp\left(-\frac{t - t^*{}_j}{\tau_m} \right) - \exp\left(-\frac{t - t^*{}_j}{\tau_s} \right) \right] \Theta(t - t^*{}_j) \tag{2}$$

where τ_m is the membrane time constant (10 ms), τ_s is the synapse time constant (5 ms), Θ is the Heaviside step function, and K is a multiplicative constant chosen so that the maximum value of the kernel is 1, as in [4,16]. When the membrane potential $h_i(t)$ exceeds the spiking threshold θ_{th}, a spike is scheduled, and the membrane potential is reset to the resting value zero. The STDP learning rule used to determine the synaptic efficacy has been introduced in [7,10,11,15,21,12,17,19,18,20]. After learning a spatiotemporal periodic pattern μ, the synaptic efficacies are:

$$J_{ij}^{\mu} = \frac{1}{N} \sum_{n=-\infty}^{\infty} A(t_j^{\mu} - t_i^{\mu} + nT^{\mu}) \tag{3}$$

where $t_i^{\mu} = (\phi_i^{\mu}/2\pi)T^{\mu}$ is the spike time of unit i in the pattern μ.

The kernel $A(\tau)$ is the STDP measure of the strength of synaptic change when a time delay τ occurs between pre and post-synaptic activity. We use the kernel introduced and motivated by [23], $A(\tau) = a_p e^{-\tau/T_p} - a_D e^{-\eta\tau/T_p}$ if $\tau > 0$ and $A(\tau) = a_p e^{\eta\tau/T_D} - a_D e^{\tau/T_D}$ if $\tau < 0$, with the same parameters used in [23] to fit the experimental data of [6], $a_p = \gamma\,[1/T_p + \eta/T_D]^{-1}$ and $a_D = \gamma\,[\eta/T_p + 1/T_D]^{-1}$ with $T_p = 10.2$ ms and $T_D = 28.6$ ms, $\eta = 4$, $\gamma = 42$.

This function satisfies the balance condition $\int_{-\infty}^{\infty} A(\tau) d\tau = 0$. Notably, when $A(\tau)$ is used in Eq. (3) to learn phase-coded patterns with uniformly distributed phases, then the balance condition assures that the sum of the connections on the single neuron $\sum_j J_{ij}$ is of order $1/\sqrt{N}$, and therefore, it assures a balance between excitation and inhibition [16].

The phases ϕ_j^{μ} of the spikes in the periodic spatiotemporal pattern are randomly chosen from a uniform distribution in $[0, 2\pi)$. When multiple phase coded patterns are stored, the learned connections are simply the sum of the contributions from individual patterns, namely

$$J_{ij} = \sum_{\mu=1}^{P} J_{ij}^{\mu}. \tag{4}$$

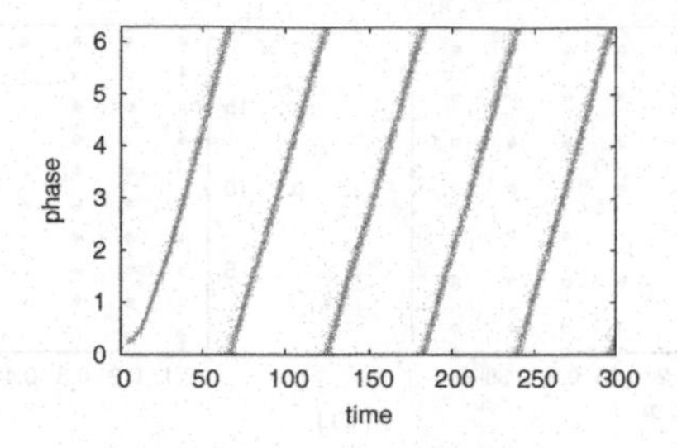

Fig. 1. Dynamics of the network with $N = 10000$ neurons, $z = 100$ positive connections per neuron, $P = 10$ stored patterns, after a stimolation with $M = 300$ consecutive spikes with the phases of one of the patterns. On the vertical axis the phase of the spiking neuron in the replayed pattern is shown.

Previous works [11,16] showed that multiple items can be memorized in such fully connected LIF model, in such a manner that the intrinsic network dynamics recall the specific phases of firing when a partial cue is presented, if spiking threshold is in a proper range of parameters. A critical value of spiking threshold exists[18,19], such that for value higher of this critical value no persistent replay is possible. Here we study the effects of a stdp-driven pruning process. After the STDP-based learning Eq. (4), both positive and negative weak synapses are removed through a competitive process.

After learning P stored patterns, we start the pruning process, deleting the weaker syapses. We delete both negative and positive synapses with low values of $|J_{ij}|$. We prune a fixed percentage of positive connections, so that each unit has z positive outgoing connections. We then prune a variable percentage of negative connections, to ensure that sum of negative outgoing connections is equal to the sum of positive outgoing connections.

Notably, even for $z \ll N$, there's a range of spiking threshold θ_{th}, in which the response of the system to a cue external stimulation, in absence of noise, shows the replay of one of the stored pattern. A short cue with $M = 300$ spikes with the proper phase relationship is able to induce the selective persistent replay of the stored pattern similar to the cue, even when $z \ll N$. For example when z is as small as $z = 100$ in a network with $N = 10000$ units and $P = 10$ patterns are stored, the dynamics after the cue is shown in Fig. 1.

2 Storage Capacity of the Pruned Network

We study here the storage capacity of the network in the case of pruning, as a function of the number z of surviving connections. The storage capacity is defined, as usual, as the maximum number of patterns that can be stored and succesfully selectively retrieved, when the network respond to a short cue stimulation. To measure quantitatively the successfull of retrieval we introduce an order parameter m^{μ}, which measure the similarity between the network activity during retrieval dynamics and the stored pattern, defined as [18,16,19]:

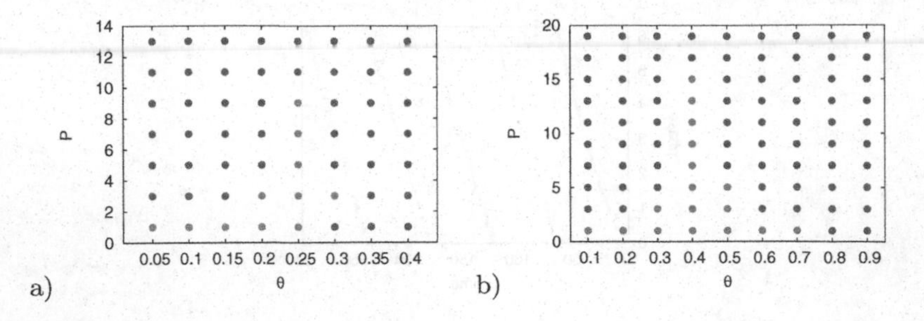

Fig. 2. The capacity $P_{\max}$ as a function of spiking threshold of the units, for different values of connections z per neuron, in a network made up of $N = 10000$ neurons. Succesfull retrieval is indicated in red, while failure in blue. (a) Pruned network with $z = 50$, and (b) pruned network with $z = 100$.

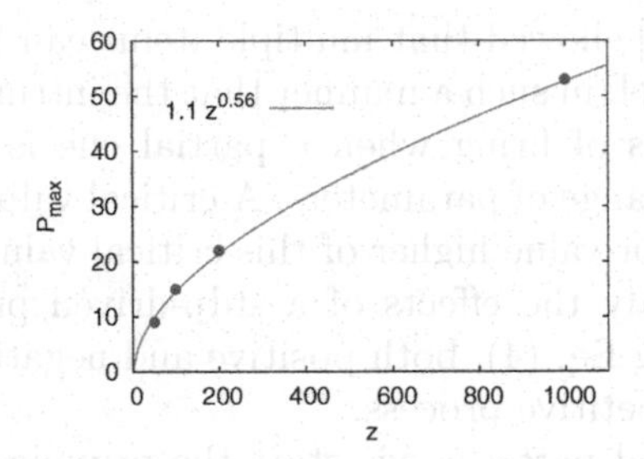

Fig. 3. The capacity $P_{\max}$ as a function of the number z of positive outgoing connections per unit, in a network with $N = 10000$ units

$$m^{\mu}(T^w) = \frac{1}{\langle N_s \rangle} \langle |M(t, T^w)| \rangle \tag{5}$$

where

$$|M^{\mu}(t, T^w)| = \left| \frac{1}{N} \sum_{\substack{j=1,\ldots,N \\ t<t_j^*<t+T^w}} e^{-i2\pi t_j^*/T^w} e^{i\phi_j^{\mu}} \right| \tag{6}$$

where t_j^* is the spike timing of neuron j during the spontaneous dynamics, and T^w is an estimation of the period of the collective spontaneous periodic dynamics, the average $\langle \cdots \rangle$ is done on the starting time t of the window, and $\langle N_s \rangle$ is the average number of spikes on a window of time T^w. This quantity is maximal (equal to one) when collective activity is periodic and the ordering of spiking times coincide with that of the stored pattern, and is order $\simeq 1/\sqrt{N}$ when the spike timings are uncorrelated with the stored ones. For instance, if the retrieval of pattern $\mu = 1$ (out of 3) is successful, the overlap values will be $|m^1| > 0, |m^2| \simeq 0, |m^3| \simeq 0$. In the following we consider a pattern perfectly retrieved when order parameter $|m| > 0.5$ for at least 3 over 5 different realizations.

We consider a network of $N = 10000$ neurons, and we study network dynamics for different degree of pruning. In Fig. 2 we plot the maximum capacity $P_{\max}$ as a function of θ_{th} for different values of z. Throuout the paper we use $N = 10000$, $\omega^{\mu} = 3Hz$, and a cue stimuation with $M = 300$ spikes with phases equals to the phases of pattern 1, and frequency $\omega^{cue} = 20Hz$.

To see how storage capacity changes as a function of the degree of pruning, we plot in Fig. 3 the storage capacity at optimal spiking threshold as a function of z. The number of patterns is well fitted by the function $P_{\max} = 1.1\, z^{0.56}$.

3 Summary and Discussion

In this paper we study the effects of synapses pruning in an oscillatory network, with particular attention to the function of storage and recall of phase-coded patterns of spikes. Oscillations are ubiquitus in the brain[2,9]. In the spatiotemporal patterns we consider the information is encoded in the phase-based timing of firing relative to the cycle[1,3,8,16].

We analyze the ability of the STDP-bases pruning strategy to memorize multiple phase-coded pattern, such that the spontaneous dynamics of the network selectively gives sustained activity which match one of the stored phase-coded patterns, depending on the initialization of the network.

We prune both negative and positive synapses that have low absolute values of synaptic efficacy in the STDP prescription. We eliminate a fixed percentage of positive weaker connections, so that each unit has exactly z positive outgoing connections. We then prune a number of the weakest negative connections, chosen to ensure that sum of negative outgoing connections is equal to the sum of positive outgoing connections.

We study the storage capacity for different degrees of pruning. The effect of spiking threshold on storage capacity have been studied previously [19,18,20]. Here we compute the storage capacity, as a function of connectivity z, at the optimal value of spiking threshold. We see that capacity of a pruned network with $z = 200 = 0.02\,N$ positive connections, already has 25% of the capacity of of the fully connected network, that is about $P_{\max} \simeq 90$ for $N = 10000$.

The task of storing and recalling phase-coded memories with sparse connectivity has been also investigated in [27] in the framework of probabilistic inference and in [24,25,13] in the framework of networks with binary units. In [24,25,27], the case of limited connectivity is studied, showing how recall performance depends to the degree of connectivity when connections are cut randomly, and how inhibition may enhance memory capacity[25]. In our work connections are not cutted randomly, but only the ones which are weaker after the learning strategy Eq. (4) are eliminated. The pruning is guided from the STDP learning, the balance between excitation and inhibition is keeped, and therefore very good capacity is achieved even at small z. In [13] it has been shown that storage capacity of spatiotemporal patterns in the binary network also depends from the topology of the connectivity, and capacity depends not only from the number of connections but also from the fraction of long range versus short range connections. While in [13] we fixed the topology of the network, and then on the survided connections we applied the learning rule Eq. (4) for synaptic efficacies, here we first apply the learning rule for synaptic efficacies and then we eliminate the weaker synapses, in a sort of learning-guided competitive pruning. So the topology of the pruned network here is determined by the patterns to be stored.

References

1. Siegel, M., Warden, M.R., Miller, E.K.: Phase-dependent neuronal coding of objects in short-term memory. PNAS 106, 21341–21346 (2009)
2. Buzsaki, G., Draguhn, A.: Neuronal Oscillations in Cortical Networks. Science 304, 1926–1929 (2004)
3. Kayser, C., Montemurro, M.A., Logothetis, N.K., Panzeri, S.: Spike-phase coding boosts and stabilizes information carried by spatial and temporal spike patterns. Neuron. 61, 597–608 (2009)
4. Gerstner, W., Ritz, R., van Hemmen, J.L.: Why spikes? Hebbian learning and retrieval of time-resolved excitation patterns. Biological Cybernetics 69(5-6), 503–515 (1993)
5. Markram, H., Lubke, J., Frotscher, M., Sakmann, B.: Regulation of synaptic efficacy by coincidence of postsynaptic APs and EPSPs. Science 275, 213–215 (1997)
6. Bi, G.Q., Poo, M.M.: Precise spike timing determines the direction and extent of synaptic modifications in cultured hippocampal neurons. J. Neurosci. 18, 10464–10472 (1998)
7. Scarpetta, S., Zhaoping, L., Hertz, J.: Hebbian Imprinting and Retrieval in Oscillatory Neural Networks. Neural Computation 14(10), 2371–2396 (2002)
8. Scarpetta, S., Marinaro, M.: A learning rule for place fields in a cortical model: Theta phase precession as a network effect. Hippocampus 15(7), 979–989 (2005)
9. Zhaoping, L., Lewis, A., Scarpetta, S.: Mathematical analysis and simulations of the neural circuit for locomotion in lampreys. Physical Review Letters 92(19), 198106 (2004)
10. Yoshioka, M., Scarpetta, S., Marinaro, M.: Spatiotemporal learning in analog neural networks using spike-timing-dependent synaptic plasticity. Phys. Rev. E 75, 051917 (2007)
11. Scarpetta, S., De Candia, A., Giacco, F.: Storage of phase-coded patterns via STDP in fully-connected and sparse network: a study of the network capacity. Frontiers in Synaptic Neuroscience 2 (2010)
12. Marinaro, M., Scarpetta, S., Yoshioka, M.: Learning of oscillatory correlated patterns in a cortical network by a STDP-based learning rule. Mathematical Biosciences 207(2), 322–335 (2007)
13. Scarpetta, S., Giacco, F., de Candia, A.: Storage capacity of phase-coded patterns in sparse neural networks. EPL (Europhysics Letters) 95(2), 28006 (2011)
14. Scarpetta, S., De Candia, A., Giacco, F.: Dynamics and storage capacity of neural networks with small-world topology. In: Proceedings of the 2011 Conference on Neural Nets WIRN10. Frontiers in Artificial Intelligence and Applications, vol. 226 (2011) ISBN: 978-1-60750-691-1
15. Yoshioka, M., Scarpetta, S., Marinaro, M.: Spike-Timing-Dependent Synaptic Plasticity to Learn Spatiotemporal Patterns in Recurrent Neural Networks. In: de Sá, J.M., Alexandre, L.A., Duch, W., Mandic, D.P. (eds.) ICANN 2007, Part I. LNCS, vol. 4668, pp. 757–766. Springer, Heidelberg (2007)

16. Scarpetta, S., Giacco, F.: Associative memory of phase-coded spatiotemporal patterns in leaky Integrate and Fire networks. Journal of Computational Neuroscience. J Comput Neurosci. 34(2), 319–336 (2013), doi:10.1007/s10827-012-0423-7; Epub (October 4, 2012)
17. Giacco, F., Scarpetta, S.: Attractor networks and memory replay of phase coded spike patterns. In: Frontiers in Artificial Intelligence and Applications, vol. 234, pp. 265–274 (2011)
18. Scarpetta, S., Giacco, F., Lombardi, F., de Candia, A.: Effects of Poisson noise in a IF model with STDP and spontaneous replay of periodic spatiotemporal patterns, in absence of cue stimulation. Biosystems 112(3), 303–2647 (2013), doi:10.1016/j.biosystems.2013.03.017, ISSN 0303-2647
19. Scarpetta, S., de Candia, A.: Critical behavior near a phase transition between retrieval and non-retrieval regimes in a LIF network with spatiotemporal patterns. AIP Conf. Proc, vol. 1510, pp. 36–43 (2013), doi:http://dx.doi.org/10.1063/1.4776499
20. Scarpetta, S., de Candia, A.: Neural avalanches at the critical point between replay and non-replay of spatiotemporal patterns. Plos One (accepted May 11) (in press, 2013), doi:10.1371/journal.pone.0064162, PONE-D-13-11021R1
21. Scarpetta, S., Yoshioka, M., Marinaro, M.: Encoding and Replay of Dynamic Attractors with Multiple Frequencies: Analysis of a STDP Based Learning Rule. In: Marinaro, M., Scarpetta, S., Yamaguchi, Y. (eds.) Dynamic Brain. LNCS, vol. 5286, pp. 38–60. Springer, Heidelberg (2008)
22. Gerstner, W., Kempter, R., van Hemmen, L., Wagner, H.: A neuronal learning rule for sub-millisecond temporal coding. Nature 383, 76–78 (1996)
23. Abarbanel, H., Huerta, R., Rabinovich, M.I.: Dynamical model of long-term synaptic plasticity. Proc. Nas. Acad. Sci. 99(15), 10132–10137 (2002)
24. Leibold, C., Kempter, R.: Memory Capacity for Sequences in a Recurrent Network with Biological Constraints. Neural Computation 18(4), 904–941 (2007)
25. Kammerer, A.A., Tejero-Cantero, A.A., Leibold Inhibition, C.C.: enhances memory capacity: optimal feedback, transient replay and oscillations. J. Comput. Neurosci. 34(1), 125–136 (2013)
26. Lengyel, M., Dayan, P.: Uncertainty, phase, and oscillatory hippocampal recall. Advances in Neural Information Processing Systems 19, 833–840 (2007)
27. Lengyel, M., Kwag, J., Paulsen, O., Dayan, P.: Matching storage and recall: hippocampal spike timing-dependent plasticity and phase response curves. Nat. Neurosci. 8, 1677–1683 (2005)
28. Thurley, K., Leibold, C., Gundlfinger, A., Schmitz, D., Kempter, R.: Phase precession through synaptic facilitation. Neural Computation 20(5), 1285–1324 (2008)
29. Latham, P.E., Lengyel, M.: Phase Coding: Spikes Get a Boost from Local Fields. Curr. Biology 18(8), R349–R351 (2008)

Predictive Analysis of the Seismicity Level at Campi Flegrei Volcano Using a Data-Driven Approach

Antonietta M. Esposito[1], Luca D'Auria[1], Andrea Angelillo[1,2],
Flora Giudicepietro[1], and Marcello Martini[1]

[1] Istituto Nazionale di Geofisica e Vulcanologia,
Sezione di Napoli Osservatorio Vesuviano, Napoli, Italy
[2] ENI S.p.A., Exploration & Production Division, GINE Dept. S. Donato Milanese, Italy
{antonietta.esposito,luca.dauria,flora.giudicepietro,
marcello.martini}@ov.ingv.it, andrea.angelillo@eni.com

Abstract. This work aims to provide a short-term tool to estimate the possible trend of the seismicity level in the area of Campi Flegrei (southern Italy) for Civil Protection purposes. During the last relevant period of seismic activity, between 1982 and 1984, an uplift of the ground (bradyseism) of more than 1.5 m occurred. It was accompanied by more than 16,000 earthquakes up to magnitude 4.2 which forced the civil authorities to order the evacuation of about 40,000 people from Pozzuoli town for several months. Scientific studies evidenced a temporal correlation between these geophysical phenomena. This has led us to consider a data-driven approach to obtain a forecast of the seismicity level for this area. In particular, a technique based on a Multilayer Perceptron (MLP) network has been used for this intent. Neural networks are data processing mechanisms capable of relating input data with output ones without any prior correlation model but only using empirical evidences obtained from the analysis of available data. The proposed method has been tested on a set of seismic and deformation data acquired between 1983 and 1985 and then including the data of the aforementioned crisis which affected the Campi Flegrei. Once defined the seismicity levels on the basis of the maximum magnitude recorded within a week, three MLP networks were implemented with respectively 2, 3 and 4 output classes. The first network (2 classes) provides only an indication about the possible occurrence of earthquakes felt by people (with magnitude higher than 1.7), while the remaining nets (3 and 4 classes) give also a rough suggestion of their intensity. Furthermore, for these last two networks one of the output classes allows to obtain a forecast about the possible occurrence of strong potentially damaging earthquakes with magnitude higher than 3.5. Each network has been trained on a fixed interval and then tested for the forecast on the subsequent period. The results show that the performance decreases as a function of the complexity of the examined task that is the number of covered classes. However, the obtained results are very promising, for which the proposed system deserves further studies since it could be of support to the Civil Protection operations in the case of possible future crises.

Keywords: Campi Flegrei volcano, seismicity forecast, MLP neural networks.

S. Bassis et al. (eds.), *Recent Advances of Neural Network Models and Applications*,
Smart Innovation, Systems and Technologies 26,
DOI: 10.1007/978-3-319-04129-2_14, © Springer International Publishing Switzerland 2014

1 Introduction

Campi Flegrei is a volcanic area located in Naples, southern Italy. Because of the high population density lying nearby, it is considered a high risk zone [7, 19]. It presents several craters and a large depression, the caldera, generated during two major explosive eruptions, the Campanian Ignimbrite eruption and Neapolitan yellow tuff eruption [20]. Thus, a probabilistic seismic hazard analysis of the area has been performed [1] and several research activities are promoted to increase the understanding of volcanic phenomena in order to mitigate the risk and support the actions of the Civil Protection Department (DPC).

During the 1982-1984 the area was affected by a bradyiseismic crisis (bradyseism means “slow ground motion”, in contrast with the “fast motion” due to an earthquake) consisting in ground deformation and a high number of earthquakes (more than 16,000 events), also of strong intensity (up to magnitude 4.2), that led to the evacuation of about 40,000 from the town of Pozzuoli [2]. The crisis began with a rapid ground deformation at the beginning of 1982, which reached its peak in late 1983 and ended in late 1984, with an observed uplift of 1.79 m [6]. The seismicity was characterized by events of significant magnitude, in particular during the year 1983, and it was very intense during 1984. At the end of 1984 the deformation began to subside and also the seismicity.

After this episode, Campi Flegrei showed a general quiescence, with rare and minor episodes of ground uplift and seismicity [6, 18, 24]. However, from November 2005, a variation of the geodetic measurements was observed indicating a new gradual progressive ground uplift in the Campi Flegrei area [5, 16].

Therefore, we proposed to examine the seismic and geodetic data of the bradyseism period in order to obtain a short-term tool for forecasting the possible trend of the seismicity level at Campi Flegrei. The analysis is realized through the use of a data-driven methodology, the MLP neural network [4], whose strength lies in its ability to create a correlation between the input and output patterns using only the information extracted from the examined data. In particular, three MLP networks have been implemented for two, three and four classes forecasting, according to the defined seismicity levels.

Other techniques exist for time series forecasting, but not applied to the seismicity level forecasting. So, a direct comparison with the neural network on this specific context is not possible. Thus, Mitrea et al. (2009) [17] compared different forecasting methods like Moving Average (MA) and Autoregressive Integrated Moving Average (ARIMA) with Neural Networks (NN) showing that NN offers better forecasting performance in comparison with these traditional methods. Moreover, Sukanesh and Harikumar (2007) [23] have compared the performance of a genetic algorithm (GA) with those of a multi-layer Perceptron (MLP) network for classification purposes for both obtaining good results, but still a superior performance of the MLP.

However, it seems appropriate to point out that the main purpose of this work is the feasibility of the seismicity level forecasting through an approach with neural networks never used before. As we shall see, the obtained results are very promising, although they depend on the complexity of the task, and then by the number of covered classes. In any case, despite its limitations, the presented network has proven to provide a tool for making decisions. The optimization of its performance is scheduled as the next step.

In the following, the description of the multiparametric dataset, composed of seismic and geodetic data of 1982-1985 crisis, is provided. Then, the encoding of the input and target vectors is illustrated. Section 4 describes the data-driven approach based on a MLP network used for the forecasting and the obtained results. Finally the conclusions are presented.

2 The Multiparametric Dataset

The surveillance of the volcanic activity of Campi Flegrei is achieved through the observation of geophysical and geochemical parameters detected through a properly designed monitoring network (Figure 1). It includes instruments for the continuous monitoring of the seismicity, ground deformation and gas emissions from soil and fumaroles. Furthermore, periodic campaigns are carried out to measure the geophysical and geochemical parameters. All collected data are then analyzed by automatic systems and interpreted by researchers from different areas of expertise.

Campi Flegrei volcano is characterized by a moderate seismicity that generally occurs in swarms. It may increase considerably, both as earthquake magnitude and as occurrence frequency of the events or seismic swarms, during periods of more intense ground deformation.

In this work we have used a multiparametric dataset composed of seismic and geodetic data recorded between January, 1, 1983 and July, 1, 1985, and so including the bradyseism previously described. Thus, the criteria applied for the data selection was to have data (seismological and geodetic) complete, with a more immediate correlation with the volcano activity. Moreover, unlike other data (e.g. those geochemical), they can be acquired in real time and with continuity.

The seismic dataset contains 15935 earthquakes recorded at the "STH" station (Figure 1), located at Agnano, within the Campi Flegrei area. The following information are provided for each event:

- the occurrence date, expressed in year, month and day;
- the **duration Magnitude** (*Md*) of the earthquakes calculated as:

$$Md = A * Log(t) + B * d + C \qquad (1)$$

where t is the event duration, d is the hypocenter-station distance and A, B, C are correction factors for the distance between the seismometer and the earthquake source and local factors of the area where the seismic station is placed. This magnitude scale is comparable with common Richter magnitude but has shown to be reliable for small local seismic networks.

- the *Benioff Strain Release*, defined as:

$$\sqrt{10^{2.05+1.96*Md}} \qquad (2)$$

This parameter, obtained from Md values, provides the cumulative strain associated with a set of events.

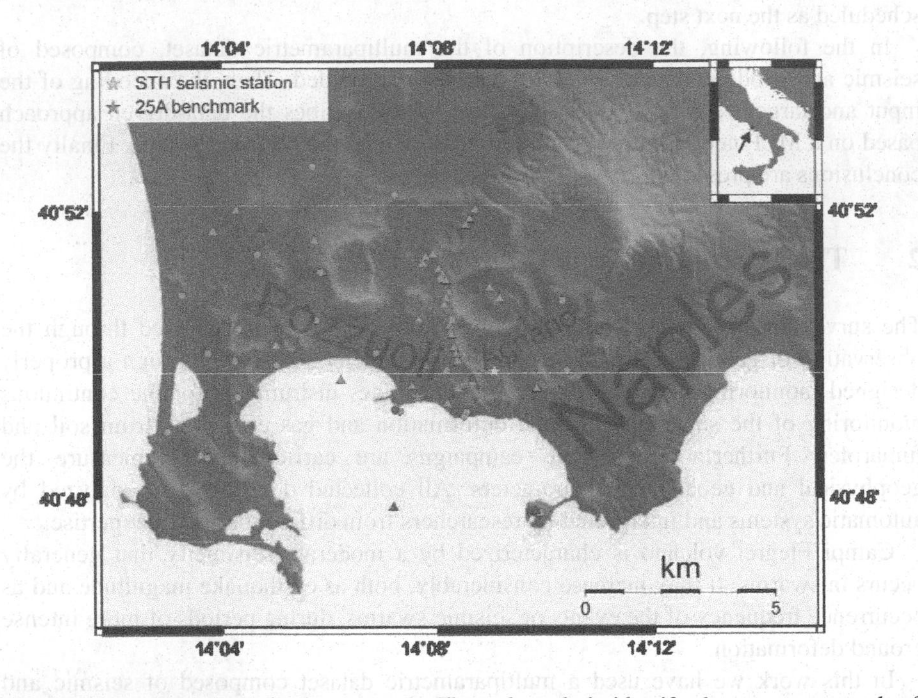

Fig. 1. Campi Flegrei monitoring network. The red symbols identify the current network of seismic stations: the circles indicate the vertical seismometers while the triangles the triaxial ones. The gray symbols refer instead to stations used in the past. The stars specify special locations relative to the legend in the upper left. Finally, the map in the upper right identifies the position of the caldera on the Italian peninsula.

Figure 2 shows the temporal distribution of *Md* values during the analyzed period. It is possible to observe the occurrence of earthquakes of significant magnitude during the crisis. Moreover, the year 1984 was characterized by an intense seismic activity that began to decline at the end of the same year.

The geodetic measurements of the Campi Flegrei area are usually made through:

- permanent GPS network for continuous recording;
- tiltmetric network recording in continuous;
- tide gauge network for continuous recording;
- leveling periodic campaigns;
- gravimetric periodic campaigns;
- SAR Interferometry data processing.

In our case, the geodetic data are measures of optical levelling made at the 25A benchmark (Figure 1), located in Pozzuoli, and recorded with a time interval ranging from about 1 year to about 15 days. For each measurement are known:

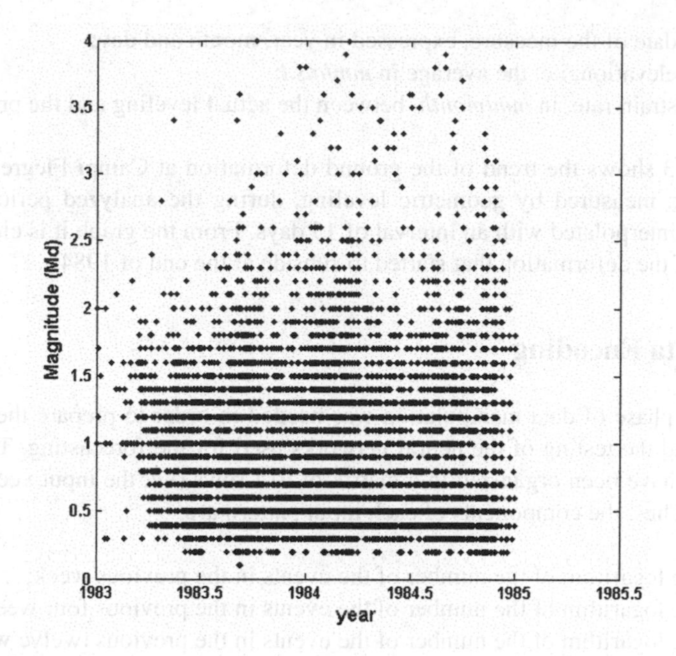

Fig. 2. The temporal distribution of *Md* values during the examined period

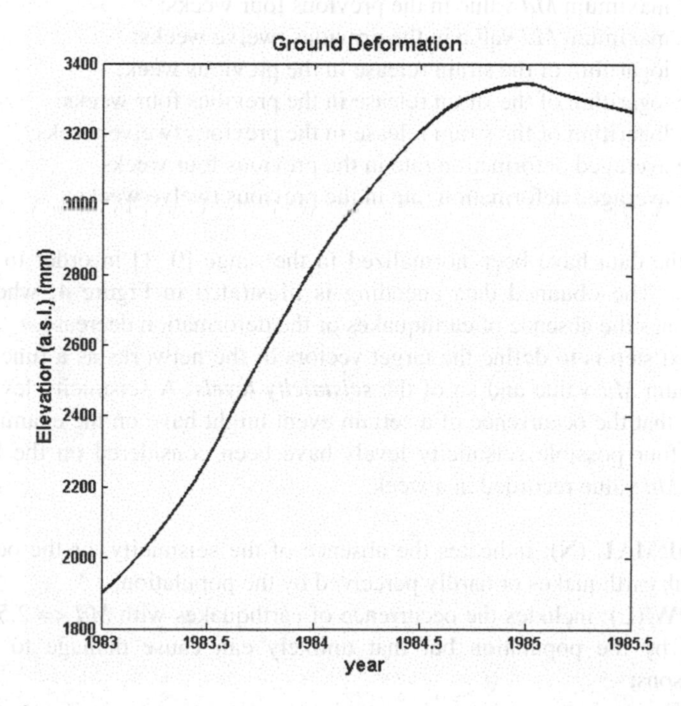

Fig. 3. Graphical representation of deformation data, interpolated with an interval of 15 days, in the analyzed period

- the date of the measure, expressed in year, month and day;
- the elevation, i.e. the average in *mm/a.s.l*;
- the strain rate, in *mm/month*, between the actual leveling and the previous one.

Figure 3 shows the trend of the ground deformation at Campi Flegrei at the 25A benchmark, measured by geometric leveling, during the analyzed period. The data have been interpolated with an interval of 15 days. From the graph it is clear the rapid increase of the deformation that started to subside at the end of 1984.

3 Data Encoding

An earlier phase of data manipulation was needed in order to prepare the sets for the training and the testing of the neural networks used for the forecasting. The available input data have been organized in a matrix of 914 rows (i.e. the input vectors) and 11 columns. Thus, the components of each input pattern are:

1. the logarithm of the number of the events in the previous week;
2. the logarithm of the number of the events in the previous four weeks;
3. the logarithm of the number of the events in the previous twelve weeks;
4. the maximum *Md* value in the previous week;
5. the maximum *Md* value in the previous four weeks;
6. the maximum *Md* value in the previous twelve weeks;
7. the logarithm of the strain release in the previous week;
8. the logarithm of the strain release in the previous four weeks;
9. the logarithm of the strain release in the previous twelve weeks;
10. the averaged deformation rate in the previous four weeks;
11. the averaged deformation rate in the previous twelve weeks;

Then, the data have been normalized in the range [0, 1] in order to make them comparable. The obtained data encoding is illustrated in Figure 4, where the blue color indicates the absence of earthquakes or the deformation decrease.

The next step is to define the target vectors of the networks as a function only of the maximum *Md* value and so of the ***seismicity levels***. A seismicity level describes the impact that the occurrence of a certain event might have on the examined area. In this work four possible seismicity levels have been considered on the basis of the maximum *Md* value recorded in a week:

- **NORMAL (N)**: indicates the absence of the seismicity, or the occurrence of weak earthquakes or hardly perceived by the population;
- **LOW (L)**: includes the occurrence of earthquakes with *Md* <= 2.5 that can be felt by the population but that unlikely can cause damage to property or persons;

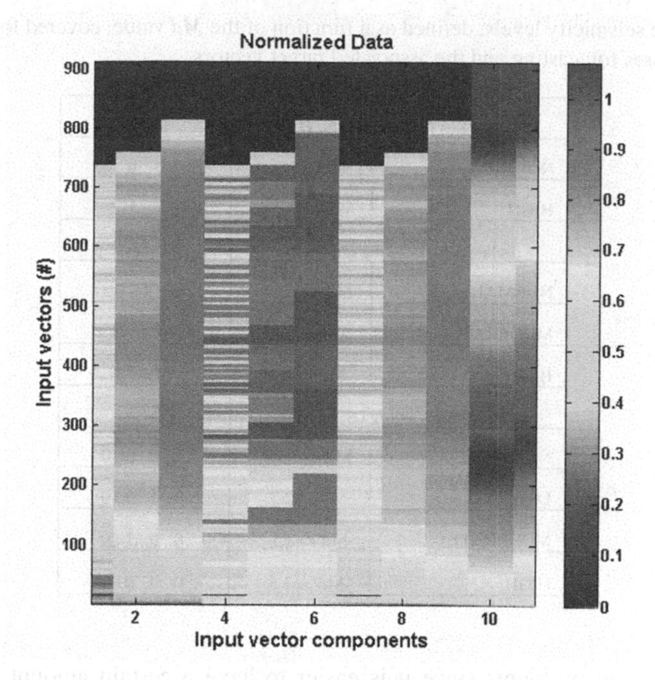

Fig. 4. The matrix of the normalized data with 914 (*input vectors*) x 11 (*components*) dimension

- **MODERATE** (**M**): contains the earthquakes with *Md* <= 3.5 that could elicit panic in the population but minor damage to property;
- **HIGH** (**H**): covers the earthquakes with *Md* > 3.5 resulting in damage to property or people or even in building collapsing.

As a result, the corresponding target vectors has been characterized as binary vectors with two, three and four dimension (columns). Each vector contain only one value equal to 1 in the position associated with the membership class.

4 Data-Driven Approach and Results

The forecasting task is performed by using a Multi-layer Perceptron (MLP) [4] network. It has been shown that neural networks are powerful pattern classifiers and recognizers, with applications in different fields [8-11, 22, 26]. In addition, they represent a valuable and attractive tool for the forecasting for several reasons. First, they are data-driven self-adaptive techniques able to adjust themselves to the data without any explicit functional requirement or distributional form for the underlying model [28]. This ability to learn from experience proves to be very important in

Table 1. The seismicity levels, defined as a function of the *Md* value, covered for the two, three and four classes forecasting and the associated target vectors

2-classes forecasting (N, H)		Target
NORMAL	Md <=1.7	1 0
HIGH	Md >1.7	0 1
3-classes forecasting (N, M, H)		**Target**
NORMAL	Md <=1.7	1 0 0
MODERATE	Md <=3.5	0 1 0
HIGH	Md >3.5	0 0 1
4-classes forecasting (N, L, M, H)		**Target**
NORMAL	Md <=1.7	1 0 0 0
LOW	Md <=2.5	0 1 0 0
MODERATE	Md <=3.5	0 0 1 0
HIGH	Md >=3.5	0 0 0 1

several practical problems since it is easier to have a certain amount of data rather than to obtain information about the system that generated them. Moreover, due to their generalization capability, they are appropriate for applications such as forecasting where the prediction of the future behavior is based on the past examples. Finally, they are considered universal approximators [12], since they can approximate any function with arbitrary accuracy [12, 13], and are nonlinear models. This last property makes them a more flexible tool in modeling real world complex relationships with respect to the traditional model-based and statistical methods [21] for forecasting [27]. The first application of neural networks for forecasting dates back to 1964, in particular for weather forecasting [14].

In this work, three MLP have been implemented, one for each of the forecasting tasks previously described (see Table 1). However, in this specific application, while the training is the usual one for neural networks, the testing or forecasting should be understood in a different manner. This means that once obtained the network by training, it is continuously modified during the testing. So it is possible to have cases where the performance of the trained network are better than those of the testing (see Figure 7).

Each network has been trained on a fixed interval by using 491 samples, about the first year and half of data, while the forecasting has been realized on the next period with a testing set of 364 samples, approximately one year of data, following that of the training. Both training and forecasting were performed on intervals of $t = 1$ day.

Although techniques for automatic calibration of parameters exist in the literature [3, 15], in this work the setting of the networks parameters has been carried out considering previous works [10, 11] and applying a trial and error procedure. This choice is dictated by the fact that the seismicity level and earthquakes forecasting with neural networks is a novel approach and the objective is to demonstrate that even if the obtained results are not optimal, the method is nevertheless feasible.

So, for all three networks were used: five hidden neurons, a nonlinear hyperbolic-tangent and a logistic sigmoidal activation function for the hidden and the output layer respectively. The Quasi-Newton algorithm [4] has been selected for the weight optimization during the training.

The results of the four MLPs are visualized by using the confusion matrix (CM) tool [25], which allows to evaluate the performance of the networks and at the same time to obtain a misclassification measure of misclassified versus correctly classified data. A confusion matrix with 0s everywhere except on the diagonal indicates a 100% correct classification, while a nonzero off-diagonal value represents the number of events of a particular class that have been "confused" by the system (wrongly classified). Thus, it basically visualizes how many actual results match the predicted results. The percentage of each class in the training and testing set for each task is the sum of the elements in each row, that is the number of the correct added to the misclassified.

Figures 5, 6 and 7 show the confusion matrices, on the training and testing data, which display the performance of the MLPs for each of the forecasting tasks illustrated in Table 1. Examining the obtained results, it should be remembered that they do not represent a true earthquake forecasting, but a cautious forecast of the seismicity level.

For the two-classes case (Figure 5) we obtained a total performance of 97.1% on the training and of 86.5% for the forecasting. In particular, a good prediction for both classes (N 81.5%, H 94.4%) is shown.

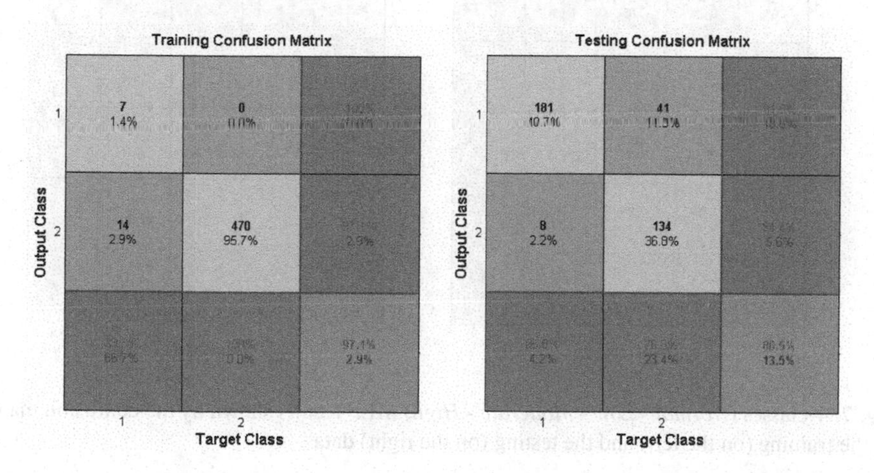

Fig. 5. 2-classes (*Normal - High*) MLP results shown by the confusion matrix on the training (on the left) and the testing (on the right) data

For the three-classes task (Figure 6) we achieved a total performance of 92.7% on the training and of 72.5% for the forecasting. In this case, however, only for two on three classes a good forecast is reached (N 91.7%, M 70.6%).

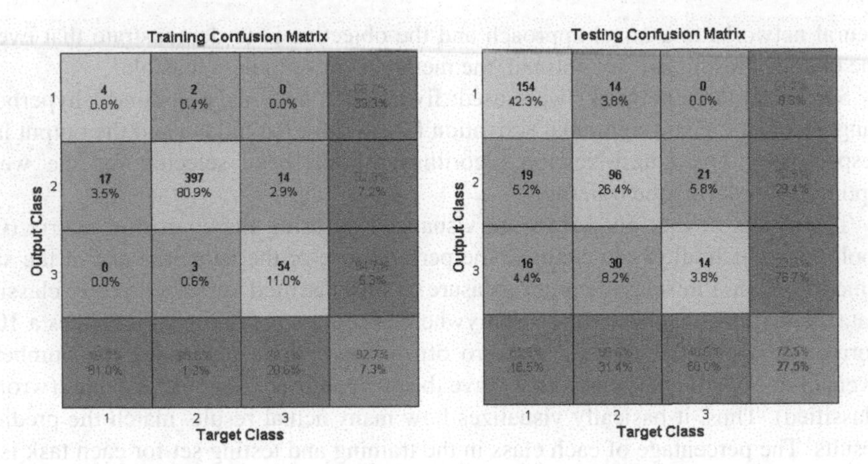

Fig. 6. 3-classes (*Normal* - *Moderate* - *High*) MLP results shown by the confusion matrix on the training (on the left) and the testing (on the right) data

Finally, for the four-classes case (Figure 7) we got a total performance of 70.7% on the training and of 63.7% for the forecasting. Given the complexity of the this last task, related to the number of examined classes, in this case only one class on the four considered has obtained a good prediction (N 95.3%).

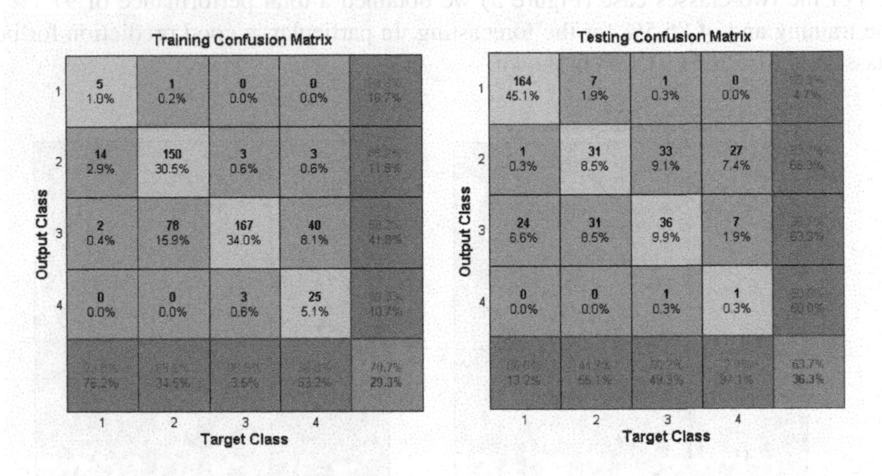

Fig. 7. 4-classes (*Normal* - *Low* - *Moderate* - *High*) MLP results shown by the confusion matrix on the training (on the left) and the testing (on the right) data

5 Conclusions

We presented a tool for the short-term forecast of the seismicity level at Campi Flegrei volcano as a support for decision-makers. Four seismicity levels have been defined as a function of the magnitude *Md* of the recorded earthquakes and related to the possible damage to things or people. A data-driven methodology has been applied that uses as only available information as one extracted from the acquired data. Three neural networks have been implemented for two, three and four classes considering the four seismicity levels above defined and so the tasks specified in Table 1. The obtained results showed a decrease of the network performance for the forecasting which depends on the increasing complexity of the examined task i.e. the number of considered classes.

However, even if the performance are lower than those usually achieved with neural networks, they can still be considered encouraging, being the "earthquake prediction task" impossible at the current state of scientific knowledge.

Finally, in order to improve these results, other tests are in progress involving the variation and optimization of the network parameters setting, changes in the dataset dimension, and so on.

The real time implementation of this procedure could provide a useful tool for the managing of a possible future volcanic crisis at Campi Flegrei.

References

1. Convertito, V., Zollo, A.: Assessment of pre-crisis and syn-crisis seismic hazard at Campi Flegrei and Mt. Vesuvius volcanoes, Campania, southern Italy. Bulletin of Volcanology 73(6), 767–783 (2011), doi:10.1007/s00445-011-0455-2
2. Barberi, F., Corrado, G., Innocenti, G., Luongo, G.: Phlegrean Fields 1982–1984: Brief chronicle of a volcano emergency in a densely populated area. Bull. Volcanol. 47(2), 1–22 (1984)
3. Birattari, M., Yuan, Z., Balaprakash, P., Stützle, T.: F-Race and Iterated F-Race: An Overview. In: Experimental Methods for the Analysis of Optimization Algorithms, pp. 311–336. Springer, Heidelberg (2010), http://link.springer.com/chapter/10.1007%2F978-3-642-02538-9_13#, doi:10.1007/978-3-642-02538-9_13
4. Bishop, C.: Neural Networks for Pattern Recognition, p. 500. Oxford University Press, New York (1995)
5. D'Auria, L., Giudicepietro, F., Aquino, I., Borriello, G., Del Gaudio, C., Lo Bascio, D., Martini, M., Ricciardi, G.P., Ricciolino, P., Ricco, C.: Repeated fluid transfer episodes as a mechanism for the recent dynamics of Campi Flegrei caldera (1989-2010). Journal of Geophysical Research 116, B04313 (2011), doi:10.1029/2010JB007837
6. Del Gaudio, C., Aquino, I., Ricciardi, G.P., Ricco, C., Scandone, R.: Unrest episodes at Campi Flegrei: A reconstruction of vertical ground movements during 1905–2009. J. Volcanol. Geotherm. Res. 195(1), 48–56 (2010), doi:10.1016/j.jvolgeores.2010.05.014
7. De Natale, G., Troise, C., Pingue, F., Mastrolorenzo, G., Pappalardo, L., Battaglia, M., Boschi, E.: The Campi Flegrei caldera: Unrest mechanisms and hazards. Geol. Soc. London Spec. Publ. 269(1), 25–45 (2006), doi:10.1144/GSL.SP.2006.269.01.03
8. Del Pezzo, E., Esposito, A., Giudicepietro, F., Marinaro, M., Martini, M., Scarpetta, S.: Discrimination of earthquakes and underwater explosions using neural networks. Bull. Seism. Soc. Am. 93(1), 215–223 (2003)

9. Esposito, A.M., Giudicepietro, F., Scarpetta, S., D'Auria, L., Marinaro, M., Martini, M.: Automatic Discrimination among Landslide, Explosion-Quake, and Microtremor Seismic Signals at Stromboli Volcano Using Neural Networks. Bull. Seism. Soc. Am. 96(4A), 1230–1240 (2006), doi:10.1785/0120050097
10. Esposito, A.M., D'Auria, L., Giudicepietro, F., Peluso, R., Martini, M.: Automatic recognition of landslide seismic signals based on neural network analysis of seismic signals: an application to the monitoring of Stromboli volcano (Southern Italy). Pure and Applied Geophysics pageoph © Springer Basel (2012), doi:10.1007/s00024-012-0614-1
11. Giacco, F., Esposito, A.M., Scarpetta, S., Giudicepietro, F., Marinaro, M.: Support Vector Machines and MLP for automatic classification of seismic signals at Stromboli volcano. In: Apolloni, B., et al. (eds.) Neural Nets WIRN 2009 Proceedings of the 19th Italian Workshop on Neural Nets, Vietri sul Mare, Salerno, Italy, May 28-30. IOS Press (2009)
12. Hornik, K., Stinchcombe, M., White, H.: Multilayer feedforward networks are universal approximators. Neural Networks 2, 359–366 (1989)
13. Hornik, K.: Approximation capabilities of multilayer feedforward networks. Neural Networks 4, 251–257 (1991)
14. Hu, M.J.C.: Application of the adaline system to weather forecasting. Master Thesis. Technical Report 6775-1, Stanford Electronic Laboratories, Stanford, CA (June 1964)
15. Hutter, F., Hoos, H.H., Leyton-Brown, K., Stuetzle, T.: ParamILS: An Automatic Algorithm Configuration Framework. Journal of Artificial Intelligence Research 36, 267–306 (2009)
16. Martini, M., Giudicepietro, F., D'Auria, L., Orazi, M., Borriello, G., Buonocunto, C., Capello, M., Caputo, A., Caputo, T., De Cesare, W., Esposito, A., Lo Bascio, D., Ricciolino, P., Peluso, R., Scarpato, G.: Seismological monitoring of Campi Flegrei caldera. Geophysical Research Abstracts, vol. 10, EGU2008-A-09610, 2008 SRef-ID: 1607-7962/gra/EGU2008-A-09610 EGU General Assembly (2008)
17. Mitrea, C.A., Lee, C.K.M., Wu, Z.: A Comparison between Neural Networks and Traditional Forecasting Methods: A Case Study. International Journal of Engineering Business Management 1(2), 19–24 (2009)
18. Orsi, G., Civetta, L., Del Gaudio, C., de Vita, S., Di Vito, M.A., Isaia, R., Petrazzuoli, S.M., Ricciardi, G.P., Ricco, C.: Short-term ground deformations and seismicity in the resurgent Campi Flegrei caldera (Italy): An example of active block-resurgence in a densely populated area. J. Volcanol. Geotherm. Res. 91, 415–451 (1999)
19. Orsi, G., Di Vito, M.A., Isaia, R.: Volcanic hazard assessment at the restless Campi Flegrei caldera. Bull. Volcanol. 66, 514–530 (2004)
20. Priolo, E., Lovisa, L., Zollo, A., et al.: The Campi Flegrei Blind Test: Evaluating the Imaging Capability of Local Earthquake Tomography in a Volcanic Area. International Journal of Geophysics 2012, Article ID 505286, 37 (2012), doi:10.1155/2012/505286
21. Richard, M.D., Lippmann, R.: Neural network classifiers estimate Bayesian a posteriori probabilities. Neural Comput. 3, 461–483 (1991)
22. Scarpetta, S., Giudicepietro, F., Ezin, E.C., Petrosino, S., Del Pezzo, E., Martini, M., Marinaro, M.: Automatic Classification of seismic signals at Mt. Vesuvius Volcano, Italy using Neural Networks. Bull. Seism. Soc. Am. 95, 185–196 (2005)
23. Sukanesh, R., Harikumar, R.: A Comparison of Genetic Algorithm & Neural Network (MLP) In Patient Specific Classification of Epilepsy Risk Levels from EEG Signals. Engineering Letters 14(1), EL_14_1_18 (2007) (Advance online publication: February 12, 2007)

24. Troise, C., De Natale, G., Obrizzo, F., De Martino, P., Tammaro, U., Boschi, E.: Renewed ground uplift at Campi Flegrei caldera (Italy): New insight on magmatic processes and forecast. Geophys. Res. Lett. 34, L03301 (2007), doi:10.1029/2006GL028545
25. Young, S.J.: HTK: Hidden Markov Model Toolkit V1.5. Cambridge University Engineering Department Speech Group and Entropic Research Laboratories, Inc., Washington, D.C (1993)
26. Widrow, B., Rumelhart, D.E., Lehr, M.A.: Neural networks Applications in industry, business and science. Communication of the ACM 37(3), 93–105 (1994)
27. Zhang, G., Patuwo, B.E., Hu, M.Y.: Forecasting with artificial neural networks: The state of the art. International Journal of Forecasting 14, 35–62 (1998); Elsevier Science B.V. PII S0169-2070(97)00044-7
28. Zhang, G.P.: Neural Networks for Classification: A Survey. IEEE Transactions on Systems, Man, and Cybernetics- Part C: Applications and Reviews 30(4), 1094–6977 (2000); Publisher Item Identifier S 1094-6977(00)11206-4

Robot Localization by Echo State Networks Using RSS

Stefano Chessa[1], Claudio Gallicchio[1], Roberto Guzman[2], and Alessio Micheli[1]

[1] Department of Computer Science, University of Pisa
Largo Pontecorvo 3, 56127 Pisa, Italy
{ste,gallicch,micheli}@di.unimi.it
[2] Robotnik Automation, SLL
C/Berni y Catala, 53 bajo
rguzman@robotnik.es

Abstract. In this paper we present an application of Reservoir Computing to indoor robot localization, based on input received signal strength signals from a wireless sensor network. The proposed localization system allows to combine good predictive performance with particularly efficient and practical solutions. Promising results are shown in preliminary experiments on a real-world scenario.

Keywords: Echo State Networks, Reservoir Computing, Wireless Sensor Networks, Robot Localization.

1 Introduction

Recently, one of the main scenarios of interest emerging in the field of robotic ecologies, concerns the autonomous robot navigation in critical environments. In this context, within the framework of the EU FP7 RUBICON [3] project[1], which addresses the development of adaptive/self-learning robotic ecologies, one of the main scenarios addressed is an application to a hospital transport problem. In this scenario, an Automated Guided Vehicle (AGV) autonomously carries out transport operations that consist in the robot moving to a given loading position, loading a trolley (e.g. with food, medicines), transporting it to a target location, and finally unloading it. This scenario, which comes from a real-world application, makes use of an AGV from Robotnik[2], a RUBICON partner.

The AGV navigation systems combine an internal gyro with a set of small 'landmark' magnets, installed along the guide path in holes in the floor. These magnets provide information about the positional deviations of the robot as well as the distance traveled. A human operator is needed to give orders to the AGV on demand by using a touch panel and by referring to fixed load/unload port.

With these solutions, costs of installation, programming and start-up of landmark-magnet based systems are high. On the other hand, the evolution experienced in Simultaneous Localization and Mapping (SLAM) algorithms

[1] The EU FP7 RUBICON project, http://fp7rubicon.eu/
[2] http://www.robotnik.es/en/

S. Bassis et al. (eds.), *Recent Advances of Neural Network Models and Applications*, Smart Innovation, Systems and Technologies 26,
DOI: 10.1007/978-3-319-04129-2_15,

(some of them successfully implemented even in domestic vacuum cleaners) allows to consider their introduction in AGV navigation in hospital applications. The most important advantage of SLAM with respect to the above mentioned localization methods is flexibility, since the programming of new paths with conventional landmarks means an extra effort in the beacon installation and mapping, while doing this in a map based solution could mean only a task reprogramming. There are, however, a number of limitations (mainly related with reliability, accuracy and speed) that may delay the effective use of SLAM algorithms in this application. For these reasons, it is interesting to investigate the possibility of ultimately enhancing conventional SLAM techniques by combining them with the use of localization systems based on Wireless Sensor Network (WSN) and received signal strength (RSS), in order to increase the overall localization system reliability and avoid the need for landmarks. Such solution requires the installation of a number of sensors in the ceiling of the hospital corridors. This results in a more flexible solution since the deployment of the sensors on the ceiling does not require the high precision required by installation of landmarks, and the localization system based on WSN measuring RSS can be easily reconfigured.

Note also that there are several options for WSN-based localization. In our solution we opted for a WSN based on the standard IEEE 802.15.4 [8] that performs localization by means of RSS measurements. Despite the fact that this kind of solution is probably not the most performing from the point of view of localization accuracy, it has three main advantages: it exploits devices that are widely available on the market and at low costs (it does not requires any special hardware), it is based on a widely known wireless standard whose electromagnetic compatibility has been tested in many different conditions, and, finally, its working frequency (2,4 GHz) is the same of WiFi, which is known to be compatible with the medical instrumentations used in the laboratories in the hospital where the RUBICON testbed is going to be experimented, a requirement that is particularly important in our scenario.

In this paper the problem of RSS robot localization in real-life indoor environment is approached using Reservoir Computing (RC) networks [14]. The RC paradigm represents a particularly efficient approach for modeling Recurrent Neural Networks, based on the separation between a recurrent non-linear *reservoir* component, and a feed-forward linear *readout* component, which is the only trained part of the network architecture. Within the RC paradigm, we take into particular consideration the Echo State Network (ESN) model [12, 9]. ESNs have been applied to a range of tasks related to autonomous system modeling (see e.g. [4]), though mainly based on simulated, artificial data. More recently, ESNs have been applied to the problem of predicting user movements in indoor environments based on real-world RSS data [5–7, 10]. Such investigations have shown that, in this applicative context, RC networks are able to generalize well to unseen environments and reach a good trade-off between the predictive performance and the implementation costs in real system deployments.

Here we present an application of an RC-based indoor robot localization system using RSS data. An experimental analysis is conducted by taking into consideration a preliminary real-world challenging testbed.

The rest of this paper is organized as follows. Section 2 describes the ESN model, Section 3 introduces the settings for the real-world scenario considered in experiments, Section 4 illustrates the experimental results and Section 5 presents the conclusions.

2 Echo State Networks

An ESN consists in an N_U-dimensional input layer, an N_R-dimensional reservoir, and an N_Y-dimensional readout. The reservoir is a large, recurrent, non-linear, typically sparsely-connected layer, used to encode the history of the driving input signals into the state space. The readout is a linear layer, which is used to compute the output of the model through a linear combination of the activation of the reservoir units. In this paper, we take into consideration a variant of the ESN model, called the Leaky Integrator ESN (LI-ESN) (see [13, 14]), which applies an exponential moving average to the reservoir state values. LI-ESNs were shown to better handle input sequences that change slowly with respect to the frequency of sampling [14], resulting in an RC model which is particularly suitable for the characteristics of the RSS input signals in [10]. At each time step t, the reservoir part of the LI-ESN computes the state transition function

$$\mathbf{x}(t) = (1-a)\mathbf{x}(t-1) + af(\mathbf{W}_{in}\mathbf{u}(t) + \hat{\mathbf{W}}\mathbf{x}(t-1)) \tag{1}$$

where $\mathbf{x}(t) \in \mathbb{R}^{N_R}$ is the reservoir state at pass t, $\mathbf{u}(t) \in \mathbb{R}^{N_U}$ is the input at pass t, $\mathbf{W}_{in} \in \mathbb{R}^{N_R \times N_U}$ is the input-to-reservoir weight matrix, $\hat{\mathbf{W}} \in \mathbb{R}^{N_R \times N_R}$ is the recurrent reservoir weight matrix, f is a component-wise applied activation function for the reservoir units (we use $f \equiv tanh$). The parameter a in equation 1 is a *leaking rate*, assuming values in $[0, 1]$, and controlling the speed of the network state dynamics. For $a = 1$, equation 1 reduces to the standard ESN state transition function. The reservoir parameters in matrices $\hat{\mathbf{W}}$ and $\mathbf{W}_{in}$ are randomly initialized, under the constraints of the *Echo State Property* (ESP) [12], and then they are left *untrained*. In applications, the ESP is generally accomplished by satisfying the necessary condition $\rho < 1$, where ρ is the *spectral radius* of $\tilde{\mathbf{W}} = (1-a)\mathbf{I} + a\hat{\mathbf{W}}$. The weight values in matrix $\mathbf{W}_{in}$ are typically chosen from a uniform distribution over the interval $[-scale_{in}, scale_{in}]$, where $scale_{in}$ is an input scaling parameter.

At each time step t, the output of the LI-ESN, i.e. $\mathbf{y}(t) \in \mathbb{R}^{N_Y}$, is computed by the readout component as $\mathbf{y}(t) = \mathbf{W}_{out}\mathbf{x}(t)$, where $\mathbf{W}_{out} \in \mathbb{R}^{N_Y \times N_R}$ is the reservoir-to-readout weight matrix. In this paper, the weight values in $\mathbf{W}_{out}$ are trained using ridge regression.

3 Robot Localization Experimental Setting

Experimental data for the real-world robot localization scenario was collected in an indoor Robotnik laboratory with some manufacturing facilities and

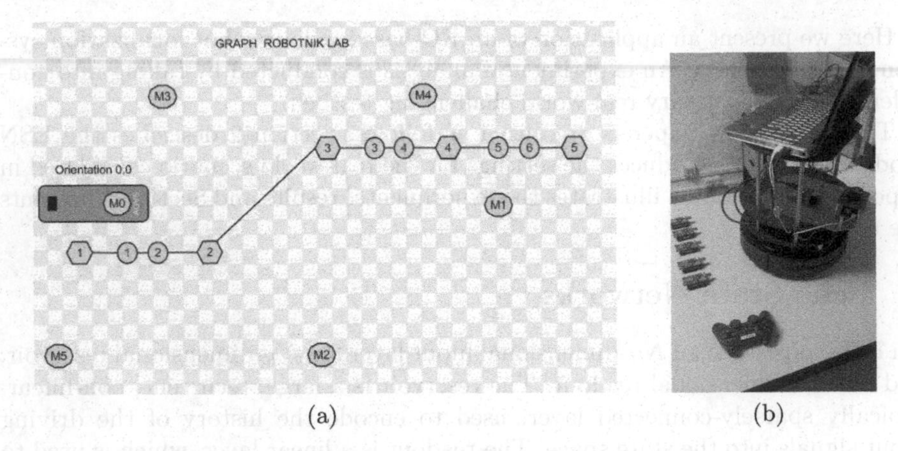

Fig. 1. (a) Graph of the Robotnik Lab, showing the position of the anchors M1, ..., M5 in the environment and the path followed by the robot. (b) Picture showing the robot, the motes of the WSN and the controller for the robot movement, used during the measurement campaign.

dimensions of 20 x 6 m (see Figure 1(a)). In such environment a small WSN was installed, using Telosb sensors [1]. In particular, 5 anchor sensors were distributed in the laboratory at a fixed height of 2 m, whereas a mobile sensor was placed on the Turtlebot robot [2]. The disposition of the motes and the hardware used in the measurement campaign are shown in Figure 1.

Experimental RSS and ground-truth localization data was collected by moving the robot in the environment according to a trajectory which goes from node 1 to node 5 and back to node 1 in the Robotnik laboratory graph (see the Figure 1(a)). Two types of trajectories were considered, i.e. *exact* trajectories, which closely follow the s-shaped path in Figure 1(a), and *zig-zag* trajectories, which deviate (even consistently) from the path in Figure 1(a). In both the cases, the sampling frequency was fixed to 10 Hz.

The ground-truth was collected using the Gmapping algorithm [11]. An additional experiment condition was the ability of the algorithm to fix the map frame in the (0,0,0) robot pose prior to the robot motion. If this does not happen, the physical map and the Gmapping map would not match and a transformation would be required. The initial orientation of the robot has been fixed manually. The accuracy in the initial position added an error of about +/-2mm. In our experiments, the location estimation provided by the localization algorithm has not been compared with other ground-truth, other than the physical localization of the robot in the ground grid, therefore there is no estimation about the orientation error. According to these measurements, the robot position was always in a ± 0.5 m error range (delta of the Gmapping map was 0.05m).

Overall, the measurement campaign resulted in the collection of 2 exact trajectories and 2 zig-zag trajectories, leading to the definition of a regression task on time-series for robot localization. The RSS data samples from the 5 anchors, are organized in 4 sequences, with lengths varying between 1040 and 2052. The

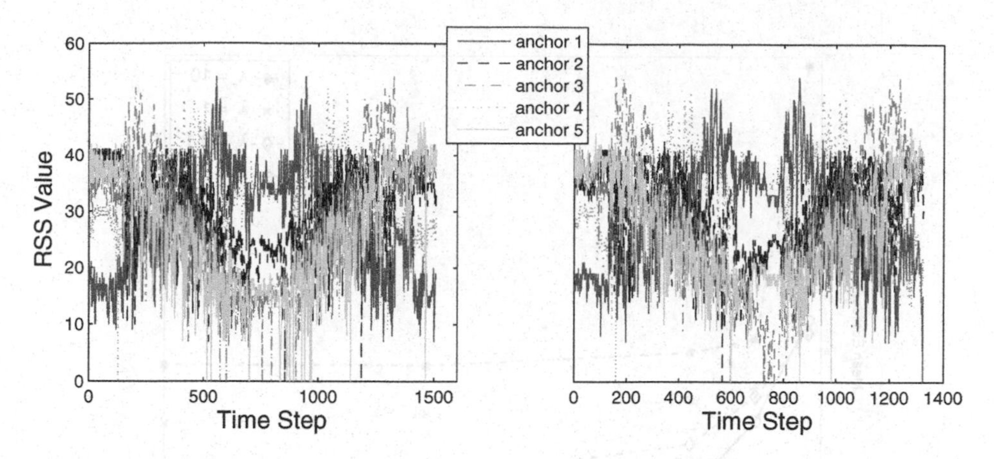

Fig. 2. RSS values for an exact (left plot) and a zig-zag (right plot) trajectory in the robot localization dataset. Each time step corresponds to 100 milliseconds (10 Hz).

corresponding ground-truth (x, y) localization data was used as bi-dimensional target data. Such task represents a suitable preliminary testbed for an experimental assessment of the proposed RC system in terms of effectiveness and efficiency. The RSS values from the 5 anchors in correspondence of two complete example trajectories in the robot localization dataset are in Figure 2, pointing out the significant noise which typically characterizes this type of measurements in real applications.

4 Experimental Results

In our experiments we used LI-ESNs with fully connected reservoir and increasing dimension $N_R \in \{10, 50, 100, 300, 500\}$. For each reservoir dimension, a number of 10 independent random guessed reservoir was considered, and the results were averaged over the 10 guesses. Other reservoir parameters (which do not have a direct influence on the computational cost) were set coherently to our previous experimental analysis reported in [5–7, 10]. In particular we considered the values of spectral radius $\rho = 0.9$, leaking rate parameter $a = 0.1$, and input scaling $scale_{in} = 0.01$. The readout was trained using ridge regression, considering a regularization parameter values $\lambda_r \in \{0.01, 0.1, 1, 10\}$. The performance of the LI-ESNs were evaluated by 4-fold cross-validation, were for each fold one of the 4 available sequences was used for the test set and the other 3 sequences were used as training set.

Figure 3 shows the mean Euclidean test error achieved by LI-ESNs with increasing reservoir dimension and different values for the readout regularization parameter. In general the mean test error varies between 1.59 m (for $N_R = 500$ and $\lambda_r = 0.01$) and 1.75 m (for $N_R = 10$ and $\lambda_r = 10$), resulting in a good predictive performance (which is close enough to the known approximation error in

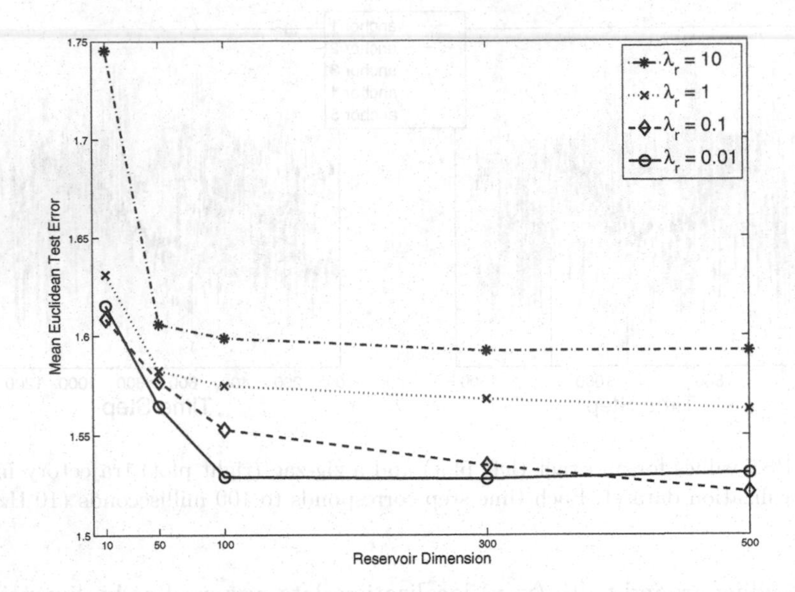

Fig. 3. Mean Euclidean test error on the robot localization dataset, for increasing reservoir dimension N_R and varying the readout regularization parameter λ_r

the ground-truth data). From Figure 3 it can be observed that smaller values for the readout regularization parameter lead to better performances. Indeed the test error decreases for decreasing the value of λ_r for every choice of the reservoir dimension N_R. At the same time we observed a general decreasing of the training error for decreasing values of λ_r, suggesting that larger values of the regularization parameter resulted in an under-fitting behavior of the RC models in these experiments. It is also evident that larger reservoirs lead to better predictive performances, although a saturation behavior can also be observed for reservoir with dimension $N_R \geq 100$.

Note that the reservoir dimension is one of the most critical parameters to be considered in view of the embedding of the RC modules on-board the motes of a WSN [5, 6]. With this point in mind, the results in Figure 3 point out that RC networks with limited dimension and suitable value for the regularization parameter, can result in a good trade-off between predictive performance and practical implementation efficiency. For example, LI-ESNs with $N_R = 50$ and $\lambda_r = 0.01$ achieved a mean test error of 1.56 m, with standard deviation (over the 4 folds) of 0.07.

We also found that the great part of the test robot trajectories were localized with small errors, whereas only a small portion of the test data led to high errors. This point is illustrated in Figure 4, through a plot of the percentiles of the Euclidean test errors (on fold number 4), corresponding to the setting with $N_R = 50$ reservoir units and readout regularization $\lambda_r = 0.01$. Each bar in the

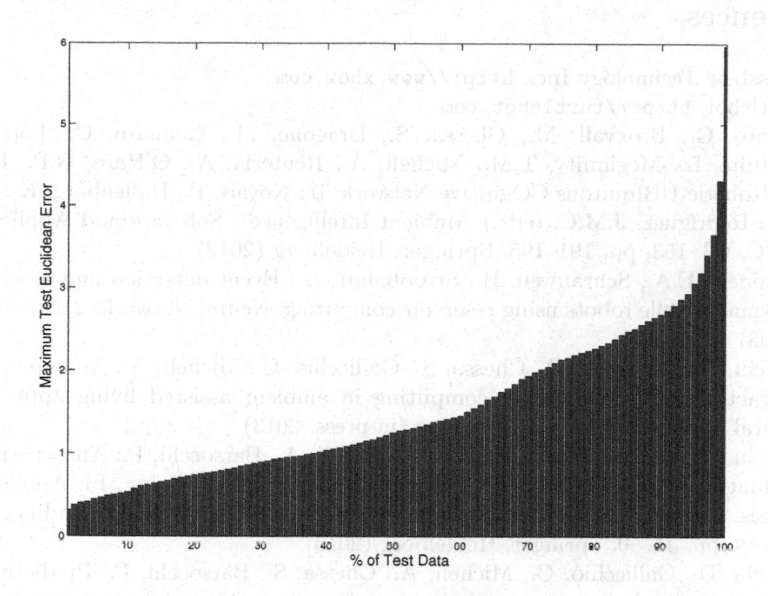

Fig. 4. Percentiles of the Euclidean test error on fold 4 of the robot localization dataset, achieved by ESNs with $N_R = 50$ reservoir units and readout regularization parameter $\lambda_r = 0.01$

plot corresponds to the maximum test error found in correspondence of a specific percentage of the test data. It can be seen that the 50%, the 75% and the 90% of the test data are localized with an error not larger than 1.25, 2.14 and 2.68 m, respectively.

5 Conclusions

We have presented an application of RC to the problem of indoor robot localization using RSS information. A real-world preliminary testbed has been set up for an experimental performance assessment of the proposed localization system, in terms of effectiveness and efficiency. The promising results obtained, pave the way for further experimental investigations based on more extensively sampled robot trajectories and more accurate ground-truth data. Overall, the proposed localization system opens the perspective of practical implementation of learning in robotic ecology, by exploiting efficient RC modules embedded in the nodes of WSNs. This will allow, on the one hand, to successfully combine SLAM algorithms with the use of WSNs, and, on the other hand, to continuously update the localization system through the learning functionalities.

Acknowledgement. This work is partially supported by the EU FP7 RUBICON project (contract n. 269914).

References

1. Crossbow Technology Inc., http://www.xbow.com
2. Turtlebot, http://turtlebot.com
3. Amato, G., Broxvall, M., Chessa, S., Dragone, M., Gennaro, C., López, R., Maguire, L., Mcginnity, T.M., Micheli, A., Renteria, A., O'Hare, G.P., Pecora, F.: Robotic UBIquitous COgnitive Network. In: Novais, P., Hallenborg, K., Tapia, D.I., Rodríguez, J.M.C. (eds.) Ambient Intelligence - Software and Applications. AISC, vol. 153, pp. 191–195. Springer, Heidelberg (2012)
4. Antonelo, E.A., Schrauwen, B., Stroobandt, D.: Event detection and localization for small mobile robots using reservoir computing. Neural Networks 21(6), 862–871 (2008)
5. Bacciu, D., Barsocchi, P., Chessa, S., Gallicchio, C., Micheli, A.: An experimental characterization of reservoir computing in ambient assisted living applications. Neural Computing and Applications (in press, 2013)
6. Bacciu, D., Chessa, S., Gallicchio, C., Micheli, A., Barsocchi, P.: An experimental evaluation of reservoir computation for ambient assisted living. In: Apolloni, B., Bassis, S., Esposito, A., Morabito, F.C. (eds.) Neural Nets and Surroundings. SIST, vol. 19, pp. 41–50. Springer, Heidelberg (2013)
7. Bacciu, D., Gallicchio, C., Micheli, A., Chessa, S., Barsocchi, P.: Predicting user movements in heterogeneous indoor environments by reservoir computing. In: Bhatt, M., Guesgen, H.W., Augusto, J.C. (eds.) Proc. of the IJCAI Workshop on Space, Time and Ambient Intelligence, STAMI 2011, pp. 1–6 (2011)
8. Baronti, P., Pillai, P., Chook, V., Chessa, S., Gotta, A., Hu, Y.F.: Wireless sensor networks: a survey on the state of the art and the 802.15.4 and zigbee standards. Computer Communications 30, 1655–1695 (2007)
9. Gallicchio, C., Micheli, A.: Architectural and markovian factors of echo state networks. Neural Networks 24(5), 440–456 (2011)
10. Gallicchio, C., Micheli, A., Barsocchi, P., Chessa, S.: User movements forecasting by reservoir computing using signal streams produced by mote-class sensors. In: Del Ser, J., Jorswieck, E.A., Miguez, J., Matinmikko, M., Palomar, D.P., Salcedo-Sanz, S., Gil-Lopez, S. (eds.) Mobilight 2011. LNICST, vol. 81, pp. 151–168. Springer, Heidelberg (2012)
11. Grisetti, G., Stachniss, C., Burgard, W.: Improved techniques for grid mapping with rao-blackwellized particle filters. IEEE Transactions on Robotics 23, 34–46 (2007)
12. Jaeger, H., Haas, H.: Harnessing nonlinearity: Predicting chaotic systems and saving energy in wireless communication. Science 304(5667), 78–80 (2004)
13. Jaeger, H., Lukosevicius, M., Popovici, D., Siewert, U.: Optimization and applications of echo state networks with leaky- integrator neurons. Neural Networks 20(3), 335–352 (2007)
14. Lukosevicius, M., Jaeger, H.: Reservoir computing approaches to recurrent neural network training. Computer Science Review 3(3), 127–149 (2009)

An Object Based Analysis Applied to Very High Resolution Remote Sensing Data for the Change Detection of Soil Sealing at Urban Scale

Luca Pugliese[1] and Silvia Scarpetta[2]

[1] IIASS Vietri sul Mare (SA) Italy
[2] Dept. of Physics University of Salerno Fisciano(SA) Italy & INFN Gr. Coll. Salerno Italy
sscarpetta@unisa.it, lucapug@gmail.com

Abstract. An object-based strategy is presented to identify soil-sealing in urban environment using Very High Resolution (VHR) remote sensing images. A first stage of segmentation has been carried out using a watershed algorithm, and then a second stage of supervised classification has been done to classify land covers. The resulting land covers have been used to discriminate between sealed and unsealed surfaces. The selection of features and of the number of land cover classes has been guided by an exploratory clustering stage. The proposed strategy has been used to classify sealed surfaces for two single-date images. The post-classification results have been used in a Change Detection Task, in order to analyze the land take problem in a periurban area of Venezia Mestre between 2005 and 2010. The change detection results are promising, considering the good capacity to reveal changes at the characteristic dimension of small man-made structures, and they can be considered a good support to a successive photointerpretation step.

Keywords: OBIA, VHR, Change Detection, Soil Sealing, Clustering, Machine learning.

1 Introduction

An object-based strategy based on segmentation followed by supervised classification is used to discriminate between sealed and unsealed surfaces in view of a land take change detection task. The problem of Land take is becoming more and more relevant nowadays. In Europe, and particularly in Italy, soil sealing determined by Urban sprawl is the main factor of transformation of Land, which leads to irremediable loss of beneficial functions exerted by natural terrains such as evapotranspiration and drainage. To cope with the negative consequences in terms of increased vulnerability to flash floods phenomena in the urban environment, an informed procedure of Urban Planning must be implemented. Soil sealing quantitative analysis has already been undertaken by European and National environmental bodies [1], with valuable results to support medium scale planning studies. Anyway, Soil sealing mapping at urban scale has still not been defined through a standardized product. Such a product would be very useful for urban planning tasks.

S. Bassis et al. (eds.), *Recent Advances of Neural Network Models and Applications*,
Smart Innovation, Systems and Technologies 26,
DOI: 10.1007/978-3-319-04129-2_16, © Springer International Publishing Switzerland 2014

In this paper a methodology which involves information extraction from Very High Resolution (VHR) aerial and satellite data to build soil sealing change maps at Urban scale is discussed. Such product can be of a valuable support in hydrologic modeling and in prediction analysis, that should be included in the VAS (Strategic Environmental Evaluation) document provided as a component of the preliminary Knowledge Framework defined in Urban planning protocols.

One of the technical challenges of the proposed application has been the harmonization of information extraction from two different sensors in view of the final change detection objective. The problem of land cover classification of multiresolution and/or multisource remote sensing data has been considered in literature with different approaches [2-7]. In this paper, a clustering guided intermediate land cover classification, has been done, involving different types and numbers of classes for the two sensors/dates. This way, a simpler supervised classification task has been assigned to the machine; in the meantime, by means of a very simple operation of aggregation of the intermediate land covers, the final identification goal has been reached.

The Area Of Interest is a peri-urban zone located in Venezia-Mestre. A full report has been done of the processing workflow, including a thorough explanation of the processing modules and of the choice of parameters setting. The main passages for the production of the Built-up Change Map are the initial step of exploratory analysis in order to select the input features to the segmentation and classification module. Segmentation is present because Object Based Image Analysis (OBIA) is a sensible choice in urban classification tasks involving VHR imagery [8]. A second important methodological choice is the aforementioned classification scheme, based on an intermediate land cover classification, with different classes for the two sensors (Spot5 and Woldview-2 for 2005 and 2010, respectively), followed by a simple class fusion in order to reach the desired soil sealing identification.

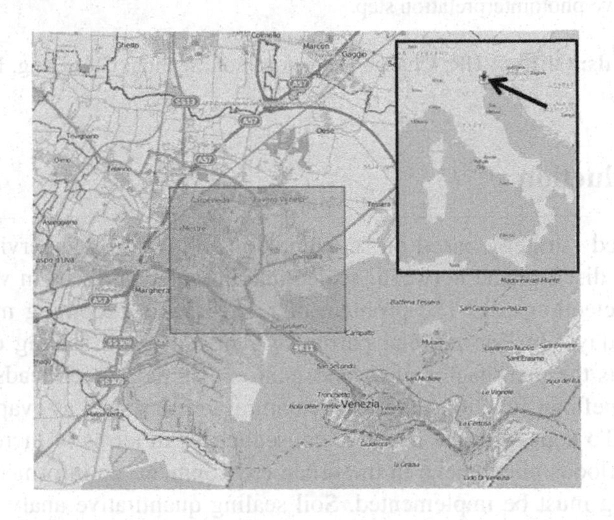

Fig. 1. The Area of Interest is located in North Italy

2 Data Set Description and Preprocessing

The Area Of Interest (AOI) is located in North Italy, inside the administrative boundaries of Venezia-Mestre (see Figure 1). A transition zone from the city center to the outskirts has been selected, due to its predisposition for the phenomenon of Land take, which is the goal of the present analysis.

In cartographic coordinates, reference system UTM, zone 33N, datum WGS84, the AOI is identified by (unit reference in meters):

UL: 283.875,0 5.043.439,0
LR: 289.343,0 5.037.947,0

The dataset is composed by a Spot 5 satellite optical image, taken in September 2005, a lidar point cloud registered during an aerial flight in September 2008 and a second satellite optical image from 2010, taken by a Worldview 2 sensor (see Table 1 for the detailed characteristics of the data).

Ancillary data have been downloaded from Veneto Region Open Data Portal (www.dati.veneto.it), both to support pre-processing tasks and to control some of the results of processing.

Table 1. Spectral and spatial characteristics of the dataset

Data name (and type)	Data Spectral (and Spatial) characteristics
Spot 5 (Passive sensor)	4 MS[1] bands (10 m): G, R, NIR, SWIR, 1 PAN band (2,5 m)
Lidar (Active sensor)	NIR channel, 3,33 points/sqm
Worldview-2 (Passive sensor)	8 MS bands (2 m): CB, B, G, Y, R, RE, NIR1, NIR2, 1 PAN band (0,50 m)

The goal of the pre-processing step has been to harmonize the two dates data in the aim of a successive change detection step. In particular, a pansharpening (i.e. panchromatic and multispectral data fusion) of Spot 5 image was done to obtain comparable spatial resolution between Spot pansharpened and Worldview (multispectral) data. Secondly, a change of projection of Worldview-2 image, involving GCPs taken from a regional Map at scale 1:5000 and a VHR Digital Elevation Model (DEM) extracted from the Lidar point cloud, has been done to register with high spatial precision the images of 2005 and 2010.

3 Object-Based Classification Strategy

The strategy is illustrated in Figure 2. A fusion of unsupervised segmentation and supervised classification has been adopted. An exploratory analysis to extract the most promising set of features for each single date has been done. A peak histogram clustering algorithm, has been adopted [9-10].

[1] MS = Multi Spectral , PAN = PANchromatic, NIR = Near InfraRed, SWIR = Short Wave InfraRed, G = Green, R = Red, CB = Coastal Blue, B = Blue, Y = Yellow, RE = Red Edge.

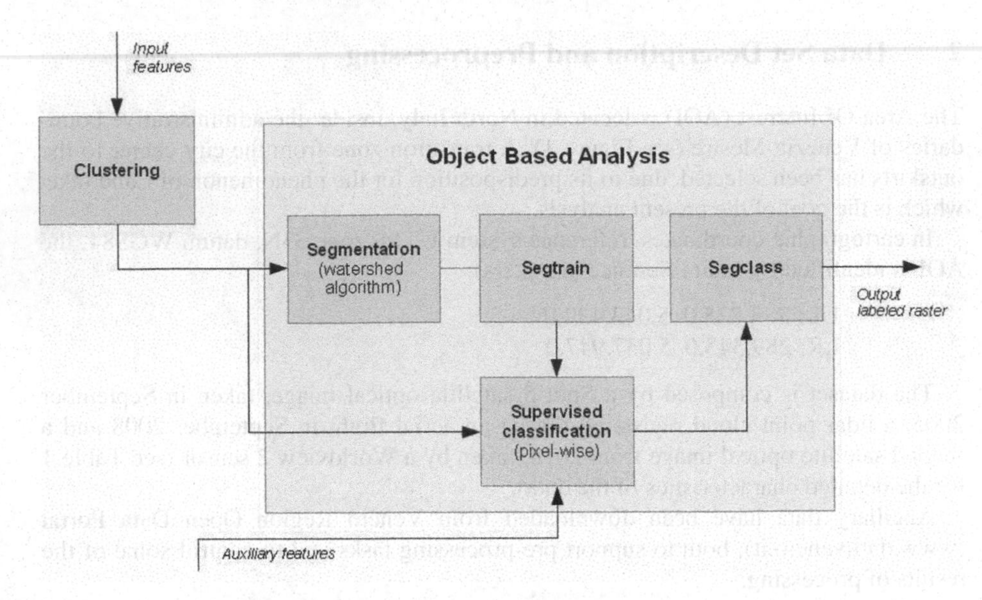

Fig. 2. Figure 1 Object Based Classification Strategy

The clusters are determined by the peaks of a multidimensional histogram whose dimensions are quantized versions of the original histograms of the input features. The result of the clustering step in terms of extracted features is reported in Table 2.

Table 2. The result of the clustering in terms of extracted features

Image (date)	Extracted features	Total number of extracted features
Spot 5 (2005)	All the bands: Green, Red, NIR, SWIR, Ratio (NIR/Red), a texture (5x5 moving window, based on a variability measure (fragmentation))	6
Worldview-2 (2010) (and Lidar (2008))	Band 2 (Blue), Band 4 (Yellow) and Band 8 (NIR 2), NDVI index, a Ratio (Band 1 /Band 8), a texture (5x5 moving window, based on fractal dimension measure) a Digital Surface Model (DSM) extracted from Lidar	7

The extracted features have been used in a segmentation procedure, based on a watershed algorithm [11-12]. For each input, a mean and a variance image are computed. The variance images are linearly combined and they are interpreted as a Digital Terrain Model (DTM) and the segments are delineated as the catch basins of this DTM, characterized by a minimum of variance in the center, and a maximum of variance at the borders. The output of this first phase is the highest possible segmented image. In dependence of the target objects, there is the possibility to generalize this "0-level" segmentation, by setting a threshold to decide the fusion of adjacent segments, if they are fairly similar. After the segmentation, the Spot 5 image has been partitioned in 31206 segments, while the Worldview-2 image has been partitioned in 26087 segments. Each segment contains hundreds of pixels, and they have been used as the units for the Object based classification. The Object based strategy consists in a pixel-wise supervised classification, whose results are then used to assign a class to each segment by means of a majority rule assignment. The selected method for the classification has been the Maximum Likelihood (ML), with equal priors and without thresholds to define minimum likelihood for the class assignment. Regarding the definition of the target classes, it has been suggested by the clustering results, that the final goal of identification of sealed and unsealed surfaces would be more easily attained if we consider an intermediate set of land covers to be found by the semi-automatic process of classification. Then, these land covers are manually aggregated to finalize the identification task. In doing so, there is a degree of freedom in the choice of the types and the number of these intermediate land covers, in the sense that we can make different choices for the two images. This is useful to obtain the common final goal, despite the disparity in spectral/spatial characteristics of the two images. In particular, the clustering outcomes has suggested to consider 10 different cover types for the Spot 5 image and 19 cover types for the Worldview-2 image.

Table 3. Intermediate land covers defined for the ML pixel-wise classification

Image type	Land cover types	Total number of Land covers	Number of training segments
Spot 5	1 water, 1 vegetation, 2 soils, 4 artificial surfaces, plus 2 shadow classes (vegetation in shadow and artificial in shadow)	10	182
Worldview-2	4 water, 4 vegetation, 3 soils, 6 artificial surfaces, plus 2 shadow classes (vegetation in shadow and artificial in shadow)	19	108

Another idea has been that of trying the sub-classification of shadowed areas into Vegetation in shadow and Artificial surfaces in shadow [13]. Indeed, due to the strong presence of shadows in urban areas, in particular if analyzed by high spatial resolution image supports, it is very important to try a classification of shadows, in order to possibly assign also these surfaces to sealed or unsealed classes. Besides the difficulties in analyzing the shadows, other challenging classification tasks have been the separation of some artificial covers by high albedo soils and the separation of shadow by shallow water. This last task has required an additional feature (a ratio between Green and NIR2 band) to be used in the Worldview-2 training phase with respect to the feature selection adopted in the segmentation task.

Fig. 3. Two examples of the Change detection results (one per column). In each column: 1) the upper image is the 2005 situation as framed by Spot 5 RGB False-color image 2) the center image is the status of the area in 2010 (Worldview-2 RGB False color image) 3) the lower image is the change detection result: in red [medium gray in the scale of grays reproduction] are the new built-up areas of each case: specifically, the enlargement of a main road and the construction of a new parking lot.

The Object based classification result has been obtained by combining the results of the pixel-wise classification with those of the segmentation through a majority rule class assignment applied to each segment. In the end the sealing identification goal has been achieved, by fusing in a unique class, named Sealed surfaces, all the artificial covers and grouping the rest of covers in a Unsealed surface class.

4 Change Detection Task and Results

For each date, after the fusion operation, the resulting map is a binary mask of Sealed surfaces. The two maps have been used for a post-classification change detection analysis [14].

The results of the application are interesting, considered the good capacity to reveal also changes at the characteristic dimension of small man-made structures (see Figure 3). Anyway, the product is more likely to be conceived as a good support to a successive photointerpretation step, than to be used as a complete automatic solution to the built-up change identification problem.

Acknowledgements. The authors acknowledge Prof. Luigi Di Prinzio's research group of IUAV - University of Venice, which have given us the opportunity to work on a valuable dataset. The first author is particularly grateful to Prof. Di Prinzio, for his suggestions and hints about the soil sealing problem, received during an High Educational activity, recently spent by the author at IUAV.

References

1. Gangkofner, U., et al.: Update of the European High-resolution Layer of Built-up Areas and Soil Sealing 2006 with Image2009 Data. In: Reuter, R. (ed.) Proceedings of the 30th EARSel Symposium, 2010, Paris, EARSel (2010)
2. Giacco, F., Colella, S., Pugliese, L., Scarpetta, S.: Application of Self Organizing Maps to multi-temporal and multi-spectral satellite images: classification and change detection. In: Frontiers in Artificial Intelligence and Applications, vol. 226, pp. 115–124. IOS Press (2011), doi:10.3233/978-1-60750-692-8-115
3. Del Frate, F., Pacifici, F., Schiavon, G., Solimini, C.: Use of Neural Networks for Automatic Classification From High-Resolution Images. IEEE Transactions on Geoscience and Remote Sensing 45(4) (April 2007)
4. Giacco, F., Scarpetta, S., Pugliese, L., Marinaro, M., Thiel, C.: Application of Self Organiz-ing Maps to multi-resolution and multi-spectral remote sensed images. In: Frontiers in Artificial Intelligence and Applications. New Directions in Neural Networks, vol. 193, pp. 245–253. IOS Press (2009), doi:10.3233/978-1-58603-984-4-245, ISSN:0922-6389
5. Giacco, F., Thiel, C., Pugliese, L., Scarpetta, S., Marinaro, M.: Uncertainty Analysis for the Classification of Multispectral Satellite Images Using SVMs and SOMs. IEEE Transactions on Geoscience and Remote Sensing 48(10), 3769–3779 (2010)

6. Pacifici, F., Del Frate, F., Solimini, C., Emery, W.J.: An Innovative Neural-Net Method to Detect Temporal Changes in High-Resolution Optical Satellite Imagery. IEEE Transactions on Geoscience and Remote Sensing 45(9), 2940 (2007), doi:10.1109/TGRS.2007.902824
7. Pugliese, L., Scarpetta, S., Giacco, F.: An application of the Self Organizing Map algorithm to computer aided classification of Aster Multispectral data. Rivista Italiana Di Telerilevamento 40, 123–129 (2008) ISSN:1129-8596
8. Blaschke, T.: Object based image analysis for remote sensing. ISPRS Journal of Photogrammetry and Remote Sensing 65, 2–16 (2010), doi:10.1016/j.isprsjprs.2009.06.004
9. Richards, J.A., Jia, X.: Remote Sensing Digital Image Analysis: An Introduction, 3rd edn. Springer-Verlag New York, Inc., Secaucus (1999) ISBN:3540648607
10. Pugliese, L., Scarpetta, S., Esposito, A., Marinaro, M.: An Application of Neural and Probabilistic Unsupervised Methods to Environmental Factor Analysis of Multi-spectral Images. In: Roli, F., Vitulano, S. (eds.) ICIAP 2005. LNCS, vol. 3617, pp. 1190–1197. Springer, Heidelberg (2005)
11. Beucher, S., Meyer, F.: The morphological approach to segmentation: the watershed transformation. In: Mathematical Morphology in Image Processing. Marcel Dekker, New York (1993)
12. Carleer, A.P., Debeir, O., Wolff, E.: Assessment of Very High Spatial Resolution Satellite Image Segmentations. Photogrammetric Engineering & Remote Sensing q(11), 1285–1294 (2005); © 2005 American Society for Photogrammetry and Remote Sensing
13. Kampouraki, M., Wood, G.A., Brewer, T.R.: Opportunities and limitations of object based image analysis for detecting urban impervious and vegetated surfaces using true-colour aerial photography. In: Blaschke, T., Lang, S., Hay, G. (eds.) Object-Based Image Analysis-Spatial Concepts for Knowledge-Driven Remote Sensing Applications, pp. 555–569. Springer, Berlin (2008)
14. Macleod, R.D., Congalton, R.G.: A Quantitative Comparison of Change-Detection Algorithms for Monitoring Eelgrass from Remotely Sensed Data. ASPRS Photogrammetric Engineering & Remote Sensing 64(3), 207–216 (1998)

EEG Complexity Modifications and Altered Compressibility in Mild Cognitive Impairment and Alzheimer's Disease

Domenico Labate[1,2,*], Fabio La Foresta[1], Isabella Palamara[1], Giuseppe Morabito[3], Alessia Bramanti[4], Zhilin Zhang[5], and Francesco C. Morabito[1]

[1] DICEAM - Mediterranea University of Reggio Calabria, Italy
[2] DIMES - University of Calabria, Cosenza, Italy
[3] University of Pavia, Italy
[4] IRCCS Centro Neurolesi bonino Pulejo, Messina, Italy
[5] Department of Electrical & Computer Engineering, University of California, San Diego, United States

{domenico.labate,fabio.laforesta,isabella.palamara,morabito}@unirc.it, peppe_mb@hotmail.it, {alessia.bramanti,zhangzlacademy}@gmail.com

Abstract. The objective of this work is to respond to the question: can quantitative electroencephalography (EEG) distinguish among Alzheimer's Disease (AD) patients, mild cognitive impaired (MCI) subjects and elderly healthy controls? In other words, are there nonlinear indexes extracted from raw EEG data that are able to manifest the background difference among EEG? The response we give here is that a synthetic index of entropic complexity (Permutation Entropy, PE) as well as a measure of compressibility of the EEG can be used to discriminate among classes of subjects. An experimental database has been analyzed to make these measurements and the results we achieved are encouraging also in terms of disease evolution. Indeed, it is clearly shown that the condition of MCI has intermediate properties with respect to the analyzed markers: thus, these markers could in principle be used to evaluate the probability of transition from MCI to mild AD.

Keywords: EEG, Alzheimer's Disease, Compressive Sensing, Permutation Entropy.

1 Introduction

Alzheimer's disease (AD) is one of the most frequent disorders among elderly population. It is characterized by neural loss and by an accumulation of neurofibrillary tangles composed of τ-amyloid fibrils and senile β-amyloid ($A\beta$) plaques [1]. The presence of synaptic loss and neurodegeneration lead to distorted signal communication among cortical circuits thus generating memory impairment and other cognitive problems. Some alterations of the patients' electroencephalogram

* Corresponding author.

S. Bassis et al. (eds.), *Recent Advances of Neural Network Models and Applications*, Smart Innovation, Systems and Technologies 26,
DOI: 10.1007/978-3-319-04129-2_17, © Springer International Publishing Switzerland 2014

(EEG) signal can be correlated to specific AD conditions, also in an earlier phase [2, 3]. AD manifests itself by several stages that provide general guidelines for understanding its progression. In the early stage, the patient suffers of frequent recent memory loss, particularly of recent event and conversation, problems expressing and understanding language and has some coordination problems like writing or using objects. In the middle stage, the patient suffers of persistent memory loss even about personal history and difficulty to recognize family and friends; furthermore, its mobility and coordination is affected by slowness and tremors. In the late stage, the patient confuses about past and present, losses the ability to remember and communicate and it is unable to take care for self. In this stage the person has need of intensive support and care.

The aim of the present study is the analysis of the EEG in terms of complexity modifications and enhanced compressibility which can be considered as the hallmark of AD. In particular, it appears that there is a gradual modification of EEG-based indexes already in the preclinical stage of the disease, commonly referred to as Mild Cognitive Impairment (MCI), when the microscopic processes that originate the disease are asymptomatic. Pharmaceutical industry is interested in how some gradual disease modifications could be detected by using a variety of biomarkers and those markers should be extracted non-invasively from possible patients. In order to achieve practical insights on these aspects, we considered a relevant number of AD patients, MCI, and normal elderly controls EEG recordings by measuring on them both an entropic index based on the recently introduced concept of Permutation Entropy (PE), and an index of compressibility based on the theory of Compressive Sensing (CS). The EEG is a synthetic description of the electromagnetic (volume conduction-filtered) brain activity: it is sparsely collected at standard locations on the patients' scalp. The related time-series contains information on correlated cortical circuits and are contaminated by artefacts and noise. The techniques here proposed aim to extract relevant information from the time-series also by exploiting their inherent redundancy. The extraction of ad hoc defined markers can be useful to differentiate among brain states basically solving an inverse problem through regularisation. EEG of AD patients reflects the reduced average complexity also at rest and it has been shown to be associated to slowing [3, 4]. To characterise the complexity of the system, the concept of entropy has been used, to take into account its nonlinearities and multiscale nature [5]. Entropy addresses randomness and predictability, with greater entropy associated to high randomness. The PE here proposed is an embedding measure that can be calculated from short time-series.

In this paper, we will also show that one novel interesting way to discriminate patients from elderly normal controls is given by a measure of compressibility. It turns out that this is also useful to apply compressive sensing strategies for therapeutic care and mental health monitoring at home through untethered wireless real-time smartphone which also alleviate the sanitary system financial overhead [6, 7]. In this study, we examined the EEG background activity in 8 patients, 8 MCI subjects, and 4 elderly control subjects.

2 Methodology

In the next paragraphs, a brief description of Compressive Sensing methodology and Permutation Entropy measure is given.

2.1 Compressive Sensing

Compressive Sensing (CS) is an emerging technique of signal processing for efficiently acquiring a signal at a rate proportional to the true information rate rather than to the Nyquist rate [8, 9]. In EEG signal, the electrical potential samples are collected in vectors (one per channel) with elements *x(n)*, with *n=1,2, ... , N*. A signal, of length *N*, denoted by $x \in \mathbb{R}^{N*1}$ could be compressed by using a random matrix, called sensing matrix, denoted by $\phi \in \mathbb{R}^{M*N}$. The signal can be compressed as: $y = \phi * x$.

Compressed Sensing (CS) algorithms use the compressed data y and the sensing matrix ϕ to recover the original signal x. The procedure is based on the assumption that most entries of the original signal x are zero (i.e., x is sparse). Current CS cannot achieve good recovery quality in case of a EEG signal because it is neither sparse in the original time domain nor sparse in transformed domains, for this reason a different approach is assumed. The signal x is expressed as $x = D * z$, where D is a *dictionary matrix*, denoted by $D \in \mathbb{R}^{M*M}$ and z is sparse. This way, the model becomes $y = \phi * D * z$.

The CS algorithms first recover z using y and $\phi * D$, and then recover the original signal x by $x = D * z$ [7]. To recover EEG signals in this study is proposed Block Sparse Bayesian Learning algorithm (BSBL) [10]: it was initially proposed for recovering a signal with block structure but it was then proposed to compress signals with no a priori known block structure. In the last case, a machine learning procedure automatically detect the underlying structure. It assumes the signal is interpreted as concatenation of g non-overlapping blocks:

$$x = [x_1, ..., x_{d_1}, ..., x_{d_{g-1}+1}, ..., x_{d_g}]^T \tag{1}$$

where the blocks are:

$$x_1 = [x_1, ..., x_{d_1}]^T, ..., x_g = [x_{d_{g-1}+1}, ..., x_{d_g}]^T \tag{2}$$

The BSBL algorithm requires users to define the block partition of x. However, this such user-defined block partition does not need to be consistent with the true block partition of the signal [7]: the user-defined block partition can be viewed as a regularization for the estimation of the signal's covariance matrix. So, even if a signal has no distinct block structure, the BSBL framework is still effective and this allows to use this model to recover an EEG signal even if it has arbitrary waveforms and the representation coefficients z generally lack block structure. An important contribution, is given by [10], that propose a learning rule based on the bound-optimization method, denoted by BSBL-BO and used in this study. The first assumption in BSBL-BO algorithm is that each block x_i, satisfies a parameterized multivariate Gaussian distribution:

$$p(x_i; \gamma_i, B_i) \sim \mathcal{N}(0, \gamma_i, B_i), \qquad i = 1, ...g \tag{3}$$

where γ_i is a non-negative parameter, if $\gamma_i = 0$ then the i-th block becomes zero. During learning procedure, because of mechanism of automatic relevance determination [11] most γ_i tend to zero. B_i is a positive definite matrix which is learned from data.

2.2 Permutation Entropy

Permutation Entropy (PE) [12] represents an alternative way of measuring similarity among patterns with respect to other types of complexity measurements, like Approximate Entropy and Sample Entropy. It has been introduced as a fast and robust method for extracting information from a time series, with special regard to its complexity [3]. With this algorithm, the one-dimensional dynamical recording corresponding to the time series is analyzed from a pure ordinal viewpoint [13]. PE is based on the counting of ordinal patterns ("motifs") and it is based on the measure of the relative frequencies of the different motifs. Ordinal patterns are not symbols ad hoc, but they actually encapsulate qualitative information about the temporal structure of the time series [4]. Symbolic representations based on motifs are rather robust against additive noise since ordinal patterns are defined by means of inequalities, and are also algorithmically simple and computationally fast [4, 12]. The PE is dependent on two-parameters: an embedding dimension, d, and a time-lag, τ.

Given a scalar time series, $y = \{y_1, ..., y_i, ..., y_N\}$, an embedding procedure forms data segments where d is the number of samples belonging to the segment, and τ represents the distance between the sample points spanned by each section of the motif. For d different samples, there will be $d!$ possible ordinal patterns, π, or "motifs". For each single motif π_j, let $f(\pi_j)$ denote its frequency of occurrence in the time series. The relative frequency is thus:

$$p(\pi_j) = \frac{f(\pi_j)}{N - d + 1} \tag{4}$$

For fixed embedding dimension $d > 2$, and fixed time-lag $\tau = \tilde{\tau}$, PE is defined as:

$$H(d, \tilde{\tau}) = -\sum_{\pi_j=1}^{d!} p(\pi_j) \, log_2 p(\pi_j) \tag{5}$$

The maximum value of $H(d)$ is $log_2(d!)$, which implies that all motifs have equal probability. The smallest value of $H(d)$ is zero, which implies the time-series is very regular, i.e., it is a mere repetition of the same basic motif.

3 Results

3.1 Data Description

The experimental EEG dataset used in this work has been extracted from a database that has been designed and acquired within a research cooperation agreement involving our NeuroLab and the "IRCSS, Centro Neurolesi Bonino-Pulejo, Messina (ITALY)". In this study three groups of subjects have been considered: 8 Alzheimer's Disease (AD), 8 Mild Cognitive Impairment (MCI) and 4 age-matched Healthy Control (HC). This section of the original database has been further processed to exclude part of the signals that evidently show the presence of artifacts. However, the CS procedure has been also checked on epochs of signal containing artifacts, which is finally the condition of home monitoring through passive electrodes. An alternative is to use active sensors which incorporates artifact cancellation automatic algorithms [14, 15]. The recordings have been collected according to the sites defined by the standard 10-20 international system, at a sampling rate of 200 Hz. The data are band-pass filtered between 0.5 and 32 Hz. In the course of the experimental activity, EEG was recorded in rest condition with closed eyes.

Table 1. Performance results of the BSBL-BO CS algorithm, in terms of both NMSE and SSIM for the HC, MCI and AD. The averaged PE value is also reported.

	NMSE			SSIM			PE		
Sensors	**HC**	**MCI**	**AD**	**HC**	**MCI**	**AD**	**HC**	**MCI**	**AD**
Fp1	0,299	0,164	0,049	0,665	0,754	0,899	0,703	0,703	0,676
Fp2	0,430	0,089	0,095	0,529	0,819	0,854	0,719	0,685	0,700
F3	0,197	0,067	0,017	0,757	0,905	0,944	0,686	0,664	0,600
F4	0,201	0,074	0,016	0,749	0,865	0,950	0,696	0,660	0,619
C3	0,159	0,062	0,017	0,771	0,876	0,939	0,682	0,661	0,615
C4	0,182	0,079	0,030	0,797	0,860	0,938	0,691	0,662	0,634
P3	0,152	0,056	0,023	0,795	0,896	0,919	0,665	0,644	0,630
P4	0,158	0,064	0,033	0,788	0,879	0,916	0,682	0,632	0,633
O1	0,133	0,070	0,034	0,788	0,890	0,918	0,649	0,648	0,622
O2	0,153	0,079	0,045	0,810	0,881	0,902	0,661	0,635	0,635
F7	0,129	0,049	0,024	0,813	0,913	0,923	0,678	0,649	0,594
F8	0,157	0,041	0,044	0,747	0,903	0,906	0,696	0,664	0,637
T7	0,189	0,058	0,035	0,748	0,893	0,900	0,668	0,633	0,628
T8	0,298	0,097	0,059	0,647	0,854	0,904	0,692	0,653	0,623
P7	0,169	0,060	0,033	0,750	0,893	0,913	0,663	0,631	0,607
P8	0,253	0,097	0,060	0,705	0,835	0,911	0,686	0,624	0,627
Fz	0,196	0,085	0,036	0,740	0,839	0,908	0,688	0,667	0,614
Cz	0,170	0,104	0,020	0,818	0,849	0,948	0,685	0,669	0,634
Pz	0,170	0,076	0,014	0,787	0,860	0,967	0,680	0,640	0,634
Fpz	0,309	0,090	0,048	0,636	0,844	0,905	0,705	0,661	0,624
Oz	0,139	0,087	0,044	0,788	0,870	0,896	0,659	0,647	0,640

3.2 Discussion

The dataset of EEG recordings include about 50 epochs each containing 384 points per category. Only artifact-free epochs have been selected. For the sake of simplicity in results presentation, all the electrodes' traces have been considered. To compress the recordings epoch by epoch, a sparse binary matrix of 192x384 as the sensing matrix has been selected (the dictionary matrix, based on standard inverse DCT, is sized 384x384). The algorithm selected for characterize the difference of compressibility among categories is the BSBL-BO described in Section 2.1. Table 1 reports the performance results of the CS algorithm, in terms of both Normalized Mean Squared Error (NMSE) and Structural SIMilarity (SSIM) [16]. NMSE is defined as:

$$NMSE = \frac{\|\hat{x} - x\|_2^2}{\|x\|_2^2} \qquad (6)$$

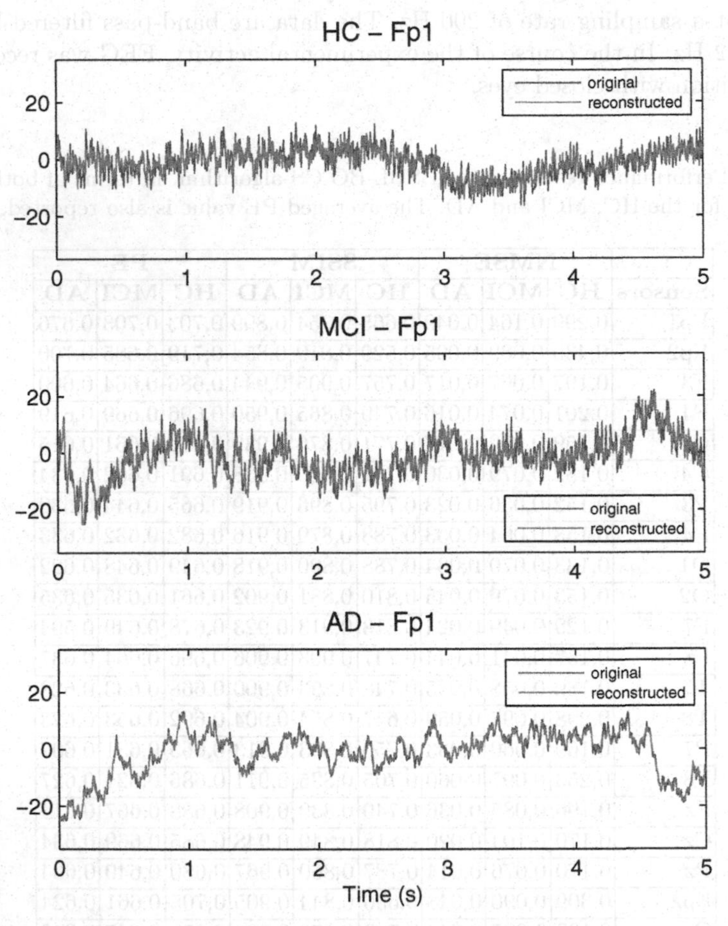

Fig. 1. AD, MCI and HC - EEG epochs (reconstructed signal in red vs. original signal in blue) for the electrode Fp1

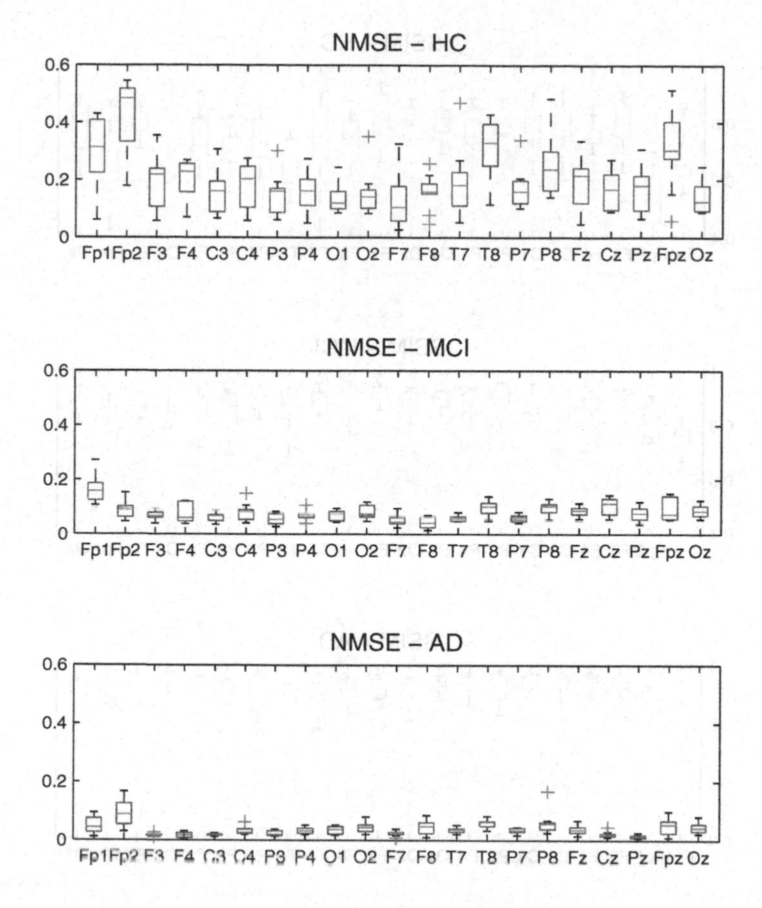

Fig. 2. Box-plots of the metric NMSE for HC, MCI and AD. Boxes and whiskers comprise 50% and 95% of values respectively. Horizontal lines within boxes are median values.

where $\hat{x}$ is the estimate of the true signal x. The averaged PE value is also reported.

The results reported in Table 1 clearly show that the estimated averaged PE of the AD EEG is typically smaller in the presence of disease, and it gradually decreases in the course of the disease evolution. Indeed, MCI subjects already show a reduced PE. The reconstruction of the original signal after compression, carried out through BSBL-BO algorithm, show the performance reported in terms of two metrics, i.e. the Normalized Mean Squared Error of reconstruction (NMSE) and the Structural Similarity Index (SSIM). The selected compression ratio is here fixed to 2:1. The compression performance indexes both confirmed the gradual transition to AD and the easier reconstruction of AD EEG signals.

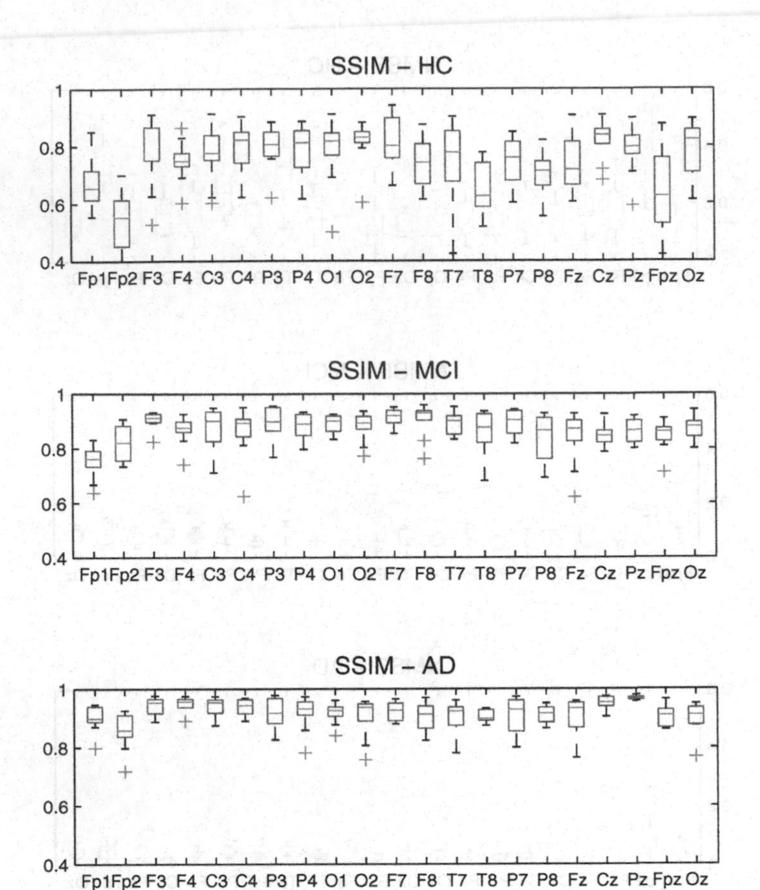

Fig. 3. Box-plots of the metric SSIM for HC, MCI and AD. Boxes and whiskers comprise 50% and 95% of values respectively. Horizontal lines within boxes are median values.

In Figure 1, the comparison between the original signal for a selected frontal electrode (Fp1) is reported. For the selected epoch, it is clearly shown the super-compressibility property of AD EEG. We then carry out a statistical analysis on the whole database and all of the electrodes' time-series. Figures 2-4 report the results achieved for the different metrics significance. Figure 2 shows that the NMSE computed for all of the channels is significantly reduced for AD EEG both in terms of median value and of standard deviations. The results achieved are good also for MCI EEG. Figure 3 illustrates that the SSIM of reconstructed signals with respect to the original ones is very good both for MCI and AD. In particular, the time-series are hardly distinguishable for some electrodes (namely, F4, Pz). Figure 4 reports the results for the estimated entropic index, PE. Since

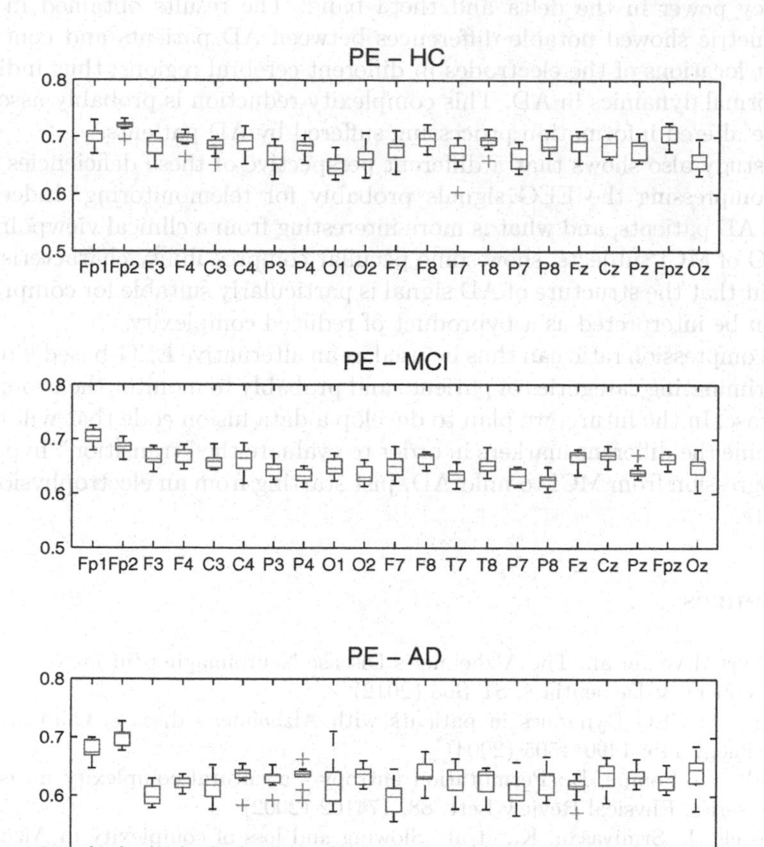

Fig. 4. Box-plots of the metric PE for HC, MCI and AD. Boxes and whiskers comprise 50% and 95% of values respectively. Horizontal lines within boxes are median values.

PE is a stiff, normalized measure, the results appears somehow difficult to interpret; however, the statistical significance of the discrimination among classes is generally good and excel for some electrodes (namely, F3, F7).

4 Conclusions

Summing all up, our study leads us to conclude that EEG background activity in AD patients shows some relevant characteristic with respect to elderly control subject. In particular, the signal is less complex, thus reflecting a distorted and reduced activity. The reduction of complexity appears to be already present in the early stage of the disease, for probable AD patients (MCI). This effect is probably correlated to the well known slowing effect which implies a reduction of the mean frequency within the alpha band and a relative increasing of the frequency power in the delta and theta band. The results obtained in terms of PE metric showed notable differences between AD patients and controls at different locations of the electrodes in different cerebral regions, thus indicating an abnormal dynamics in AD. This complexity reduction is probably associated with the altered information processing suffered by AD patients.

The study also shows that a different perspective of these deficiencies is met when compressing the EEG signals probably for telemonitoring. Indeed, the EEG of AD patients, and what is more interesting from a clinical viewpoint, also the EEG of MCI subjects, show some peculiar compressibility characteristics. It turns out that the structure of AD signal is particularly suitable for compression. This can be interpreted as a byproduct of reduced complexity.

The compression ratio can thus be used as an alternative EEG-based biomarker for discriminating categories of patients and probably to monitor the evolution of the disease. In the future, we plan to develop a data fusion code that will be able to combine the different markers in order to evaluate the "transition" hypothesis of a progression from MCI to mild AD, just starting from an electrophysiological substrate.

References

1. Weiner, M.W., et al.: The Alzheimer's Disease Neuroimaging Iniziative: A review. Alzheimer's & Dementia 8, S1–S68 (2012)
2. Jeong, J.: EEG Dynamics in patients with Alzheimer's disease. Clinical Neurophysiology 115, 1490–1505 (2004)
3. Bandt, C., Pompe, B.: Permutation entropy - a natural complexity measure for time series. Physical Review Lett. 88, 174102 (2002)
4. Dauwels, J., Srinivasan, K., et al.: Slowing and loss of complexity in Alzheimer's EEG: two sides of the same coin? Intl. J. of Alzheimer's Disease (2011)
5. Labate, D., La Foresta, F., Morabito, G., Palamara, I., Morabito, F.C.: Entropic Measures of EEG Complexity in Alzheimer's Disease through a Multivariate Multiscale Approach. IEEE Sensors Journal 13(9), 3284–3292 (2013)
6. Morabito, F.C., Labate, D., Bramanti, A., La Foresta, F., Morabito, G., Palamara, I., Szu, H.H.: Enhanced Compressibility of EEG Signal in Alzheimer's Disease Patients. IEEE Sensors Journal 13(9), 3255–3262 (2013)
7. Zhang, Z., Jung, T.P., Makeig, S., Rao, B.D.: Compressed Sensing of EEG for Wireless Telemonitoring with Low Energy Consumption and Inexpensive Hardware. IEEE Trans. Biomed. Eng. 60(1), 221–224 (2013)
8. Cands, E.J., Wakin, M.B.: An introduction to compressive sensing. IEEE Signal Processing Magazine 25(2), 14–20 (2008)
9. Donoho, D.: Compressed sensing. IEEE Trans. Inform Theory 52(4), 1289–1306 (2006)
10. Zhang, Z., Rao, B.D.: Extension of SBL algorithms for the recovery of block sparse signals with intra-block correlation. IEEE Trans. on Signal Processing 61(8), 2009–2015 (2013)
11. Tipping, M.: Sparse Bayesian learning and the relevance vector machine. Journal of Machine Learning Research 1, 211–244 (2001)

12. Morabito, F.C., Labate, D., La Foresta, F., Bramanti, A., Morabito, G., Palamara, I.: Multivariate Multi-Scale Permutation Entropy for Complexity Analysis of Alzheimer's Disease EEG. Entropy 14(7), 1186–1202 (2012)
13. Keller, K., Sinn, K.: Ordinal analysis of time series. Physica A 356, 114–120 (2005)
14. Szu, H., et al.: Smartphone household wireless electroencephalogram hat. Applied Computational Intelligence and Soft Computing, ID 241489 (2013)
15. Mammone, N., La Foresta, F., Morabito, F.C.: Automatic artifact rejection from multichannel scalp by EEG by wavelet ICA. IEEE Sensors Journal 12(3), 533–542 (2012)
16. Wang, Z., Bovik, A.: Mean squared error: Love it or leave it? a new look at signal fidelity measures. IEEE Signal Processing Magazine 26(1), 98–117 (2009)

Smart Home Task and Energy Resource Scheduling Based on Nonlinear Programming

Severini Marco, Stefano Squartini, Gian Piero Surace, and Francesco Piazza

Università Politecnica delle Marche
{m.severini,s.squartini,g.surace,f.piazza}@univpm.it

Abstract. The computational intelligence community has invested many efforts in the last few years on the challenging problem of automatic task and energy resources scheduling in smart home contexts. Moving from a recent work of some of the authors, jointly considering the electrical and thermal comfort needs of the user, in this paper a nonlinear optimization framework, namely "Mixed-Integer Nonlinear Programming", is proposed on purpose. It allows dealing with nonlinearities resulting from the constraints imposed by the involved building thermal model, which was not feasible in the original linear approach. Performed computer simulations related to a realistic domestic scenario have shown that a certain improvement is attainable in terms of satisfaction of user thermal requirements, attaining at the same time an enhanced overall energy cost reduction with respect to the non-optimized scheduling strategy.

Keywords: Optimal Home Energy Management, Task and Energy Resource Scheduling, Mixed-Integer Nonlinear Programming, Thermal Comfort, Smart Grid.

1 Introduction

Smart grid technology, nowadays, is regarded as the cornerstone of the next generation power grids [1], with improved robustness, energy routing abilities and energy production plant redundancy. Differently from the current generation grids, a smart grid has multiple links among its nodes, hence, it can balance the load over the links. In fact, by routing the energy through different paths, it is possible to prevent link overloads, or bypass interrupted links, resulting in the increased robustness of the grid [2]. Clearly, to take fully advantage of the energy routing abilities, a distributed energy production system may be necessary.

From a wider perspective, the introduction of micro grids [3] can further improve the main grid performance. In fact, if local production and local consumption match, only a minimal surplus, to be stored [4] or routed towards the main grid, is produced. Also, the energy supplied by the main power grid is a minimal part of the actual energy demand, and thus the grid can attain better voltage and frequency regulation. Additionally, a dynamic pricing scheme can be applied to energy of the main grid. Under these circumstances, the optimization of the monetary balance may, also, modulate the energy sale and purchase, depending

S. Bassis et al. (eds.), *Recent Advances of Neural Network Models and Applications*, Smart Innovation, Systems and Technologies 26,
DOI: 10.1007/978-3-319-04129-2_18,

on the energy sale price and purchase price. As a result, the micro grid energy demand can adapt to the main grid energy offer. In monetary terms, this means that a lower cost is borne by the customer. In term of energy management, this means, instead, that the micro grid activity can be coordinated, thus equalizing the energy demand across the main grid.

To account the multiple aspects of the management of a micro grid, a high level of automation is mandatory, so that the needs of the users can be satisfied. Among these, the thermal management of the building is also to be accounted, since it affects the comfort of the customers. Naturally, the complexities of such a problem cannot be handled by mean of simple task scheduling algorithms [5], and therefore more powerful computational means are necessary.

The efforts of the computational intelligence community prove that, by modelling the energy management as an optimization problem, different techniques may be applied [6], such as Particle Swarm Optimization (PSO) [7], Artificial Neural Networks [8], Fuzzy Logic [9], Adaptive Dynamic Programming [10–12], but also Linear Programming [13], Mixed Integer Linear Programming (MILP) [14] and Mixed Integer Nonlinear Programming (MINLP) [15]. A previous attempt, based on MILP optimization approach [16,17], proves that Mixed Integer Linear Programming is quite effective, when dealing with constrained problems such as task scheduling. At the same time, however, it involves linear approximation of the thermal model to take the temperature constraints into account.

Thus, if a nonlinear thermal model is used, an accuracy gain can be achieved. In this case, the optimal schedule search assumes the form of a MINLP global optimization problem. Clearly, different optimization schemes may be effective, as well. However, up to the authors knowledge, very few optimization methods are able to manage highly constrained scenarios. In particular, the approach proposed in [16, 17] appears to be a suitable reference since, differently from other schemes, it addresses the household energy management, including both task scheduling and thermal comfort, as an integrated problem.

Therefore, to evaluate the enhanced optimization scheme, a scenario is modelled. By defining the monetary balance of the energy management, an objective function is obtained, and the contributions resulting from production, storage and consumption are accounted. In the case studies, the production provided by a photovoltaic power plant is considered, the storage system is modelled including both batteries and plug-in hybrid vehicles (PHV), while the consumption accounts both task execution and thermal regulation, the latter provided by a heat pump.

The implemented model is discussed in Section 2, whereas in Section 3 the evaluated case studies are presented along with the conducted simulations and their results. Section 4 draws the work conclusions.

2 The Optimization Problem

Presently, although many efforts are made to promote and enhance the thermal efficiency of buildings, old constructions still require a great deal of energy to achieve the user thermal comfort. Hence, an accurate characterization of the thermal behaviour may improve the automation of the thermal management of the building, with respect to an approximate model. Even more so if the thermal management is part of the overall energy management of the building, which includes the user tasks and needs.

As such, as a first step, the thermal characterization provided in [16, 17] has been revised to accommodate the inherent nonlinearities of the thermal behaviour of the building. As a result, the optimization problem now takes the form of a "Mixed-Integer Nonlinear Programming" (MINLP) problem.

2.1 Building Thermal Model

To model the heat balance of the buildings [18–20], the overdot notation can be used to denote the time derivative. Therefore, the heat fluxes can be defined as:

$$\dot{Q}_{hp} = \Phi c_p(\theta_{hp} - \theta_i)\,, \tag{1}$$

$$\dot{Q}_{loss} = k_l(\theta_i - \theta_o)\,. \tag{2}$$

In particular, equation (1) expresses the thermal energy provided (heating), or removed (cooling), by the heat pump. Instead, equation (2) represents the thermal energy losses (heating), or gains (cooling), through the walls.

As already suggested in [16], by simplifying the building model, the heat loss factor can be addressed as k_l, whereas M_{air} and c_p can be used to address, respectively, the air mass within the building and its heat capacity at NTP. Also, the indoor temperature variation, over time, can be calculated as the net heat flux on the thermal capacity of the air mass inside the room, which leads to the following:

$$\dot{\theta}_i = \frac{\dot{Q}_{hp} - \dot{Q}_{loss}}{M_{air}c_p}\,. \tag{3}$$

Differently from [16] however, in this case no approximation is used. Thus, since the scheduling process is based on a discrete time domain, from (3) a difference equation can be obtained. The indoor temperature, then, may be defined as:

$$\theta_i(t+1) = \theta_i(t) + \frac{(\theta_{hp} - \theta_i(t))\Phi(t)c_p - k_l(\theta_i(t) - \theta_o(t))}{M_{air}c_p} \tag{4}$$

being t the discrete time variable. By algebraic manipulation, the temperature as function of the thermal flux is obtained:

$$\theta_i(t+1) = \theta_i(t)\left[1 - \frac{\Phi(t)}{M_{air}} - \frac{k_l}{M_{air}c_p}\right] + \frac{\Phi(t)\theta_{hp}}{M_{air}} + \frac{k_l\theta_o(t)}{M_{air}c_p}\,. \tag{5}$$

By replacing the time-continuous heat flux $\Phi(t)$ with its discrete counter part $\Phi_k \cdot b_{k,t}$, the building internal temperature can be computed by the framework, and compared against the temperature boundaries to determine the thermal needs of the user.

2.2 Schedule Cost

The objective function is represented by the monetary balance of the user activity and its comfort needs. The one used in the present work is almost identical to the one proposed in [16,17], with the exception that the time discrete variable, in this case, has been explicitly noted as Δ:

$$Q = \sum_{t=1}^{T} \left\{ \sum_{j=1}^{houses} \left[\sum_{i=1}^{tasks} (w_{j,i,t} \Delta P_{j,i} - S_{j,i,t}) C_t \right] + \right.$$

$$+ \sum_{m=1}^{s\_items} \left[(\Delta Pc_{m,t} \eta_m - Sst_{m,t}) C_t - \frac{\Delta Pd_{m,t}}{\eta_m} (C_t - Cmb) \right] +$$

$$\left. - Sex_t Cs_t + \left[\sum_{k=1}^{flows} (\Delta \Gamma_k \Phi_k b_{k,t}) - Shp_t \right] C_t \right\}.$$

The constraints, on the other hand, remain unchanged with respect to the ones proposed in [16,17], thus they are not addressed in the present manuscript.

3 Smart Home Task and Energy Resource Scheduling: Case Studies and Simulation Results

In order to evaluate the performance of the proposed scheduling scheme, the suggested model is implemented with the help of the 64-bit version of the MathWorks MatLab 2011a[1] numerical computing environment. The software is hosted on a notebook PC, based on the Intel Core i7 CPU series, with 8GB of ram and Microsoft Windows Seven 64-bit OS on board. In this environment, to retrieve the schedule meant to minimize the costs, the MINLP problem is solved by means of the SCIP solver, obtained as a part of the academic version of Opti Toolbox 1.75[2] for MatLab.

The resulting framework accounts the consumption of tasks (white appliances), the outdoor temperature, the temperature constraints and the model of the thermal behaviour of the building. At the end of the scheduling process, it returns the timetable corresponding to the minimum overall cost for the given scenario. Based on the computed timetable, a simulation of the task execution

[1] `http://www.mathworks.com`
[2] `http://www.i2c2.aut.ac.nz/Wiki/OPTI/index.php/Main/HomePage`

is carried out. During the simulation, at the beginning of each time slot, the simulator check if the actual scenario matches the one within the schedule. If a mismatch is found, the task execution is halted, and a new scheduling process is carried out prior to resuming the task execution. As such, to evaluate the scheduling process performance, the order of execution of the tasks and the allocation of the energy contributions, resulting at the end of the entire simulation process, are used to compute the overall energy cost. The same operation is carried out regarding the MILP scheme. A baseline model, also suggested in [16], which lacks any kind of optimization or energy storage, is used as an additional reference.

The scheduling problem spans over a time interval of 10 hours, from 8 a.m to 6 p.m. Presently, the list of used white appliances includes a washing machine cycle, a drying machine cycle , oven cooking and a dish washer cycle. Concerning the washing machine and the drying machine, since their power consumption changes over time, the washing cycle has been divided in subtask, each lasting an hour, and requiring a specific power amount. Critical conditions, for the system, are also accounted. The state of tasks based on the user presence, such as the PC activity, and specific configuration of the system, such as the connection of a PHV to the building wirings, are supposed to remain unchanged until an event marks its change occurrence.

Concerning the framework set-up, in regard to the building structure, the PHV characterization, the solar plant and energy storage, the solar irradiation and the energy prices, the same parameters adopted in [16] are used. The outdoor temperature values, albeit part of the same data set used in [16], refer to January the 4th in the area of San Francisco and are presented in Fig. 1. Also the set of tasks suggested in the current evaluation, slightly differs from the one used in [16], thus is detailed here in Table 1. Specifically, Table 1 describes the accounted tasks, reports their critical or non critical nature, the time frame within which each shall occur. Also, for each task, Table 1 reports the number of related subtasks and the consumption, ID and duration of each of them.

For the sake of clearness, the information reported in Table 1 is also presented in Tables 2 - 3. In particular, with the same notation of [16], the time slots are denoted in sequence from "s1" to "s15". Also, for each task, a time slot falling within the time frame of the task is denoted with an empty circle ○, whereas a cross × is used if the time slot falls outside the assigned time frame. A bullet •, on the other hand, marks a time slot allocated to the task execution. In addition, while the first column on the left reports the addressed tasks, the task parameters block reports, for each task, ID, priority, dependency, power, and duration (in time slots). The remaining two block present two consecutive time tables generated, at different times, by the scheduler.

For instance, the first time table, reported in Table 2, is obtained when the simulation starts. At the beginning of time slot "s3" (underlined in Table 2), however, because of the PC, which results turned on, a new scheduling process occurs, and the second time table, also reported in Table 2, is obtained. Similarly, at the beginning of time slot "s5" (underlined in Table 2), an external event

Table 1. Assigned tasks, upcoming events and their respective timings

	Crit.	T_{start}	T_{end}	Tasks	$P_{(W)}$	ID	$T_{(hour)}$
			Predictable tasks				
Washing machine	×	8:00 a.m.	3:00 p.m.	3	800	11	1
	×				1000	12	1
	×				900	13	1
Drying Machine	×	11:00 a.m.	5:00 p.m.	1	2500	21	2
Oven	×	10:00 a.m.	12:00 a.m.	1	2000	31	1
Dishwasher	×	1:00 p.m.	5:00 p.m.	2	1000	41	1
	×				2000	42	1
			Unpredictable events				
PHV	-	10:00 a.m.	6:00 p.m.	-	-	2	8
PC	✓	10:00 a.m.	1:00 p.m.	1	150	51	-

highlights that the PHV is plugged into the building wirings. As a result a third schedule, reported in Table 3, is produced by the framework. The last event, occurs at the beginning of time slot "s6" (underlined in Table 3), and results from the PC being turned off. In this case, as well, a new time table is obtained and is presented in Table 3.

Concerning the scheduling process, it may be useful to point out that, each time an event occurs, the completed subtasks are removed from scheduling pool. Also, since after the fourth time table is generated no additional event is recorded, all the remaining tasks are executed to completion following the fourth schedule. Although the reported schedules concern the MINLP scheme, they are identical to the ones produced by the MILP scheme, thus only one set of tables is reported.

Table 2. First and Second Time Table

	Task parameters					First Time Table										Second Time Table									
	ID	Priority	Follows	P	T	s1	s2	s3	s4	s5	s6	s7	s8	s9	s10	s3	s4	s5	s6	s7	s8	s9	s10	s11	s12
Task 1	11	3	-	800	1	○	○	○	●	○	○	○	×	×	×	○	○	○	○	●	×	×	×	×	×
Task 2	12	3	ID 11	1000	1	○	○	○	○	●	○	○	×	×	×	○	○	○	○	●	×	×	×	×	×
Task 3	13	3	ID 12	900	1	○	○	○	○	○	●	○	×	×	×	○	○	○	○	●	×	×	×	×	×
Task 4	21	1	ID 13	2500	2	×	×	×	○	○	○	●	●	○	×	×	○	○	○	●	●	○	×	×	×
Task 5	31	3	-	2000	1	×	×	○	●	×	×	×	×	×	×	○	×	×	×	×	×	×	×	×	×
Task 6	32	3	ID 31	1000	1	×	×	×	×	×	○	●	○	○	×	×	×	×	○	○	○	●	×	×	×
Task 7	41	3	-	2000	1	×	×	×	×	×	○	●	○	○	×	×	×	×	○	○	●	○	×	×	×
Task 8	51	1	-	150	9	×	×	○	○	○	○	○	○	○	○	●	●	●	●	●	●	●	●	●	×

Although both scheduling approaches yield the same task execution order, each thermal model affects the heat pump and energy management in a different way, and this is responsible of the enhanced overall costs reduction which will be commented in the following. The heat pump and energy allocation table referring to the MILP case is reported in Table 4, whereas the one referring to

Table 3. Third and Fourth Time Table

	Task parameters					Third Time Table										Fourth Time Table									
	ID	Priority	Follows	P	T	s5	s6	s7	s8	s9	s10	s11	s12	s13	s14	s6	s7	s8	s9	s10	s11	s12	s13	s14	s15
Task 1	11	3	-	800	1	○	○	●	×	×	×	×	×	×	×	○	●	×	×	×	×	×	×	×	×
Task 2	12	3	ID 11	1000	1	○	○	●	×	×	×	×	×	×	×	○	●	×	×	×	×	×	×	×	×
Task 3	13	3	ID 12	900	1	○	○	●	×	×	×	×	×	×	×	○	●	×	×	×	×	×	×	×	×
Task 4	21	1	ID 13	2500	2	○	○	●	●	○	×	×	×	×	×	○	●	●	○	×	×	×	×	×	×
Task 5	32	3	ID 31	1000	1	×	○	○	●	○	×	×	×	×	×	○	○	○	●	×	×	×	×	×	×
Task 6	41	3	-	2000	1	×	○	○	○	●	×	×	×	×	×	○	○	●	○	×	×	×	×	×	×
Task 7	51	1	-	150	9	●	●	●	●	●	●	●	×	×	×	○	○	○	○	○	○	×	×	×	×

the MINLP case is reported in Table 5. Again, the entries are denoted with the same symbolism proposed in [16], thus, for each time slot, the rows, from top to bottom, report respectively: heat pump demand (heat pump), charge rate (Ch), discharge rate(Dh), and residual storage level (SL) of the solar plant storage (with ID 1) and of the PHV (with ID 2). The next rows report the amount of renewable energy allocated, respectively, to the tasks (R_{task}) and to the energy storage (R_{store}), the amount being sold R_{sell}, the renewable energy allocated to the heat pump (R_{hp}), and the total solar energy yield (R_{tot}). The remaining two rows report, respectively the energy input from the grid (I_{en}), and the overall load of the building.

Table 4. MILP Energy Balance Contributions

	ID	Priority	s1	s2	s3	s4	s5	s6	s7	s8	s9	s10
Heat pump	-	-	400	400	400	400	400	0	400	400	400	0
Ch	1	-	0	0	0	0	51	0	0	0	150	0
Dh	1	-	397	261	238	0	0	0	5	0	0	0
SL	1	-	1603	1342	1104	1104	1155	1155	1150	1150	1301	1301
Ch	2	2	-	-	-	-	0	540	18	442	0	0
Dh	2	2	-	-	-	-	0	0	0	0	0	0
SL	2	2	-	-	-	-	4000	4540	4558	5000	5000	5000
R_{task}	-	-	0	0	0	0	150	0	614	0	0	0
R_{store}	-	-	0	0	0	0	51	540	0	0	150	0
R_{sell}	-	-	0	0	0	0	0	0	0	0	0	0
R_{hp}	-	-	3	140	236	363	400	0	0	342	85	0
R_{tot}	-	-	3	139.5	235.5	363	601.5	540	613.5	342	235.5	0
En_{cost}	-	-	6.58	5.70	5.61	5.60	6.12	5.99	5.09	5.37	5.53	6.53
I_{en}	-	-	0	0	76	2187	0	0	5000	5000	1315	0
Load	-	-	400	400	550	2550	550	0	5600	4900	1400	0

In both cases, concerning the energy management, by looking at the rows marked with "SL", it is possible to notice a limited change in the stored energy level over time. In fact, since the efficiency of both charge and discharge processes is accounted, the scheduler routes the energy directly to either the tasks and the heat pump whenever possible. Moreover, by looking at the row marked with "R_{tot}", it is possible to observe that, accordingly to the winter scenario considered, the amount of renewable energy is quite limited and the solar plant is not able to keep its own energy storage fully charged. Under these circumstances, the energy storage has a low influence in the energy allocation process.

Table 5. MINLP Energy Balance Contributions

	ID	Priority	s1	s2	s3	s4	s5	s6	s7	s8	s9	s10
Heat pump	-	-	400	400	400	400	0	400	0	400	400	0
Ch	1	-	0	0	0	0	452	0	0	0	150	0
Dh	1	-	397	261	315	324	0	5	0	0	0	0
SL	1	-	1603	1342	1027	704	1155	1150	1150	1150	1301	1301
Ch	2	2	-	-	-	-	0	145	414	442	0	0
Dh	2	2	-	-	-	-	0	0	0	0	0	0
SL	2	2	-	-	-	-	4000	4145	4558	5000	5000	5000
R_{task}	-	-	0	0	150	150	150	0	200	0	0	0
R_{store}	-	-	0	0	0	0	452	145	414	442	0	0
R_{sell}	-	-	0	0	0	0	0	0	0	0	0	0
R_{hp}	-	-	3	140	86	213	0	395	0	342	85	0
R_{tot}	-	-	3	139.5	235.5	363	601.5	540	613.5	342	235.5	0
En_{cost}	-	-	6.58	5.70	5.61	5.60	6.12	5.99	5.09	5.37	5.53	6.53
I_{en}	-	-	0	0	0	1863	0	0	5000	5000	1315	0
Load	-	-	400	400	550	2550	150	400	5200	4900	1400	0

In other words, the time slot allocation for each task is affected by the task constraints and the hourly energy cost, but it does not depends on the hourly energy demand, which explains why both schemes provide the same task schedule.

Nonetheless, the improvement of the MINLP scheme, over the MILP one, is clear. Indeed, the improved control over temperatures allows an enhanced management of the energy demand of the heat pump, which is evident by looking at Tables 4 and 5. In details, in the MILP case, the linear model predicts an accentuate temperature descent, thus the heat pump activity is programmed, among others, at time slots "s5" and "s7", whereas at time slot "s6" the heat pump is turned off. In the MINLP case, on the other hand, a more accurate heat loss is predicted, thus the heat pump is turned on at time slot "s6", whereas it remains switched off at time slots "s5" and "s7". As a result, the MINLP scheme lower the extra activity of the heatpump, and thus its additional energy consumption. This improvement can be detailed in terms of monetary balance. In particular, with respect to the baseline scheduler, the MILP achieves a saving of 30.4 $cents corresponding to the 28.1 % of the baseline costs, while the MINLP model allows to save 34.34 $cents corresponding to the 31.7 % of the baseline costs.

From the thermal comfort perspective, the MINLP approach leads to a better indoor temperature management, as shown by the comparison of the MILP temperature profile (dot-slash line in Fig. 1) against the MINLP temperature profile (slash line in Fig. 1). In this case, the improvement can be measured by means of the mean square error (MSE), revealing that the MILP achieves an MSE equal to 23.14 while for the MINLP the MSE equals 15.71, with respect to the assigned target temperature (diamond marked line in Fig. 1).

Based on these results, it can be concluded that the MINLP characterization can achieve a better performance, either in terms of assigned target temperature and of overall costs minimization. Further simulations have been carried out and similar results were obtained: they have not been detailed here for the sake of conciseness.

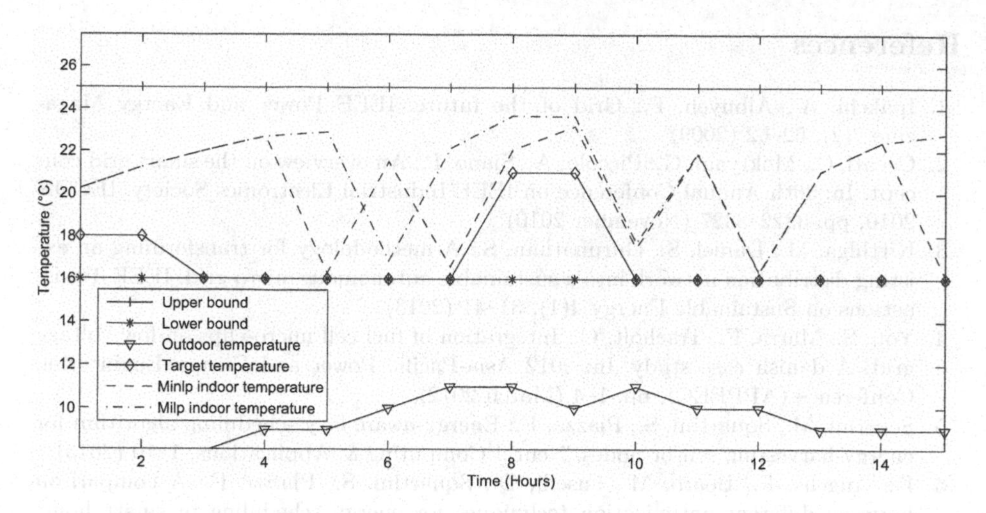

Fig. 1. Indoor temperature profile: MILP against MINLP comparison

4 Conclusions

The interest toward a smart home environment, capable of an efficient energy management through the integration of power sources into the scenario, promotes the development of several optimization techniques. Among them, the theoretical approach proposed in [16], that relies on a MILP characterization of the problem, is deemed particularly interesting. On this basis, a further improvement is proposed, by replacing the constraints, resulting from the linear approximation of the building thermal model, with a nonlinear version, more adherent to the theoretical model. The resulting MINLP optimization framework is then used, on purpose, to perform the task and energy resource scheduling.

To evaluate the improvement of the proposed solution, and to prove that the scheduling problem can benefit from an accurate modelling, a simplified domestic scenario is suggested and simulated. The results point out that, not only the obtained temperature profile is closer to the assigned target, but also the energy cost is reduced, with respect to the MILP characterization of the problem. Obviously, the choice of a rather long time slot impairs the energy management, and therefore a shorter Δ may lead to better performance.

By introducing different types of energy sources, such as combined heat and power generators, and consequently different types of energy storage, it is possible to increase the degrees of freedom of the problem, moreover if even thermal or cooling energy can be sold. In these circumstances, the suggested MINLP approach represents a natural scheme to suitably deal with these contexts.

Further improvements can be obtained by enhancing the model of the energy storage [21]. Additionally, since in real case scenarios information such as solar power availability is known solely through forecasting [22], a forecaster can be accounted, in order to evaluate how it affects the scheduling performance [23].

References

1. Ipakchi, A., Albuyeh, F.: Grid of the future. IEEE Power and Energy Magazine 7(2), 52–62 (2009)
2. Cecati, C., Mokryani, G., Piccolo, A., Siano, P.: An overview on the smart grid concept. In: 36th Annual Conference on IEEE Industrial Electronics Society, IECON 2010, pp. 3322–3327 (November 2010)
3. Kirthiga, M., Daniel, S., Gurunathan, S.: A methodology for transforming an existing distribution network into a sustainable autonomous micro-grid. IEEE Transactions on Sustainable Energy 4(1), 31–41 (2013)
4. You, S., Marra, F., Traeholt, C.: Integration of fuel cell micro-chps on low voltage grid: A danish case study. In: 2012 Asia-Pacific Power and Energy Engineering Conference (APPEEC), pp. 1–4 (March 2012)
5. Severini, M., Squartini, S., Piazza, F.: Energy-aware lazy scheduling algorithm for energy-harvesting sensor nodes. Neural Computing & Applications, 1–10 (2013)
6. De Angelis, F., Boaro, M., Fuselli, D., Squartini, S., Piazza, F.: A comparison between different optimization techniques for energy scheduling in smart home environment. In: Apolloni, B., Bassis, S., Esposito, A., Morabito, F.C. (eds.) Neural Nets and Surroundings. SIST, vol. 19, pp. 311–320. Springer, Heidelberg (2013)
7. Soares, J., Morais, H., Sousa, T., Vale, Z., Faria, P.: Day-ahead resource scheduling including demand response for electric vehicles. IEEE Transactions on Smart Grid (in press)
8. Hernandez, L., Baladron, C., Aguiar, J., Carro, B., Sanchez-Esguevillas, A., Lloret, J., Chinarro, D., Gomez-Sanz, J., Cook, D.: A multi-agent system architecture for smart grid management and forecasting of energy demand in virtual power plants. IEEE Communications Magazine 51(1), 106–113 (2013)
9. Cui, H., Dai, W.: Multi-objective optimal allocation of distributed generation in smart grid. In: 2011 International Conference on Electrical and Control Engineering (ICECE), pp. 713–717 (September 2011)
10. Fuselli, D., De Angelis, F., Boaro, M., Squartini, S., Wei, Q., Liu, D., Piazza, F.: Action dependent heuristic dynamic programming for home energy resource scheduling. International Journal of Electrical Power & Energy Systems 48, 148–160 (2013)
11. Fuselli, D., De Angelis, F., Boaro, M., Liu, D., Wei, Q., Squartini, S., Piazza, F.: Optimal battery management with adhdp in smart home environments. In: Wang, J., Yen, G.G., Polycarpou, M.M. (eds.) ISNN 2012, Part II. LNCS, vol. 7368, pp. 355–364. Springer, Heidelberg (2012)
12. Boaro, M., Fuselli, D., Angelis, F.D., Liu, D., Wei, Q., Piazza, F.: Adaptive dynamic programming algorithm for renewable energy scheduling and battery management. Cognitive Computation (in press)
13. Tham, C.-K., Luo, T.: Sensing-driven energy purchasing in smart grid cyber-physical system. IEEE Transactions on Systems, Man, and Cybernetics: Systems (in press)
14. Bozchalui, M., Sharma, R.: Analysis of electric vehicles as mobile energy storage in commercial buildings: Economic and environmental impacts. In: 2012 IEEE Power and Energy Society General Meeting, pp. 1–8 (July 2012)
15. Nagasaka, K., Ando, K., Xu, Y., Takamori, H., Wang, J., Mitsuta, A., Saito, O., Go, E.: A research on operation planning of multi smart micro grid. In: 2012 International Conference on Advanced Mechatronic Systems (ICAMechS), pp. 351–356 (September 2012)

16. De Angelis, F., Boaro, M., Fuselli, D., Squartini, S., Piazza, F., Wei, Q.: Optimal home energy management under dynamic electrical and thermal constraints. IEEE Transactions on Industrial Informatics 9(3), 1518–1527 (2013)
17. De Angelis, F., Boaro, M., Fuselli, D., Squartini, S., Piazza, F., Wei, Q., Wang, D.: Optimal task and energy scheduling in dynamic residential scenarios. In: Wang, J., Yen, G.G., Polycarpou, M.M. (eds.) ISNN 2012, Part I. LNCS, vol. 7367, pp. 650–658. Springer, Heidelberg (2012)
18. Kazanavičius, E., Mikuckas, A., Mikuckienė, I., Čeponis, J.: The heat balance model of residential house. In: Information Technology and Control, vol. 35(4), pp. 391–396 (November 2006)
19. Qela, B., Mouftah, H.: Simulation of a house heating system using C# – an energy conservation perspective. In: 2010 23rd Canadian Conference on Electrical and Computer Engineering (CCECE), pp. 1–5 (May 2010)
20. Fux, S.F., Ashouri, A., Benz, M.J., Guzzella, L.: Ekf based self-adaptive thermal model for a passive house. Energy and Buildings (2012), http://www.sciencedirect.com/science/article/pii/S0378778812003039
21. Squartini, S., Fuselli, D., Boaro, M., De Angelis, F., Piazza, F.: Home energy resource scheduling algorithms and their dependency on the battery model. In: IEEE Symposium Series on Computational Intelligence (April 2013)
22. Ciabattoni, L., Grisostomi, M., Ippoliti, G., Longhi, S., Mainardi, E.: Online tuned neural networks for PV plant production forecasting. In: 38th IEEE Photovoltaic Specialists Conference (PVSC), Austin, TX, pp. 2916–2921 (June 2012)
23. Squartini, S., Boaro, M., De Angelis, F., Fuselli, D., Piazza, F.: Optimization algorithms for home energy resource scheduling in presence of data uncertainty. In: International Conference on Intelligent Control and Information Processing (June 2013)

Part III
Applications

Data Fusion Using a Factor Graph for Ship Tracking in Harbour Scenarios*

Francesco Castaldo and Francesco A.N. Palmieri

Dipartimento di Ingegneria Industriale e dell'Informazione
Seconda Universitá di Napoli (SUN)
via Roma 29, 81031 Aversa (CE)-Italy
france.castaldo@gmail.com, francesco.palmieri@unina2.it

Abstract. Data coming from cameras deployed along an harbour coastline are fused to extract the state of unknown vessels framed by the sensors. We embed the ship dynamic model into a Factor Graph that through probability propagation provides a very flexible merge of sensory data and inferences. Preliminary results and experiments from videos gathered in the Gulf of Naples are reported with a discussion on future trends.

Keywords: Data Fusion, Harbour Management, Factor Graph, Ship Model, Camera.

1 Introduction

State estimation of moving objects in a complex scene is a research topic much explored in recent years and in constant evolution as such technology can provide improvements to safety and management in various application areas (Harbours, Airports, Public areas, etc.). Intelligent and automatic fusion of data coming from different sensor modalities [1], that provide information that differ in type, quality and reliability, requires a framework that is capable of handling in real time previous knowledge about the model together with environmental constraints and measurements' uncertainty.

In this paper we consider an harbour scenario, and the state estimation problem for tracking one or more vessels appearing in the field of view of cameras deployed along the coastline [2].

Tracking of multiple targets can be achieved using several and well-known techniques [3] [4] [5], such as Kalman Filters, Particle Filters, Bayesian Networks, Factor Graphs, etc. We have chosen Factor Graphs, and in particular their normal realizations, also called Forney-Style Factor Graphs (FFG) [6]. Graphical models such as FFGs are very attractive for two main reasons: simplicity and rigorousness. The first feature allows to map into a nice and comprehensible

* This work has been partially sponsored by Ministero Infrastrutture e Trasporti, PON01-01936, Harbour Traffic Optimization System (HABITAT) with Consorzio Nazionale Interuniversitario per le Telecomunicazioni (CNIT)- Italy.

S. Bassis et al. (eds.), *Recent Advances of Neural Network Models and Applications*, Smart Innovation, Systems and Technologies 26,
DOI: 10.1007/978-3-319-04129-2_19, © Springer International Publishing Switzerland 2014

structure very complex problems and algorithms, while the second one guarantees robustness and optimality of the resulting estimates. Factor graphs are also modular, a feature very important in mutable scenarios as harbours are.

The paper is organized as follow. In Section 2 we present the ship model used in the algorithm. In Section 3 the factor graph is introduced, with particular attention to the type of messages exchanged between the nodes. Section 4 discusses preliminary results obtained in a test in the Gulf of Naples and Section 5 draws conclusions with suggestions for future developments.

2 Ship Model

The discrete-time mathematical model we use for the state of ships sailing near the harbour is based on a single point (usually the ship's barycenter) that moves on the 2D sea surface. We use a modified version of the model presented in [7] [8], considering vessels sailing at almost constant speed and with moderate maneuvering capabilities.

Defining $\mathbf{x}_k = (x_k, y_k, v_{x_k}, v_{y_k})^T$ the ship state at time k, with (x_k, y_k) the GPS ship coordinates (longitude and latitude) and (v_{x_k}, v_{y_k}) ship velocity components, we have the following discrete-time model

$$\begin{aligned} x_k &= x_{k-1} + T v_{x_{k-1}} + w_{x_k}, \\ y_k &= y_{k-1} + T v_{y_{k-1}} + w_{y_k}, \\ v_{x_k} &= v_{x_{k-1}} + w_{v_{x_k}}, \\ v_{y_k} &= v_{y_{k-1}} + w_{v_{y_k}}, \end{aligned}$$

where T is the sampling time and $\mathbf{w}_k = (w_{x_k}, w_{y_k}, w_{v_{x_k}}, w_{v_{y_k}})^T$ is a white-gaussian noise that allows slight changes in ship's position, speed and course. We can write the model in matrix form as

$$\begin{pmatrix} x_k \\ y_k \\ v_{x_k} \\ v_{y_k} \end{pmatrix} = \begin{pmatrix} 1 & 0 & T & 0 \\ 0 & 1 & 0 & T \\ 0 & 0 & 1 & 0 \\ 0 & 0 & 0 & 1 \end{pmatrix} \begin{pmatrix} x_{k-1} \\ y_{k-1} \\ v_{x_{k-1}} \\ v_{y_{k-1}} \end{pmatrix} + \begin{pmatrix} w_{x_k} \\ w_{y_k} \\ w_{v_{x_k}} \\ w_{v_{y_k}} \end{pmatrix},$$

or equivalently as

$$\mathbf{x}_k = A\mathbf{x}_{k-1} + \mathbf{w}_k, \tag{1}$$

with $A = \begin{pmatrix} 1 & 0 & T & 0 \\ 0 & 1 & 0 & T \\ 0 & 0 & 1 & 0 \\ 0 & 0 & 0 & 1 \end{pmatrix}$. Initial conditions are in the vector $\mathbf{x}_0 = (x_0, y_0, v_{x_0}, v_{y_0})^T$.

3 Data Fusion Using Factor Graph

Sensor data and ship model are embedded here into a Forney-Style Factor Graph [6] [9] [10]. Factor graphs are an emerging framework that allows to map many problems of different kind (signal processing, statistics, data fusion, etc) into a graph. With both simplicity and rigor the model can be manipulated with the information flow that travels bidirectionally for prediction and inference.

A factor graph is composed of *edges* (through which information is propagated) and *nodes* (through which information is processed). Forward and backward messages are denoted as f_x and b_x, where x is the edge variable. Either forward or backward messages coming out of a node depend on messages entering the node. For the reader not familiar with this framework, we would like to emphasize that, even though the propagating equations may appear somewhat unjustified, they are rigorous translation of marginalization and Bayes' rule [6]. In this paper we assume that messages are gaussian pdfs, i.e. fully describable by a mean vector m and a covariance matrix Σ. Clearly the state mean at time k represents the state expected value, while the covariance represents how that value is dispersed around the mean. Our confidence may evolve in time as the result of information gathered and fused in various parts of the graph. Forward and backward messages for a variable $\mathbf{x}$ are defined as the sets

$$f_{\mathbf{x}} = \{m_{f_{\mathbf{x}}}, \Sigma_{f_{\mathbf{x}}}\},$$
$$b_{\mathbf{x}} = \{m_{b_{\mathbf{x}}}, \Sigma_{b_{\mathbf{x}}}\}.$$

Each graph edge corresponds to a variable and forward/backward messages can be combined along the same edge to get the variable's best knowledge pdf by the product rule [6] [9]. Given the hypothesis of gaussianity, after the application of the product rule we get again a mean and a covariance matrix that fully describe the result. More specifically, at an edge $\mathbf{x}$ from knowledge of forward $f_{\mathbf{x}}$ and backward $b_{\mathbf{x}}$ messages, we can compute mean and covariance of $\mathbf{x}$ as

$$m_{\mathbf{x}} = \Sigma_{\mathbf{x}}(\Sigma_{f_{\mathbf{x}}}^{-1} m_{f_{\mathbf{x}}} + \Sigma_{b_{\mathbf{x}}}^{-1} m_{b_{\mathbf{x}}}),$$
$$\Sigma_{\mathbf{x}} = (\Sigma_{f_{\mathbf{x}}}^{-1} + \Sigma_{b_{\mathbf{x}}}^{-1})^{-1}.$$

More details about the sum-product rule can be found in [9].

In our tracking application, the factor graph used in this paper, for each discrete time k, has the structure shown in Figure 1.

The linear A block represents the first part of the model defined in (1). The initial forward message is $f_{\mathbf{x}_0} = \{\mathbf{x}_0, \Sigma_0\}$, with $\mathbf{x}_0$ initial condition and Σ_0 a fixed covariance. The smaller the covariance's values are chosen, the greater our confidence on initial values are. The following *sum* block adds the white gaussian noise $\mathbf{w}_k$ through the forward message $f_{\mathbf{w}_k} = \{\mathbf{0}, \Sigma_{f_{\mathbf{w}_k}}\}$ that represents our model's uncertainty. Low values of covariance's norm imply great confidence on model's evolution, while high values imply strongly reliance on sensory data. The *equal* block implements the actual fusion between ship predictions and sensory measurements. The information that converges into this node can be sharp, partial or null: the rules of message propagation and combination guarantee that we always utilize all the information at best.

Block g combines cameras' 2D information (called obs_k) to provide on-line measurements of the ship state at any time k, as shown in the simplified diagram

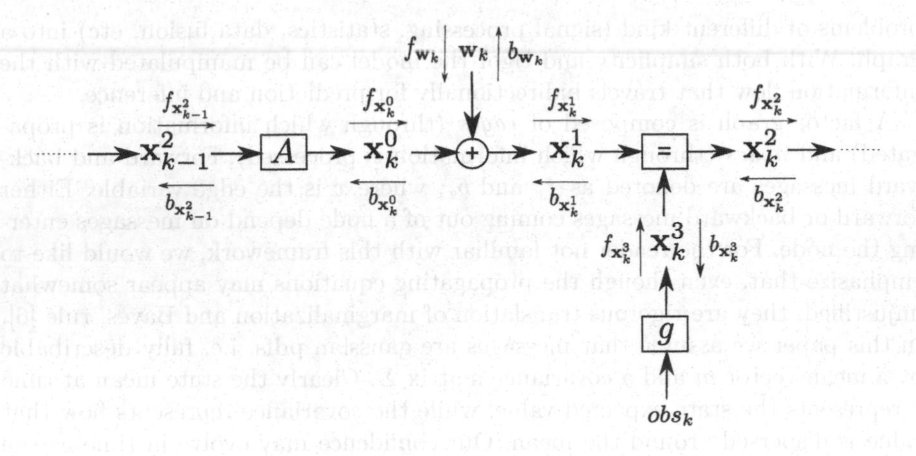

Fig. 1. The k-th stage in the factor graph

of Figure 2. The camera system, not detailed here, but discussed in [2], consists of a set of N cameras that have been deployed and calibrated using AIS data. Calibration is equivalent to obtaining a prospective homography matrix for each camera. The vision system contains many details, some of which have been reported in [2], that will be discussed elsewhere for space reasons. Essentially, image points are mapped into geographical positions via a 2D reconstruction procedure; after pattern recognition and pixel association the estimates are fused to provide an instantaneous estimate of the ship's parameters. To the purpose of this paper we assume that the camera system at each time k provides a message $f_{\mathbf{x}^3_k}$ on the ships state. Note that the camera system can improve its tracking capabilities using the available $b_{\mathbf{x}^3_k}$ coming from the factor graph (not discussed here). The block g can work even with one single camera, but obviously more cameras can cover a larger area, providing more reliable and general estimates.

In the message $f_{\mathbf{x}^3_k} = \{m_{f_{\mathbf{x}^3_k}}, \Sigma_{f_{\mathbf{x}^3_k}}\}$, the covariance matrix $\Sigma_{f_{\mathbf{x}^3_k}}$ carries information on how reliable the data is at that moment. Note this model allows time-dependent fusion in which sensory data may carry a very large covariance, which means that at that time camera information may be very unreliable (lost target), or essentially unavailable (poor guess).

The nice feature of the factor graphs is that messages can be propagated bi-directionally providing predictions into the future and estimation improvements (smoothing) into the past. To infer into the *past*, we also need a backward message coming from the end of the graph. We can model it as $b_{\mathbf{x}^2_k} = \{\mathbf{0}, \Sigma_{b_{\mathbf{x}^2_k}}\}$, with $\Sigma_{b_{\mathbf{x}^2_k}}$ a very high-valued covariance matrix. Message fusion can be accomplished along each edge of the graph, even though the most relevant information is at the end of each graph's element, that is $\mathbf{x}^2_i = \{m_{\mathbf{x}^2_i}, \Sigma_{\mathbf{x}^2_i}\}$, with $i = 1, ..., k$.

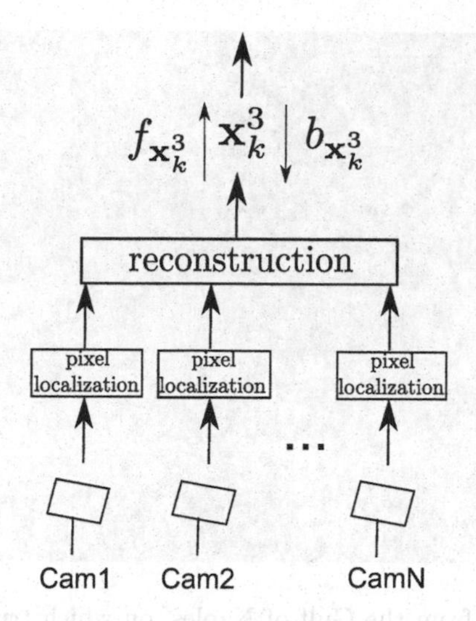

Fig. 2. A simplified diagram of the multiple-camera system (g block)

Given inputs $f_{\mathbf{x}_{k-1}^2}$, $b_{\mathbf{x}_k^2}$, $f_{\mathbf{x}_k^3}$ and $f_{\mathbf{w}_k}$, forward and backward messages can be computed as

$$\begin{aligned}
f_{\mathbf{x}_k^0} &= \{Am_{f_{\mathbf{x}_{k-1}^2}}, A\Sigma_{f_{\mathbf{x}_{k-1}^2}} A^T\}, \\
f_{\mathbf{x}_k^1} &= \{m_{f_{\mathbf{x}_k^0}} + m_{f_{\mathbf{w}_k}}, \Sigma_{f_{\mathbf{x}_k^0}} + \Sigma_{f_{\mathbf{w}_k}}\}, \\
f_{\mathbf{x}_k^2} &= \{(\Sigma_{\mathbf{x}_k^1}^{-1} + \Sigma_{\mathbf{x}_k^3}^{-1})^{-1}(\Sigma_{\mathbf{x}_k^1}^{-1} m_{\mathbf{x}_k^1} + \Sigma_{\mathbf{x}_k^3}^{-1} m_{\mathbf{x}_k^3}), (\Sigma_{\mathbf{x}_k^1}^{-1} + \Sigma_{\mathbf{x}_k^3}^{-1})^{-1}\}, \\
b_{\mathbf{x}_k^1} &= \{(\Sigma_{f_{\mathbf{x}_k^3}}^{-1} + \Sigma_{b_{\mathbf{x}_k^2}}^{-1})^{-1}(\Sigma_{f_{\mathbf{x}_k^3}}^{-1} m_{f_{\mathbf{x}_k^3}} + \Sigma_{b_{\mathbf{x}_k^2}}^{-1} m_{b_{\mathbf{x}_k^2}}), (\Sigma_{f_{\mathbf{x}_k^3}}^{-1} + \Sigma_{b_{\mathbf{x}_k^2}}^{-1})^{-1}\}, \\
b_{\mathbf{x}_k^0} &= \{m_{b_{\mathbf{x}_k^1}} - m_{f_{\mathbf{w}_k}}, (\Sigma_{b_{\mathbf{x}_k^1}}^{-1} + \Sigma_{f_{\mathbf{w}_k}}^{-1})^{-1}\}, \\
b_{\mathbf{x}_{k-1}^2} &= \{(A^T \Sigma_{b_{\mathbf{x}_k^0}}^{-1} A)^{-1}(A^T \Sigma_{b_{\mathbf{x}_k^0}}^{-1} m_{b_{\mathbf{x}_k^0}}), (A^T \Sigma_{b_{\mathbf{x}_k^0}}^{-1} A)^{-1}\}.
\end{aligned}$$

4 Preliminary Results

Our tracking framework has been tested on real data coming from two cameras deployed along the Gulf of Naples coastline and calibrated with AIS data [11] of known ships [2].

The test regards state reconstruction of a large ship appearing in the cameras' fields of view. The ship under consideration is one of the vessels used for calibration, therefore its position, speed and course are already known from AIS. This information is compared with the state provided by the algorithm, in order to assess estimation accuracy. Assuming known initial state $\mathbf{x}_0$ and, for each discrete step k, measures from the cameras $f_{\mathbf{x}_k^3}$, the algorithm fuses altogether forward and backward messages, getting ship state estimate, $k = 1, ..., 7$.

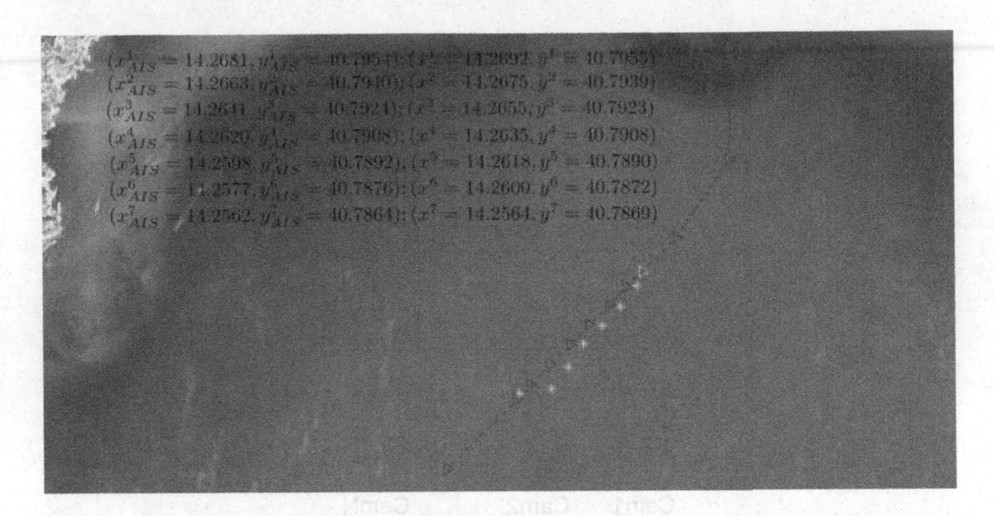

Fig. 3. A Google Maps from the Gulf of Naples, on which true (red) AIS and reconstructed (green) trajectories are superimposed. On the left there is the confrontation between longitude and latitude coordinates obtained from the AIS and the ones computed from our algorithm.

The covariances are diagonal matrices with standard deviations chosen to be as follows: for the initial conditions $\sigma_0 = 0.1$; for the prediction model $\sigma_w = 0.5$; for the backward information at the end of the chain $\sigma_{b_{\mathbf{x}_7^2}} = 1000$; for the information coming from the camera system $\sigma_{f_{\mathbf{x}_k^3}} = 0.5$.

In Figure 3 we plot ship GPS positions (red triangles) coming from AIS and the reconstructed positions obtained from our procedure (green asterisks). Data from both cameras and from AIS have been interpolated as they are all available asynchronously. As we can see from the figure, discrepancies are very limited.

Figure 4 is a more detailed version of Figure 3, where green crosses represent state estimates calculated at each step k. In the algorithm, at each time step a backward message $b_{\mathbf{x}_k^2}$ is injected into the graph, and previous states are recomputed, in order to improve the estimation accuracy.

Green circles represent approximately covariances calculated at the last step, and indicate how confident we are on the value they represent. They are enlarged 100-times in order to have a comprehensible figure. As stated before, backward messages enhance previous states' reconstruction, therefore going backwards estimation uncertainty and covariances' values decrease. The values go from $\sigma_{\mathbf{x}_7^2} \cong 0.5$ (standard deviation at step $k = 7$) to $\sigma_{\mathbf{x}_1^2} \cong 0.27$ (standard deviation at step $k = 1$).

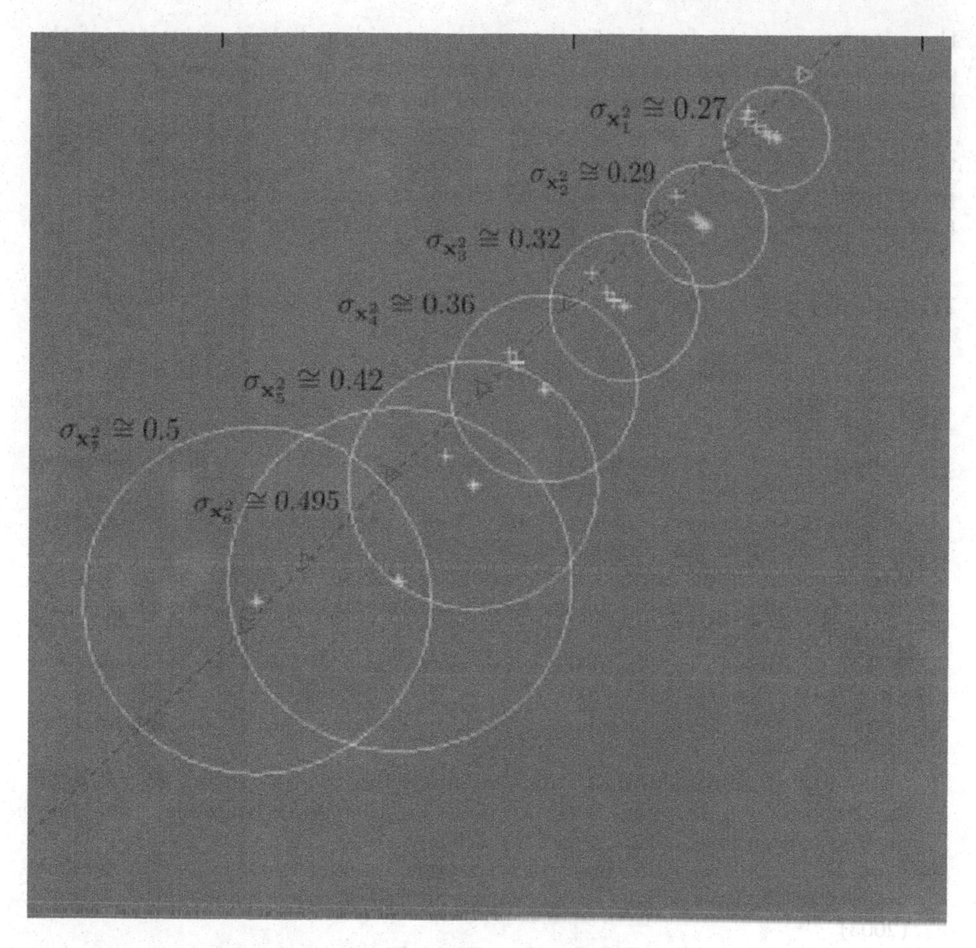

Fig. 4. Zoomed version of Figure 3, with green circles that represent a 100-fold magnified indication of the dispersion around the mean (covariance)

5 Conclusions and Future Developments

The framework described in this paper is intended to be the "milestone" around which construct an autonomous and intelligent system that provide state information about vessels in movement near the coast. Information about the behaviours of moving targets into the scene can be used to undertstand whether the situation of the harbour is into a normal or a possibly dangerous state. Camera sensory data is fused with the dynamic model and it will be extended to account for multiple targets. Furthermore positional estimates may be available at selective times from different sensor modalities and be fused into the factor graph.

The flexibility of the framework also permits integration of other sensor modalities (e.g. Radar, Lidar, etc.) without much efforts. The estimates coming from these sensors can be fused into the graph, providing better estimates of the object state. This last feature is crucial in complex and changing scenarios as harbours, where sensors can be substituted and modified anytime.

Current effort is also being devoted to provide automatic covariance learning through the factor graph's blocks. More specifically, block g extracts 2D GPS points using a non-linear operation based on singular values decomposition [2] (not discussed here) and the issue of data dispersion is still an open issue.

References

1. Hall, D.L.: An introduction to multisensor data fusion. Proceedings of the IEEE 85, 6–23 (1997)
2. Palmieri, F.A.N., Castaldo, F., Marino, G.: Harbour Surveillance with Cameras Calibrated with AIS Data. In: Proceedings of the 2013 IEEE Aerospace Conference, March 2-9. Big Sky, Montana (2013)
3. Liggins, M.E., Hall, D.L., Llinas, J.: Handbook of Multisensor Data Fusion: Theory and Practice, 2nd edn. (September 2008)
4. Arulampalam, S.M., Maskell, S., Gordon, N., Clapp, T.: A tutorial on particle filters for online nonlinear/non-gaussian bayesian tracking. IEEE Transactions on Signal Processing 50, 174–188 (2002)
5. Welch, G., Bishop, G.: An Introduction to the Kalman Filter. In: Proceedings of SIGGRAPH, Course, USA (2001)
6. Forney Jr., G.D.: Codes on graphs: Normal realizations. IEEE Trans. Inform. Theory 47(2), 520–548 (2001)
7. Semerdijev, E., Mihaylova, L.: Adaptive interacting multiple model algorithm for manouvering ship tracking. In: Proceedings of the 1998 International Conference on Information Fusion, Las Vegas, NV (1998)
8. Rong Li, X., Jilkov, V.P.: Survey of maneuvering target tracking. part I: dynamic models. IEEE Transactions on Aerospace and Electronic Systems 39, 1333–1364 (2003)
9. Kschischang, F.R., Frey, B.J., Loeliger, H.-A.: Factor graphs and the sum-product algorithm. IEEE Trans. Inform. Theory 47, 498–519 (2001)
10. Loeliger, H.-A., Dauwels, J., Korl, S., Ping, L., Kschischang, F.R.: The factor graph approach to model-based signal processing. Proceedings of the IEEE 95, 1295–1322 (2007)
11. IALA Guideline No. 1082 On an Overview of AIS. International Association of Marine Aids to Navigation and Lighthouse (June 2011)

Reinforcement Learning for Automated Financial Trading: Basics and Applications

Francesco Bertoluzzo[1] and Marco Corazza[1,2]

[1] Department of Economics, Ca' Foscari University of Venice
[2] Advanced School of Economics of Venice
Sestiere Cannaregio n. 873, 30121 Venice, Italy
{fbertoluzzo,corazza}@unive.it

Abstract. The construction of automated financial trading systems (FTSs) is a subject of high interest for both the academic environment and the financial one due to the potential promises by self-learning methodologies. In this paper we consider Reinforcement Learning (RL) type algorithms, that is algorithms that real-time optimize their behavior in relation to the responses they get from the environment in which they operate, without the need for a supervisor. In particular, first we introduce the essential aspects of RL which are of interest for our purposes, second we present some original automatic FTSs based on differently configured RL-based algorithms, then we apply such FTSs to artificial and real time series of daily stock prices. Finally, we compare our FTSs with a classical one based on Technical Analysis indicators. All the results we achieve are generally quite satisfactory.

Keywords: Financial trading system, Reinforcement Learning, stochastic control, *Q*-learning algorithm, Kernel-based Reinforcement Learning algorithm, financial time series, Technical Analysis.

1 Introduction

According to the well-known weak form of the *Efficient Market Hypothesis* (EMH), it is not possible to systematically make profitable trading in financial markets. In fact, following this theory, the economic agents operating in these markets are rational, that is, through the law of the demand and the supply, they are able to instantaneously and appropriately vary the prices of the financial assets on the basis of the past and the current information. In this theoretical framework, the only source of (unpredictable) variations of the financial asset prices between two consecutive time instants can be the arrival of unexpected new information (see for more details [7]).

But, as the common sense suggests, human beings (and therefore economic agents) are often non rational when making decisions under uncertainty. In fact, since the 80s of the past century experimental economists have documented several departures of the real investors' behaviours from the ones prescribed by the EMH (see [12] and the references therein). The main implication coming from these departures consists in the fact that financial markets are not so rarely inefficient, and consequently that they offer, more or less frequently, possibilities of profitable trading.

S. Bassis et al. (eds.), *Recent Advances of Neural Network Models and Applications*, Smart Innovation, Systems and Technologies 26,
DOI: 10.1007/978-3-319-04129-2_20,

At this point, a (crucial) question naturally arises: How to take advantage of these possibilities of trading? Of course, the answer depends on the chosen reference theoretical framework. In our opinion, the currently most convincing attempt to reconcile the EMH with the empirical departures from it is given by the so-called *Adaptive Market Hypothesis* (AMH). Following this theory, a financial market can be viewed as an evolutionary environment in which different "species" (for instance, hedge funds, market makers, pension funds, retail investors, ...) interact among them in accordance with unknown and structurally time-varying dynamics in order to achieve the efficiency (see for more details [8] and [12]). Note that this evolutionary tending towards efficiency is not instantaneous and that it generally does not imply appropriate variations of the financial asset prices. Because of that, the AMH entails that ≪[f]*rom an evolutionary perspective, the very existence of active liquid financial markets implies that profit opportunities must be present. As they are exploited, they disappear. But new opportunities are also constantly being created as certain species die out, as others are born, and as institutions and business conditions change*≫ (from [12], page 24). So, coming back to the above question, it should be clear that an effective financial trading system (FTS) has to be a new specie:

- Able to real-time interact with the considered financial market in order to learn its unknown and structurally time-varying dynamics;
- Able to exploit this knowledge in a not emotive way in order to real-time detecting profitable financial trading policies.

Therefore, given the reference theoretical framework we have chosen (i.e. the AMH) and given the features of the FTS we have required in the two previous points, in this paper we resort to a self-adaptive machine learning known as *Reinforcement Learning* (RL) (see [1]), also known as *Neuro-Dynamic Programming* (see [4]), in order to develop a fully automated FTS. In particular, to this end we consider two different RL-based policy evaluation approaches: the *Temporal Difference* (TD) (see subsection 3.3) and the *Kernel-based Reinforcement Learning* (KbRL) (see subsection 3.4). Note also that, from a more traditional standpoint, the implementation of such a FTS can be viewed as a stochastic control problem in which the RL-based approaches have to discover the optimal financial trading strategies directly interacting with the financial market (so, any need to build a priori formal models for the description of the dynamics of the prices and/or the returns is eliminated).

The remainder of the paper is organized as follows. In the next Section we give a brief review of the prominent literature on the RL-based FTSs, and we describe the elements of novelties we present in our paper with respect to this literature. In Section 4 we present the essential aspects of the RL methodology which are of interest to our purposes. In Section 3 we present our RL-based FTSs, we provide the results of their applications to an artificial time series and to five real ones, and we compare these results with those coming from a classical FTS. In section 5 we propose some concluding remarks.

2 A Brief Review of the Literature

In this Section we provide a few review of the most influential papers about the RL-based FTSs. Our purpose is not primarily to be exhaustive, but rather that to highlight the main research directions. Then we describe the elements of novelties present in our paper.

Among the first contributions in this research field, we recall [13], [14], and [9]. In general, the respective Authors show that RL-based financial trading policies perform better than those based on supervised learning methodologies when market frictions are considered. In [13] a simplified version of the RL methodology, called *Direct Learning*, is proposed and used in order to set a FTS that, taking into account transaction costs, maximizes an appropriate investor's utility function based on a differential version of the well-known Sharpe ratio. Then, it is shown by controlled experiments that the proposed FTS performs better than standard FTSs. Finally, the Authors use the so developed FTS to make profitable trades with respect to assets of the U.S. financial markets. In [14], the Authors mainly compare FTSs developed by using various RL-based approaches with FTSs developed by using stochastic dynamic programming methodologies. In general they show by extensive experiments that the former approaches are better than the latter ones. In [9] the Author considers a FTS similar to the one developed in [13] and applies it to the financial high-frequency data, obtaining profitable performances. Also in [3] the Authors consider a FTS similar to the one developed in [13], but they consider as investor's utility function based on the differential version of the returns weighted direction symmetry index. Then, they apply this FTS to some of the most relevant world stock market indexes achieving satisfactory results. In [16] the Authors proposes a RL-based asset allocation strategy able to utilize the temporal information coming from both a given stock and the fund over that stock. Empirical results attained by applying such asset allocation strategy to the Korean stock market show that it performs better than several classical asset allocation strategies. In [10] two stock market timing predictors are presented: an actor-only RL and and actor-critic RL. The Authors show that, when both are applied to real financial time series, the latter generally perform better than the former. Finally, in [2] an actor-critic RL-based FTS is proposed, but in a fuzzy version. The Authors show that, taking into account transaction costs, the profitability of this FTS when apply to important world stock market indexes is consistently superior to that of other advanced trading strategies.

With respect to the this literature, in this paper we do the following:

- As introduced in Section 1, we consider two different RL-based trading policy evaluation approaches: the TD and the KbRL ones. Generally, the *squashing function* used in the former is the well-known S-shaped logistic. In this paper we have substitute it with another well-known S-shaped squashing function, the hyperbolic tangent, in order to check its performances. Notice that, to the best of our knowledge, the hyperbolic tangent has been rarely used in this context. Then, as regards the KbRL, again to the best of our knowledge, it has never been used before for developing FTSs;
- In several papers the differential Sharpe ratio is used as performance indicator. In this paper we utilize the classical Sharpe ratio computed on the last $L \in \mathbb{N}$ trading days. This surely reduce the computational effort;

- Usually, the very very big majority of the classical FTSs and of the advanced ones consider two signals or actions: "sell" or equivalently "stay-short-in-the-market", and "buy" or equivalently "stay-long-in-the-market". In this paper we consider also a third signal or action: "stay-out-from-the-market". By doing so, we give to our RL-based FTS the possibility to take no position, or to leave a given position, when the buy/sell signal appears weak. Note that the set of the action is finite and discrete;
- As state descriptors of the considered financial market we consider the simple current and some past returns. Generally, this is not the case for several FTSs. Anyway, we have made this choice in order to check the performance capability of our FTS also starting from not particularly sophisticated information. Notice that any state descriptor is a continuous variable over $\mathbb{R}$.

3 Some Basic on the RL

In this Section we present the basic aspects of the RL methodology which are of interest to our purposes.

To learn by directly interacting with the environment[1] without the need of a supervisor is likely the more immediate idea about the nature of the learning itself. The consequences of the actions[2] of the learning agent[3] lead she/he to choose what are the actions that allow to obtain the desired results and what are the ones to avoid. RL formalizes this kind of learning by maximizing a numerical reward[4] (see [1]). The agent has to discover which actions yield the most reward by trying them. RL is different from Supervised Learning (SL) in which the agent learns from past examples provided by an external supervisor. SL is an important kind of learning, but in interactive problems it is often impractical to obtain examples of desired current and future behavior that are representative of the situation in which the agent will have to act.

To formalize these first ideas, let us consider a system observed at discrete time in which the state[5] at time t, $s_t \in \mathscr{S}$, summarizes all information concerning the system available to the agent. In the RL framework it is assumed that the system satisfies the Markov property, that is that the probability of transition from the actual state s_t to the next one s_{t+1} depends only on the current state s_t. On the basis of s_t, the agent selects an action $a_t \in \mathscr{A}(s_t)$, where $\mathscr{A}(s_t)$ is the set of all the possible actions the agent can take given the state s_t. At time $t+1$ the agent receives a reward, $r(s_t, a_t, s_{t+1}) \in \mathbb{R}$, as consequence of her/his actions a_t and of the new state s_{t+1} in which she/he finds herself/himself. The reward is a numerical representation of the satisfaction of the agent. Generally, at time t the agent wish to maximize the expected value of some global return, $R(s_t)$, which is defined as function of the actual reward and of the future suitably discounted ones. This function can be written as:

---

[1] In our case a financial market.
[2] In our case stay-short-in-the-market, stay-out-from-the-market, and stay-long-in-the-market.
[3] In our case a FTS.
[4] In our case the performances of the FTS.
[5] In our case the current and some past returns.

$$R(s_t) = r(s_t, a_t, s_{t+1}) + \gamma r(s_{t+1}, a_{t+1}, s_{t+2}) + \gamma^2 r(s_{t+2}, a_{t+2}, s_{t+3}) + \ldots,$$

where $\gamma \in (0,1)$ is the discount factor.

In RL, a policy $\pi(s_t) = a_t$ is a mapping from states to actions defining the choice of action a_t given state s_t. In order to maximize the expected $R(s_t)$, RL searches for a suitable policy. Considering also the policy $\pi(\cdot)$, we can rewrite the global return as:

$$R^\pi(s_t) = r(s_t, \pi(s_t), s_{t+1}) + \gamma r(s_{t+1}, \pi(s_{t+1}), s_{t+2}) + \gamma^2 r(s_{t+2}, \pi(s_{t+2}), s_{t+3}) + \ldots.$$

3.1 The Value Functions

Generally, RL-based approaches[6] need the estimation of the so-called *value functions*. These functions of states (or of state-action pairs) attribute a value to each state s_t (or to each state-action pairs) proportional to the rewards achievable in the future from the current state s_t (or from the current state-action pairs). They evaluate how good is for the agent to be in a given state (or how good to perform a given action in a given state). The notion of "how good" is defined in terms of R_t. In particular, the value of a state $s_t = s$ following policy $\pi(\cdot)$ is the expected sum of the current and of the future discounted rewards when starting in state $s_t = s$, and thereafter following policy $\pi(\cdot)$, that is:

$$V^\pi(s) = \mathbb{E}\left[R^\pi(s_t) | s_t = s\right].$$

Similarly, the value of tacking action $a_t = a$ being in state $s_t = s$ under policy $\pi(\cdot)$ is the expected sum of the current and of the future discounted rewards starting from state $s_t = s$, taking action $a_t = a$, and thereafter following policy $\pi(\cdot)$, that is:

$$Q^\pi(s,a) = \mathbb{E}\left[R^\pi(s_t) | s_t = s, a_t = a\right].$$

A fundamental property of value functions is that they satisfy particular recursive relationships. In particular, for any policy $\pi(\cdot)$ and any state $s_t = s$, the following consistency condition holds between the value of s_t and the value of any possible successor state s_{t+1}:

$$V^\pi(s) = \mathbb{E}\left[r(s_t, \pi(s_t), s_{t+1}) + \gamma V^\pi(s_{t+1}) | s_t = s\right]. \tag{1}$$

Equation (1) is the so-called *Bellman equation* for $V^\pi(s_t)$. It is possible to prove that the value $V^\pi(s)$ is the unique solution to its Bellman equation.

3.2 The Generalized Policy Iteration

The task of the RL methodologies consists in finding an optimal policy. The optimal policy identifies the values $V^*(s)$ and $Q^*(s,a)$ such that

$$V^*(s) = \max_\pi V^\pi(s) \text{ and } Q^*(s,a) = \max_\pi Q^\pi(s,a) \;\forall\, s \in \mathcal{S} \text{ and } \forall\, a \in \mathcal{A}(s). \tag{2}$$

[6] In our case the financial trading strategies.

Since $V^*(s)$ is the value function for a policy, it satisfy the Bellman equation (1). As it is also the optimal value function, $V^*(s)$'s Bellman condition can be written in a special form without reference to any specific policy. This form is the Bellman equation for $V^*(s)$, or the *Bellman optimality equation*, which expresses the fact that the value of a state under an optimal policy must equal the expected global return for the best action from the state itself, that is:

$$V^*(s) = \max_a Q^*(s,a) = \mathbb{E}[R^*(s_t)|s_t = s, a_t = a]. \tag{3}$$

With equivalent arguments, the Bellman optimality equation for $Q^*(s,a)$ is:

$$Q^*(s,a) = \mathbb{E}\left[r(s_t, a_t, s_{t+1}) + \gamma \max_{a'} Q^*(s_{t+1}, a')|s_t = s, a_t = a\right].$$

At this point, it is possible to iteratively calculate the value function for a state (or for a state-action pair). Let $V_0^\pi(s_t)$ for all $s_t = s \in \mathscr{S}$ be an arbitrarily initialization of the state value function. Each successive approximation is obtained by using as follows the Bellman equation in an update rule:

$$\widehat{V}_{k+1}^\pi(s_t) = \mathbb{E}\left[r(s_t, a_t, s_{t+1}) + \gamma \widehat{V}_k^\pi(s_{t+1})\right] \; \forall \, s_t = s \in \mathscr{S}. \tag{4}$$

If the expectation $\widehat{V}_{k+1}^\pi(s_t)$ exists, then $\lim_{k \to +\infty} \widehat{V}_k^\pi(s) = V^\pi(s)$.

The reason for computing the value functions for a policy is to find better policies in order to increase the expected value of the global returns. This process is called *policy improvement*.

As described, the policy improvement process requires the evaluation of the previous policy. This evaluation can be made by (4), which is itself an iterative process that converges in the limit. Fortunately, there is no need to wait for the exact convergence or for a particularly high level of convergence. In fact one can stop the policy evaluation iteration in several ways without losing the convergence (see [1]). An important special case is when policy evaluation is stopped after just one step. In particular, it is possible to combine the evaluation process and the improvement one by stopping the policy evaluation process at each step and then by improving the policy itself. This mixed algorithm is called *generalized policy iteration*. It can be written as:

$$\widehat{V}_{k+1}^\pi(s_t) = \max_a \mathbb{E}\left[r(s_t, a, s_{t+1}) + \gamma \widehat{V}_k^\pi(s_{t+1})\right], \tag{5}$$

where $\widehat{V}_{k+1}^\pi(s_t)$ is the update estimate with the improved policy at step $k+1$ with respect to the previous estimate and the previous policy at step k.

In order to improve the policy we choose an approach, among the ones presented in literature, which may produce increasing of the global return in the long run. Following such an approach, the choice of the action at each time t is given by:

$$a_t = \begin{cases} \pi'(s_t) & \text{with probability } 1 - \varepsilon \\ a \in \mathscr{A}(s_t) & \text{with probability } \varepsilon \end{cases},$$

where $\varepsilon \in (0,1)$ and $\pi'(s_t)$ is the candidate action which maximizes $Q^\pi(s,a)$.

In the next subsections we shortly introduce the two different approaches we use in our FTSs for calculating the expected value (5). Notice that we can not take into account other kinds of approaches like the *Dynamic Programming*-based ones and the *Monte Carlo*-based ones. In fact:

- The former need models to calculate the true probabilities of transition from a state to another one, whereas in financial trading such a model is generally not known or not available;
- The latter, in order to improve the policy, need to wait for until the end of all the trades, whereas a real FTS trades an a priori indefinite number of times.

3.3 TD Methods and the Q-Learning Algorithm

In this subsection we present a class of (trading) policy evaluation algorithms which update step by step the estimate of $V(s_t)$. First of all one puts in evidence that it is possible to write $\widehat{V}_{k+1}(s_t)$ in the following recursive way:

$$\widehat{V}_{k+1}(s_t) = \frac{1}{k+1}\sum_{j=1}^{k+1} R_j(s_t) = \frac{1}{k+1}\left[R_{k+1}(s_t) + \sum_{j=1}^{k} R_j(s_t)\right] = \cdots =$$
$$= \widehat{V}_k(s_t) + \alpha_k\left[R_{k+1}(s_t) - \widehat{V}_k(s_t)\right],$$

where $\alpha_k = 1/(k+1)$ is the so-called learning rate.

The TD methods can update the estimate $\widehat{V}_{k+1}(s_t)$ as soon as the quantity

$$d_k = R_{k+1}(s_t) - \widehat{V}_k(s_t) = r(s_t, s_{t+1}) + \gamma\widehat{V}_k(s_{t+1}) - \widehat{V}_k(s_t),$$

becomes available. Therefore the above recursive relationship can be rewritten as:

$$\widehat{V}_{k+1}(s_t) = \widehat{V}_k(s_t) + \alpha_k\left[r(s_t, s_{t+1}) + \gamma\widehat{V}_k(s_{t+1}) - \widehat{V}_k(s_t)\right].$$

With regard to the convergence of the TD methods, it is possible to prove that, if the sufficient conditions $\sum_{k=1}^{+\infty}\alpha_k = +\infty$ and $\sum_{k=1}^{+\infty}\alpha_k^2 < +\infty$ hold, then $\lim_{k\to+\infty}\Pr\left\{\left|\widehat{V}_k(s_t) - -V(s_t)\right| < \varepsilon\right\} = 1$, for any $\varepsilon > 0$ (see [4]).

The TD methods are "naturally" developed in an incremental and on-line fashion which make them particularly appealing for building FTSs.

In literature there exist several different TD methods, the most widespread of whom is the *Q-Learning algorithm* (QLa). The QLa is an *off*-policy control method, where "off" indicates that two different policies are used in the policy improvement process: A first one is used to estimate the value functions, and another one is used to control the improvement process. It is possible to prove that the so-obtained state-action value function is given by:

$$\widehat{Q}_{k+1}(s_t, a_t) = \widehat{Q}_k(s_t, a_t) + \alpha_k\left[r_{t+1} + \gamma\max_a \widehat{Q}_k(s_{t+1}, a_{t+1}) - \widehat{Q}_k(s_t, a_t)\right]. \quad (6)$$

At this point we recall that the states of interest taken into account for the development of our FTSs are continuous variables over $\mathbb{R}$. In this case it is possible to prove that the value function at step k, $\widehat{V}_k(s_t)$, can now be approximated by a parameterized functional form with parameter vector θ_k at step k. It involves that the associated value function $\widehat{V}_k(s_t) = \widehat{V}_k(s_t;\theta_k)$ totally depends on θ_k, which varies step by step. In order to determine the optimal parameter vector, θ^*, which minimizes the "distance" between the unknown $V^\pi(s_t)$ and its estimate $\widehat{V}^\pi(s_t;\theta_k)$, in most machine learning approaches the minimization of the mean square error is used, that is:

$$\min_{\theta_k} \sum_s \left[V^\pi(s) - \widehat{V}^\pi(s;\theta_k)\right]^2.$$

The convergence of θ_k to θ^* is proven for approximators characterized by simple functional forms like the affine ones (see [4]). Among the linear functional forms, in building our FTSs we use the following one:

$$\widehat{V}^\pi(s;\theta) = \sum_{i=1}^{n} \theta_i \phi_i(s_i) = \theta' \phi(s), \tag{7}$$

where n is the number of considered states, and $\phi_i(\cdot)$ is a suitable transformation of the i-th state.

Under mild assumptions it is possible to prove that the update rule to use for estimating the state-action value function in the case of continuous states becomes:

$$\begin{aligned} d_k = r(s_t,a_t,s_{t+1}) + \gamma \max_a \widehat{Q}(s_{t+1},a;\theta_k) - \widehat{Q}^\pi(s_t,a_t;\theta_k) \text{ and } \theta_{k+1} = \\ = \theta_k + \alpha d_k \nabla_{\theta_k} \widehat{Q}^\pi(s_t,a_t;\theta_k). \end{aligned} \tag{8}$$

3.4 The KbRL Algorithm

Another method for approximating $Q^\pi(s,a)$, alternative to the QLa and also usable in case of continuous states, is the nonparametric regression-based one known as KbRL algorithm (see [17] and [18]). Given a kernel $K(\cdot)$ and defined $q_t = (s_t\ a_t)$, the KbRL algortihm estimates $Q(s,a)$ as follows:

$$\widehat{Q}_k(q_t) = \sum_{i=1}^{t-1} p_i(q_t) \widehat{Q}_k(q_i),$$

where $p_i(q_t) = K\left(\frac{q_t - q_i}{h}\right) \Big/ \sum_{i=1}^{t-1} K\left(\frac{q_t - q_i}{h}\right)$.

Note that the choice of the kernel $K(\cdot)$ is not crucial, whereas the choice of the bandwidth h has to be done carefully (see [6]).

In this new reference methodological framework, the approximation relationship for $\widehat{Q}_{k+1}(q_t)$ is given by:

$$\widehat{Q}_{k+1}(q_i) = \widehat{Q}_k(q_i) + p_i(q_t)\left[\widehat{Q}_{k+1}(q_t) - \widehat{Q}_k(q_i)\right],\ i = 1,2,\ldots,t-1. \tag{9}$$

This approach is particularly interesting because constitutes a kind of minimization method without derivatives (see [5]). Further, it is also usable in non-stationary contexts as the update relationship (9) is based on the current values of state-action pairs.

4 The RL-Based FTSs

In this Section we present our RL-based FTSs, we provide the results of their applications to an artificial time series and to five real ones, and we compare these results with those coming from a classical FTS.

4.1 Our RL-Based FTSs and the Applications

In this Subsection we use the QLa and the KbRL algorithm for developing daily FTSs. First one has to identify the quantities which specify the states, the possible actions of the FTSs, and the reward function.

With regards to the states, recalling that we are interested in checking the performance capability of our FTSs also starting from not particularly sophisticated information, we simply use as states the current and the past four percentage returns of the asset to trade. So, given the current price of the asset, p_t, the state of the system at the time t is given by the following vector:

$$s_t = (e_{t-4},\, e_{t-3},\, e_{t-2},\, e_{t-1},\, r_t),$$

where $e_\tau = \dfrac{p_\tau - p_{\tau-1}}{p_{\tau-1}}$.

Further, as we are also interested in checking the applicability of the considered RL-based methods to the development of effective FTSs, we do not consider frictional aspects.

Concerning the possible action of the FTS, we utilize the three following ones:

$$a_t = \begin{cases} -1 \text{ (sell or stay-short-in-the-market signal)} \\ 0 \text{ (stay-out-from-the-market signal)} \\ 1 \text{ (buy or stay-long-in-the-market signal)} \end{cases},$$

in which the stay-out-from-the-market implies the closing of whatever previously open position (if any)[7].

Finally, with reference to the reward function, following [14] we take into account the well known Sharpe ratio, that is:

$$Sr_t = \frac{\mathrm{E}_L\left[g_{t-1}\right]}{\sqrt{\mathrm{Var}_L\left[g_{t.1}\right]}},$$

where $\mathrm{E}_L(\cdot)$ and $\mathrm{Var}_L(\cdot)$ are, respectively, the sample mean operator and the sample variance one calculated in the last L stock market days, and $g_t = a_{t-1}e_t$ is the gained/lost percentage return obtained at time t as a consequence of the action undertaken by the FTS at time $t-1$.

In particular, as we wish an indicator that reacts enough quickly to the consequences of the actions of the considered FTSs, we calculated Sr_t only by using the last $L = 5$ stock market days (a stock market week), and the last $L = 22$ stock market days (a stock market month).

[7] Notice that in most of the prominent literature, like for instance in [14], only the sell and the buy signals are considered.

Now let us pass to the two RL-based approaches we consider: The QLa and the KbRL algorithm. With respect the QLa, the kind of linear approximator of the state-action value function we choose is:

$$Q(s_t, a_t; \theta_k) = \theta_{k,0} + \sum_{n=1}^{5} \theta_{k,n} \tanh(s_{t,n}) + \theta_{k,6} \tanh(a_t),$$

in which $\tanh(\cdot)$ plays the role of transformer of the state.
Then, the ε-greedy function we consider is:

$$a_t = \begin{cases} \arg\max_{a_t} Q(s_t, a_t; \theta_k) & \text{with probability } 1 - \varepsilon \\ u & \text{with probability } \varepsilon \end{cases},$$

in which $\varepsilon \in \{2.5\%, 5.0\%, 7.5\%\}$, and $u \sim \mathscr{U}_d(-1,1)$.

Concerning the KbRL algoritm, we follow [6] and [17] (a reasonable setting of the bandwidth h has be done through some trial-and-error experiments).

Summarizing, we consider two different RL-based approaches (QLa and KbRL algorithm), three different values of ε (2.5%, 5.0% and 7.5%), and two different values of L (5 and 22), for a total of twelve different configurations.

We apply the so specified RL-based FTSs to six different time series of daily prices: An artificial one and five real ones. With reference to the artificial time series, as in [14] we generate log-price series as a random walks with autoregressive trend processes. In particular, we used the model:

$$p_t = \exp\left\{\frac{z_t}{\max z - \min z}\right\},$$

where $z_t = z_{t-1} + \beta_{t-1} + 3a_t$, in which $\beta_t = 0.9\beta_{t-1} + b_t$, $a_t \sim \mathscr{N}(0,1)$, and $b_t \sim \mathscr{N}(0,1)$. The length of the so-generated series is $T = 5{,}000$.
This artificial price series shows features which are often present in real financial price series. In particular, it is trending on short time scales and has a high level of noise.

As far as the real time series regards, we utilize the closing prices of Banca Intesa, Fiat, Finmeccanica, Generali Assicurazioni, and Telecom Italia (from the Italian stock market), from January 1, 1973 to September 21, 2006. The length of these series is $T = 5{,}400$, more than 21 stock market years.

At this point we can present the results of the applications of the variously configured FTSs. In all the applications we set $\alpha = 0.8$ and $\gamma = 0.7$.

In figure 1 we graphically report the results of the application of the QLa-based FTS to the real price series of Banca Intesa, with $\varepsilon = 5\%$ and $L = 5$. In particular: The first panel shows the price series, the second panel shows the actions taken by the FTS at each time, the third panel shows the rewards, that is the Sharpe ratios, at each time, and the fourth panel shows the cumulative return one should obtain by investing the same monetary amount at each time. At the end of the trading period, $t = T$, the cumulative return is 271.42%.

It is important to note that at the beginning of the trading period, $t = 0$, the vector of the parameters used in the linear approximator, θ_k, is randomly initialized. Because of it, by repeating more times the same application we observe a certain variability in the final cumulative return. With regards to the application to the real price series of Banca

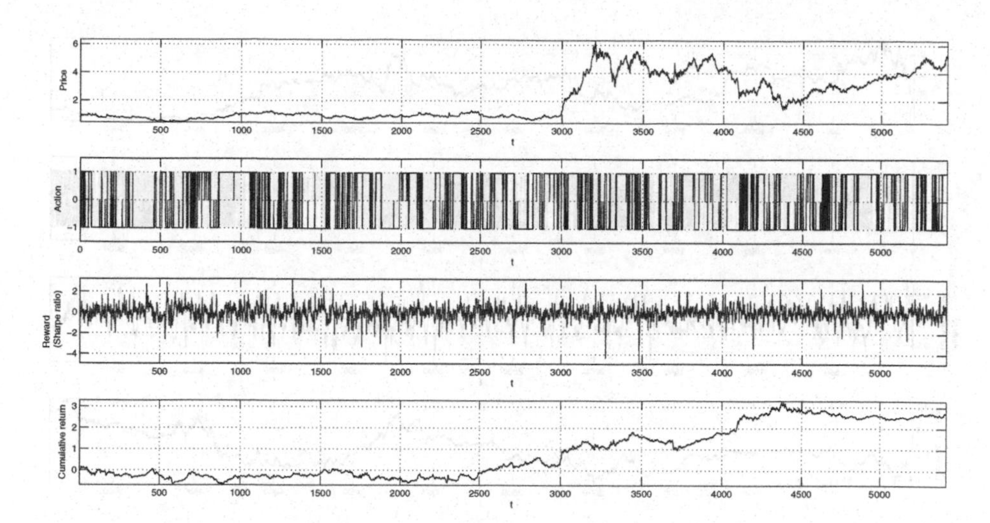

Fig. 1. Results of the *Q*La-based FTS applied to the Banca Intesa price series, with $\varepsilon = 5\%$ and $L = 5$. Final cumulative returns: 271.42%.

Intesa, some of the final cumulative returns we obtain are: -158.52%, 174.91%, and -26.38%. This shows that the influence of the random initialization heavily spreads overall the trading period instead to soften as time step increases. To check the effects of this random initialization, we repeated 1,000 times the application of each of the investigated configurations. When the application was to the artificial time series, the series has been taken the same in all the repetition. In tables 1 to 4 we report some statistics concerning the final cumulative returns.

With reference to the KbRL algorithm, we obtain results similar to the one related to the *Q*La, although a bit less performing. In figure 2 we graphically report the results of the application of the KbRL-based FTS to the real price series of Fiat, with $\varepsilon = 7.5\%$ and $L = 22$ (at the end of the trading period, $t = T$, the cumulative return is 135.47%). In particular, note that for the KbRL algorithm there is not parameter vector to randomly initialize, but at the beginning of the trading period, $t = 0$, it is necessary to randomly initialize a suitable set of state-action pairs (see [17]). So, also in this approach one puts the question of the variability of the results. As for the *Q*La, for the KbRL algorithm too we repeated 1,000 times the application of each of the investigated configurations (see tables 1 to 4 for some statistics about the final cumulative returns).

The main facts detectable from tables 1 to 4 are the following ones:

- Given the values of the means:
  - Generally, the investigated configurations appears to be profitable, in fact the most of the means (88.89%) are positive. Further, the *Q*La approach seems more performing than the KbRL one;
  - It appears that the value of L has a significant impact on the performances of the FTSs. In particular, with reference to the artificial time series, all the checked FTS configurations are better performing when $L = 5$, whereas, with reference

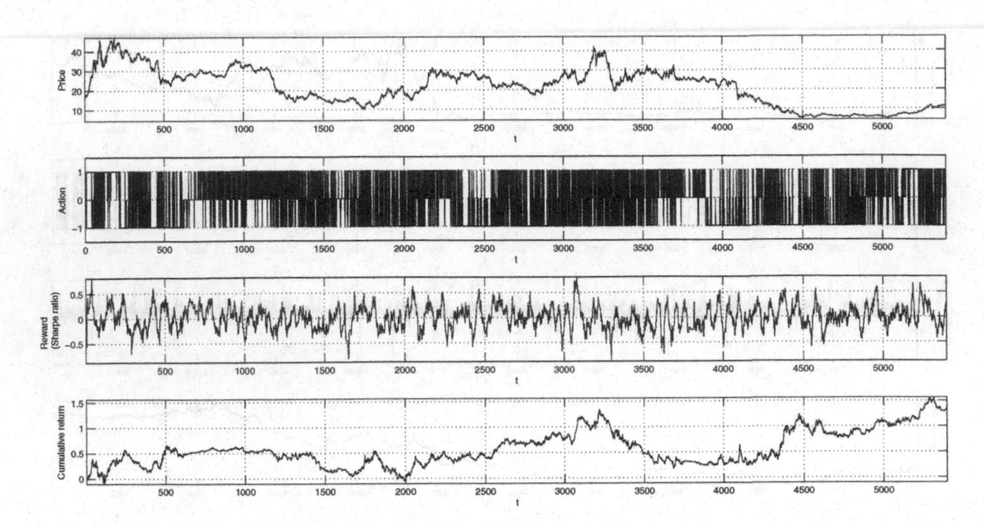

Fig. 2. Results of the KbRL-based FTS applied to the Fiat price series, with $\varepsilon = 7.5\%$ and $L = 22$. Final cumulative returns: 135.47%.

to all the real time series, the most of the checked FTS configurations (75.00%) are better performing when $L = 22$;

∘ It appears that the value of ε too has a significant impact on the performances of the FTSs. In particular, although with reference to the artificial time series empirical regularities does not come into view, with reference to all the real time series, the most of the checked FTS configurations (85.00%) are better performing when $\varepsilon = 7.5\%$;

- Given the values of the standard deviations, the results of all the considered variously configurated FTSs are characterized by a certain level of variability. In particular, such values emphasize that the question of the influence of the random initialization on the results is mainly true for the real financial time series.

4.2 The Comparison

In this Subsection we compare the results coming from our RL-based FTSs with those coming from a classical FTS based on Technical Analysis (TA) indicators. The choice of the latter FTS as term of comparison is due to (at least) two reasons: first, TA, together with Fundamental Analysis, is one of the most used "toolbox" among professionals and practitioners for the building of FTSs; second, in the last 10-15 years also the academic world has begun to recognize the soundness of some of the features related to the TA (see [11]).

Table 1. Some statistics about the final cumulative returns

Method	ε	L	Statistics	Artificial time series	Banca Intesa time series	Fiat time series
QLa	2.5%	5	μ	443.62%	−3.77%	40.58%
			σ	37.83%	127.48%	139.35%
			Conf. interval	[369.48%, 517.76%]	[−253.53%, 246.08%]	[−232.55%, 313.70%]
QLa	2.5%	22	μ	307.03%	70.85%	101.21%
			σ	43.02%	146.34%	144.27%
			Conf. interval	[222.71%, 391.35%]	[−215.97%, 357.68%]	[−181.55%, 383.98%]
QLa	5.0%	5	μ	472.02%	40.00%	91.26%
			σ	32.79%	139.09%	137.30%
			Conf. interval	[407.76%, 536.29%]	[−226.73%, 306.73%]	[−177.85%, 360.37%]
QLa	5.0%	22	μ	337.68%	92.92%	124.69%
			σ	40.80%	149.28%	144.89%
			Conf. interval	[257.71%, 417.64%]	[−199.66%, 385.50%]	[−159.29%, 408.68%]
QLa	7.5%	5	μ	467.48%	49.15%	114.41%
			σ	31.11%	139.71%	142.42%
			Conf. interval	[406.50%, 528.46%]	[−224.67%, 322.97%]	[−164.73%, 393.56%]
QLa	7.5%	22	μ	339.04%	99.66%	133.81%
			σ	39.60%	146.74%	147.53%
			Conf. interval	[261.43%, 416.66%]	[−187.95%, 387.28%]	[−155.34%, 422.96%]

Table 2. Some statistics about the final cumulative returns

Method	ε	L	Statistics	Artificial time series	Banca Intesa time series	Fiat time series
KbRL	2.5%	5	μ	483.71%	−26.16%	73.33%
			σ	60.10%	141.37%	144.00%
			Conf. interval	[365.91%, 601.51%]	[−303.25%, 250.93%]	[−208.90%, 355.56%]
KbRL	2.5%	22	μ	237.64%	15.90%	75.37%
			σ	63.55%	159.92%	131.08%
			Conf. interval	[113.09%, 362.20%]	[−297.54%, 329.34%]	[−181.55%, 332.28%]
KbRL	5.0%	5	μ	435.42%	−7.99%	77.32%
			σ	41.13%	131.26%	130.42%
			Conf. interval	[354.81%, 516.02%]	[−265.26%, 249.28%]	[−178.30%, 332.95%]
KbRL	5.0%	22	μ	216.61%	13.20%	71.98%
			σ	49.98%	153.34%	136.04%
			Conf. interval	[118.64%, 314.58%]	[−287.34%, 313.74%]	[−194.66%, 338.61%]
KbRL	7.5%	5	μ	401.76%	−0.63%	67.55%
			σ	39.80%	130.40%	128.00%
			Conf. interval	[323.75%, 479.77%]	[−256.23%, 254.96%]	[−183.33%, 318.42%]
KbRL	7.5%	22	μ	197.78%	35.36%	74.82%
			σ	42.55%	143.51%	132.06%
			Conf. interval	[114.38%, 281.18%]	[−245.91%, 316.63%]	[−184.02%, 333.65%]

Table 3. Some statistics about the final cumulative returns

Method	ε	L	Statistics	Finmeccanica time series	Generali Assicurazioni time series	Telecom Italia time series
*Q*La	2.5%	5	μ	177.03%	1.04%	288.79%
			σ	171.97%	99.12%	168.71%
			Conf. interval	[−160.03%,514.09%]	[−193.24%,195.31%]	[−41.88%,619.46%]
*Q*La	2.5%	22	μ	161.77%	31.70%	246.44%
			σ	176.86%	103.50%	165.71%
			Conf. interval	[−184.87%,508.40%]	[−171.15%,234.55%]	[−78.35%,571.23%]
*Q*La	5.0%	5	μ	269.77%	2.69%	334.86%
			σ	174.92%	101.54%	167.66%
			Conf. interval	[−73.08%,612.62%]	[−196.33%,201.70%]	[6.25%,663.48%]
*Q*La	5.0%	22	μ	217.58%	30.60%	256.90%
			σ	173.14%	110.94%	179.84%
			Conf. interval	[−121.77%,556.93%]	[−186.85%,248.05%]	[−95.59%,609.39%]
*Q*La	7.5%	5	μ	295.33%	21.69%	335.63%
			σ	163.18%	104.54%	163.96%
			Conf. interval	[−24.50%,615.16%]	[−183.21%,226.59%]	[14.26%,657.00%]
*Q*La	7.5%	22	μ	227.32%	47.56%	279.09%
			σ	168.45%	110.49%	168.09%
			Conf. interval	[−102.85%,557.49%]	[−169.00%,264.12%]	[−51.54%,609.73%]

Table 4. Some statistics about the final cumulative returns

Method	ε	L	Statistics	Finmeccanica time series	Generali Assicurazioni time series	Telecom Italia time series
KbRL	2.5%	5	μ	61.20%	−13.35%	77.46%
			σ	165.83%	122.98%	190.70%
			Conf. interval	[−263.81%,386.22%]	[−254.40%,227.69%]	[−296.31%,451.22%]
KbRL	2.5%	22	μ	6.41%	−26.78%	26.82%
			σ	170.46%	126.01%	183.79%
			Conf. interval	[−327.68%,340.50%]	[−273.76%,220.20%]	[−333.40%,387.05%]
KbRL	5.0%	5	μ	80.34%	10.77%	93.43%
			σ	162.01%	120.89%	187.08%
			Conf. interval	[−237.19%,397.87%]	[−226.17%,247.71%]	[−273.25%,460.12%]
KbRL	5.0%	22	μ	29.58%	−9.44%	42.94%
			σ	170.01%	120.57%	188.15%
			Conf. interval	[−303.65%,362.81%]	[−245.75%,226.88%]	[−325.82%,411.71%]
KbRL	7.5%	5	μ	107.97%	20.62%	102.27%
			σ	155.40%	119.64%	189.44%
			Conf. interval	[−196.61%,412.55%]	[−213.89%,255.12%]	[−269.04%,473.57%]
KbRL	7.5%	22	μ	38.24%	−3.31%	34.85%
			σ	158.94%	121.46%	190.09%
			Conf. interval	[−273.28%,349.75%]	[−241.38%,234.75%]	[−337.73%,407.43%]

The TA-based FTS we consider as benchmark utilizes the following five classical IF-THEN-ELSE rules:

Rule 1

IF $EMA3 > EMA12$ THEN (buy OR stay-long-in-the-market)
ELSE IF $EMA3 < EMA12$ THEN (sell OR stay-short-in-the-market)
ELSE stay-out-of-the-market

where $EMAm$ is the exponential moving average calculated by using the last m daily stock prices;

Rule 2

IF ($MACD > 0$ AND $MACD > Signal\ line$) THEN (buy OR stay-long-in-the-market)
ELSE IF ($MACD < 0$ AND $MACD < Signal\ line$) THEN (sell OR stay-short-in-the-market)
ELSE stay-out-of-the-market

where $MACD = EMA12 - EMA26$ is the moving average convergence/divergence, and $Signal\ line = EMA9$;

Rule 3

IF $RSI < 30$ THEN (buy OR stay-long-in-the-market)
ELSE IF $RSI > 70$ THEN (sell OR stay-short-in-the-market)
ELSE stay-out-of-the-market

where RSI is the relative strength index calculated by using the last 14 daily stock prices;

Rule 4

IF $ROC < -1$ THEN (buy OR stay-long-in-the-market)
ELSE IF $ROC > 1$ THEN (sell OR stay-short-in-the-market)
ELSE stay-out-of-the-market

where ROC is the rate of change calculated by using the last 12 daily stock prices;

Rule 5

IF $TSI > 0$ THEN (buy OR stay-long-in-the-market)
ELSE IF $TSI < 0$ THEN (sell OR stay-short-in-the-market)
ELSE stay-out-of-the-market

where TSI is the true strength index calculated by using the last 13 and the last 25 daily stock prices.

If three or more of the listed rules "propose" the same action then the considered TA-based FTS takes that action, else it stays-out-of-the-market (see for more details on TA [15]).

In table 5, in the second row we report the results (in terms of final returns) of the applications of the specified TA-based FTS to the six time series of daily stock prices considered in Subsection 4.1, and in the third row we report, for each of the same time series, the percentage of times in which the variously configured RL-based FTSs have over-performed the TA-based FTS.

Table 5. Results about the comparison

	Artificial time series	Banca Intesa time series	Fiat time series	Finmeccanica time series	Generali A. time series	Telecom Italia time series
Final return	2.11%	2.06%	2.40%	1.74%	0.53%	4.57%
%	100.00%	66.67%	100.00%	100.00%	66.67%	100.00%

Given these results:

- Only for two time series of daily stock prices, the ones related to Banca Intesa and to Generali Assicurazioni, the TA-based FTS over-performs some of the configurations of the RL-based FTSs;
- Considering all the results as a whole, the RL-based FTSs over-perform the TA-based FTS in the 88.89% of the cases.

All this seems to show the goodness of our approaches, at least with respect to the chosen TA-based benchmark.

5 Some Concluding Remarks

In this paper we have developed and applied some original automated FTSs based on differently configured RL-based algorithms. Here we have presented the results coming out from the current phase of our research on this topic. Of course, many questions have again to be explored. In particular:

- The choice of percentage returns as states is a naive choice. Now we are beginning to work to specify some new indicators to use as states (in the first experimentations they have provided interesting results);
- As known, the Sharpe ratio as performance measure suffers several financial limits. Currently, as reward function we are considering alternative and more realistic performance measures;
- The management of the learning rate, α, we have used here is appropriate for stationary systems. But generally financial markets are non-stationary. Because of that, we are beginning to work to develop methods for the dynamic management of the learning rate in non-stationary contexts;
- In order to deepen the valuation about the capabilities of our FTSs, we wish to apply them to more and more financial price series coming from different markets;
- Finally, when all the previous questions will be explored, transaction costs and other frictions will be considered.

Acknowledgements. The authors wish to thank the Department of Economics of the Ca' Foscari University of Venice for the support received within the research project *Machine Learning adattativo per la gestione dinamica di portafogli finanziari* [*Adaptative Machine Learning for the dynamic management of financial portfolios*]).

References

1. Barto, A.G., Sutton, R.S.: Reinforcement Learning: An Introduction. Adaptive Computation and Machine Learning. The MIT Press (1998)
2. Bekiros, S.D.: Heterogeneous trading strategies with adaptive fuzzy Actor-Critic reinforcement learning: A behavioral approach. Journal of Economic Dynamics & Control 34, 1153–1170 (2010)
3. Bertoluzzo, F., Corazza, M.: Making financial trading by recurrent reinforcement learning. In: Apolloni, B., Howlett, R.J., Jain, L. (eds.) KES 2007/WIRN 2007, , Part II. LNCS (LNAI), vol. 4693, pp. 619–626. Springer, Heidelberg (2007)
4. Bertsekas, D.P., Tsitsiklis, J.N.: Neuro-Dynamic Programming. Athena Scientific (1996)
5. Brent, R.P.: Algorithms for Minimization without Derivatives. Prentice-Hall (1973)
6. Bosq, D.: Nonparametric Statistics for Stochastic Processes. Estimation and Prediction, vol. 110. Springer (1996)
7. Cuthbertson, K., Nitzsche, D.: Quantitative Financial Economics. Wiley (2004)
8. Farmer, D., Lo, A.W.: Market force, ecology and evolution. Industrial and Corporate Change 11, 895–953 (2002)
9. Gold, C.: FX trading via recurrent Reinforcement Learning. In: Proceedings of the IEEE International Conference on Computational Intelligence in Financial Engineering, pp. 363–370 (2003)
10. Li, H., Dagli, C.H., Enke, D.: Short-term stock market timing prediction under reinforcement learning schemes. In: Proceedings of the 2007 IEEE Symposium on Approximate Dynamic Programming and Reinforcement Learning, pp. 233–240 (2007)
11. Lo, A.W., Mamaysky, H., Wang, J.: Foundations of technical analysis: Computational algorithms, statistical inference, and empirical (2000)
12. Lo, A.W.: The Adaptive Markets Hypothesis. Market efficiency from an evolutionary perspective. The Journal of Portfolio Management 30, 15–29 (2004)
13. Moody, J., Wu, L., Liao, Y., Saffel, M.: Performance functions and Reinforcement Learning for trading systems and portfolios. Journal of Forecasting 17, 441–470 (1998)
14. Moody, J., Saffel, M.: Learning to trade via Direct Reinforcement. IEEE Transactions on Neural Network 12, 875–889 (2001)
15. Murphy, J.J.: Technical Analysis of the Financial Markets. A Comprehensive Guide to Trading Methods and Applications. New York Institute of Finance (1999)
16. Jangmin, O., Lee, J., Lee, J.W., Zhang, B.-T.: Adaptive stock trading with dynamic asset allocation using reinforcemnt learning. Information Sciences 176, 2121–2147 (2006)
17. Ormonet, D.: Kernel-Based Reinforcement Learning. Machine Learning 49, 161–178 (2002)
18. Smart, W.D., Kaelbling, L.P.: Practical Reinforcement Learning in continuous spaces. In: Proceedings of the 17th International Conference on Machine Learning, pp. 903–910 (2000)

A Collaborative Filtering Recommender Exploiting a SOM Network

Giuseppe M.L. Sarnè

DICEAM, University "Mediterranea" of Reggio Calabria
Loc. Feo di Vito, 89122 Reggio Calabria, Italy
sarne@unirc.it

Abstract. Recommender systems are exploited in many fields for helping users to find goods and services. A collaborative filtering recommender realizes a knowledge-sharing system to find people having similar interests. However, some critical issues may lead to inaccurate suggestions. To provide a solution to such problems, this paper presents a novel SOM-based collaborative filtering recommender. Some experimental results confirm the effectiveness of the proposed solution.

1 Introduction

To provide users with attractive suggestions, recommender systems can adopt *Content-based* CB (exploiting past users' interests [15]), *Collaborative Filtering*, CF (using knowledge-sharing techniques to find people similar for interests [2]) approaches or a their combination (*Hybrid recommenders* [4]). In particular, CF recommenders generate suggestions by computing similarities between users or items based on: (*i*) users' ratings or automatic elicitation (memory-based approach [9]); (*ii*) data mining or machine learning algorithms (model-based approach [32]). A common solution, to save computational resources for adopting more complex CF algorithms, is the off-line computation of users' similarity and suggestions [27] but recurring updating are required to avoid mismatching.

In such a context, this paper proposes the Social Relevance-based CF (SRCF) recommender (*i*) to pre-compute suggestions based on a novel CF algorithm using the concept of "social relevance" of an item and (*ii*) a *Self-Organizing Map* (SOM) network to cluster similar users in the respect of their privacy. Some experiments confirm the effectiveness of the proposed solution.

2 The Knowledge Representation

SRCF adopts a compact knowledge representation (see Figure 1) to describe user's interests and preferences in a *User Profile* UP [8,19,20] on the basis of a common *Dictionary*, called $\mathcal{D}$, implemented by an XML Schema. $\mathcal{D}$ is organized in *categories* and *instances*, with each instance belonging to only one category.

The profile of a user u stores: (*i*) Cid_c (resp. Iid_i), an identifier of the category c (resp. instance i); (*ii*) LCA_c^u (resp. LIA_i^u), the date of the last access of u to

S. Bassis et al. (eds.), *Recent Advances of Neural Network Models and Applications*,
Smart Innovation, Systems and Technologies 26,
DOI: 10.1007/978-3-319-04129-2_21, © Springer International Publishing Switzerland 2014

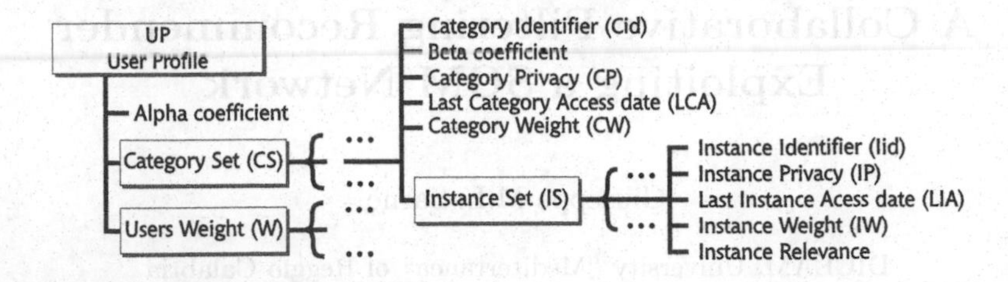

Fig. 1. The structure of the *User Profile*

c (resp. i); (*iii*) CW_c^u (resp. IW_i^u), a real value, ranging in the interval $[0, 1]$, to measure the interest of u about c (resp. i); (*iv*) CP_c^u (resp. IP_i^u), a flag set by u to 0 or 1 to make public or private his interest about c (resp. i); (*v*) α^u, a real coefficient set by u in $]0, 1[$ and used in computing the IW measures; (*vi*) β_c^u, a real coefficient ranging in $[0, 1]$ to consider the u's expertise about the category c; (*vii*) ρ_i^u, the *Instance Relevance*, a real coefficient that measures in $[0, 1]$ the relevance assigned to the instance i; (*viii*) *Category Set* CS, *Instance Set* IS and *Users Weight* W, three sets storing data on categories, instances and weights (these later are real values, ranging in $[0, 1]$, that measure the reliability of the other users in providing their relevance measures to u).

IW represents the u's history in visiting i. Based on the time t_i^u (in seconds) spent by u on the page containing the instance i [21], IW_i^u is updated as $IW_i^u = \alpha^u \cdot t_i^u + (1 - \alpha^u) \cdot \mathcal{F}(current, past) \cdot IW_i^u$, where the coefficient α^u, set in $]0, 1[$, weights the contribution due to the u's visit and $(1 - \alpha^u)$ weights the current IW value, suitably decreased by using the function $\mathcal{F}(current, past)$ in which *current* and *past* (i.e., LIA_i^u) are the dates of the actual and the last visits of u to i. In particular, the function $\mathcal{F}(\,)$ returns $1 - \frac{curren-past}{365}$ if $(current-past) \leq 365$, otherwise it returns 0. Similarly, CW_c^u measures the u's interest in a public category c as the sum of all the IW_i^u of its public instances. After an IW_c^u (i.e., CW_c^u) updating the parameter LIA_i^u (i.e., LCA_c^u) is set to the current date (i.e., the most recent LIA_i^u of the public instances associated with c).

3 The Social Relevance Collaborative Filtering Algorithm

The SRCF algorithm is derived by [3,30] and adapted to a recommender context. It exploits the concept of *relevance* (ρ) of the instance i belonging to the category c in the users' community $\mathcal{A}$. In other words, for each user $u \in \mathcal{A}$ the relevance ρ_i^u is the measure of how much i is significant for him (*individual relevance* μ_i^u) and for his community (*social relevance* γ_i^u, built with the opinions of each user $j \neq u \in \mathcal{A}$). More formally, ρ_i^u is calculated as $\rho_i^u = \beta_c^u \cdot \mu_i^u + (1 - \beta_c^u) \cdot \gamma_i^u$, where the coefficient β_c^u takes into account the expertise level of u about c which the instance i belongs to (with ρ, μ, γ and β real values ranging in $[0, 1]$).

The *individual relevance* μ_i^u is computed as $\mu_i^u = IW_i^u / \sum_{t=1}^{r} IW_{i_t}^u$ by using the updated IW values, with *past* $= LIA^u$ in $\mathcal{F}(\,)$, and normalized on all the r instances stored in UP^u that belong to the same category of i. The *social*

```
RecList R=SRCFRecommender(user u, community A, int M, int N)
  { RelevanceSet RSet_A=Relevance(community A, int M, int N)
    RecList R Recommendations(user u, community A, RelevanceSet RSet_A, int M, int N)
    return R; }

RelevanceSet RSet_A=Relevance(community A, int M, int N)
  { void Updating-A(userProfile UP^A, date current_date);
    for(j = 0; j < n; j++)
    {   InstanceSet ISet_j=SelectTop(user j, userProfileUP^j, int M, int N);
        ISet_A = ISet_A ∪ ISet_j;}
    {   for(k = 0; k < p; k++)
        Relevance R_k=InstanceRelevance(Instance k, userProfile UP^A);
        RSet_A = RSet_A ∪ R_k;}
    return RSet_A; }

RecList R Recommendations(user u, community A, RelevanceSet RSet_A, int M, int N)
  { void Updating-u(userProfile UP^u, date current_date);
    InstanceSet ISet_u=Visited(user u, userProfile UP^u);
    RecList R = RSet_A - RSet_A ∩ ISet_u;
    RecList R=Sort(RecList R);
    RecList R=Extract(RecList R, int N);
    return R; }
```

Fig. 2. The Social Relevance Collaborative Filtering Recommender Algorithm

relevance γ_i^u is the weighted mean of the ρ measures about i asked by u to the other users of $\mathcal{A}$ and computed as $\gamma_i^u = \sum_{j=1}^{n-1} W_j^u \cdot \rho_i^j/(n-1)$, $\forall j \neq u \in \mathcal{A}$. W_j^u, a real value ranging in $[0,1]$, weights j's reliability in providing his ρ measures to u, it is computed based on the last d differences (with $d = 5$) between the ρ measures computed by u and those provided by j as $W_j^u = 1 - \left|\sum_{d=1}^{v}(\rho_d^u - \rho_d^j)\right|/v$. The real coefficient $\beta_c^u = IC_c^u/\sum_{k=1}^{p} IC_k^u$ measures in $[0,1]$ the u's expertise on the category c which i belongs to. The IC_u values will be updated by using $\mathcal{F}(\,)$ with *past* $= LCA^u$, and then normalized. Therefore:

$$\rho_c^u = \frac{CW_c^u}{\sum_{k=1}^{p} CW_k^u} \cdot \frac{IW_c^u}{\sum_{t=1}^{r} IW_t^u} + \left(1 - \frac{CW_c^u}{\sum_{k=1}^{p} CW_k^u}\right) \cdot \frac{\sum_{j=1}^{n-1} W_j^u \cdot \rho_c^j}{n-1} \tag{1}$$

where n, p and r are the number of users of $\mathcal{A}$, instances and categories. To compute the relevance measures of each instance, a system of n equations of the type (1) in n variables needs to be solved [3,30] and it admits only one solution.

The function `SRCFRecommender( )` (see Figure 2) describes the SRCF algorithm. This function requires as input the current user u, a community (cluster) $\mathcal{A}$ of n users with their profiles and the integers M and N (with $M, N > 0$); a list R of suggestions computed for u is returned as output. In turn the function `SRCFRecommender( )` calls the functions `Relevance( )` and `Recommendations( )`.

The function `Relevance( )` receives in input the community $\mathcal{A}$ (with all the public data of its users' profiles, pointed out by $UP^{\mathcal{A}}$) and the integers N, while it returns the set $RSet_{\mathcal{A}}$ storing the relevance measures of the most significant instances in $\mathcal{A}$. Firstly, $LCA, LCI \in UP^{\mathcal{A}}$ are updated to the current date by `Updating-A( )`. Then for each user of $\mathcal{A}$, `SelectTop( )` receives in input a user with his profile and the integers M and N, while a set containing the first N Top-rank public instances of interest for each one of the M Top-rank public categories for that user is returned. Each one of these sets is incrementally added to the global set $ISet_{\mathcal{A}}$ of the most relevant instances in $\mathcal{A}$. After this, for each instance the function `InstanceRelevance( )` is called; it receives the instance k and the users' profiles $UP^{\mathcal{A}}$ as input and returns in R_k the relevance measure of k in $\mathcal{A}$ as output. Finally, each R_k is added to the global set $RSet_{\mathcal{A}}$ returned by `Relevance( )`. Note that `Relevance( )` is the most time consuming computation.

`Recommendations( )` receives the user u, the community $\mathcal{A}$, the set $RSet_{\mathcal{A}}$ of the most relevant instances in $\mathcal{A}$, the integers M and N and returns a list of suggestions. Firstly $LCA, LCI \in UP^u$ are updated to the current date by `Updating-`u`( )`. Then `Visited( )`, for each user u and his profile, returns $ISet_u$, a set storing those public instances already visited by u. The instances in $ISet_u$ are deleted from $RSet_{\mathcal{A}}$ by `Recommendation( )`, that is ordered by `Sort( )`, based on the relevance values of the instances. Finally, `extract( )` returns the list R of suggestions for u with the first N instances of $RSet_{\mathcal{A}}$ having the highest score.

4 The SOM Neural Network Clustering

A high-quality clustering of users might imply a high computational burden. Self-Organizing Maps (SOM) networks [12] can provide to an effective solution to such a problem. SOMs are unsupervised neural networks able to map high dimensional data in low dimensional spaces [12]. Two different learning algorithms are available: (*i*) the *sequential* or *stochastic learning algorithm* updates the synaptic weights immediately after a single input vector is presented; (*ii*) the *batch learning algorithm* updates the synaptic weights after all the input vectors have been presented. The first one is less likely stopped to a local minimum, the other one does not suffer for convergence problems. To save computational resources, SRCF exploits a SOM to cluster the users' profiles by using the public information stored therein by using an off-line strategy. More in detail, each user is represented by an input vector consisting of his public CW measures of interest updated to the current date and normalized among them.

5 Related Work

A wide number of recommender systems have been proposed in the literature (see [4,13] for an overview). Recommenders can adopt Centralized (CR) or Distributed (DR) architectures. CRs are easy to be implemented but suffer for lack scalability, failure risks, privacy and security [11]. DRs share information and computation tasks among more entities but time and space complexities quickly grow [11], while design, setting and privacy can be critical to be obtained [18].

In particular, CF systems search for similar users to suggest items popular among them. An earlier centralized CF recommender is GroupLens [24] that searches for agreed news that could be again agreed. Another personal CF system is PocketLens [16], a DR system that adopts a variant of the item-to-item algorithm [31] and works free by connectivity and device constraints, while in [6,29] the CF component also considers the device characteristics in generating suggestions. Also many e-Commerce platforms (i.e.,*Amazon*, eBay, etc.) drive visitors' purchases with centralized CF by exploiting the past visitors' behaviours.

To save computational resources, clustering algorithms are exploited [17]. In [1,5] two CF agent-based systems exploit a dynamic catalogue of products, based on categories and attributes and users' profiles, respectively. Social and trust information are exploited in [22], while a two-level clustering is adopted in [10]. In EVA [26,28], recommender agents can "migrate" among users with a cloning mechanism based on a "genealogic" reputation model. When generating suggestions for a new user, the agents also consider the preferences of their past owners similarly to a CF contribution. Often DRs also take advantage from P2P networks to exchange data locally stored on each peer in a decentralized domain by using their efficient, scalable and robust routing algorithms as in [7].

In [14,25] SOMs are combined with CF techniques. In [14] demographic information are exploited to segment users in clusters by means of a SOM and in [25] a SOM is combined with a Case Based Reasoning . Then, similarly to SRCF, a CF algorithm is applied on each cluster to generate recommendations.

Finally, some of the cited systems adopt the agent technology but this could be easily adopted also from the remaining part of these systems including SRCF.

6 Experiments

The test of the proposed SRCF recommender system involved: (*i*) a community of 10.000 simulated users; (*ii*) a common *Dictionary* of 20 categories, each one provided with 50 instances, implemented by a unique XML Schema; (*iii*) 20 XML Web sites to simulate the users' behaviour, each one dealing with only two categories; (*iv*) 20 clusters of users; (*v*) 2 Top-rank categories ($M = 2$) and 5 Top-rank istances ($N = 5$). The first 10 Web sites have been exploited for generating the user's profiles, while the other 10 Web sites for measuring the recommenders performances. For sake of simplicity, all the information stored in the profiles have been considered as public. Finally, SRCF has been compared with the CF components of the MWSuggest [29] and EC-XAMAS [6] recommender systems.

The simulated users have been clustered on 20 clusters comparing, on the basis of the information stored in their profiles, two different clustering approaches, (*i*) a SOM neural network [12] and (*ii*) a partitional clustering [23] based on the Jaccard measure of similarity, respectively. To this purpose, each user has been represented by a pattern consisting of the *CW* values of the 20 categories of interest, stored in his profile, normalized among them.

Then for each user u belonging to a given cluster, each of the three recommender systems generated a set of recommendations stored in a different list L_g^u (with $g = 1, \cdots, 3$) including N (i.e., $N = 5$) suggestions $l_{g,s}^u$ (with $s = 1, \cdots, N$), ordered based on their supposed relevance for u. To measure the quality of each suggestion generated for u, a rate $r_{g,s}^u$ (i.e. an integer ranging in $[0, 5]$)

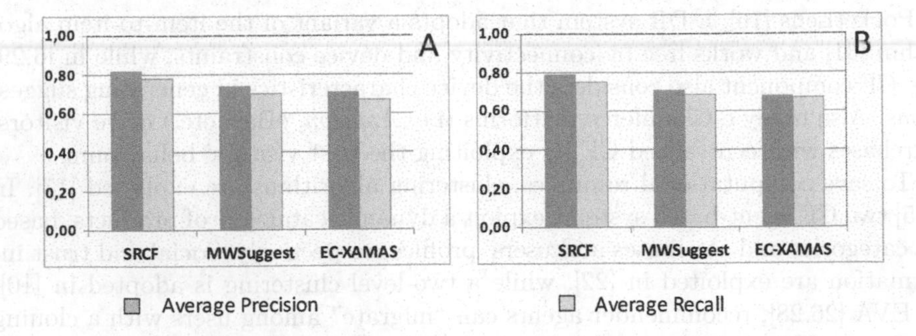

Fig. 3. Average Precision and Average Recall of the SRCF, MWSuggest and EC-XAMAS recommender systems exploiting *A*) a SOM and *B*) a partitional clustering based on the Jaccard measure of similarity

is computed based on the information stored in his profile. If, for u and for each suggestion $l^u_{g,s} \in L^u_g$, the computed rate is $r^u_{g,s} \geq 4$, it is considered as a *true positive* and inserted in the set TP^u_g containing all the true positives of L^u_g. If it is $0 < r^u_{g,s} \leq 3$, it is considered as a *false positive* and inserted in the set FP^u_g of that user. Finally, if the rate is 0, it is considered as a *false negative* (i.e., the user performed a choice not belonging to L^u_g) and inserted in the set FN^u_g.

Therefore, for the user u the standard measures *Precision* (P) and *Recall* (R) for the sets L^u_g, generated for him, have been computed. Precision can be interpreted by the probability that a suggestion is considered as relevant by the user, while Recall can be considered as the probability that it is relevant. Formally, for the user u and for the recommender system g, the P^u_g is computed as $P^u_g = |TP^u_g|/|TP^u_g \cup FP^u_g|$ and R^u_g as $R^u_g = |TP^u_g|/|TP^u_g \cup FN^u_g|$. The Average Precision $\overline{P}^{\,u}_{\,g}$ (resp. the Average Recall $\overline{R}^{\,u}_{\,g}$) of each system is defined as the average of the P^u_g (resp. R^u_g) values of all the users belonging to a cluster.

Using the clustering performed by the SOM, the maximum advantage of SRCF on a single cluster in terms of Average Precision with respect to the second best performer, that is MWSuggest, is of the 13,23%, while in the average with respect to all the clusters it is of the 9,13%. In terms of Average Recall with respect to the second best performer, that is always MWSuggest, the maximum advantage on a single cluster is of the 14,62% and in the average on all the clusters it is of the 9,86%. In this context, the performances of these systems adopting the SOM clustering have an advantage around 1÷2% with respect to the partitional clustering based on the Jaccard measure of similarity. These results are represented in Figure 3 in terms of Average Precision and Average Recall for the three recommender systems with respect to the two tested clustering techniques. Summarizing, these good SRCF performances are surely due to the proposed novel CF algorithm but also to the contribute provided by the high quality of the clustering performed by the SOM network.

7 Conclusions

In this paper a novel collaborative filtering recommender, called SRCF, introducing the concept of relevance of each item among the users of a community has been presented. SRCF generates high-quality suggestions exploiting light users' profiles, built by monitoring their interests and preferences and preserving the users' privacy. SRCF takes advantage from the use of a SOM network to cluster users based on their profiles. The results of some experiments show a significant improvement in terms of effectiveness of SRCF with respect to the other tested recommenders. The keys of these good results are due both to the proposed recommender and to the accurate clustering performed by a SOM network.

References

1. Braak, P., Abdullah, N., Xu, Y.: Improving the Performance of Collaborative Filtering Recommender Systems through User Profile Clustering. In: Proc. IEEE/WIC/ACM Int. Joint Conf. on Web Intelligence and Intelligent Agent Technologies, 2009, pp. 147–150. IEEE (2009)
2. Breese, J., Heckerman, D., Kadie, C.: Empirical Analysis of Predictive Algorithms for Collaborative Filtering. In: Proc. 14th Int. Conf. on Uncertainty in Artificial Intelligence, pp. 43–52. Morgan Kaufmann (1998)
3. Buccafurri, F., Palopoli, L., Rosaci, D., Sarné, G.M.L.: Modeling Cooperation in Multi-Agent Communities. Cognitive Systems Research 5(3), 171–190 (2004)
4. Burke, R.D.: Hybrid Recommender Systems: Survey and Experiments. UMUAI 12(4), 331–370 (2002)
5. Castro-Schez, J.J., Miguel, R., Vallejo, D., López-López, L.M.: A Highly Adaptive Recommender System Based on Fuzzy Logic for B2C e-Commerce Portals. Expert Systems with Applications 38(3), 2441–2454 (2011)
6. De Meo, P., Rosaci, D., Sarnè, G.M.L., Terracina, G., Ursino, D.: EC-XAMAS: Supporting e-Commerce Activities by an XML-Based Adaptive Multi-Agent System. Applied Artificial Intelligence 21(6), 529–562 (2007)
7. Draidi, F., Pacitti, E., Kemme, B.: P2Prec: A P2P Recommendation System for Large-Scale Data Sharing. In: Hameurlain, A., Küng, J., Wagner, R. (eds.) TLDKS III. LNCS, vol. 6790, pp. 87–116. Springer, Heidelberg (2011)
8. Garruzzo, S., Rosaci, D., Sarné, G.M.L.: ISABEL: A Multi Agent e-Learning System That Supports Multiple Devices. In: Proc. of the 2007 Int. Conf. on Intel. Agent Technology (IAT 2007), pp. 485–488. IEEE (2007)
9. Hofmann, T.: Latent Semantic Models for Collaborative Filtering. ACM Transaction on Information Systems 22(1), 89–115 (2004)
10. Hoseini, E., Hashemi, S., Hamzeh, A.: SPCF: a Stepwise Partitioning for Collaborative Filtering to Alleviate Sparsity Problems. Journal of Information Science 38(2), 578–592 (2012)
11. Jogalekar, P., Woodside, M.: Evaluating the Scalability of Distributed Systems. IEEE Trans. Parallel Distributed Systems 11(6), 589–603 (2000)
12. Kohonen, T.: Self-Organizing Maps, 3rd edn. Springer (2001)
13. Konstan, J., Riedl, J.: Recommender Systems: from Algorithms to User Experience. User Modeling and User-Adapted Interaction 22(1), 101–123 (2012)

14. Lee, M., Choi, P., Woo, Y.: A hybrid recommender system combining collaborative filtering with neural network. In: De Bra, P., Brusilovsky, P., Conejo, R. (eds.) AH 2002. LNCS, vol. 2347, pp. 531–534. Springer, Heidelberg (2002)
15. Lops, P., Gemmis, M., Semeraro, G.: Content-based Recommender Systems: State of the Art and Trends. In: Recommender Systems Hand, pp. 73–105. Springer, Heidelberg (2011)
16. Miller, B.N., Konstan, J.A., Riedl, J.: PocketLens: Toward a Personal Recommender System. ACM Transaction on Information Systems 22(3), 437–476 (2004)
17. Mobasher, B., Dai, H., Luo, T., Nakagawa, M.: Discovery and Evaluation of Aggregate Usage Profiles for Web Personalization. Data Mining Knowledge Discovery 6, 61–82 (2002)
18. Olson, T.: Bootstrapping and Decentralizing Recommender Systems. Ph.D. Thesis, Dept. of Information Technology, Uppsala Univ. (2003)
19. Palopoli, L., Rosaci, D., Sarné, G.M.L.: A Multi-tiered Recommender System Architecture for Supporting e-Commerce. In: Fortino, G., Badica, C., Malgeri, M., Unland, R. (eds.) IDC 2012. SCI, vol. 446, pp. 71–80. Springer, Heidelberg (2012)
20. Palopoli, L., Rosaci, D., Sarné, G.M.L.: Introducing Specialization in e-Commerce Recommender Systems. Concurrent Engineering: Research and Applications 21(3), 187–196 (2013)
21. Parsons, J., Ralph, P., Gallagher, K.: Using Viewing Time to Infer User Preference in Recommender Systems. In: AAAI Work. on Semantic Web Personalization, pp. 52–64. AAAI (2004)
22. Pham, M.C., Cao, Y., Klamma, R., Jarke, M.: A Clustering Approach for Collaborative Filtering Recommendation Using Social Network Analysis. Journal of Universal Computer Scienc 17(4), 583–604 (2011)
23. Postorino, M.N., Sarné, G.M.L.: Cluster analysis for road accidents investigations. In: Advances in Transport - Proc. of Urban Transport VIII, Urban Transport and the Environment in the 21st Century, 2002, pp. 785–794. WIT Press (2002)
24. Resnick, P., Iacovou, N., Suchak, M., Bergstrom, P., Riedl, J.: GroupLens: an Open Architecture for Collaborative Filtering of Netnews. In: Proc. 1994 ACM Conf. on Computer Supported Cooperative. Work, pp. 175–186. ACM (1994)
25. Roh, T.H., Oh, K.J., Han, I.: The collaborative filtering recommendation based on som cluster-indexing cbr. Expert Systems with App. 25(3), 413–423 (2003)
26. Rosaci, D., Sarné, G.M.L.: Supporting Evolution in Learning Information Agents. In: Proc. of the 12th Work. on Objects and Agents. CEUR Workshop Proceedings, vol. 741, pp. 89–94. CEUR-WS.org (2011)
27. Rosaci, D., Sarnè, G.M.L.: Efficient Personalization of e-Learning Activities Using a Multi-Device Decentralized Recommender System. Computational Intelligence 26(2), 121–141 (2010)
28. Rosaci, D., Sarnè, G.M.L.: Cloning Mechanisms to Improve Agent Performances. Journal of Network and Computer Applications 36(1), 402–408 (2012)
29. Rosaci, D., Sarnè, G.M.L.: Recommending Multimedia Web Services in a Multi-Device Environment. Information Systems (2012)
30. Rosaci, D., Sarnè, G.M.L., Garruzzo, S.: Integrating Trust Measures in Multiagent Systems. International Journal of Intelligent Systems 27(1), 1–15 (2012)
31. Sarwar, B., Karypis, G., Konstan, J., Reidl, J.: Item-based Collaborative Filtering Recommendation Algorithms. In: Proc. 10th Int. Conf. on WWW 2001, pp. 285–295. ACM (2001)
32. Shani, G., Brafman, R., Heckerman, D.: An MDP-based Recommender System. In: Proc. 18th Conf. on Uncertainty in Artificial Intelligence, UAI 2002, pp. 453–460. Morgan Kaufmann Pub. (2002)

SVM Tree for Personalized Transductive Learning in Bioinformatics Classification Problems

Maurizio Fiasché

Department of Computer Science, University of Milan, Italy
maurizio.fiasche@ieee.org

Abstract. Personalized modelling joint with Transductive Learning (PTL) uses a particular local modelling (personalized) around a single point for classification of each test sample, thus it is basically neighbourhood dependent. Usually existing PTL methods define the neighbourhood using a (dis)similarity measure, in this paper we propose a new transductive SVM classification tree (tSVMT) based on PTL. The neighbourhood of a test sample is built over the classification knowledge modelled by regional SVMs, and a set of such SVMs adjacent to the test sample are aggregated systematically into a tSVMT. Compared to a normal SVM/SVMT approach, the proposed tSVMT, with the aggregation of SVMs, improves classifying power in terms of accuracy on bioinformatics database. Moreover, tSVMT seems to solve the over-fitting problem of all previous SVMTs as it aggregates neighbourhood knowledge, significantly reducing the size of the SVM tree.

Keywords: Personalized Modelling, SVM, SVMT, transductive learning, transductive reasoning.

1 Introduction

Most learning models and systems in artificial intelligence developed and implemented thus far are based on inductive reasoning approach, where a model (a function) is derived from all training data representing the available information from the problem space (induction), and then the application of the model to new coming data to predict the property of interest (deduction or generalization).

In contrast, the transductive method, first introduced by Vapnik [1], is defined as a technique which estimates the value of an unknown model for a single point in the problem space (a sample vector) employing information relevant to that vector.

While the inductive approach is useful when a global model of the problem is needed in an approximate form, the transductive approach is more appropriate for applications where the focus is not on the overall precision of the model, but rather on every individual case. In this sense, it well fits the cases in clinical and medical applications where the interest is on the preciseness of the prediction for an individual patient.

Contrary to an inductive learner which conducts learning only on training data $h_L = L_i(Strain)$, a transductive learner runs from both on training and testing data,

S. Bassis et al. (eds.), *Recent Advances of Neural Network Models and Applications*,
Smart Innovation, Systems and Technologies 26,
DOI: 10.1007/978-3-319-04129-2_22, © Springer International Publishing Switzerland 2014

$h_I = L_t(S_{train}, S_{test})$. With respect to how the unlabelled data S_{test} can be used for learning, personalized transductive learning methods use 'personalized' learning by creating a unique model for each test sample based on its neighbouring samples. A typical personalized learning approach usually consists of two steps: (1) Neighbour-sample filtering: for each new input sample $x_j \in S_{test}$ that needs to be processed for a prognostic/classification task, its N_j nearest neighbours are selected from the training set to form a neighbourhood set $D_j \in S_{train}$. (2) Extremely local decision making: a personalized model M_j is created on D_j to approximate the function value y_j at point x_j. The simplest personalized transductive model is the k nearest neighbour classifier (kNN), it dynamically creates a prediction model for each test sample by learning upon its k nearest neighbours. Neuro-Fuzzy Inference (NFI) is a relative newly introduced personalized transductive learning method [2]. By employing a fuzzy inference model for regional decision making, NFI outperforms kNN on some benchmark data sets. However, since the personalized prediction models for NFI and kNN are based on the same neighbour-sample filtering, their generalization abilities are not significantly different. It is worth noting that for data collected from real world applications, the neighbourhood directly defined by inter-sample dissimilarities is subjected to high noise or ill-posed conditions, which in turn renders the personalized prediction models unreliable. A neighbourhood based on inter-sample dissimilarity is unable to approximate such neighbourhood data distribution precisely, because samples adjacent to the query sample might be just noise data. Thus for a better neighbourhood modelling, it is desirable to take into consideration, besides the distance metric, also the classification information/knowledge to enable constructed neighbourhood knowledgeable. About this issue, we analyze in this paper a transductive SVM tree (t-SVMT), which implements personalized multi-model cooperative learning in a transductive approach.

2 Personalized Neighbourhood Filtering

Given a query sample x in data space D, the neighbourhood of x is defined as a subset Zx adjacent to x, possibly containing samples from different classes.

$$Zx = z1, \cdots, zk, x, d(zi, x) < \theta, \tag{1}$$

where zi represents a neighbouring sample, $d(zi, x)$ is the distance between neighbouring sample and the query sample x, θ is a predefined neighbourhood up bound. Personalized transductive learning assumes that $f(x, Zx)$ approximates the ground truth $f(x)$ better than $f(x,D)$ for the reason that noise data is removed due to neighbouring instance selection. Thus, a personalized transductive function can be learned just from the neighbourhood data set Zx, but not from the full data set D. However, it is noticeable that (1) exclusively relies on inter-sample dissimilarity evaluation. The obtained neighbourhood, depending on what kind of dissimilarity measure is used, is presented either as a circle-type scope or as a rectangle shown in Fig. 1 (c). Such neighbourhood calculation is improper when a dynamic neighbourhood is required for more accurate personalized modelling. Moreover, the neighbourhood obtained from (1) is merely in the sense of dissimilarity measure, ignoring discriminative knowledge between different classes, which in some cases leads to degeneration in the generalization performance. A personalized neighbourhood Zx is modelled based on the following two policies: (1) class label information is incorporated for defining the neighbourhood, and (2) a neighbourhood is composed of flexible-size multi-instance packages instead of individual instances. Formally, the proposed personalized neighbourhood model is defined as

$$Z^*_x = Z1 \cup Z2 \cup \cdots \cup Zk \cup \{x\}, \textit{ Subject to } d(Zi, x) < \theta, \textit{ and } Zi \leftarrow \wp(D), \quad (2)$$

where Zi is a subset containing neighbouring samples to x, $d(Zi, x)$ the distance between Zi and query sample x, and θ a predefined dissimilarity upper bound. It is noticeable that Zi is a flexible-size multi-instance package, which is a subset of D, or sometimes only a single instance. Also, $Z{*}\,x$ considers the class label information as every Zi is obtained from a supervised clustering $\wp(D)$. Therefore, $Z{*}\ x$ is personalized for x in terms of the membership of the neighbourhood. The last illustration of Fig. 1 gives an aggregated neighbourhood by a set of multi-instance packages represented as circles in the figure. It is obvious that this last one presents a more accurate neighbourhood approximation than previous images of the same Fig.1.

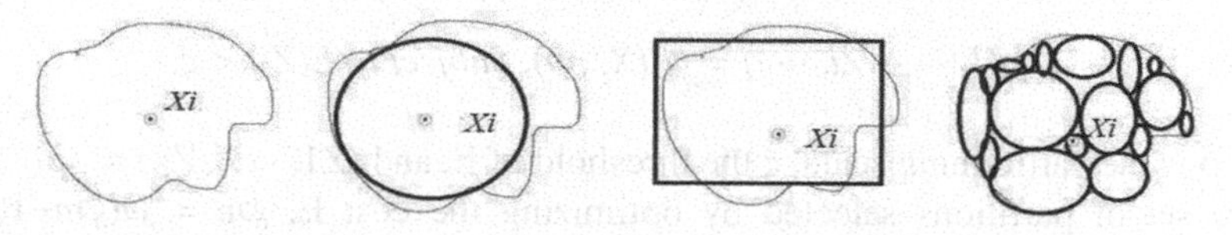

Fig. 1. Dynamic neighbourhood modelling. (a) Truth personalized neighbourhood of xi; (b) spherical neighbourhood approximation; (c) cubic neighbuorhood approximation; (d) the proposed subset aggregated neighbourhood.

3 Personalized- Transductive- Spanning- SVM Tree

SVM tree (SVMT) [3] is a type of inductive SVM aggregation. It overcomes the difficulty of SVM ensemble by determining the number of SVMs automatically during learning. The disadvantage of existing SVMTs is that the spanning of SVMT is completely data-driven (called DDS SVMT here afterward): the tree grows easily into a large-size SVM decision tree with terminal SVM nodes built on extremely small dataset, which results in the generated SVMT over-fitting to the training data. Here we present a new SVM aggregating method which constructs an SVMT in a transductive way [4]. The proposed approach is capable of preventing the over-fitting of SVMT learning as it aggregates only neighbourhood knowledge and hence largely reduces the size of generated SVM tree. Towards a multi-model transductive SVM aggregation classifier with high generalization ability, in this section we address three aspects of the model:

(1) problem decomposition, i.e. data partitioning,
(2) modelling of local knowledge, i.e. individual SVM training on regional multi-instance package,
(3) individual SVMs aggregation, i.e. transductive aggregation.

3.1 Data Partitioning

In the spirit of searching for better partitions with good discriminative ability, we are aiming at partitioning data into geographically adjacent subsets while taking class label information into account. Let $X = \{x1, \ldots, xN\}$ be the training set, and the associative labels given by $\{y1, \ldots, yN\}$, a new attribute zi can be formed out of the original attributes as $Zi = (xi, yi)$, embodying the information contained in X as a set of data partitions $\{Z1, \ldots, Zk\}$. A supervised data decomposition can be accomplished by introducing cost function based on a regional SVM approximation,

$$E(Zk) = \sum_{x \in Zk} (x - xk) + \alpha \sum_{x \in Zk} (fsvm(xk) - fsvm(x)), \tag{3}$$

where $fsvm$ is a standard SVM approximation given later, and α is a balancing weight. and α is a balancing weight determined in practice by cross-validation tests. The default value of α is 1. Given a partitioning function ℘ on X with adjustable partitioning scale, (3) can be used for a supervised recursive partitioning procedure as,

$$[Z1, \cdots, Zi, \cdots] = \wp^{n}(X, \rho 0), \text{ Subject to } E(Zi) < \xi, \tag{4}$$

where ρ is the partitioning scale, ξ the threshold of E, and $(Z1, \cdots, Zk) = (g_1, \ldots, g_k)$ are the set of partitions selected by optimizing the cost E. $\wp n = \wp(\wp n-1(Xn-1, \rho n-1), \rho n)$, $\rho n = \rho n-1 + \Delta$, and Δ is the partitioning scale interval normally set to a small value in the range of ℘ definition. For example, for K-means, Δ can be an integer greater-equal to 1. By (4), X can be partitioned into a set of data partitions in different sizes $\{Z1, Z2, \cdots, Zi, \cdots\}$, such that $Z1 \cup Z2 \cup \cdots \cup Zi \cup \cdots \approx D$, and each partition Zi is associated with the partitioning scale ρ, in which Zi is produced.

3.2 SVM Particles

To formalize classification as the task of finding partitions with maximum likelihood, a local SVM is associated with each partition obtained by the above data partitioning procedure. An SVM particle is defined as a structure that combines the dataset Zi, and the trained SVM f^{i}_{svm} on Zi, $Vi = \{Zi, f^{i}_{svm}\}$. In the case that Zi has data from two classes, a regular two-class SVM model $f_{svm} = sign(\sum^{l}_{i=1} yi(wTi\, \phi(x \cdot i) + bi^*))$, where ϕ is the kernel function, l is the number of training samples, and wi and b_i^* are optimized by

$$min\ \frac{1}{2}(w^{T}_{i}\, w_i) + C\sum^{l}_{t=1} \eta^{2}_{i}, \text{ Subject to } y_i(w^{T}_{i}\, \phi(x_t) + b_i) \geq 1 - \eta_i, \tag{5}$$

where C is the margin parameter to weight the error penalties η_i, and $\phi(.)$ is the mapping function implemented by the kernel function. On the other hand, when *Zi* only contains data from one class, a one-class SVM is applied. Following [5], a class SVM function can be modelled as an outlier classifier by labelling samples in *Zi* as positive and samples in the complement set $\overline{Z}_i$ as negative. Then the SVM is trained on a dataset $Z'i = \{x_i | i = 1, \cdots, N\}$, and

$$y'_i = \begin{cases} +1 \text{ if } xi \in Zi, \\ -1 \text{ if } xi \in \overline{Z}_i. \end{cases} \tag{6}$$

Consequently, the following two types of SVM particles are obtained: (1) oneclass SVM particles $V^{<1>} = \{g, \rho, f_{svm}{}^{<1>}\}$ for class 1, $V^{<2>} = \{g, \rho, f_{svm}{}^{<2>}\}$ for class 2, and (2) two-class SVM particle $V^{<2>} = \{g, \rho, f_{svm}\}$. Given a data partition belonging to one class, a one-class SVM is applied to model the data by separating the samples in the partition from the outliers. In the case that the data belong to two classes, satisfying $E(g) > \xi$, then a two-class SVM is applied to the partition, and a standard SVM particle is created.

3.3 Personalized Mode Transductive Aggregation

Given a test sample x, and a set of SVM particles {Vi} derived over Strain, the distance between x and a particle Zi is measured by the normalized Euclidean distance defined as follows,

$$d(x, Zi)^2 = (x - (\textstyle\sum_{xk \in Zi} xk)/|Zi|)^T (x - (\textstyle\sum_{xk \in Zi} xk)/|Zi|), \tag{7}$$

where |Zi| denotes the cardinal number of the particle.

Then, all selected SVM particles compose the knowledgeable neighbourhood for the new input instance x, and the decision of classification is made transductively by an aggregation $\hat{f}$ of those SVM particles in the neighbourhood as,

$$\hat{f}(x) = \begin{cases} f_{svm}<1> \text{ if } x \to V^{<1>}, \\ f_{svm}<2> \text{ if } x \to V^{<2>}, \\ f_{svm} \text{ otherwise,} \end{cases} \tag{8}$$

where $f_{svm}<1>$ and $f_{svm}<2>$ are one-class SVM decision makers for class 1 and 2, respectively, and f_{svm} is a 2-class SVM decision maker. Taking each SVM particle as a tree node and the partitioning scale ρ as the level of tree structure, $\hat{f}$ can be represented as an SVM tree. Thus, '→ ' in (8) indicates that x is branched to a certain node of the SVM tree. Clearly, there may be errors in the classification of above constructed $\hat{f}$, as $\hat{f}$ may differ from the true classification function *f*. Thus, a suitably chosen real-value loss function $L = L(\hat{f}, f)$ is used to capture the extent of this error. Therefore loss L is data dependent: $L = |\hat{f} - y|$. As $\hat{f}$ is applied to datasets drawn from the data domain D under a distribution of $\theta(x)$. The expected loss can be quantified as,

$$L = \sum_{i=1}^{I} |f_{svm}i - fi|\, \varrho_i + \sum_{j=1}^{J} |f_{svm}j\langle 1\rangle - fj|\, q_{1j} + \sum_{k=1}^{K} |f_{svm}j\langle 2\rangle - fk|\, q_{1k}, \quad (9)$$

where *fi*, *fj*, and *fk* are the regional true classification function. ϱ_i is the probability that the *ith* particle contains two class data. q_r(r = 1, 2) represent the probability that a partition contains data from only one class. I, J, and K represent the number of two-class SVM particles, one-class SVM particles for class 1, and class 2, respectively. Here, I, J, K and N are determined automatically after $\hat{f}$ is created. From (7), the aggregation risk is determined by 2-class SVM and one-class SVM classifiers over all the particles/regions. Also, the risk from $\wp$ has already been minimized during the process of data partitioning. Thus an optimized SVM aggregation is achieved as long as the risk from every regional SVM is minimized.

4 The Proposed t-SVMT Algorithm

In this section, we describe algorithm presented in [3] where the above personalized transductive modelling is interpolated as an algorithm of transductive SVM tree (t-SVMT). First, the training data is divided in a recursive data partitioning procedure. Then, regional knowledge of the input data is approximated as a set of SVM particles. Finally, the selected transductive particles (i. e. particles neighbouring to a test instance x) are aggregated into a personalized SVM tree model. Below is a step by step explanation of the operations taken by the t-SVMT Algorithm:

Step 1: Partitioning. The input training data is decomposed and modelled into a set of SVM particles. Algorithm 1 gives the partitioning function, where ρ0 is a predefined initial resolution for t-SVMT to start analyzing. Default ρ0 is normally set as a scale that gives the biggest size data partitions. For example, for K-mean, ρ0 is set as 2. If some prior knowledge of the data is known, e.g. serious class-imbalance and class-overlap, a finer scale is suggested to enable t-SVMT to analyse the data with a finer resolution. In our experiments, we adopt a standard K-mean clustering approach with ρ = 2.

Step 2: Spanning transductive SVMT. Given a new input instance x, a personalized t-SVMT *Ti* is constructed by transductive SVM particle training followed by SVMT aggregation. Algorithm 2 describes the t-SVMT training function. In our experiments, two-class SVMs use a linear kernel and one-class SVMs employ RBF kernel with parameter adjusted via cross-validation.

Step 3: Testing the constructed t-SVMT y = T (x). Test sample x is first judged by the test function T0(x) at the root node in the SVM tree. Depending on the decision made by the root node, x will be branched to one of the children of the root node. This procedure is repeated until a leaf node or an SVM node is reached, then the final classification decision is made for the test sample by a node one-class classifier or a regular SVM classification.

```
Function: Partitioning(X_train, ρ0)
 X_train; /* training dataset */
 ρ; /* initial partition scale */
 P; /* output data partition set*/
begin
1  P = ø ; /* initialize output */
2  ρ = ρ0 ; /* initialize partition scale*/
3  if X_train is empty
4      return P; /* Iteration stops when X_train is empty */
5  [Z1, ···, Zk] = Partition(X_train, ρ0);
6  for each Zk {
7      if all x ∈ Zk is in one-class
8          P = P ∪ [Zk, ρ];
9      if Zk in two-class, and E(Zk) > ξ
10         P = P ∪ [Zk, ρ];
11     X_train = X_train − Zk; }
12 if X_train size is not decreasing
13     ρ = ρ − Δ;
14 Partitioning(X_train, ρ);/* zooming in */
end
```

Algorithm 1. Partitioning

```
Function: t-SVMTtraining(P, x)
 x; /* a test instance */
 K; /* SVM Kernel for constructing SVM Tree*/
 T; /* output t-SVMT of x */
begin
1  Vt = ø;/* initialize transductive particle set */
2  for each Zk ∈ P {
3      if Zk is close to x
4          Vt = Vt ∪ [Zk, ρp]; }
5  T = ø ; /* initialize t-SVMT as a root node */
6  for each Zk ∈ Vt {
7      if Zk is one-class
8          Mk=Train_SVMone(Zk, K);/* one-class SVM*/
9      else
10         Mj=Train_SVM(Zk, K);}
11     T = T ∪ [Zk, Mk, ρk];/* Add a tree node at level ρj */
end
```

Algorithm 2. Transductive Spanning SVM Classification Tree

5 Experimental Setup

We applied this technique on bioinformatics problems, datasets has been provided by http://www-genome.wi.mit.edu/. In our experiments, we compare the results of t-SVMT with those of standard inductive SVM, and previous SVM aggregation methods: SVM ensemble [6] and DDS SVMT [7]. On the other hand, we also compare t-SVMT with other transductive methods including transductive SVM [4][6], kNN and NFI [2], but for limit of pages we won't describe these last analysis having these last one worst performances. In the comparison, we set all SVMs with a linear kernel, and aggregation parameters for SVMTs (e.g. ρ0 and ξ) as well as parameters for kNN and NFI to be determined by cross validation experiments. For performance evaluation, a consistent 10-fold cross-validation policy is applied for all experiments by using one tenth data as a testing set, and the rest of data as a training set. To see if the t-SVMT can prevent over-fitting to the training dataset, we compared the proposed t-SVMT with DDS SVMT on the scale of the tree structures, i.e. the number of nodes in the tree, and the classification performance, i.e. the classification accuracy with 10-fold cross validation. The experiments are done on eight well-known two-class cancer datasets. We check the performance of the algorithms under two conditions: For the first case, the raw gene features are used as input, and for the second case, 100 genes selected by a standard t-test gene selection algorithm are used as input. Table 1 lists the details and the experimental results of the datasets. As seen from the table, over-fitting happens for DDS SVMT as it yields big tree structures that have about 30 nodes for datasets with less than 80 instances, such as CNS Tumour, Breast Cancer, and Lymphoma(2). In these cases, the classification accuracies are lower than 50%. On the contrary, the trees created by t-SVMT are normally 2-3 times smaller than those by DDS SVMT, and outperforms DDS SVMT in classification accuracy on 12 out of 16 case studies. This indicates that t-SVMT is capable of preventing over-fitting by reducing the size of SVMT, whilst maintains a superior classification accuracy on datasets with significant class imbalance. By computational cost point of view we confirm results present in [6], our simulations on 5 fold cross validation computed over all datasets in mean value highlight that our technique runs 35% faster than DDS SVMT (an AMD Opteron Dual Core 2.2 GHZ has been used).

Table 1. Average classification accuracies for 8 cancer datasets, based on 10-fold cross validation, values are means of 10 runs. Numbers in boldface indicate the best results.

Cancer dataset	Genes/with selection	Bias ratio class 1/2	DDS SVMT /Tree size	t-SVMT /Tree size
	7129/		**84.4%**/15	77.9%/6
Lymphoma(1)	100	19/58=0.33	80.5%/12	**84.4%**/8
	7219/		64.7%/24	**78.3%**/12
Leukemia*	100	11/27=0.41	**91.2%**/12	**91.2%**/7
	7129/		50.0%/34	**72.0%**/8
CNS Tumour	100	21/39=0.53	63.0%/26	**78.3%**/8
Colon	2000/		71.3%/21	**80.7%**/10
Cancer	100	22/40=0.55	75.8%/31	**86.5%**/9
	15154/		**97.3%**/13	75.0%/4
Ovarian	100	91/162=0.56	96.4%/12	**98.4%**/6
Breast*	24482/		52.6%/38	**73.7%**/4
Cancer	100	34/44=0.77	68.4%/14	**78.9%**/6
	6431/		51.7%/27	**60.3%**/10
Lymphoma(2)	100	26/32=0.81	58.6%/26	**66.7%**/15
Lung*	12533/		64.4%/15	**75.0%**/8
Cancer	100	16/16=1.0	**77.8%**/12	73.8%/7

* Independent validation dataset was used for the accuracy evaluation.

6 Discussions and Conclusions

In this paper, we approach a new type of classification SVM-tree based that performs effective personalized transductive learning for new test instances implementing a new type of SVM aggregating intelligence [4] for transductive learning on bioinformatics datasets. The proposed t-SVMT is a new type of personalized learning, different from previous PTL methods such as kNN and NFI, as it uses SVM particle based knowledgeable neighbourhood instead of the simple neighbourhood defined by distance metric. On the other hand, from the viewpoint of SVM aggregating intelligence [3][4], t-SVMT follows the same route of recursive data partitioning plus aggregating as the previous SVM tree methods. Moreover it doesn't aggregate the knowledge of the overall dataset, but only the regional knowledge (i.e. knowledge related to the test instance), therefore the proposed t-SVMT solves successfully the over-fitting problem of previous SVMTs as described by our tests on cancer datasets.

References

1. Vapnik, V.N.: The Nature of Statistical Learning Theory, 2nd edn., pp. 237–240, 263–265, 291–299. Springer, Berlin (1999)
2. Verma, A., Fiasché, M., Cuzzola, M., Iacopino, P., Morabito, F.C., Kasabov, N.: Ontology based personalized modeling for type 2 diabetes risk analysis: An integrated approach. In: Leung, C.S., Lee, M., Chan, J.H. (eds.) ICONIP 2009, Part II. LNCS, vol. 5864, pp. 360–366. Springer, Heidelberg (2009)
3. Pang, S., Ban, T., Kadobayashi, Y., Kasabov, N.: Personalized mode transductive spanning SVM classification tree. Information Sciences 181(11), 2071–2085 (2011)
4. Chen, Y., Wang, G., Dong, S.: Learning with progressive transductive support vector machine. Pattern Recogn. Lett. 24(12), 845–1855 (2003)
5. Schölkopf, J.C., Platt, J.C., Shawe-Taylor, J., Smola, A.J., Williamson, R.C.: Estimating the support of a high-dimensional distribution. Technical report, Microsoft Research, MSR-TR-99-87 (1999)
6. Pang, S., Kim, D., Bang, S.Y.: Face membership authentication using SVM classification tree generated by membership-based LLE data partition. IEEE Trans. Neural Network 16(2), 436–446 (2005)
7. Joachims, T.: Transductive Inference for Text Classification using Support Vector Machines. In: Procs of the Sixteenth International Conference on Machine Learning, pp. 200–209 (1999)

Multi-Country Mortality Analysis Using Self Organizing Maps

Gabriella Piscopo and Marina Resta

Department of Economics,
University of Genoa, via Vivaldi, Genova, Italy
{piscopo,resta}@economia.unige.it

Abstract. In this paper we introduce the use of Self Organizing Maps (SOMs) in multidimensional mortality analysis. The rationale behind this contribution is that patterns of mortality in different areas of the world are becoming more and more related; a fast and intuitive method understanding the similarities among mortality experiences could therefore be of aid to improve the knowledge on this complex phenomenon. The results we have obtained highlight common features in the mortality experience of various countries, hence supporting the idea that SOM may be a very effective tool in this field.

Keywords: Clustering, Mortality, Self Organizing Maps.

1 Introduction

In the recent actuarial literature the interest in the development of common mortality models for different countries has been growing. As highlighted in Hatzopoulos and Haberman [5], populations in different world countries are becoming more closely related, due to the globalization phenomenon, which affects communication, transportation, trade, technology and disesase. The investigation of similarities among different countries could then provide interesting information about the factors driving changes in mortality, in particular across ages and times. The literature on this topic is fervent. Wilson [18] notes a global convergence in mortality experience; White [17] studies the similarity in life experience among twenty–one industrialized countries. Li and Lee [8] introduce the concept of coherent mortality forecasting, where forecasting for a simple country are based on patterns exploited from a larger basket. Tuljapurkar et al.[16] identify a *universal pattern* of mortality decline by analysing together the seven richest world countries. Cairns et al. [1] represent the joint development over time of mortality rates in a pair of related populations. Fiig Jarner and Masotti Kryger [3] introduce a proposal for robust forecasting based on the existence of a larger reference population.

This paper aims to provide fast and intuitive mortality data analysis through Self Organizing Maps –SOMs (Kohonen [7]). In particular, we employ the SOM algorithm in order to discover clusters in mortality data among different countries. As widely known, clustering is a useful data mining technique for unbiased

S. Bassis et al. (eds.), *Recent Advances of Neural Network Models and Applications*, Smart Innovation, Systems and Technologies 26,
DOI: 10.1007/978-3-319-04129-2_23, © Springer International Publishing Switzerland 2014

exploration of a data set; SOM, in turn, is an unsupervised clustering algorithm that has been widely used, with applications which virtually span over all fields of knowledge, from robotics, medical imaging, characters recognition, ([6], [12] and [11]), to economics and financial markets ([9] and [2], or more recently [10], and [14]). In the analysis of mortality data, clustering has been already used within the medical context, in order to separate the causes of death for a given pathology, like cancer, malformations and so on, but in the actuarial context the application of both clustering and SOM is a very innovative tool. In particular, Resta and Ravera [15] introduced it to extract similar mortality patterns from different countries; running a separate SOM for each country's lifetable. Nevertheless, the approach followed in our work is different: we run SOM on a unique multidimensional dataset composed by the mortality data of a bunch of thirty–seven countries, in order to test whether the SOM technique induces a clusterization by countries, or just by ages and years; in the latter case, we could highlight common feature between all countries, which are experiencing the same mortality evolution for groups of ages in a given period, hence tracing the route towards which building coherent mortality forecasts (Li and Lee[8]). The structure of the paper is as follows. In Section 2 we present some technical aspects about SOM. In Section 3 we describe the multidimensional mortality dataset, run SOM and show numerical and graphical results. Concluding remarks are offered in Section 4.

2 Some Brief Remarks on the Self Organizing Map

The Self-Organizing Map –SOM (Kohonen [7]), also known as Kohonen's network, is a computational method for the visualization, low-dimensional approximation and analysis of high-dimensional data. It works with two spaces: a low-dimensional space with a regular grid of nodes, and the higher-dimensional space of data: the SOM learning procedure puts them into contact improving the approximation of data by the nodes but at the same time, approximately preserving the neighbourhood relations between nodes: if nodes are neighbors then the corresponding input vectors are maintained relatively close to each other. To improve the approximation of data, each vector node moves during learning and it involves in this movement its neighbors to preserve neighbourhood relations between nodes as well. From a mathematical viewpoint, the characterization of the learning algorithm may be given as follows. Assume $\mathbf{X}$ to denote a finite input dataset, merged into a d–dimensional space ($d >> 2$), and indicate by $\mathbf{x}_i \in \mathbf{X}$ the i–th input pattern. Let M a bi–dimensional $h \times m$ grid whose nodes $n_{i,j}$ (i=1,. . .,h, and j=1,. . .,m) in turn, are associated to arrays $\mathbf{w}_{i,j}$ of dimension d. Then:

1. Initialize each node's weights.
2. Choose a vector $\mathbf{x}$ from $\mathbf{X}$, and present it to the SOM.
3. Find the Best Matching Unit (BMU) by calculating the distance between the input vector and the weights of each node.

4. The radius of the neighborhood around the BMU is calculated. The size of the neighborhood decreases with each iteration.
5. Each node in the BMUs neighborhood has its weights adjusted to become more like the BMU. Nodes closest to the BMU are altered more than the nodes furthest away in the neighborhood.
6. Repeat from Step 2 for enough iterations for convergence

3 The Application of SOM to Multidimensional Mortality Datasets

We are going to discuss the application of SOM to multidimensional lifetables, as an innovative way to analyse multi–country mortality experiences. The interest in this research vein is very recent: so far we have found only the works of Resta and Ravera [15], and Hatzopoulos and Haberman [5]. With respect to the former, the main difference of our work is in the use of a single SOM trained on a multidimensional lifetable (whose composition will be explained later in next rows), whereas [15] employ as many SOMs as the countries they examine. On the other hand [5] use a Fuzzy C-Means clustering as starting point to derive an index of the time trend of mortality through the fitting of an age-period parametric model ([4]). In this case the originality of our contribution is to find *a-priori* common mortality features between apparently different countries, without applying any parametric model, and hence disregarding any forcing due to the application of either a parametric framework or to the choice of a model rather than another.

3.1 The Multidimensional Dataset

The multidimensional dataset employed as SOM training set is composed by thirty–seven lifetables of the countries included in the Human Mortality Database (HMD)[1], that contains detailed calculations of death rates and lifetables for thirty-seven countries that are listed in Table 1.

Single country lifetable is nothing but a matrix whose components are: the collection's year (t), the age (x), the central death rate m_x between age x and $x+1$; the probability of death q_x between age x and $x+1$, the average length of survival a_x between ages x and $x+1$ for person dying in that interval; the number l_x of survivors at age x, assuming $l_0 = 100,000$, the number of deaths d_x between ages x and $x+1$, the number of persons per years L_x lived between x and $x+1$; the number of persons per years T_x remaining after exact age x, and the life expectancy e_x in years at exact age x. Note that generally it is $x \in [0, 110]$, since all ages from birth ($x = 0$) to extremal age (i.e. the highest age at which someone in the population is still living, e.g.: $x = 110$) are represented; moreover t depends on the year from which the country's demographic bureau began to collect data. In the case of Sweden, for instance, data began to be collected since 1751, so

[1] http://www.mortality.org

Table 1. Countries included in the Human Mortality Database and related abbreviations

Country & ID	Country & ID	Country & ID
Australia (AUS)	Germany (GER)	Norway (NOR)
Austria (AUT)	Hungary (HUN)	Poland (POL)
Belarus (BIE)	Iceland (ICE)	Portugal (POR)
Belgium (BEL)	Ireland (EIRE)	Russia (RUS)
Bulgaria (BUL)	Israel (ISR)	Slovakia (SLK)
Canada (CAN)	Italy (ITA)	Slovenia (SLO)
Chile (CHI)	Japan (JAP)	Spain (SP)
Czech Rep. (CR)	Latvia (LAT)	Sweden (SWE)
Denmark (DEN)	Lithuania (LIT)	Switzerland (SWI)
Estonia (EST)	Luxembourg (LUX)	Taiwan (TW)
Finland (FIN)	Netherlands (NL)	United Kingdom (UK)
France (FRA)	New Zealand (NZ)	U.S.A. (USA)
		Ukraine (UKR)

that the available lifetable has more than 28,000 entries (obtained as 111×258, i.e. 111 years for each collection time $t = 1751, \ldots, 2009$). Moving to Russia and Ukraine, on the other hand, the dataset is sensitively smaller (approximately 6,000 rows), because data began to be collected after 1953. In order to make meaningful comparisons, we considered $x \in [0, 110]$ and $t \in [1992, T]$ where T is the final year of available data in the HMD (it can vary from 2007 to 2011). Finally, although lifetables for males and females are available separately, here we use those for the whole population. Our dataset is therefore composed by 74,394 entries. We run SOM on it and the results are shown hereinafter.

3.2 The Results

The SOM procedure, applied to the whole HMD dataset, divided it into seven clusters, as shown in Figure 1.

Figure 2 represents how the variables of the lifetable are distributed in each cluster.

The procedure has collected the data by ages into different groups, while each cluster contains data from all thirty-seven countries. The interpretation of this result is intuitive: even though the countries are different from each other for socio–political and economics reasons, they have experienced similar patterns in the evolution of life expectancy.

In order to better understand both the common features of mortality and the differences between the countries, we then analysed the composition of each cluster. We discuss with full details the first cluster (S_1). S_1 includes 27.02% of the data and covers ages in the interval [26,76] for each $t \in [1992, T]$. Table 2 shows the different countries ages range, as it results in S_1, for the initial time $t = 1992$, and the conclusive time $t = T$.

The main effect one can easily note is a kind of *ages shifting forward* from t to T for almost all countries. This is due to an improvement in life expectancy

Table 2. Composition of S_1

Country	$t - T$	Ages in t	Ages in T
AUS	1992-2009	41-68	45-73
AUT	1992-2010	39-67	45-73
BEL	1992-2009	39-68	42-71
BIE	1992-2010	28-63	28-62
BUL	1992-2009	37-64	39-65
CAN	1992-2007	40-69	45-72
CHI	1992-2005	39-66	39-69
CR	1992-2009	39-64	41-67
DEN	1992-2011	39-65	44-71
EIRE	1992-2009	42-66	42-71
EST	1992-2009	27-62	30-66
FIN	1992-2009	37-66	40-72
FRA	1992-2010	39-70	41-74
GER	1992-2009	39-67	44-72
HUN	1992-2009	31-62	39-65
ICE	1992-2010	46-69	45-76
ISR	1992-2009	43-67	46-72
ITA	1992-2009	42-69	46-73
JAP	1992-2009	43-71	45-74
LAT	1992-2009	27-62	30-65
LIT	1992-2010	29-64	32-65
LUX	1992-2009	31-67	43-71
NL	1992-2009	42-68	45-72
NOR	1992-2009	42-68	46-72
NZ	1992-2008	40-67	43-72
POL	1992-2009	35-63	39-67
POR	1992-2009	37-67	39-72
RUS	1992-2010	26-62	26-62
SLK	1992-2009	39-63	39-66
SLO	1992-2009	36-66	40-71
SP	1992-2009	39-69	45-73
SWE	1992-2011	42-69	47-73
SWI	1992-2011	40-69	46-74
TAI	1992-2010	37-66	39-70
UK	1992-2009	41-66	42-71
UKR	1992-2009	28-62	28-63
USA	1992-2009	38-67	39-70

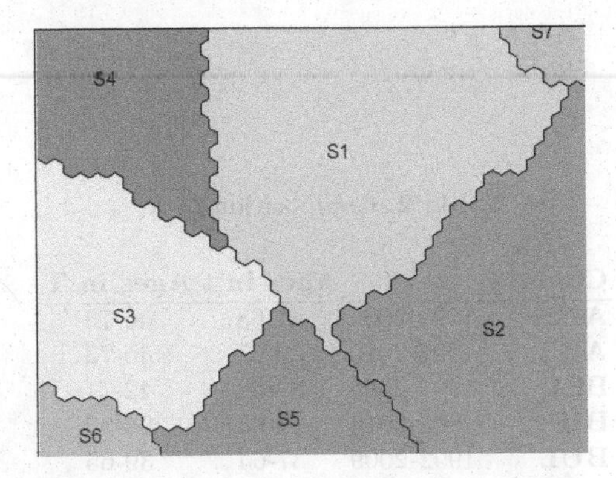

Fig. 1. Clusters exploited by SOM on mortality dataset

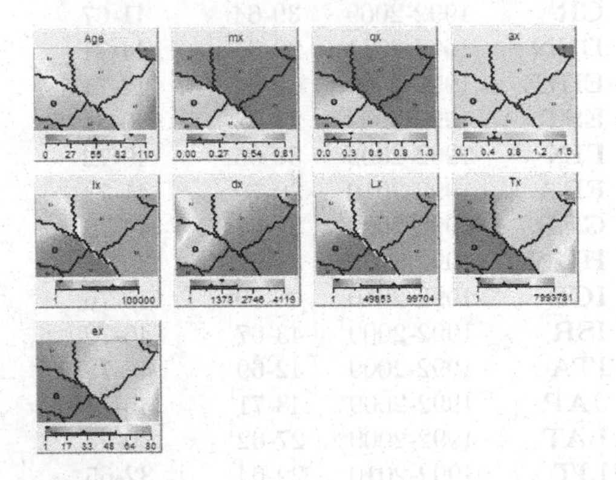

Fig. 2. Distribution of the variables of lifetable in each cluster

documented in actuarial literature as *longevity risk.* We register three exceptions to this well–known phenomenon: BIE, RUS and UKR, but this is probably due to socio–political rather then demographic factors. The youngest ages displayed in S1 are a consequence of the inclusion in the cluster of: BIE, EST, HUN, LAT, LIT, RUS and UKR. This fact has a quite trivial explanation for all cited countries: in the starting time $t = 1992$ those countries just opened to free market, and living standards were significantly far from those of other industrialised nations.

The composition of remaining clusters is as follows. S_2 includes ages between 1 and $y - 1$, where y's are the lower bounds of the age intervals collected in S_1; S_6 and S_7 include respectively the two extreme ages 110 and 0; S_3, S_4, S_5 contain the remaining data for all examined countries.

4 Final Remarks

In this paper we have introduced the use of SOM in the analysis of a multi-country lifetable. The main advantage of the procedure is to offer intuitive and fast computing results. The clusters we have obtained highlight common features in mortality experience of different countries, avoiding any forcing due to the application of a parametric framework. This approach could then represent a starting point to develop a coherent model to forecast mortality across many areas of the world, taking into account and sharing the common information available on the evolution of the life expectancy. Further researches will focus on the relationship between many lifetables, trying to create a well instructive network, whose nodes are the different countries.

References

1. Cairns, A.J.G., Blake, D., Dowd, K., Coughlan, G.D., Khalaf-Allah, M.: Bayesian Stochastic Mortality Modelling for Two Populations. ASTIN Bull. 41(1), 29–59 (2011)
2. Deboeck, G., Kohonen, T.: Visual Explorations in Finance: with Self-Organizing Maps. Springer Finance, New York (1998)
3. Fiig Jarner, S., Masotty Kryger, E.: Modelling adult mortality in small populations: the Saint model. Pensions Institute Discussion Paper PI-0902 (2009)
4. Hatzopoulos, P., Haberman, S.: A parameterized approach to modelling and forecasting mortality. Insur.: Math. and Econ. 44(1), 103–123 (2009)
5. Hatzopoulos, P., Haberman, S.: Common mortality modeling and coherent forecasts. An empirical analysis of worldwide mortality data. Insur.: Math. and Econ. 52, 320–337 (2013)
6. Kaski, S., Kangas, J., Kohonen, T.: Bibliography of Self-Organizing Map (SOM) Papers: 1981-1997. Neur. Comp. Surveys 1, 102–350 (1998)
7. Kohonen, T.: Self-Organizing Maps, 3rd extended edn. Springer, Berlin (2001)
8. Li, N., Lee, R.: Coherent mortality forecasts for a group of populations: An extension of the Lee-Carter method. Demography 42(3), 575–594 (2005)
9. Martn, B., Serrano Cinca, C.: Self Organizing Neural Networks for the Analysis and Representation of Data: some Financial Cases. Neu. Comp. & Appl. 1(2), 193–206 (1993)
10. Montefiori, M., Resta, M.: A computational approach for the health care market. Health Care Man. Sc. 12(4), 344–350 (2009)
11. Oja, M., Kaski, S., Kohonen, T.: Bibliography of Self-Organizing Map (SOM) Papers: 1998-2001 Addendum. Neur. Comp. Surveys 3, 1–156 (2003)
12. Polla, M., Honkela, T., Kohonen, T.: Bibliography of Self-Organizing Map (SOM) Papers: 2002-2005 Addendum. TKK Reports in Information and Computer Science, Helsinki University of Technology, Report TKK-ICS-R23 (2009)
13. Resta, M.: Early Warning Systems: an approach via Self Organizing Maps with applications to emergent markets. In: Proc. of the 2009 Conference on New Directions in Neural Networks: 18th Italian Workshop on Neural Networks, WIRN 2008, pp. 176–184. IOS Press, Amsterdam (2009)
14. Resta, M.: Assessing the efficiency of Health Care Providers: A SOM perspective. In: Laaksonen, J., Honkela, T. (eds.) WSOM 2011. LNCS, vol. 6731, pp. 30–39. Springer, Heidelberg (2011)

15. Resta, M., Ravera, M.: A model for mortality forecasting based on Self Organizing Maps. In: Estevez, P.A., Principe, J.C., Zegers, P. (eds.) Advances in Self-Organizing Maps. AISC, vol. 198, pp. 335–343. Springer, Heidelberg (2013)
16. Tuljapurkar, S., Li, N., Boe, C.: A universal pattern of mortality decline in the G7 countries. Nature 40, 789–792 (2000)
17. White, K.M.: Longevity Advances in High-income Countries, 1955-96. Pop. and Dev. Rev. 28, 59–76 (2002)
18. Wilson, C.: On the Scale of Global Demographic Convergence 19502000. Pop. and Dev. Rev. 24, 593–600 (2001)

A Fuzzy Decision Support System for the Environmental Risk Assessment of Genetically Modified Organisms

Francesco Camastra[1], Angelo Ciaramella[1], Valeria Giovannelli[2], Matteo Lener[2], Valentina Rastelli[2], Antonino Staiano[1,⋆], Giovanni Staiano[2], and Alfredo Starace[1]

[1] Dept. of Applied Science, University of Naples "Parthenope", Isola C4, Centro Direzionale, I-80143, Napoli (NA), Italy
{camastra,angelo.ciaramella,staiano}@ieee.org, alfredo.starace@gmail.com
[2] Nature Protection Dept., Institute for Environmental Protection and Research (ISPRA), via v. Brancati 48, 00144 Roma
{valeria.giovannelli,matteo.lener,valentina.rastelli, giovanni.staiano}@isprambiente.it

Abstract. Aim of the paper is the development of a Fuzzy Decision Support System (FDSS) for the Environmental Risk Assessment (ERA) of the deliberate release of genetically modified plants. The evaluation process permits identifying potential impacts that can achieve one or more receptors through a set of migration paths. ERA process is often performed in presence of incomplete and imprecise data and is generally yielded using the personal experience and knowledge of the human experts. Therefore the risk assessment in the FDSS is obtained by using a Fuzzy Inference System (FIS), performed using jFuzzyLogic library. The decisions derived by FDSS have been validated on real world cases by the human experts that are in charge of ERA. They have confirmed the reliability of the fuzzy support system decisions.

Keywords: Fuzzy Decision Support Systems, Risk Assessment, Genetically Modified Organisms, Fuzzy Control Language, jFuzzyLogic library.

1 Introduction

The development of genetic engineering in the last years produced a very high number of genetically modified organisms (GMOs). Whereas in USA the use of GMOs is widely spread in agriculture, in Europe there are discordant policies w.r.t. GMO usage. For instance, commercialization of food and feed containing or consisting of GMOs is duly approved in European Community (EC), while cultivation of new genetically modified crops are not adopted. The maize MON 810, approved by the old EC legislation framework, is currently the unique GMO cultivated in the EC (e.g., Czech Republic, Poland, Spain, Portugal, Romania

⋆ Corresponding author.

S. Bassis et al. (eds.), *Recent Advances of Neural Network Models and Applications*, Smart Innovation, Systems and Technologies 26,
DOI: 10.1007/978-3-319-04129-2_24,

and Slovakia). According to EC, the environmental release of GMOs is ruled by Directive 200118EC and Regulation 18292003EC. The Directive refers to the deliberate release into the environment of GMOs and sets out two regulatory regimes: Part C for the placing on the market and Part B for the deliberate release for any other purpose, i.e. field trials [17]. In both legislations the notifier, i.e., the person who requests the release into the environment of GMO, must perform an Environmental Risk Assessment (ERA) on the issue. The ERA is formally defined as "the evaluation of risks to human health and the environment, whether direct or indirect, immediate or delayed, which the deliberate release or the placing on the market of GMOs may pose". ERA should be carried out case by case, meaning that its conclusion may depends on the GM plants and trait concerned, their intended uses, and the potential receiving environments. The ERA process should lead to the identification and evaluation of potential adverse effects of the GMO, and, at the same time, it should be conducted with a view for identifying if there is a need for risk management and it should provides the basis for the monitoring plans. The aim of this work is the development of a decision system that should advise and help the notifier in performing the ERA about the cultivation of a specific genetically modified plant (GMP). ERA process is often performed in presence of incomplete and imprecise data. Moreover, it is generally yielded using the personal experience and knowledge of the notifier. Therefore the usage of fuzzy reasoning in the ERA support decision system is particularly appropriate as witnessed by the extensive application of fuzzy reasoning to the risk assessment in disparate fields [1, 2, 6–9, 12, 14, 16, 18]. Having said that, the fuzzy decision support system, object of the paper, is inspired by the methodological proposal of performing ERA on GMP field trials [17] and it has been developed as a Web Application[1]. The methodology would allow to describe the relationships between potential receptors and the harmful characteristics of a GMP field trial, leading to the identification of potential impacts. The paper is organized as follows: In Section 2 the methodological proposal that has inspired the system is described; The FDSS structure of the Fuzzy System is discussed in Section 3; Section 4 describes how the system validation has been performed; Finally some conclusions are drawn in Section 5.

2 The Methodological Approach

The methodological proposal, that has inspired the system object of the paper, is based on a conceptual model [17]. The schema, shown in Figure 1, illustrates the possible paths of the impact from a specific source to a given receptor through disparate diffusion factors and migration routes. The model implies that the notifier fills an electronic questionnaire. The notifier answers are collected in a relational database management system and, in a second time, become input of a fuzzy decision support engine that is the system core and provides to the notifier the overall evaluation of risk assessment related to a specific GM plant. The questionnaire can be grouped in specific sets of questions where each set

[1] Available at this address: `http://www.ogm-dss.isprambiente.it/`

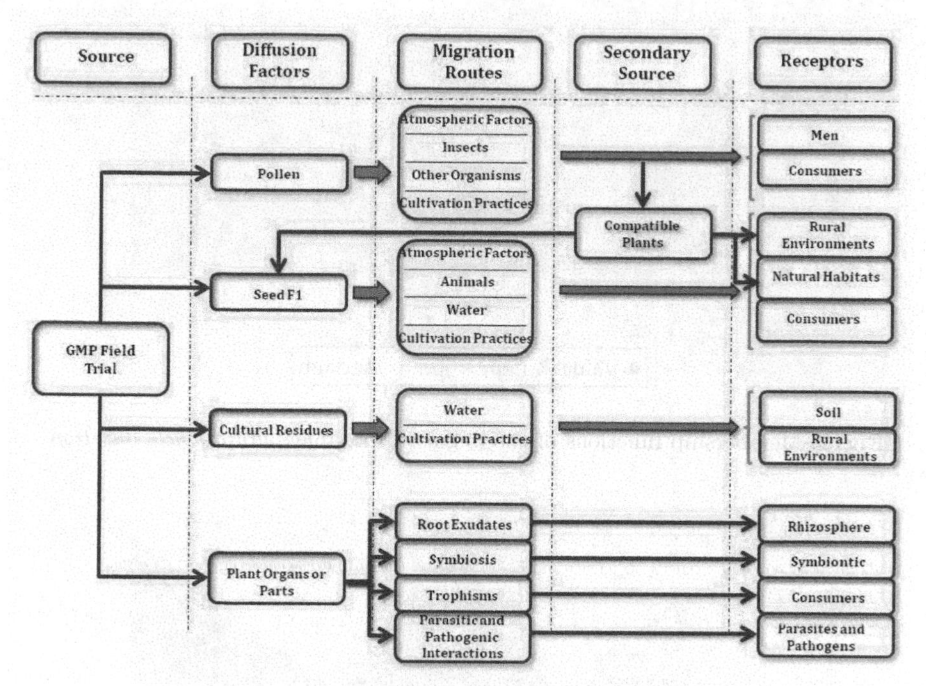

Fig. 1. Conceptual model

corresponds to a specific box of the diagram of the conceptual model. For each block the potential effects are calculated by using fuzzy concepts and a fuzzy reasoning system. The questions can be of two different types, e.g., qualitative and quantitative. The former is typically descriptive and it is not used by fuzzy decision support system in the reasoning process. On the contrary, the latter is used by the fuzzy engine and can be an item chosen within a limited number of possible replies or a numeric or a boolean value.

3 The Fuzzy Decision Support System

The FDSS, object of the paper, has the same architecture of a Fuzzy Logic Control System. Moreover, the Fuzzy Inference System (FIS) of FDSS has been implemented using the jFuzzyLogic library. Therefore the section is organized in two subsections. In the former subsection we discuss the architecture of a generic Fuzzy Logic Control System and, hence, of our system. In the latter subsection jFuzzyLogic is described, in detail.

3.1 A Fuzzy Logic Control System

A Fuzzy Logic Control (FLC) system incorporates the knowledge and experience of a human operator, the so-called expert, in the design of a system that

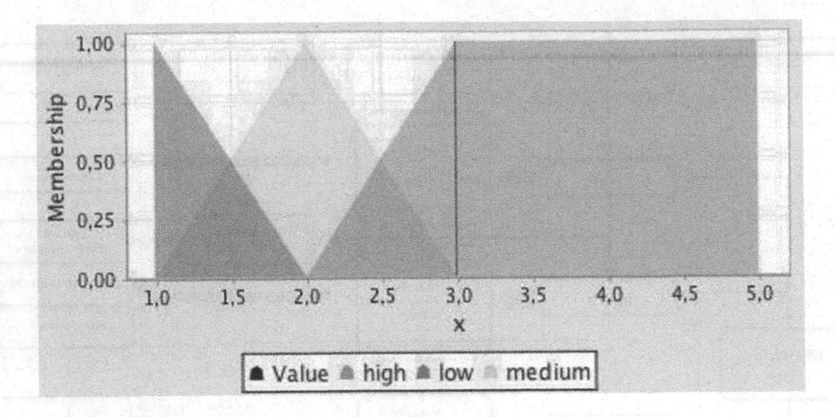

Fig. 2. Membership functions of the linguistic variable *cultural cycle duration*

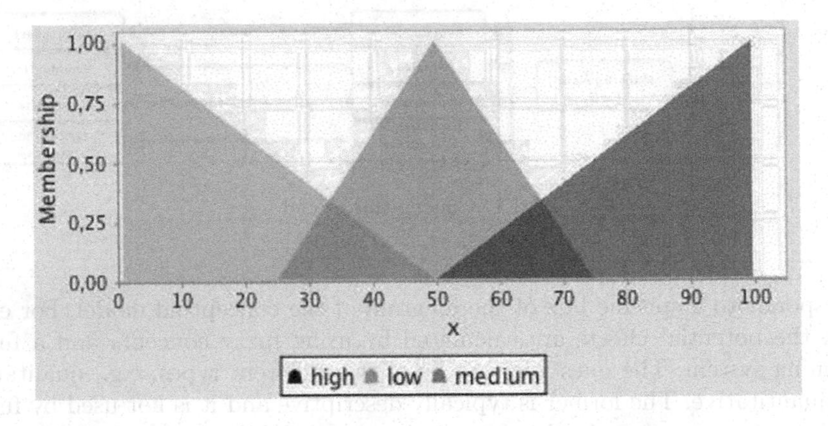

Fig. 3. Membership functions of the linguistic variable *phenological risk*

controls a process whose input-output relationships are described by a set of fuzzy control rules, e.g., IF-THEN rules. We recall that the *antecedent* is the part of rule delimited by the keywords IF and THEN. Whereas the *consequent* is the part of the rule that follows the keyword THEN. The rules involve linguistic variables (*LV*s) that express qualitative high level concepts. A typical FLC architecture is composed of four principal components: a *fuzzifier*, a *fuzzy rule base*, an *inference engine* and a *defuzzifer* [13]. The fuzzifier has the task of transforming crisp measured data (e.g., *vegetative cycle duration* is 1 year) into suitable linguistic values, namely the data becomes, for instance, *vegetative cycle duration* is *low*. The fuzzy rule base stores the operation knowledge of the process of the domain experts. The inference engine is the FLC core, and it can simulate human decision making process by performing approximate reasoning in order to achieve a desired control strategy. The defuzzifier is used for yielding a control action inferred by the inference engine. In the inference engine the generalized modus ponens [13] plays an important role. For the application of fuzzy reasoning in FLCs, the generalized modus ponens can written as

$$\begin{array}{l} \text{Premise 1: IF } x \text{ is } A \text{ THEN } y \text{ is } B \\ \text{Premise 2: } x \text{ is } A' \\ \hline \text{Conclusion } y \text{ is } B' \end{array}$$

where A, A', B and B' are fuzzy predicates i.e., Fuzzy Sets or Fuzzy Relations. In general, a fuzzy control rule, e.g., premise 1, is a fuzzy relation expressed by a fuzzy implication. In most general cases, the fuzzy rule bases has the form of a Multi-Input-Multi-Output (MIMO) system. In this case the inference rules are combined by using the connectives AND and ELSE that can be interpreted as the intersection and the union for different definitions of fuzzy implications, respectively. For instance, if we consider the LV *cultural cycle duration* as shown in Figure 2, the fuzzy inference system of the LVs *cultural cycle duration* and *vegetative cycle duration* could be represented by:

IF vegetative cycle duration is *Low* AND cultural cycle duration is *Low*
THEN phenological risk is *High*
ELSE
IF vegetative cycle duration is *High* AND cultural cycle duration is *Low*
THEN phenological risk is *Low* ELSE
IF vegetative cycle duration is *High* AND cultural cycle duration is *High*
THEN phenological risk is *High*

In Figure 3 an example of fuzzy memberships for the output LV *phenological risk* is presented. Finally, a crisp (i.e., non fuzzy) output is obtained considering a *Center of Area* defuzzifier [13].

3.2 jFuzzyLogic

jFuzzyLogic [4] is an open source software library for fuzzy systems which allows to design Fuzzy Logic Controllers supporting the standard *Fuzzy Control Programming* [10], published by the International Electrotechnical Commission. The library is written in Java and permits FLC design and implementation, following IEC standard for Fuzzy Control Language (FCL). The standard defines a common language to exchange portable fuzzy control programs among different platforms. The aim of jFuzzyLogic is to bring the benefits of open source software and standardization to the researchers and practitioners of fuzzy systems community, namely:

- *Standardization*, which reduces programming work.
- *Extensibility*, i.e., the object model and the Application Programming Interface (API) allow to create a wide range of different applications.
- *Platform independence*, that permits the software development on whatever hardware platform and operating system supporting Java.

Moreover, jFuzzyLogic allows to implement a *Fuzzy Inference System* (FIS). A FIS is usually composed of one or more *Function Block*s (FBs). Each FB has variables (input, output or instances) and one or more *Rule Block*s (RBs). Each RB is composed of a set of rules, as well as Aggregation, Activation and Accumulation methods. Having said that, jFuzzyLogic is based on *ANTLR* [15] that generates a lexer and a parser based on a FCL grammar defined by the user. The parser, written in Java, uses a left to right leftmost derivation recursive strategy, formally known as *LL*(*). Using the lexer and the parser generated by ANTLR, it can parse FCL files by creating an *Abstract Syntax Tree* (AST), a well known structure in compiler design. The AST is converted into an *Interpreter Syntax Tree* (IST), which is capable of performing the required computations. This means that the IST can represent the grammar, in a similar way of AST, but it is also capable of performing computations. Moreover, all methods defined in the IEC norm are implemented in jFuzzyLogic. It should be noted that jFuzzyLogic fulfills the definitions of Aggregation, Activation and Accumulation as defined in [10]. Aggregation methods define the *t-norms* and *t-conorms* correspond to the familiar AND, OR and NOT operators [3]. These can be *Minimum*, *Product* or *Bounded difference* operators. Each set of operators fulfills *De Morgan*'s laws [5]. The Activation method establishes how rule antecedents modify the consequent. Activation operators are Minimum and Product. Finally, accumulation methods determines the way the consequents from multiple rules in the same RB are combined. Accumulation methods defined in the IEC norm include: Maximum, Bounded sum, Normed sum, Probabilistic OR and Sum [10]. Only two membership functions are defined in the IEC standard, i.e., *singleton* and *piece-wise linear*. jFuzzyLogic implements other membership functions such as *trapezoidal*, *sigmoidal*, *gaussian*, *generalized bell*, *difference of sigmoidal*, and *cosine*, too. Moreover, jFuzzyLogic permits making arbitrary memberships by means of a combination of mathematical functions. Thanks to the flexibility in defining membership functions that can be sampled in a fixed number of points, called samples. The number, that is by default one thousand, can be modified on the basis of the precision-speed trade-off required for a specific application. Inference is performed by evaluating membership functions at these samples. To perform the sampling, the domain, called *universe*, of each variable, must be estimated. The variable universe is defined as the range where the variable assumes non-negligible value. For each variable, both membership function and term are considered during the universe computation. After the examination of all rules is terminated, the accumulation for each variable is complete. The last step in the evaluation of a FIS is the *defuzzification* [13]. The value for each variable is computed by means of the defuzzification method selected by the user within the following set: *Center of gravity*, *Rightmost Max*, *Center of area*, *Leftmost Max*, *Mean max* (for the continuous membership functions), or *Center of gravity* (for the discrete membership functions).

4 System Validation

The FDSS, object of the work, has a knowledge base organized in 123 FBs and it consists of 6215 rules of the type described in Section 3[2]. FDSS was tested producing about 150 ERAs related to GM plants (e.g., Bt-maize and HT oil seed rape). We adopted a Mamdani fuzzy inference system using min and max for *t-norm* and *t-conorm* operators, respectively, and a *Center of area* defuzzification mechanism. The ERAs, yielded by FDSS, were submitted to a pool of ISPRA experts, not involved in the FDSS knowledge base definition, in order to assess the consistency and completeness of FDSS evaluations. The tables 1 and 2 show the outputs produced applying the FDSS in two different scenarios. The assessed scenarios are described in [11]. The former, related to Bt maize cultivation in the Lazio Region estimates that the risks deriving from gene flow and seed dispersal (see Table 1 of scenario 1, rows 1 and 2) are negligible, while risks on soil system are higher. Instead, in case of the HR oilseed rape cultivation, potentially high risks deriving from gene flow and seed dispersal (see Table 2 of scenario 2) have been identified. FDSS has been developed as a Web Application and uses MySql database to store the data required to perform the ERA process. FDSS is implemented in Java2EE under Linux UBUNTU on a PC with Intel dual core 2.6GHz, 2GB RAM. It requires a CPU time lower than 2 sec to perform the whole ERA.

Table 1. Estimation of the risks for scenario 1. Value is in the [0, 100] interval and LV are the Linguistic Variables.

Risk	Bt maize (scenario 1)	
	value	LV
Potential genetic pollution of natural resources	0	low
Potential increase in weed population	0	low
Potential changes to soil microbe and fungal biodiversity	100	high
Potential changes to structure of symbiotic populations	50	medium
Potential changes to structure of rhizosphere populations	50	medium

Table 2. Estimation of the risks for scenario 2. Value is in the [0, 100] interval and LV are the Linguistic Variables.

Risk	HT oil seed rape (scenario 2)	
	value	LV
Potential genetic pollution of natural resources	100	high
Potential increase in weed population	0	high
Potential changes to soil microbe and fungal biodiversity	50	medium
Potential changes to structure of symbiotic populations	0	low
Potential changes to structure of rhizosphere populations	0	low

[2] The database of the rules is available on request or can be download from the web site of the Web Application.

5 Conclusions

In this paper a FDSS for the ERA of the deliberate release of GMPs has been presented. The evaluation process permits identifying potential impacts that can achieve one or more receptors through a set of migration paths. The risk assessment in the FDSS is obtained by using a FIS, implemented by means of jFuzzyLogic library. The decisions yielded by FDSS have been validated on real world cases by the human experts of ISPRA, confirming, in this way, the FDSS reliability. In the next future we plan to develop machine learning algorithms that allow to learn automatically the knowledge base of FDSS.

Acknowledgements. This research has been partially funded by the LIFE project MAN-GMP-ITA (Agreement n. LIFE08 NAT/IT/000334).

References

1. Chen, Y.-L., Weng, C.-H.: Mining fuzzy association rules from questionnaire data. Knowledge-Based Systems 22, 46–56 (2009)
2. Chen, Z., Zhao, L., Lee, K.: Environmental risk assessment of offshore produced water discharges using a hybrid fuzzy-stochastic modeling approach. Environmental Modelling & Software 25, 782–792 (2010)
3. Ciaramella, A., Tagliaferri, R., Pedrycz, W.: The genetic development of ordinal sums. Fuzzy Sets and Systems 151(2), 303–325 (2005)
4. Cingolani, P., A.-Fdez, J.: jFuzzyLogic: A Robust and Flexible Fuzzy-Logic Inference System Language Implementation. In: Proceedings of IEEE World Congress on Computational Intelligence 2012, June 10-15 (2012)
5. Cormen, T.H., Leiserson, C.E., Rivest, R.L., Stein, C.: Introduction to Algorithms, 3rd edn. MIT Press (2009)
6. Davidson, V.J., Ryks, J., Fazil, A.: Fuzzy risk assessment tool for microbial hazards in food systems. Fuzzy Sets and Systems 157, 1201–1210 (2006)
7. Guimara, A.C.F., Lapa, C.M.F.: Fuzzy inference to risk assessment on nuclear engineering systems. Applied Soft Computing 7, 17–28 (2007)
8. Kahraman, C., Kaya, I.: Fuzzy Process Accuracy Index to Evaluate Risk Assessment of Drought Effects in Turkey. Human and Ecological Risk Assessment 15, 789–810 (2009)
9. Karimi, I., Hullermeier, E.: Risk assessment system of natural hazards: A new approach based on fuzzy probability. Fuzzy Sets and Systems 158, 987–999 (2007)
10. International Electrotechnical Commission technical committee industrial process measurement and control2. IEC 61131 - Programmable Controllers. Part 7: Fuzzy Control Programming. IEC (2000)
11. Lener, M., Giovannelli, V., Arpaia, S., Baldacchino, F., Benedetti, A., Burgio, G., Canfora, L., Dinelli, G., Manachini, B., Marotti, I., Masetti, A., Sbrana, C., Rastelli, V., Staiano, G.: Applying an operating model for the environmental risk assessment in Italian Sites of Community Importance (SCI) of the European Commission Habitats Directive (92/43/EEC). Bulletin of Insectology 66(2), 257–267 (2013)

12. Li, W., Zhou, J., Xie, K., Xiong, X.: Power System Risk Assessment Using a Hybrid Method of Fuzzy Set and Monte Carlo Simulation. IEEE Transactions on Power Systems 23(2) (2008)
13. Lin, C.-T., Lee, C.S.: Neural Fuzzy Systems: A Neuro-Fuzzy Synergism to Intelligent Systems. Prentice Hall (1996)
14. Ngai, E.W.T., Wat, F.K.T.: Design and development of a fuzzy expert system for hotel selection. Omega 31, 275–286 (2003)
15. Parr, T.J., Quong, R.W.: Software: Practice and Experience, vol. 25(7), pp. 789–810. Wiley & Sons (1995)
16. Sadiqa, R., Husain, T.: A fuzzy-based methodology for an aggregative environmental risk assessment: a case study of drilling waste. Environmental Modelling & Software 20, 33–46 (2005)
17. Sorlini, C., Buiatti, M., Burgio, G., Cellini, F., Giovannelli, V., Lener, M., Massari, G., Perrino, P., Selva, E., Spagnoletti, A., Staiano, G.: La valutazione del rischio ambientale dell' immissione deliberata nell' ambiente di organismi geneticamente modificati. Tech. Report (2003), `http://bch.minambiente.it/EN/Biosafety/propmet.asp` (in Italian)
18. Wang, Y.-M., Elhag, T.M.S.: An adaptive neuro-fuzzy inference system for bridge risk assessment. Expert Systems with Applications 34, 3099–3106 (2008)

Adaptive Neuro-Fuzzy Inference Systems vs. Stochastic Models for Mortality Data

Valeria D'Amato[1], Gabriella Piscopo[2], and Maria Russolillo[1]

[1] Department of Economics and Statistics,
University of Salerno, Campus Fisciano, Italy
{vdamato,mrussolillo}@unisa.it

[2] Department of Economics,
University of Genoa, via Vivaldi,Genoa, Italy
piscopo@economia.unige.it

Abstract. A comparative analysis is done between stochastic models and Adaptive Neuro–Fuzzy Inference System applied to the projection of the longevity trend. The stochastic models provides the heuristic rule for obtaining projections. In the context of ANFIS models, the fuzzy logic allows for determining the learning algorithm on the basis of the relationship between inputs and outputs. In other words the rule is here deducted by the actual mortality data, because this allows for fuzzy systems to learn from the data they are modelling. This is possible by computing the membership function parameters that best allow the associated fuzzy inference system to track the input/output data. The literature indicates that the self-predicting model of ANFIS is better than other models in a lot of fields. Shortcomings and advantages of both approaches are here highlighted.

Keywords: Adaptive Neuro-Fuzzy Inference System, Stochastic Models, Longevity Projections.

1 Introduction

The life insurance products are mostly characterized by long durations and companies performances largely depend on demographic changes. In particular, insurance business is negatively affected by longevity risk, resulting from systematic deviation of realized survival probabilities from the expected probabilities over time. Also most private pension systems as well as financial stability of the governments through public pension plans and social security schemes are endangered by longevity risk. The size of the phenomenon is meaningful and in the past century survival probabilities have increased for each age group. In the light of this consideration, accurate mortality forecasts become necessary to avoid failing consequences on solidity of insurance companies and public institutes. Useful instruments for controlling longevity risk are the projected mortality tables, constructed forecasting future mortality trends. Various approaches have been proposed to improve mortality forecasts. In the actuarial literature and practice, the

S. Bassis et al. (eds.), *Recent Advances of Neural Network Models and Applications*, Smart Innovation, Systems and Technologies 26,
DOI: 10.1007/978-3-319-04129-2_25,

widely used is the Lee Carter (LC) model [14]. In 1992, Lee and Carter designed a model to extrapolate the trends of mortality and described the secular change in mortality as a function of a time index. The model is fit to historical data through the singular value decomposition of the matrix of the log-mortality data and then the time-varying parameter is modeled and forecast as an ARIMA process using standard Box-Jenkins methods. The Authors suggest to refer to [14] for an in-depth analysis of the LC model. According to [2], the Lee-Carter-based approach is widely considered because it produces fairly realistic life expectancy forecasts, which are used as reference values for other modelling methods. During the last two decades, there have been several extensions of the standard LC method, retaining some of its basic properties, but adding additional statistical features too (see [3], [4]). Many Authors have proposed approaches to mortality forecasting based on smoothing procedure ([8]; [7]). [12] show a particular version of the LC methodology, the so-called Functional Demographic Model, based on the combination of smoothing techniques and forecasting robust to outliers. All these contributions are based on time series analysis and the trend of mortality is forecasted through stochastic processes.

In the last years, a different approach is being experimenting. Very recent contributions introduce the fuzzy logic in mortality projection ([1], [10]). In the field of data modelling, regression models based on fuzzy data were first developed by [17] and [9]. Fuzzy logic has gained recognition and was intensively applied in mathematics and computer sciences, but its use in mortality projection is still dawning. [10] introduce fuzzy logic to clusterize mortality experience in different countries, using a fuzzy cluster analysis on the main time effect of the LC model for each countries. [1] combine fuzzy logic and artificial neural network for mortality forecasts.

In this paper we show a comparative study of stochastic model versus Adaptive Neuro–Fuzzy Inference System (ANFIS) in the context of mortality data. The structure of the paper is as follows. In Section 2 we briefly present the LC and the FDM models. Section 3 offers some details on ANFIS algorithm. In Section 4 we show a comparative application of stochastic model vs ANFIS algorithm to Italian mortality data. Concluding remarks are presented in Section 4.

2 Brief Remarks on Stochastc Mortality Models

In this section we retrace the main feature of two widely used stochastic mortality models: the LC and the FDM.

The LC model describes the log of the observed mortality rate for age x and year t, $m_{x,t}$, as the sum of an age-specific component α_x, that is independent of the time, and another component that is the product of a time-varying parameter k_t, reflecting the general level of mortality, and an age-specific component β_x, that represents how mortality at each age varies when the general level of mortality changes:

$$ln(m_{x,t}) = \alpha_x + k_t\beta_x \tag{1}$$

The model is fit to historical data through the singular value decomposition of the matrix of the log-mortality data and then the time-varying parameter is modeled and forecast as an ARIMA process using standard Box-Jenkins methods.

Hyndman and Ullah show a particular version of the LC methodology, the so-called Functional Demographic Model; they propose a methodology to forecast age-specific mortality rates, based on the combination of functional data analysis, nonparametric smoothing and robust statistics. Let $y_t(x)$ denote the log of the observed mortality rate for age x and year t, $f_t(x)$ the underlying smooth function, $\{x_i, y_t(x_i)\}, t = 1, ..., n, i = i, ..., p$ the functional time series where

$$y_t(x_i) = f_t(x_i) + \sigma_t(x_i)\varepsilon_{t,i} \tag{2}$$

with $\varepsilon_{t,i}$ an iid standard normal random variable and $\sigma_t(x_i)$ allowing for the amount of noise to vary with x.

The dataset is smoothed for each t by applying penalized regression splines. Using a nonparametric smoothing with constraint the functions $f_t(x_i)$ is estimated for each t. Then the fitted curves are decomposed by using a basis function expansion:

$$f_t(x) = \mu(x) + \sum_{k=1}^{K} \beta_{t,k}\varphi_k(x) + e_t(x) \tag{3}$$

To each coefficient $\beta_{t,k}$ univariate time series models are fitted and then are forecasted for $t = n+1, ..., n+h$ period. Finally, from eq. 3 $y_t(x)$ can be projected for $t = n + 1, ..., n + h$ period, with h the forecasted periods ahead.

3 The ANFIS Procedure

A Neuro–Fuzzy system is defined as a combination of Neural Networks and Fuzzy Inference System (FIS). [13] introduced an Adaptive Neuro-Fuzzy Inference System (ANFIS) where the membership function (MF) parameters are fitted to a dataset through a hybrid–learning algorithm.

The difference between FIS and ANFIS is that in the former case the rule structure is essentially predetermined by the user's interpretation, while in the latter case the procedure learn information from the data. In particular, the basic structure of FIS is a model that maps input characteristics to input MF, input MF to rules, rules to a set of output characteristics, output characteristics to output MF, and the output MF to a single-valued output or a decision associated with the output. The MF are chosen arbitrarily and the rule structure is essentially predetermined by the user's interpretation of the characteristics of the variables in the model. The steps for creating a FIS model are:

1. Fuzzification: the input variables are compared with the MFs on the premise part of the fuzzy rules to obtain the probability of each linguistic label (we refer to [5],[6],[15] for the specification of the fuzzy rules to obtain coherent probabilities) .

2. Combination (through logic operators) the probability on the premise part to get the weight (fire strength) of each rule.
3. Application of firing strength to the premise MF of each rule to generate the qualified consequent of each rule depending on their weight.
4. Defuzzification: Aggregate the qualified consequents to obtain the output.

On the other side, the ANFIS procedure provide a method for the fuzzy modeling procedure to involve information about a data set. Using a given data set, the ANFIS algorithm constructs a FIS whose MF parameters are adjusted using either a backpropagation algorithm alone or in combination with a least squares type of method. This adjustment allows the fuzzy systems to learn from the data they are modeling.The MF parameters changes through the learning process. The computation of theme takes place through a gradient vector, that measures how well the FIS is modeling the input/output data for a given set of parameters. When the gradient vector is obtained, any of several optimization routines can be applied in order to adjust the parameters to reduce some error measure, such as the sum of the squared difference between actual and desired outputs.

The ANFIS can be implemented using a first–order Sugeno type FIS. The advantages of the Sugeno Method are the efficiency in the computation because it works well with linear techniques; the continuity of the output surface and the possibility to adapt it for modelling nonlinear systems by interpolating between multiple linear models.

One application of this procedure to mortality data is offered by [16]. They implement an ANFIS which uses a first order Sugeno-type FIS to forecast mortality. The model predicts the yearly mortality in a one step ahead prediction scheme. The method of trial and error is used in order to decide the type of membership function that describe better the model and provides the minimum error. The least-squares method and the backpropagation gradient descent method are used for training the Fuzzy Inference System (FIS) membership function parameters.

4 Numerical Application

In this section we compare the results obtained by Artificial Intelligence-based prediction model, particularly ANFIS, with the ones obtained by the following stochastic models: the LC and the FDM one. The data used are real word data and have been taken from the site `www.mortality.org` [11] which provides data of mortality for many countries. In our case, data regard the death rates for Italian total population, collected from 2001 up to 2006, from ages 0 up to 110. The data, considered by single calendar year and by single year of age, are split into two divisional ranges: ages 0-60 and ages 61-110.The analysis is run separately on the two datasets and a comparative investigation is done between the LC model, the FDM model and an Adaptive Neuro Fuzzy Inference System.

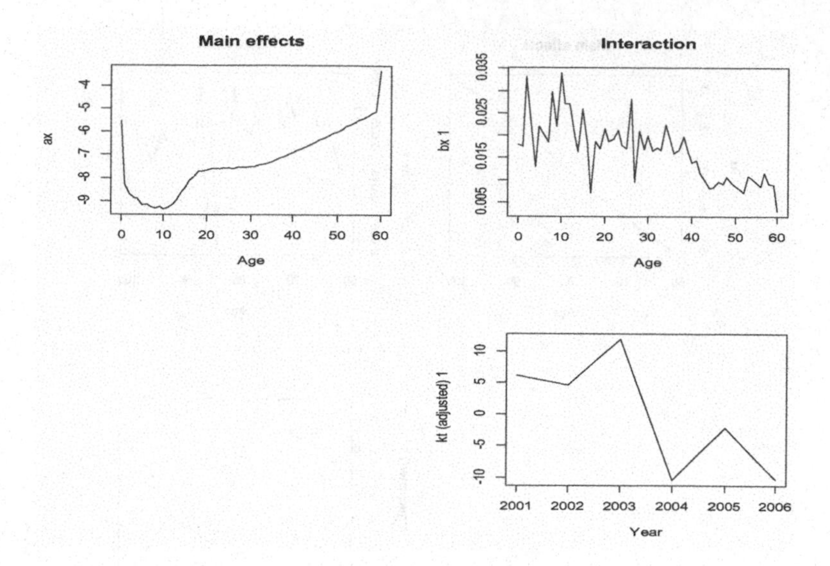

Fig. 1. The parameter estimates of basic LC model on Italian Total mortality data for years 2001 - 2006 and ages 0 - 60

Firstly, we fitted the basic LC model and the FDM version to the two subsets under consideration. For sake of simplicity, in Fig. 1 and Fig. 2, are shown the fitting results in the case of the basic LC model.

The percentage of variation explained by the LC model is 76.9% in the case of ages 0 - 60 and 76.8% the case of ages 61 - 110. As regards the Functional Demographic Model the basis functions respectively explain 79.4%, 10.6%, 3.9%, 3.6%, 2.5% of the variation and 80.0%, 12.9%, 3.7%, 2.3%, 1.2% of the variation either in the case of ages 0 - 60 either in the case of ages 61 - 110 for years 2001 - 2006. From a first comparison between the traditional LC model and the FDM one, we can notice that in terms of variance explained the goodness of fitting is better for FDM. This result is confirmed also looking at the following error measures as shown in Table 1 and Table 2: mean error (ME), mean square error (MSE), mean percentage error (MPE) and mean absolute percentage error (MAPE).

In the second part of the application, we trained and tested different types of ANFIS models. The parameters of interest include the number of epochs, maximum error, and number of neurons at the input and output. In our case, the number of epochs was set to 500. The maximum error has been set to $1e^{-}10$. We initialize with a fuzzy Sugeno system. Then we generate the FIS and obtain the graphical representation of its input/output structure. We test the data against the FIS model, trained according to the optimization method. This generates error plots in the plot region. The gbell membership function provides better outcomes than the other functions like gauss2mf, trapezodial, triangular and gauss. For training the Fuzzy Inference System (FIS) membership function parameters we used the hybrid optimization method, which is a combination of

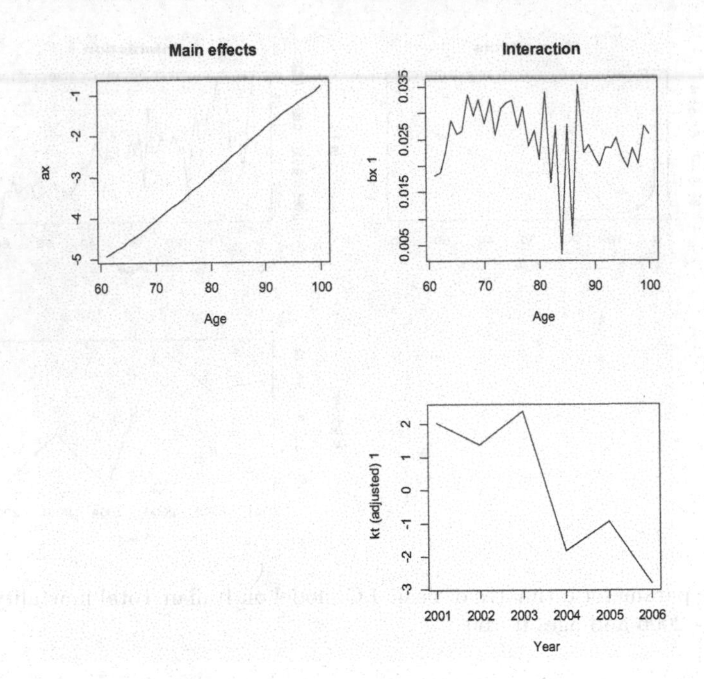

Fig. 2. The parameter estimates of basic LC model on Italian Total mortality data for years 2001 - 2006 and ages 61 - 110

Table 1. Basic LC model error measures on Italian data

Averages across ages (0-60)			
ME	MSE	MPE	MAPE
0.00000	0.00000	0.00718	0.08855
Averages across ages (61-110)			
ME	MSE	MPE	MAPE
0.00006	0.00002	0.00141	0.02164

least-squares and backpropagation gradient descent method. In order to compare the Adaptive Neuro Fuzzy Inference System to the LC and FDM model, we considered the training error, given by the difference between the training data output value and the output of the fuzzy inference system corresponding to the same training data input value. In particular, the training error records the root mean squared error (RMSE) of the training data set at each epoch. The results, in terms of RMSE, among the three models under consideration are compared in Table 3.

Table 2. FDM model error measures on Italian data

Averages across ages (0-60)			
ME	MSE	MPE	MAPE
0.00009	0.00054	0.00000	0.00222
Averages across ages (61-110)			
ME	MSE	MPE	MAPE
0.00000	0.00017	0.00008	0.00384

Table 3. RMSE - Comparison between models

RMSE (0-60)		
LC	FDM	ANFIS
0.1208718	0.0232379	2.724e-005
RMSE (61-110)		
LC	FDM	ANFIS
0.09055385	0.0130384	3.6742e-005

5 Concluding Remarks

In this contribute we have run a comparative analysis between stochastic models and Adaptive Neuro-Fuzzy Inference System applied to the projection of the longevity trend. We expect that additional research can better explore the actuarial significance of the observed deviations. Possible extensions could include independent variables in the model, for instance health status, working condition, gender, location, etc. Our final conclusion is that the system of mortality tables used to value longevity phenomenon in actuarial practice can be improved by using Artificial Intelligence-based prediction models.

References

1. Atsalakis, G., Nezis, D., Matalliotakis, G., Ucenic, C.I., Skiadas, C.: Forecasting Mortality Rate Using a Neural Network with Fuzzy Inference System. Working Papers 806 from University of Crete, Department of Economics (2007)
2. Booth, H.: Demographic forecasting: 1908 to 2005 in review. International Journal of Forecasting 22, 547–581 (2006)
3. Booth, H., Maindonald, J., Smith, L.: Applying Lee-Carter under conditions of variable mortality decline. Population Studies 56(3), 325–336 (2002), doi:10.1080/00324720215935

4. Butt, Z., Haberman, S.: A comparative study of parametric mortality projection models (Report No. Actuarial Research Paper No. 196). London, UK: Faculty of Actuarial Science & Insurance, City University London (2010)
5. Coletti, G., Scozzafava, R.: Characterization of Coherent Conditional Probabilities as a Tool for their Assessment and Extension. International Journal of Uncertainty, Fuzziness and Knowledge-Based Systems 4, 103–127 (1996)
6. Coletti, G., Scozzafava, R., Vantaggi, B.: Soft Computing: State of the Art Theory and Novel Applications. STUDFUZZ, vol. 291, pp. 193–208. Springer, Heidelberg (2013)
7. Currie, I.D., Durban, M., Eilers, P.H.C.: Smoothing and forecasting mortality rates. Statistical Modelling 4(4), 279–298 (2004), doi:10.1191/1471082X04st080oa
8. Delwarde, A., Denuit, M., Eilers, P.: Smoothing the Lee-Carter and Poisson log-bilinear models for mortality forecasting: a penalised log-likelihood approach. Statistical Modelling 7, 29–48 (2007)
9. Diamond, P.: Fuzzy least-squares. Information Sciences 3, 141–157 (1988)
10. Hatzopoulos, P., Haberman, S.: A dynamic parameterization modelling for the age-period-cohort mortality. Insurance: Mathematics and Economics 49(2), 155–174 (2011)
11. Human Mortality Database. University of California, Berkeley (USA), and Max Planck Institute for Demographic Research (Germany), `http://www.mortality.org`, `http://www.mortality.org` (referred to the period 2001 to 2006)
12. Hyndman, R.J., Ullah, S.: Robust forecasting of mortality and fertility rates: a functional data approach. Computational Statistics and Data Analysis 51, 4942–4956 (2007)
13. Jang, J.S.R.: ANFIS: Adaptive-Network-based Fuzzy Inference Systems. IEEE Trans. on Systems, Man, and Cybernetics 23, 665–685 (1993)
14. Lee, R.D., Carter, L.R.: Modelling and Forecasting U.S. Mortality. Journal of the American Statistical Association 87, 659–671 (1992)
15. Scozzafava, R., Vantaggi, B.: Fuzzy Inclusion and Similarity through Coherent Conditional Probability. Fuzzy Sets and Systems 160, 292–305 (2009)
16. Skiadas, C., Matalliotakis, G., Skiadas, C.: An extended quadratic health state function and the related density function for life table data. In: Skiadas, C. (ed.) Recent Advances in Stochastic Modeling and Data Analysis, pp. 360–369. World Scientific, Singapore (2007)
17. Tanaka, H., Uejima, S., Asai, K.: Linear Regression Analysis with Fuzzy Model. IEEE Transactions on Systems, Man and Cybernetics 12(6), 903–907 (1982)

Part IV
Special Session on "Social and Emotional Networks for Interactional Exchanges"

Recent Approaches in Handwriting Recognition with Markovian Modelling and Recurrent Neural Networks

Laurence Likforman-Sulem

Institut Mines-Telecom/Télécom ParisTech & CNRS LTCI, 46 rue Barrault, 75013 Paris, France
laurence.likforman@telecom-paristech.fr
http://perso.telecom-paristech.fr/~lauli/

Abstract. Handwriting recognition is challenging because of the inherent variability of character shapes. Popular approaches for handwriting recognition are markovian and neuronal. Both approaches can take as input, sequences of frames obtained by sliding a window along a word or a text-line. We present markovian (Dynamic Bayesian Networks, Hidden Markov Models) and recurrent neural network-based approaches (RNNs) dedicated to character, word and text-line recognition. These approaches are applied to the recognition of both Latin and Arabic scripts.

Keywords: Word recognition, Text-line recognition, Hidden Markov Models, Recurrent Neural Networks, BLSTMs.

1 Introduction

Handwriting recognition started more than 20 years ago with the recognition of bank check numerical amounts, mail postal codes and handwprinted characters in forms. Feedforward neural networks and template matching were very popular, as well as structural and syntaxic pattern recognition approaches. Cursive word recognition followed [1,2] in order to recognize street or town names, and litteral bank check amounts. The most popular approach for word recognition has been till 2009 Hidden Markov Models (HMMs). The main advantage of HMMs, in particular the segmentation-free analytical approach, is that they do not need a segmentation of a word into its compound characters, which is prone to errors, neither for training nor recognition.

Word recognition has been soon followed by text-line recognition. Text-line recognition, by recognizing all words in a document, allows companies and administrations to automatically sort or search through the large amount of handwritten mails they receive. From word recognition systems, text-line recognition systems can be derived, adding a language model.

The neural network approach has been recently revisited [3] with recurrent neural networks (RNNs) associated to LSTM blocks. Such association produces BLSTM (bi-directionnal LSTM networks) and MDLSTM (multi-dimensional

S. Bassis et al. (eds.), *Recent Advances of Neural Network Models and Applications*, Smart Innovation, Systems and Technologies 26,
DOI: 10.1007/978-3-319-04129-2_26,

LSTM) architectures. BLSTMs can process the same frame sequences as HMMs, with the ability to take into account the past and future contexts of a frame. BLSTM modelling is discriminative in constrast to HMM generative modelling. This yields high recognition accuracies even at the character level.

We present in the following, our recent research on handwriting recognition with DBNs, HMMs and RNNs. Experiments have been conducted on publicly available databases.

1.1 Character Recognition with Dynamic Bayesian Networks

Discriminative approaches for character recognition (convolutional neural networks, support vector machines) are powerful since they learn to discriminate classes through frontier. But they are sensible to degradations: additional noise, shape deformation, breaks within characters. In contrast, generative approaches (HMMs) better cope with degradation since they rely on scores, provided for each character and each class.

We have applied the generative DBN (dynamic bayesian networks) approach to broken and degraded character recognition [4]. This consists in the coupling of 2 hidden Markov processes through the DBN framework, one observing the rows, the other the columns (Fig. 1). State transition matrices encode the dependencies between rows and columns. In the example provided in Fig. 2, the states are denoted by X_i^j with i being the time-step and j the observation stream. The states of the markovian process observing the columns are linked to the states of the process observing the rows. This model is auto-regressive since observation variables are linked through time.

We have applied this approach to character recognition. Ten networks are trained, one for each character class. All character classes share the same DBN architecture but their parameters differ. The training phase consists in learning network parameters [5]. Characters [6] are normalized in size, then rows and columns are input in the bayesian network as observation variables Y. Inference algorithms are performed on the network to compute the likelihood of observation sequences.

These models provide a certain robustness to degradation due to their ability to cope with missing information. They have the ability to exploit spatial correlations between observations. Thus a corrupted observation (row or column) in the image can be compensated by an uncorrupted one. The improvement brought by coupled models increases as the level of degradation increases.

1.2 Word and Text-Line Recognition with Hidden Markov Models

Hidden Markov models are convenient for handwriting recognition. They can cope with non-linear distortions and offer a character-based representation of words: words models are obtained by the concatenation of compound character models which is convenient for enlarging the vocabulary.

We have chosen the segmentation-free strategy. This avoids the error-prone segmentation of text-lines into words, and words into characters. Moreover this

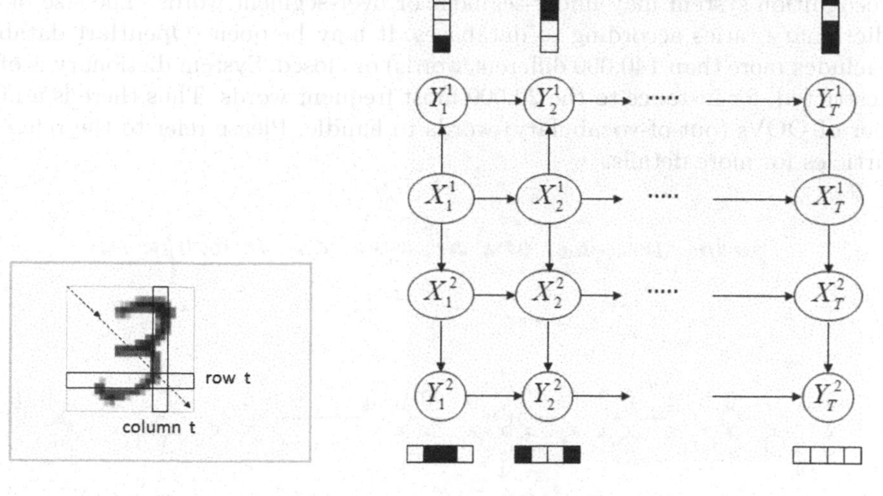

Fig. 1. Columns are the observations for the first stream, rows for the second stream

Fig. 2. Auto-regressive coupled DBN architecture: Xs are the hidden state variables, Ys the observation variables

strategy is closer to the human reading process which jointly segments handwriting into words and recognize them. A sliding window extracts a sequence of frames from the word or text-line image from left to right (for Latin script) or right to left (for Arabic). A generative markovian modelling of compound characters is built from such frame sequences along with their textual transcription. Word models are built from character models and text-line models from word models (cf. Fig. 3). Frames are composed of a set of features (directional, geometric, statistical and derivative). Image preprocessing is necessary to reduce writing variability and remove fragments from neighboring text-lines. Recognition within the HMM framework requires a dictionary, and a language model when recognizing text-lines.

We have proposed several contributions to the recognition of handwriting with HMMs: the definition of a set of efficient features and the use of slanted sliding windows to cope with shifted diacritical marks [7], the use of context-dependent character models to take into account the variations of character shapes according to the preceding and following characters. Such models require parameter tying, performed in our work with decision trees [8]. We have also developed specific text-line preprocessings [9] and language models adapted to a mail reading task [10]. We have built context-independent and context-dependent recognition systems for three major databases: IFN/ENIT [11,12], Rimes [13,14] and OpenHart [15].

Recognition rates depend on task type, image quality, and size of the dictionary. It is easier to recognize isolated words than text-lines: a text-line recognition system may under-segment or over-segment words. The size of the dictionary varies according to databases. It may be open (OpenHart database includes more than 140,000 different words) or closed. System dictionary is often restricted, for instance to the 20,000 most frequent words. Thus there is a number of OOVs (out-of-vocabulary) words to handle. Please refer to the reference articles for more details.

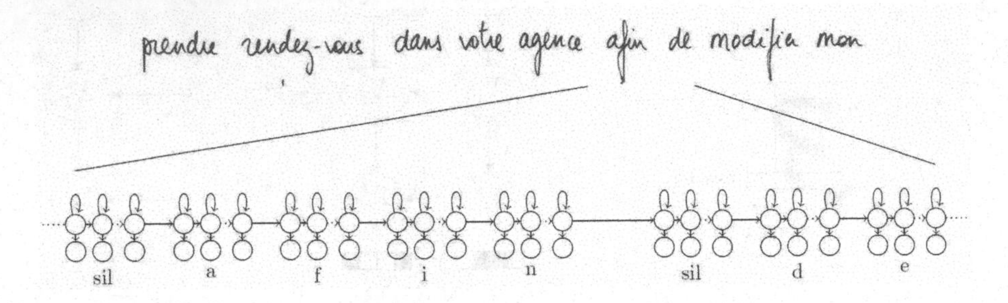

Fig. 3. Text-line modelling with Hidden Markov Models

1.3 Text-Line Recognition with Recurrent Neural Networks

Recurrent neural networks (RNNs) consist of an input layer, an output layer and an hidden layer. RNNs differ from classical feedforward networks by the fact that outputs of hidden layers, at time $t-1$, are fed as inputs to hidden layers at time t. This is the so-called recurrent links within hidden layers. BLSTM (Bi-directional long short-term memory) is recurrent neural network architecture which consists in the coupling of 2 RNNs and the use of memory blocks called LSTMs (Long Short-Term Memory).

Each recurrent network takes as input a frame sequence as for HMMs, one network (called the forward network) takes the original frame sequence as input (from $t = 1$ to $t = T$) while the backward network takes as input the reversed sequence (from $t = T$ to $t = 1$) (Fig. 4). At each time step t, the input layer consists of the frame extracted at t. The output layer consists of 91 units, each unit being associated with a given character (A to Z, a to z, numerals, symbols). The two networks, forward and backward, share the same input and output layers. They differ in their hidden layers. Hence, the value of an output unit at time-step t is the linear combination of the outputs of the forward and backward hidden layers at this time-step t. The outputs of the forward hidden layer, at time-step t, depend on the inputs at t and the outputs of the hidden layer at $t-1$ and, subsequently, depend on all previous frames (1 to $t-1$). Similarly, the backward hidden layer at time t depends on all following frames ($t+1$ to T). Thus long-term dependencies are taken into account.

To avoid the decay in the error signals during RNN learning (this issue is referred to vanishing gradient), LSTM blocks in the hidden layer include a memory cell and multiplicative logical gates which are specifically designed to memorize or forget information through time. Those gates can pass or block signals.

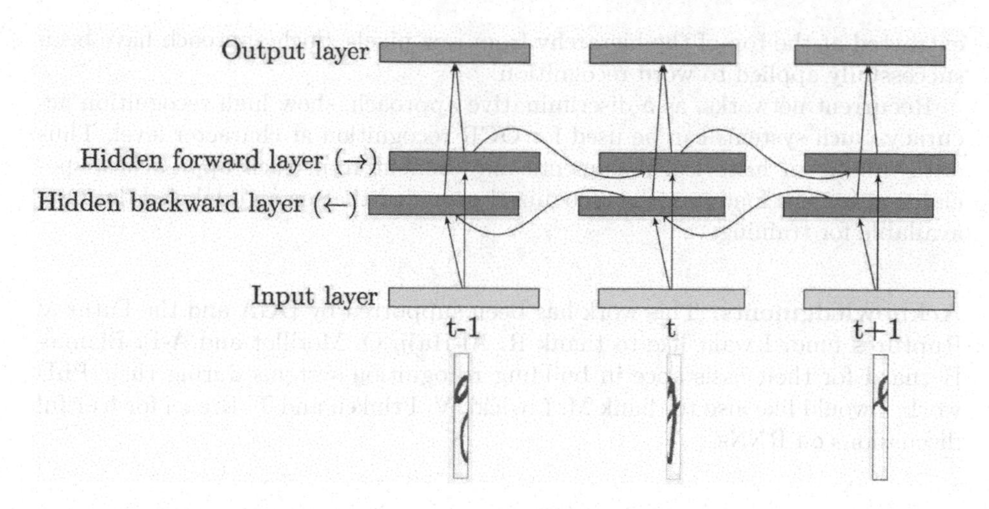

Fig. 4. Bidirectional network unfolded through 3 time steps. Inputs are features extracted from a sliding window shifted at each time step.

A BLSTM computes, for each frame its outputs, each associated to a character class. Then, a backward-forward token passing algorithm, referred to as connectionist temporal classification (CTC), provides the best sequence of words given the dictionary and the language model. Dictionary and language model are based on training set transcriptions or external resources.

BLSTMs, when correctly trained, outperform HMMs. On the Rimes text-line database the word error rate (WER) equals 16.1% for BLSTMs and 31.2% for HMMS. This shows the interest of BLSTMs over HMMs. Even without dictionary and language model, BLSTMs outputs at character level include a number of real words: most errors are doubled or missing characters. However BLSTM training is currently much longer than that of HMMs and needs more training data.

A RNN recognizer developed by T. Breuel is available through OCRopus project [16]. Another version is released by A. Graves [17].

2 Conclusion

We have presented several approaches for recognizing handwritten characters, words and text-lines. They are based on the markovian framework (Dynamic Bayesian Networks, Hidden Markov Models) and Recurrent Neural Networks.

We have presented the BLSTM recurrent network architecture, useful for text-line recognition. BLSTMs take as input the same frame sequences as HMMs. Other RNN architectures exist: the MDLSTM (multi-dimensional LSTM) architecture [18] which extends the 2-pass BLSTM framework to 4-passes. MDLSTM combined with hierarchical subsampling produces the hierarchical subsampling recurrent neural network (HSRNNs) architecture. Features are automatically extracted at the top of the hierarchy from raw pixels. Such approach have been successfully applied to word recognition.

Recurrent networks, as a discriminative approach, show high recognition accuracy. Such systems can be used for OCR recognition at character level. Thus data mining in historical documents may benefit from such approaches: special and ancient fonts may be recognized provided that enough labeled data are available for training.

Acknowledgments. This work has been supported by DGA and the Futur & Ruptures fund. I want like to thank R. Al-Hajj, O. Morillot and A-L. Bianne-Bernard for their assistance in building recognition systems during their PhD work. I would like also to thank M. Liwicki, V. Frinken and T. Breuel for fruitful discussions on RNNs.

References

1. Vinciarelli, A.: Online and offline handwriting recognition: A comprehensive survey. Pattern Recognition 35(7), 1433–1446 (2002)
2. Plamondon, R., Srihari, S.: Online and offline handwriting recognition: A comprehensive survey. IEEE Transactions on Pattern Analysis and Machine Intelligence 22(1), 63–84 (2000)
3. Graves, A., Liwicki, M., Fernandez, S., Bertolami, R., Bunke, H., Schmidhuber, J.: A novel connectionist system for unconstrained handwriting recognition. IEEE Transactions on Pattern Analysis and Machine Intelligence 31(5) (2009)
4. Likforman-Sulem, L., Sigelle, M.: Recognition of degraded characters using dynamic bayesian networks. Pattern Recognition 41(10), 3092–3103 (2008)
5. Murphy, K.: BayesNet Toolbox for Matlab (2003), `http://www.ai.mit.edu/~murphyk/Bayes/bnintro.html`
6. Likforman-Sulem, L., Sigelle, M.: ENST ancient character database (2008), `http://perso.telecom-paristech.fr/~lauli/ENST_ANCIENT_CHAR/`
7. Al-Hajj-Mohamad, R., Likforman-Sulem, L., Mokbel, C.: Combining slanted-frame classifiers for improved HMM-based arabic handwriting recognition. IEEE Transactions on Pattern Analysis and Machine Intelligence 31(7), 1165–1177 (2009)
8. Bianne-Bernard, A.L., Menasri, F., El-Hajj, R., Mokbel, C., Kermorvant, C., Likforman-Sulem, L.: Dynamic and contextual information in HMM modeling for handwritten word recognition. IEEE Transactions on Pattern Analysis and Machine Intelligence 33(10) (2011)
9. Morillot, O., Likforman-Sulem, L., Grosicki, E.: New baseline correction algorithm for text-line recognition with bi-directional recurrent neural networks (2013)
10. Morillot, O., Likforman-Sulem, L., Grosicki, E.: Construction of language models for an handwritten mail reading system. In: Proceedings of Electronic Imaging - Document Recognition and Retrieval (DRR) XIX (2012)
11. Pechwitz, M., Maddouri, S.S., Märgner, V., Ellouze, N., Amiri, H.: IFN/ENIT-database of handwritten arabic words. In: Colloque International Francophone sur l'Ecrit et le Document (2002)
12. Märgner, V., El-Abed, H.: ICDAR 2009 Arabic Handwriting Recognition Competition. In: International Conference on Document Analysis and Recognition (2009)

13. Grosicki, E., El-Abed, H.: ICDAR 2009 handwriting recognition competition. In: International Conference on Document Analysis and Recognition, pp. 1398–1402 (2009)
14. Grosicki, E., El-Abed, H.: ICDAR 2011: French handwriting recognition competition. In: International Conference on Document Analysis and Recognition (2011)
15. NIST: OpenHart 2013 Evaluation (2013), `http://www.nist.gov/itl/iad/mig/hart2013.cfm`
16. Breuel, T.: OCRopus - Open Source Document Analysis and OCR system (2013), `https://code.google.com/p/ocropus`
17. Graves, A.: RNNlib (2009), `http://www6.in.tum.de/Main/Graves`
18. Graves, A., Schmidhuber, J.: Offline handwriting recognition with multidimensional recurrent neural networks. In: Neural Information Processing Systems, pp. 545–552 (2008)

Do Relationships Exist between Brain-Hand Language and Daily Function Characteristics of Children with a Hidden Disability?

Sara Rosenblum[1] and Miri Livneh-Zirinski[2]

[1] Department of Occupational Therapy, Faculty of Social Welfare and Health Sciences, University of Haifa, Haifa, Israel
rosens@research.haifa.ac.il

[2] Child Development Center, Kupat Holim Meuhedet, Haifa, Israel (retired)

Abstract. Objective: To discover whether children with a hidden disability such as Developmental Coordination Disorders (DCD) have unique brain-hand language (handwriting) and daily function characteristics and whether there are relationships between these characteristics.

Method: 20 children diagnosed with DCD and 20 typically developed controls aged 7-10 performed the Alphabet writing task on a page affixed to an electronic tablet, a component of the ComPET which documented their handwriting process. Further, their organizational ability was evaluated through daily function events as reported by their parents.

Results: Significant group differences (DCD versus controls) were found in the coefficient of variance of spatial, temporal and pressure writing process measures. Specific handwriting measures predicted the level of children's organization abilities through daily function.

Conclusions: These results emphasize the need for further development of sophisticated computerized methods so as to gain deeper insight concerning daily function characteristics of children with hidden disabilities.

Keywords: handwriting, DCD, daily function.

1 Introduction

Approximately 5-10% of children are faced with a Hidden disability (HD) such as a Learning Disability (LD), Attention Deficit Hyperactivity Disorder (ADHD) and Developmental Coordination Disorders (DCD).

In the literature, this population is included under the definition of neuro-developmental disabilities or Atypical Brain Development (ABD), as result of large comorbidity between these phenomena (LD, DCD, ADHD) [1]. Children with HD represent a host of functional limitations, but they and their families may for years be unaware that their deficits in daily function indeed have a defined title [2].

Developmental Coordination Disorder (DCD), labeled in the past as 'Clumsiness' is one of the most common HD affecting school-aged children, characterized by

S. Bassis et al. (eds.), *Recent Advances of Neural Network Models and Applications*,
Smart Innovation, Systems and Technologies 26,
DOI: 10.1007/978-3-319-04129-2_27, © Springer International Publishing Switzerland 2014

motor impairment that interferes with the child's activities of daily living and academic achievements [3-4]. The prevalence of DCD ranges from 1.4% to 19% among school aged children, depending on the definition and used methods of evaluation [5-6].

DCD is diagnosed according to four criteria of the DSM-4. Criterion A: A marked impairment in the development of motor coordination. Criterion B: The diagnosis is made only if this impairment interferes with academic achievement or activities of daily living. Criteria C and D restrict that the coordination disorders are not due to a medical condition or beyond difficulties bounded by mental retardation [7]. According to criteria A and especially B, handwriting difficulty may be a manifestation of motor coordination deficits among school aged children and hence consist a key point for their diagnosis as DCD [8].

Handwriting as a brain-hand language is a complex human activity, compounded of cognitive, kinesthetic, perceptual, and motor components [9-10]. It is considered an "over-learned" skill involving particularly rapid sequencing of movements.

In this study, handwriting performance characteristics of children with DCD and controls were evaluated with the Computerised Penmanship Evaluation Tool (ComPET) [11]. The purpose of the study was to discover whether children aged 7-9 with DCD indeed master the Alphabet sequence which constitutes the building blocks required for written communication and academic achievements [12-14].

Similarly, based on previous results related to handwriting performance characteristics among children with DCD, in the current study the coefficient of variance of temporal spatial and pressure measures of the Alphabet writing task were analysed. Coefficient of variance (CV) is defined as the ratio of the standard deviation to the mean and is a normalized measure of dispersion of a probability distribution which in fact may reflect the variability of the individual's performance. Motor control theories have indicated that the more automatic the performance is and does not require cognitive efforts, the more consistent and less variant it will be [15-16].

A further aim of the study was to discover whether relationships exist between these handwriting measures and daily performance characteristics of children with DCD. Specifically, this study focused on 'organizational ability' [17-18] as reflected in daily function which was evaluated through the Daily Activities Performance scale for children aged 5 to 12 (DAP) developed by Dewey, Summers, & Larkin [19]. Zentall, Harper, and Stormont-Spurgin (1993)[20] defined organizational ability as the ability to plan and execute an activity within limited time, place objects where they will be easily found and plan how to execute the activity.

Hence, the research hypotheses were as follows:

1. Significant differences will be found between children with DCD and controls in:
   (a) Handwriting process measures of the alphabet sequence writing.
   (b) Organizational ability as reflected through daily function characteristics.
2. Significant correlations will be found between handwriting performance measures and organizational ability scores.
3. Several handwriting measures will predict organizational ability as reflected through daily function.

2 Methods

2.1 Participants

Twenty children with DCD (2 girls, 18 boys), ranging from 7-9 years old, and 20 age- and gender-matched control children participated in the study.

The children with DCD were recruited from developmental centers in Northern Israel. All the children in the DCD group were initially diagnosed by a pediatrician and met the criteria of DSM-4 for DCD [7]. Subsequent testing with the Movement Assessment Battery for Children (M-ABC) [21] was performed in order to verify their status as children with DCD. Children who scored below the 15th percentile for their age on the M-ABC were included in the DCD group.

The control group was recruited from the same local schools where the children included in the DCD group attended. The 20 controls had no symptoms of DCD, as indicated by the children's parents and teachers, as well as within norm M-ABC scores [21]. Detailed description of the M-ABC scores and further demographic characteristics of both groups are presented in Table 1.

Children with known neurotic/emotional disorders, autistic disorders, physical disabilities, or neurological diseases were excluded from the study. All subjects were native Hebrew speakers, attended school, and reported no hearing or vision problems.

Table 1. Background Characteristics of participants in both groups (DCD versus control)

		DCD group	Control group	T (79)	P
		M (SD) *n=20*	*M (SD)* *n=20*		
Age (in months)		7.9(.88)	7.8(.86)	.29	NS
Years of education – mother		13.64 (2.80)	14.75 (2.46)	-1.89	NS
M-ABC Fin. Score		15.08 (6.58)	8.87 (3.09)	6.13	<.0001
		Percentile	Percentile	$\chi 2$	p
Gender	Male	90% (18)	90% (18)	.064	NS
	Female	10% (2)	10% (2)		
Handwriting difficulty (parents report)	Yes	75% (15)	5% (1)	27.56	<.001
	No	25% (5)	95% (19)		

2.2 Instruments

Computerized Penmanship Evaluation Tool (ComPET, previously referred to as POET) [11].

In the current study, children were asked to write the alphabet sequence from memory on an A4-sized lined sheet of paper affixed to the surface of a WACOM Intuos II x-y digitizing tablet (404 X 306 X 10 mm), using a wireless electronic pen with a pressure-sensitive tip (Model GP-110). This pen is similar in size and weight to regular pens commonly used by children and therefore does not require a change in grip that might affect their writing performance (see Fig. 1). Displacement, pressure and pen tip angle are sampled at 100 Hz via a laptop computer. The primary outcome measures are comprised of temporal, spatial and pressure measures for each writing stroke, as well as performance over the entire task. Further data regarding the pen azimuth and tilt is also supplied by the system.

Based on previous analysis and the literature [22], the Coefficient of Variance of the following measures was analyzed:

1. Writing velocity on paper of the entire task in cm/second.
2. Maximal stroke's velocity level
3. Number of peak velocity per strokes- A measure for handwriting movement regularity, with the assumption being that the more peaks in one stroke, the less regular the movement will be [23].
4. Writing pressure applied by the pen towards the writing surface during the entire task measured in non scaled units from 0-1024.
5. Pen tilt while writing – measured at 0-90 degrees.
6. Stroke height (on the Y axis), which measures in millimeters the direct distance from the lower point of the stroke to the highest point.

The present study focused on writing the Hebrew alphabet. As demonstrated in Figure 1 some of the letters are constructed of two separate, unconnected components or strokes. For example, the first letter (Alef א) and the fifth letter (Hey ה), (underlined) are each formed by two separate components.

To enable comparison of study results in Hebrew writing with those of other languages, the measures for the writing strokes are obtained for the entire writing task performance and are not based on single letters.

Fig. 1. The Hebrew Alphabet sequence (22 letters written from right to left)

Daily Activities Performance for children aged 5 to 12 – parent questionnaire (DAP) [20].

The DAP was designed to explore the impact of children's motor coordination deficits on their daily functioning characteristics as reported by their parents.

The questionnaire is comprised of three sections, the first includes one page of demographic questions (e.g., mother/father age, education, number of people living at home etc.).

The second section includes 45 items that cover daily functions: dressing ("Requires physical assistance with clothing and fastenings"), personal hygiene ("Bathes without supervision"), eating behavior ("Uses a spoon and fork together efficiently"), homework ("Does his/her homework on time with very little prompting"), and play and academic tasks ("Has ball skills similar to age-mates"). The parent is required to score each item in relation to his child, on a Likert scale from 1 to 5 (1=not at all like my child – 5= A lot like my child). In fact, performing daily activities such as these by the child requires good organization in space and time skills.

The third section of the DAP consists of 8 questions about to the child's enjoyment level related to varied play situations with other children (e.g., running, computer games, crafts etc.). The parent is requested to mark on a five point Likert scale whether the statement that appears in each item represents his/her child very much (5) or does not represent his/her child at all (1).

Internal reliability for the DAP in the current study was found to be α=.946.

2.3 Procedure

The study design conformed to the instructions stipulated by the University of Haifa's Ethics Committee. Parents of children diagnosed by the pediatrician with DCD, were asked to sign an informed consent and to complete a demographic questionnaire. Further, parents were asked to respond to whether their child has handwriting difficulties (yes or no).

Parents were requested to complete the Daily Activities Performance (DAP).The children were evaluated at a developmental center by an occupational therapist who administrated the M-ABC test and the handwriting (ComPET) evaluation. Children who did not meet the M-ABC criteria were excluded from the study. For each child in the DCD group, an age and gender-matched control child was chosen from the same school and the same procedure described above was administered.

2.4 Statistical Analysis

Descriptive statistics of the dependent variables were tabulated and examined. T-tests were calculated to compare differences in the M-ABC and DAP final scores and children age in both groups.

Chi square tests were performed in order to discover dependency between groups (DCD versus controls) and gender as well as handwriting difficulties based on parents report.

MANOVA analyses were then used to test for group differences across the kinematic handwriting measures of the Alphabet writing. Univariate ANOVA analyses were used to determine the source for the group differences.

Pearson correlations were calculated to investigate the associations between handwriting performance measures and the daily function scales scores (DAP) in each group (DCD versus controls).

A stepwise regression analyses was applied in order to determine whether kinematic handwriting measures (ComPET) predict organization ability through daily function as measured by the DAP, beyond group membership (DCD versus typical).

3 Results

Background Characteristics

As expected and presented in Table 1, significant differences were found between the groups (DCD versus controls) on the M-ABC final score (t(78)=9.05 p<.00) while children with DCD revealed significantly higher scores, meaning lower motor coordination proficiency.

Consequently, significantly more children with DCD were labeled by their parents as dealing with handwriting difficulties.

3.1 Between Group Differences in Handwriting Kinematic Measures (ComPET)

The MANOVA yielded statistically significant differences between the groups for the handwriting kinematic measures (Wilks Lambda= .57 $F_{(6,32)}$=4.05, p=.004 η^2=.43). As presented in Table 2, the subsequent Univariate ANOVA analyses revealed that the significance was due to differences between the groups in five kinematic handwriting performance measures (CV of: velocity, stroke height, pressure, pen tilt and number of peak velocity).

Table 2. A comparison of the kinematic handwriting process measures of the Alphabet Task in both groups

	DCD group *M (SD)* *n=20*	**Control group** *M (SD)* *n=20*	**F(1,39)**	**P**	η^2
Velocity on paper CV	114.83 (25.18)	92.61 (10.30)	13.25	.001	.26
Maximal velocity per strokes	7.52 (1.04)	8.21 (.67)	.92	.34	.024
Number of peak velocity per strokes	113.63 (37.98)	84.18 (24.20)	8.42	.006	.18
Pressure CV	29.14 (5.91)	24.87 (5.83)	5.16	.029	.12
Pen Tilt CV	11.06 (3.24)	8.66 (3.20)	5.37	.026	.12
Strokes height CV	47.82 (10.13)	39.46 (9.21)	7.27	.010	.16

3.2 Between Group Differences in the Daily Function Scale (DAP) Final Score

Next, analyses were performed to draw a comparison between the daily function scale score of the study groups and the results are presented in Table 3. The DAP yielded significant group differences. (t(38)= -5.34 p<.001) with a considerably large gap between the groups.

Table 3. A comparison of organization ability through daily function

	DCD group *M (SD)* *n=20*	Control group *M (SD)* *n=20*	t(38)	P
Daily Activities Performance (DAP)	188.94 (26.84)	227.05(16.33)	-5.34	<.001

3.3 Correlations between ComPET Measures and the Daily Function Scale Score in Each Group (DCD versus Controls)

Significant medium correlations were found among the DCD group between the DAP final score and the CV of the velocity on the paper: r=.664 p<.01).

3.4 Handwriting Kinematic Measures (ComPET) As Predictors of Daily Function Ability (DAP) [20]

In the hierarchical regression, the DAP final score was entered as a dependent variable in order to determine whether the 6 handwriting kinematic measures (detailed in table 2) predicts daily function ability, beyond the group accountability (DCD versus control).

The results presented in Table 4 indicate that the group accounted for 46% of the variance of the DAP score (F (1, 36) = 30.25, β= .68, p<.0001), while among the handwriting kinematic measures only the CV of the velocity on paper added 12% of prediction (F (2, 36) =23.39 β= . 39 p<.05). All in all, this handwriting performance measure accounted for 12% of the variance of the DAP final score above group membership (DCD versus controls).

Table 4. Predicting daily function abilities based on the Daily Activities Performance (DAP) questionnaire by handwriting kinematic measures

	Model 1			Model 2		
Variable	B	SE B	Β	B	SE B	Β
Group	39.58	7.20	.68***	51.45	7.54	.88***
Velocity on paper CV				.59	1.19	.40**
R^2 (Adj.rsq)	.46 (.45)			.58 (.55)		
F change in R^2	30.25***			9.33**		

**p<.01 ***p<.0001

4 Discussion

The aim of this study was to discover whether the Computerized Penmanship Evaluation Tool (ComPET) will expose unique brain-hand language performance characteristics of children with a hidden disability such as Developmental Coordination Disorders (DCD), previously labeled as clumsiness. A further aim was to find whether relationships exist between their handwriting and daily function characteristics with a focus on organizational abilities.

Results indicated significant differences between children with DCD and those with typical development in all the five kinematic handwriting performance measures (CV of: velocity, stroke height, pressure, pen tilt and number of peak velocity).

In fact, the results of the current study indicated that the performance of children with DCD is more variable, less consistent and less sequential.

Latash [24] remarked that automatic handwriting movements increase efficacy and reduce redundancy. The more skilled and automatic the handwriting act, less variability in temporal (performance time), and spatial (length, height, width) measures will exist, and greater consistency will be evident [25].

The current results manifested that unlike children of their age, children with DCD have not acquired the ability to automatically produce the Alphabet letters which constitute the building blocks of the written language [26].

It seems that such evidence may explain previous findings about significantly longer performance time and less temporal and spatial consistency among children with DCD compared to controls while copying a sentence or a paragraph [27]. Researchers have emphasized that the ability to fluently construct correctly formed single letters represents a necessary lower level building block for the development of proficient handwriting, and for the higher-order processes involved in composing text [13], [28], thus contributing to the enhancement of overall learning potential [28-29]. A child's mastery of lower level writing skills, considered to be the transcription phase, is a basic and critical phase in the functional development of handwriting and substantially influences the quantity and quality of text composition [30].

Such deficits in the transcription phase among children with DCD may interfere with their availability to the higher level of producing the written content [31], as they need to invest extra energy in the actual on paper writing production.

The significant differences found between the groups related to their daily function (DAP questionnaire) support previous findings. Researchers reported that children with DCD have difficulties with activities of daily living such as dressing, personal hygiene, cutting, eating and playing [32-36]. The novelty in the current results is thatthe DAP focuses on organizational abilities as required in daily function and these abilities were found to be deficient among the children with DCD.

Organization ability is a component of Executive Functions (EF). These results support previous results as regards to the significant differences between children with DCD and typical controls in EF of organization and decision-making [37] or working memory [38].

Further to finding significant differences in ADL performance as related to organizational abilities, the next question was whether the objective measures of handwriting performance will correlate with the daily function questionnaire scores and which of the handwriting measures will predict organizational abilities through daily function.

Among the DCD group, significant medium correlations were found between the DAP final score and the CV of the velocity on the paper.

Furthermore, the handwriting measure means the CV of the velocity added considerable percent of variance to prediction of the of the DAP score.

As previously mentioned, the CV is a normalized measure of dispersion of a probability distribution which in fact may reflect the variability of the individual's performance. Therefore, the meaning of the results is that velocity profile of performance of actual activity (handwrting) of children with DCD may predict their organizational abilities in daily performance.

These findings are in accordance with handwriting transcription models that have indicated the involvement of EF in the writing production process, with an emphasis on planning and organization in time and space [39-40]. Such results indicate that handwriting is indeed a complex human activity that may serve as a sensitive measure for organization ability, one of the EF components, as reflected in daily function in varied activities.

These current and previous results indicate that the motive of time is of importance among this population and indeed there is evidence in the literature of timing deficits among children with DCD. Findings indicated that children with DCD have internal deficits in sense of time that affect their ability to perform precise, synchronized movements at a reasonable pace and that their movements are more temporally variable in comparison to controls [41-45]. Functional, sequential, continuous everyday tasks involving timing ability are dynamic and rapidly changing, their performance is demanding and requires complex cortical activation [46-47].

Thus, combining previous and present results, manifests that the handwrting measures indeed reflect hand function and daily function abilities among children with DCD, and it seems that the principle standing behind is their ability to organize in space and time while performing an action or a certain task.

Some explanations to the underlying brain mechanism behind function including handwriting have often been related to children with DCD. At the level of receiving and encoding varied stimuli, visual-spatial and kinesthetic processing deficits were found to be connected to their functioning [48-51]. Their timing deficits may be associated with cerebellar function deficits [52]. It was found that children with DCD demonstrated under-activation in cerebellar–parietal and cerebellar–prefrontal networks and in brain regions associated with visual-spatial learning [53].

Hence, poorer visual-spatial short term memory and problems in processing and storing temporal and spatial information among children with DCD as reflected through handwriting performance may underpin learning and daily function difficulties among this population [54].

Both deficits in handwriting production and in daily function found among children with DCD involve EF and reflect the importance of considering EF components such as attention planning and organization, reasoning and problem solving, conceptual thought, self-correction, judgment, and decision making among this population.

In summary, handwriting performance reflects brain-hand secrets related to further global deficits in EF required for daily function. Difficulties in daily function may cause this population frustration and have secondary emotional and social implications directly influencing their self-image [32], [55-57].

Thus, identification of the deficit using standardised tools, supplying knowledge to the individual about the phenomena, and finding appropriate strategies in order to confront the problem and improve quality of life among this population is of considerable importance.

References

1. Kaplan, B.J., Wilson, B.N., Dewey, D.M., Crawford, S.G.: DCD not be a discrete disorder. Human Movement Science 17, 471–490 (1998)
2. Josman, N., Rosenblum, S.: A meta-cognitive model for children with atypical brain development (ABD). In: Naomi, K. (ed.) Cognition, Occupation and Participation Across the Life Span, 3rd edn. AOTA Press, Bethesda (2011)
3. Cairney, J., Hay, J.A., Faught, B.E., Flouris, A., Klentrou, P.: Developmental coordination disorder and cardiorespiratory fitness in children. Pediatric Exercise Science 19(1), 20–28 (2007)
4. Wann, J.: Current approaches to intervention in children with developmental coordination disorder. Developmental Medicine & Child Neurology 49(6), 467–471 (2007)
5. Kadesjo, B., Gillberg, C.: Developmental coordination disorder in Swedish 7-year-old children. Journal of the American Academy of Child & Adolescent Psychiatry 38(7), 820–828 (1999)
6. Lingam, R., Hunt, L., Golding, J., Jongmans, M., Emond, A.: Prevalence of developmental coordination disorder using the DSM-IV at 7 years of age: A UK population-based study. Pediatrics 123(4), 693–700 (2009)
7. American Psychiatric Association. Diagnostic and statistical of mental disorders– DSM-IV, 4th edn. Author, Washington (1994)
8. Barnett, A.: Seminar 2: Assessment. Handwriting: Its assessment and role in the diagnosis of Developmental Coordination disorder. In: Sugden, D.A. (ed.) Developmental Coordination Disorder as a Specific Learning Difficulty, pp. 18–21. University of Leeds (2006)
9. Bonny, A.M.: Understanding and assessing handwriting difficulties: Perspective from the literature. Australian Occupational Therapy Journal 39, 7–15 (1992)
10. Reisman, J.: Development and reliability of the research version of the Minnesota Handwriting Test. Physical and Occupational Therapy in Pediatrics 13(2), 41–55 (1993)
11. Rosenblum, S., Parush, S., Weiss, P.L.: Computerized temporal handwriting characteristics of proficient and non-proficient handwriters. The American Journal of Occupational Therapy 57(2), 129–138 (2003)
12. McBride-Chang, C.: The ABCs of the ABCs: The development of letter-name and letter-sound knowledge. Merrill-Palmer Quarterly 45, 285–308 (1999)
13. Rosenblum, S.: Using the Alphabet task to differentiate between proficient and non-proficient handwriters. Perceptual and Motor Skills 100, 629–639 (2005)
14. Wasik, B.A.: Teaching the alphabet to young children. Young Children 56, 34–40 (2001)
15. Wilson, C., Simpson, S.E., Van Emmerik, R.E.A., Hamill, J.: Coordination variability and skill development in expert triple jumpers. Sports Biomechanics 7, 2–9 (2008)
16. Hiley, M.J., Zuevsky, V.V., Yeadon, M.R.: Is skilled technique characterized by high or low variability? An analysis of high bar giant circles. Human Movement Science 32, 171–180 (2013)
17. Godefroy, O.: Frontal syndrome and disorders of executive functions. Journal of Neurology 250, 1–6 (2003)

18. Godefroy, O., Aithamon, B., Azouvy, P., Didic, M., Le Gall, D., Marie´, R.M., et al.: Groupe de re´flexion sur l'evaluation des fonctions executives. Syndromes frontaux et dysexe´cutifs 160, 899–909 (2004)
19. Dewey, Summers, Larkin: Daily Activities Performance for children aged 5 to 12 – parent questionnaire (in process)
20. Zentall, S.S., Harper, G.W., Stormont-Spurgin, M.: Children with hyperactivity and their organizational abilities. Journal of Educational Research 87, 112–117 (1993)
21. Henderson, S.E., Sugden, D.: The movement assessment battery for children. The Psychological Corporation, London (1992)
22. Lacquaniti, F., Ferringo, G., Pedotti, A., Soechting, J.F., Terzuolo, C.: Changes in spatial scale in drawing and handwriting: Kinematic contributions by proximal and distal joints. Journal of Neuroscience 7, 819–828 (1987)
23. Mavrogiorgou, P., Mergl, R., Tigges, P., El Husseini, J., Schrter, A., Juckel, G., Zaudig, M., Hegerl, U.: Kinematic analysis of handwriting movements in patients with obsessive-compulsive disorder. J. Neurol. Neurosurg. Psychiatry 70, 605–612 (2001)
24. Latash, L.P.: Automation of movement: Challenges to the notions of the orienting reaction and memory. In: Mark, M.L. (ed.) Progress in Motor Control, pp. 51–88. Human Kinetics, Champaign (1998)
25. Smits-Engelsman, B.C.M., Van Galen, G.P.: Dysgraphia in children: Lasting psychomotor deficiency or transient developmental delay? Journal of Experimental Child Psychology 67, 164–184 (1997)
26. Dixon, R.A., Kurzman, D., Friesen, I.C.: Handwriting performance in younger and older adults: Age, familiarity, and practice effects. Psychology and Aging 8, 360–370 (1993)
27. Chang, S.H., Yu, N.Y.: Characterization of motor control in handwriting difficulties in children with or without developmental coordination disorder. Developmental Medicine & Child Neurology 52, 244–250 (2010)
28. Medwell, J., Wray, D.: Handwriting: what do we know and what do we need to know? Literacy 41, 10–15 (2007)
29. Christensen, C.A.: The critical role handwriting plays in the ability to produce high-quality written text. In: Roger, B., Debra, M., Jeni, R., Martin, N. (eds.) The SAGE Handbook of Writing Development, pp. 284–299. Sage, London (2009)
30. Graham, S., Weintruab, N., Berninger, V.: Which manuscript letters do primary grade children write legibly. Journal of Educational Psychology 93, 488–497 (2001)
31. Berninger, V., Swanson, H.L.: Modifying Hayes & Flower's model of skilled writing to explain beginning and developing writing. In: Earl, B. (ed.) Children's Writing: Toward a Process Theory of the Development of Skilled Writing, pp. 57–81. JAI Press, Greenwich (1994)
32. Mandich, A.D., Polatajko, H.J., Rodger, S.: Rites of passage: Understanding participation of children with developmental coordination disorder. Human Movement Science 22, 583–595 (2003)
33. May-Benson, T., Ingolia, P., Koomar, J.: Daily living skills and developmental coordination disorder. In: Sharon, C.A., Dawne, L. (eds.) Developmental Coordination Disorder, pp. 140–156. Delmar, Albany (2002)
34. Missiuna, C., Moll, S., Law, M., King, G., King, S.: Mysteries and mazes: Parents' experiences of developmental coordination disorder. Canadian Journal of Occupational Therapy 73, 7–17 (2007)

35. Rosenblum, S.: The development and standardization of the Children Activity Scales (ChAS-P/T) for the early identification of children with Developmental Coordination Disorders (DCD). Child Care Health and Development 32(6), 619–632 (2006)
36. Summers, J., Larkin, D., Dewey, D.: Activities of daily living in children with developmental coordination disorder: Dressing, personal hygiene, and eating skills. Hum. Mov. Sci. 27, 215–229 (2008)
37. Alizadeh, H., Zahedipour, M.: Executive functions in children with and without Developmental Coordination Disorder. Advances in Cognitive Science 6(3-4), 49–56 (2005)
38. Piek, J.P., Dyck, M.J., Nieman, A., Anderson, M., Hay, D., Smith, L.M., et al.: The relationship between motor coordination, executive functioning and attention in school aged children. Archives of Clinical Neuropsychology 19, 1063–1076 (2004)
39. Graham, S., Struck, M., Santoro, J., Berninger, V.W.: Dimensions of good and poor handwriting legibility in first and second graders: motor programs, visual-spatial arrangement, and letter formation parameter setting. Developmental Neuropsychology 29, 43–60 (2006)
40. Van Galen, G.P.: Handwriting: Issues for a psychomotor theory. Human Movement Science 10, 165–191 (1991)
41. Ben-Pazi, H., Kukke, S., Sanger, T.D.: Poor penmanship in children correlates with abnormal rhythmic tapping: A broad functional temporal impairment. Journal of Child Neurology 22(5), 543–549 (2007)
42. Geuze, R.H., Kalverboer, A.F.: Inconsistency and adaptation in timing of clumsy children. Journal of Human Movement Studies 13, 421–432 (1987)
43. Geuze, R.H., Kalverboer, A.F.: Tapping a rhythm: A problem of timing for children who are clumsy and dyslexic? Adapted Physical Activity Quarterly 11, 203–213 (1994)
44. Johnston, L.M., Burns, Y.R., Brauer, S.G., Richardson, C.A.: Differences in postural control and movement performance during goal directed reaching in children with development coordination disorder. Human Movement Science 21, 583–601 (2002)
45. Mackenzie, S., Getchell, N., Deutsch, K., Annemiek, W.F., Clark, J., Waitall, J.: Multi-limb coordination and rhythmic variability under varying sensory availability conditions in children with DCD. Human Movement Science 27(2), 256–269 (2008)
46. Zelanznik, N.H., Spencer, R.M., Ivry, R.B.: Dissociation of explicit and implicit timing in repetitive tapping and drawing movement. Journal of Experimental Psychology: Human Perception and Performance 28, 575–588 (2002)
47. Schaal, S., Sternad, D., Osu, R., Kawato, M.: Rhythmic arm movement is not discrete. Nature Neuroscience 7, 1136–1143 (2004)
48. Ameratunga, D., Johnston, L., Burns, Y.: Goal-oriented upper limb movements by children with and without DCD: A window into perceptuomotor dysfunction? Physiotherapy Research International 9(1), 1–12 (2004)
49. Coleman, R., Piek, J.P., Livesey, D.J.: A longitudinal study of motor ability and kinaesthetic acuity in young children at risk of Developmental Coordination Disorder. Human Movement Science 20, 95–110 (2001)
50. Wilson, P.H., McKenzie, B.E.: Information processing deficits associated with Developmental coordination disorder: A meta-analysis of research findings. Journal of Child Psychology and Psychiatry 39(6), 829–840 (1998)
51. Piek, J., Pitcher, T.A.: Processing deficits in children with movement and attention deficits. In: Dewey, D., Tupper, D.E. (eds.) Developmental Motor Disorders, Guilford Press, New York (2004)

52. Lundy-Ekman, L., Ivry, R., Keele, S., Woollacott, M.: Timing and force control deficits in clumsy children. Journal of Cognitive Neuroscience 3, 367–376 (1991)
53. Zwicker, J.G., Missiuna, C., Harris, S.R., Boyd, L.A.: Brain activation of children with developmental coordination disorder is different than peers. Pediatrics, e678–e686 (2010)
54. Alloway, T.P., Rajendran, G., Archibald, L.M.D.: Working memory in children with developmental disorders. Journal of Learning Disabilities 42(4), 372–882 (2009)
55. Kaplan, B., Dewey, D., Crawford, S., Wilson, B.: The term Comorbidity is of questionable value in reference to Developmental Disorders. Journal of Learning Disabilities 34(6), 555–565 (2001)
56. Skinner, R.A., Piek, J.P.: Psychosocial implications of poor motor coordination in children and adolescents. Human Movement Science 20, 73–94 (2001)
57. Segal, R., Mandich, A., Polatajko, H., Cook, J.V.: Stigma and its management: A pilot study of parental perceptions of the experiences of children with Developmental Coordination Disorder. American Journal of Occupational Therapy 56, 422–428 (2002)

Corpus Linguistics and the Appraisal Framework for Retrieving Emotion and Stance – The Case of Samsung's and Apple's Facebook Pages*

Amelia Regina Burns[1], Olimpia Matarazzo[2], and Lucia Abbamonte[2]

[1] Department of Law, Second University of Naples – Italy
ameliarburns@yahoo.it
[2] Department of Psychology, Second University of Naples – Italy
{lucia.abbamonte,olimpia.matarazzo}@unina2.it

Abstract. The study investigated the situated linguistic interactions of the users of the Samsung and Apple Facebook pages, with a focus on the attitudinal/affectual values they displayed towards these brands and their products, in a comparative perspective. Following Corpus Linguistics (CL) methodology, two corpora were created, named AppleCorpus (7337 tokens) and SamsungCorpus (5216 tokens), consisting in the wall posts on Apple Inc. and Samsung Mobile pages collected over a period of four days. These corpora were scrutinized both in a CL quantitative perspective and in a qualitative perspective by using the resources of the Appraisal Framework (AF) for discourse analysis to better identifying these social network users' stance and attitudinal positioning. The findings of this pilot study showed that Samsung's users display a more positive attitude toward the brand than Apple's users. Results are discussed in the text.

Keywords: Appraisal Framework, Corpus Linguistics, Word-Smith Tools, Keywords, Emotion and Stance, Facebook, Apple, Samsung.

1 Introduction

In our contemporary semiosphere, where web-mediated data capacities apparently double every two years, an amplified awareness of the importance of 'extracting' opinions, appraisals and emotions from texts has currently spread to several domains. Indeed, in fields such as advertising, marketing, and tourism, the importance of 'tracking' the consumer's voice is not easily overvalued, and social networks provide a useful medium both to retrieve and to disseminate essential information. Along these lines, our pilot study investigated the situated linguistic interactions of the users

* While the design of this pilot research is common, A. Burns and L. Abbamonte are respectively responsible for the quantitative and qualitative linguistic data retrieval and analysis, and O. Matarazzo for the study of emotions and cognition.

S. Bassis et al. (eds.), *Recent Advances of Neural Network Models and Applications*, Smart Innovation, Systems and Technologies 26,
DOI: 10.1007/978-3-319-04129-2_28, © Springer International Publishing Switzerland 2014

of the Samsung and Apple[1] Facebook (FB) pages, with a focus on the different attitudinal/affectual values they displayed towards such brands and their products. In order to obtain more fine-grained results, resources from both quantitative and qualitative applied linguistics approaches were utilized, by integrating corpus linguistics (CL) tools and the more recent resources of the appraisal studies (the Appraisal Framework 2012 version).

2 The Appraisal Framework

The Appraisal Framework (AF) can be defined as an approach to exploring and explaining the way language is used to evaluate, to adopt stances, to construct textual personas and to manage interpersonal positioning and relationships. Developed by P.R.R. White [14] in its present form, the AF provides rich analytical resources for the in-depth comprehension of a great variety of discourses and texts, by taking into account both their text-internal organizational features, and several interrelating contextual factors (participants, culture, group dynamics, textual identity), in order to interpret their local meanings as well – in other terms, the speakers'/writers' *Engagement* and dialogistic positioning in discourse. The AF may be especially useful where *Attitudinal positioning* and *lexis* play in the foreground, as in the present study, i.e., when speaker/writers pass judgement and associate emotional/affectual responses with participants, processes and things, also through the attribution of evaluative responsibility (ER) to themselves or to other actors. Attitudinal lexis can be either emotional (Affect), or ethical (Judgement) or aesthetic (Appreciation), to varying degrees of FORCE and FOCUS (i.e., *Graduation*, or the semantics of scaling). Among the AF main advantages we can list its virtually unlimited potential for fine-tuned, granular lexico-grammar and semantic analysis.[2]

The AF has been implemented to deal with issues including media discourse, education, scientific discourse, translation studies, doctor patient communication, service encounters, intercultural communication, *attitudinal data mining and sentiment extraction through corpora* – especially useful for *social network-based commercial applications* – corporate communication, covert evaluation in political communication, and much more.

3 Attitude and the Key Role of Emotions in the AF

A pivotal role in the AF is played by the effort to map the expression of feelings and attitudes as they are construed in English texts. Over the last few decades, the key role of emotions both in human evolutionary history and in cognition has been recognized in several areas of research [3], [4], [7]. Both adults and children can explain/describe when and why emotions occur and, additionally, their descriptions show important

[1] The two leading companies and innovators in the market of mobile technology.
[2] Owing to space constraints, for more information on the AF classifying frames, see: www.grammatics.com/appraisal/

similarities across individuals and cultures [9], [1]. Emotion knowledge, at least as expressed in the English language, seems to be structured in hierarchically organised, fuzzy categories, and the nature of mental representations that individuals use when accessing their knowledge about emotion is still being investigated. These deeply ingrained resources for expressing emotions are at the heart of AF elaborations, *attitudinal positioning* in particular. *Attitude* can be either explicit (overtly evaluative/attitudinal words, phrases or sentences) or *implicit* – the latter being not easy to locate. Indeed, it is the readers'/listeners' particular set of beliefs and expectations which will lead them to interpret and consider the writing as un/true, un/acceptable, or un/attractive. Analogously, the semantics of scaling is not confined to cases where the value is explicitly carried by some independent, isolating lexical item, such as *very* or *somewhat*. Implicit scaling also needs to be considered. Apparently, neither quantitative corpus-linguistics based analyses nor other current data mining techniques can easily cope with these levels of analysis as yet – not to mention the registers of irony or jocularity.

4 Corpus Linguistics

Essentially consisting in the use of a corpus of language as the basis for language description[3], CL has seen a fast growth from the 1960s onwards with an increasing variety of fields of application. It facilitates the investigation of language structures, usage and uses, (patterns of) lexis and syntax, idiolects and speech communities, discourse, rhetoric, etc... Collected according to specific criteria, a corpus is representative of a given language and can be a very useful resource for pursuing various research agendas in domains including lexicography, first/second language acquisition/learning [5]; prediction of speakers' syntactic choices, cross/inter-cultural investigations [10], syntactic priming/persistence, stylistics, sociolinguistics, forensic linguistics (e.g. authorship attribution).

Based on the evaluation of frequencies, CL analysis starts from the assumption that formal differences reflect functional differences: *if two constructions are syntactically distinct, they must be semantically or pragmatically distinct.* The notion of *frequency* involves not only the relevant linguistic phenomenon that occurs a great number of times, but also the phenomena that occur only once or not at all [12]. In brief, a corpus is made up of files and needs to be queried through a statistical software in order to retrieve valuable data by extrapolating distributional (or quantitative) information [13]. Such information –e.g., how often morphemes, words, grammatical patterns etc. occur in (parts of) a corpus – is usually represented in *frequency lists*, whereas the listing of each occurrence of a word in a text or corpus, presented with the words surrounding it (KWIC – key word in context), is shown in *concordances.*

[3] In contemporary acceptation a corpus is an electronically-stored and machine readable collection of naturally-occurring (written and/or transcribed) texts, made up of (hundreds of) millions of running words, which is designed for (general or specific) linguistic analysis (see [13] among many others).

5 The Present Study

This pilot study aimed at investigating and comparing the attitudes of the Apple's and Samsung's users towards such brands and their products, through the linguistic analysis of their posts on the companies' FB pages. The focus was on the attitudinal/affectual values the users displayed, in order to better track their voices. To this purpose both CL quantitative approach and AF multi-layered frames for qualitative analysis were utilized.

6 Methods

6.1 Tools

The present research was carried out through the support of Mike Scotts's **WordSmith Tools** (version 5.0) software, i.e. an integrated suite of programs (keywords, concord, wordlist, etc.) for looking at how words behave in texts. WST was used here to create the wordlists[4] of the two considered corpora, and to compare them through the utility ***KeyWords*** (KW), which carries out comparisons between word lists and is extremely helpful to investigate differences and similarities both across and within texts. The procedure for identifying a keyword has several stages,[5] and utilizes the chi-square statistic: a *word will be **key** if its **frequency** is either atypically high or low* in a given text by comparison to a reference corpus. When all potentially key items have been identified, they are ordered in terms of their relative keyness.

6.2 Procedure

For our analysis two comparable corpora (in English language) were built from the collection over a given period[6] of all the comments posted by the users on the walls of the Samsung Mobile FB page (henceforth, SamsungCorpus) and of the Apple Inc. FB page (AppleCorpus). In addition, a *reference corpus* (of 93,747 tokens) was built from the comments posted on the walls of the ten most-clicked companies on FB over the same four-days span (hence, FBCorpus), which according to Business Insider are: CocaCola, Disney, Converse, Starbucks, Red Bull, Oreo, Skittles, McDonald's, Pringles, Victoria's Secrets.[7]

[4] The tool WordList "generates word lists based on one or more plain text or web text files. Word lists are shown both in alphabetical and frequency order" [11]

[5] First, a word list is computed, containing all the different types in the reference corpus and the frequencies of each. Next, the same kind of word list is computed for the text whose key words one wishes to find. Finally, the software identifies the key words in comparison with the reference corpus (see also footnote 10).

[6] 17-20 April 2013.

[7] A reference corpus needs to be considerably larger than the target corpora in order to be comparable and, also, to be created from texts sharing time span, language and contexts of communicative exchanges with the target texts. Thus, it can be queried in order to retrieve the keywords.

While Samsung Mobile FB page is mostly dedicated to customers, and functions as a costumer care service with Samsung support team replying to users question and enquiries, Apple Inc. FB page is instead mainly used to share information among users.

The gathered texts were first cleared of all the notes and references not expressly written by users (e.g. share, comment, like) and then converted in .txt format. The files – divided in two groups, AppleCorpus and SamsungCorpus – were then processed through WST. By using the tool "wordlist", we queried both sub-corpora, thus obtaining *two comparable wordlists.* The second part of the data processing consisted in the extraction of their *keywords* through the comparison to the reference corpus.

6.3 Results

In Table 1 the general features of the AppleCorpus and SamsungCorpus are shown.

Table 1. AppleCorpus and SamsungCorpus overall[8]

AppleCorpus		**SamsungCorpus**	
Tokens	7145	Tokens	4980
Types	1812	Types	1199
Type/Token Ratio	25,36	Type/Token Ratio	24,07
Standardised Type/Token	44,87	Standardised Type/Token	40,20

Since they are consistently different in extension,[9] it was important to work on percentages instead of absolute figures. Another interesting datum is the higher standardised type/token ratio in AppleCorpus, which indicates the greater variety of vocabulary of the users of this FB page.

Then, we compared each sub-corpus to the reference corpus (FBCorpus), in order to retrieve (through the KW tool of WST) their **keywords**. In Table 2 we can read the first 20 keywords of each with their keyness value[10].

[8] Type refers to all different types of words of a text. Token refers to all words of a piece of text. For example, *iphone* occurs 51 times in AppleCorpus, therefore it counts 1 type and 51 tokens. The standardised type/token ratio is the average type/token ratio based on consecutive 1,000- word chunks of text.

[9] See first line of Table 1 - Tokens: AppleCorpus 7145 vs. SamsungCorpus 4980.

[10] To compute the keyness of an item, WST "computes its frequency in the small word-list, the number of running words in the small word-list, its frequency in the reference corpus, the number of running words in the reference corpus and cross-tabulates these" (Scott, 2007) applying the classic chi-square test of significance. The higher the score of keyness value, the stronger the keyness of that word.

Table 2. AppleCorpus and SamsungCorpus Keywords

	AppleCorpus		SamsungCorpus	
N	Keywords	Keyness	Keywords	Keyness
1	APPLE	375,1482239	SAMSUNG	412,9162598
2	TO	283,0169983	GALAXY	389,272644
3	IPHONE	267,8778076	THE	281,8278198
4	YOU	188,512619	I	272,3925781
5	PHONE	146,9696198	PHONE	219,2950592
6	THE	146,016098	TO	168,2855225
7	AND	140,7210541	FOR	135,3243866
8	I	105,5957336	NOTE	127,268898
9	FOR	90,08646393	WILL	109,0653534
10	YOUR	89,4288559	MY	93,00803375
11	HTTP	88,40182495	AND	88,7545929
12	WWW	85,8077774	PROBLEM	64,53414154
13	IT	76,82676697	PHONES	63,71566391
14	ITUNES	76,73320007	GET	60,55189514
15	APP	74,20905304	HAVE	57,73073959
16	ON	71,73426819	BE	55,37960434
17	SUPPORT	68,21994019	VERIZON	52,9889946
18	OF	67,27710724	PRE	52,28457642
19	A	67,06533051	UPDATE	50,54430771
20	CAN	63,86418152	BATTERY	47,09987259

As we can notice, both the FB pages put the brands themselves to the fore, ‘Apple’ and ‘Samsung’ are found at the top of the keyword lists, closely followed by the type of smartphone they produce, ‘IPhone’ and ‘Galaxy’ respectively. Of interest is the different preference for personal pronouns as it arises from the keyword lists: Apple users tend to use the pronoun ‘you’ more frequently, while Samsung users prefer the pronoun ‘I’.

Further, from the two complete keyword and concordance lists[11], we selected and compared the items which most frequently occurred in evaluative contexts towards the two brands and their products, and were 'key' in both corpora (listed in table 3). The occurrences of each item were broken up into positive (Pos.), negative (Neg.) and neutral (Neut.) connotations according to the observation of their concordance lines.

Table 3. Positive, Negative, and Neutral occurrences of 'brand-endorsing' words

	AppleCorpus					SamsungCorpus				
Keyword	KW	Tot	Pos.	Neg.	Neut.	KW	Tot.	Pos.	Neg.	Neut.
Iphone	3	51	14	14	23	31	7	1	1	5
I	8	193	19	60	114	4	228	26	14	188
Your	10	63	6	24	33	39	33	10	4	19
Support	17	13	2	8	3	46	5	1	3	1
My	44	83	8	14	61	10	94	12	16	66
Samsung	53	6	5	-	1	1	70	14	7	49
Problem	78	9		4	5	12	16	4	10	2

In order to evaluate whether, for each word, there was a significant difference in its categories of occurrence between the AppleCorpus and SamsungCorpus, the chi-square is an appropriate statistical test. However, since its use is inadequate if the expected frequency is less than 5 in more than 20% of the cells, their results are valid only for 3 of the 7 keywords: *I, Your* and *My*. The keyword *I* (χ^2 = 45.22, df =2, p<.001) has more negative occurrences in AppleCorpus than in SamsungCorpus and more neutral occurrences in the latter than in the former. The keyword *Your* (χ^2 = 10.73, df =2, p<.01) has more positive occurrences in SamsungCorpus than in AppleCorpus and more negative occurrences in the latter than in the former. No significant difference between the two subcorpora emerged for the keyword *My*. Although the results of the remaining keywords cannot be considered as valid, it is noteworthy that the frequency of *Iphone* is analogously distributed in the two sub-corpora categories, whereas Samsung has more positive occurrences than expected in AppleCorpus rather than in SamsungCorpus.

The concordance lists showed that whereas in the AppleCorpus *your* is used mainly in the context of disagreement (e.g., "your customer service … failed us this time"), in the SamsungCorpus, *your* is mainly *inclusive*, (e.g., "you guys manage your precious time in replying all of your fans' comments"). Moreover, in the latter, there is a greater prominence of the first person singular pronoun, (e.g.: "I really want a Samsung Galaxy S4"). Indeed, *I* is used very differently in the two corpora: mainly to convey the lack of interaction in the AppleCorpus (e.g., "I have read the support forum and still no luck"), and a sense of belongingness and fruitful interaction in the SamsungCorpus (e.g., "I love what you guys are doing").

[11] Space constraints do not allow us to show them here (but they are available on demand), nor to extensively comment every datum.

As will be apparent from the tables 4,5,6 (showing a sample of qualitative data), the use of these (and other) words in context construe different costumers' identities.

Starting from the keywords listed in Table 3, we selected three small sets of textual samples to illustrate how such keywords are used in context to convey the consumers' attitudes to the brands and their products, along the dimensions of Affect, Judgment and Appreciation as classified in the AF.[12]

LEGEND:

- *Italics* = words with *positive* emotion-value
- **Bold** = words with **negative** emotion-value
- Underlined = positive expressions with negative value, or negative expressions with positive value, i.e., ambiguity (irony, jocularity, etc.)
- FORCE = FR ; FOCUS = FC
- SER= self attribution of evaluative responsibility[13]
- OER = attribution of evaluative responsibility to others

Table 4. AF – Attitudinal positioning: qualitative data – AFFECT

Samsung Mobile Facebook page	Apple Inc. Facebook page
1. I absolutely *love* it! [SER]	1. *Love* my Apple Products. It's *the BEST*. [SER] Waiting for the new iPhone
2. I have the Note now and I love it!	2. *I love* my iPhone. Sorry your having so many **problems.** [OER]
3. Hi. Just wanted to say that your products are *amazing*! And I can't believe your smartphones are *blazing fast* with *ultra fast* processors. [SER]	3. I *love* my iPhone 5! Great job Apple [FC]
4 *My boyfriend*, SGS III, suffered *a terrible accident* [FR] last night. His gorilla glass screen shattered into a million pieces. *I am extremely depressed.* [FC] I don't know why bad things happen to good people. [SER] On the bright side *his hot brother SGS IV is coming home soon.* [SER]	4. - so I was wondering about how much would it cost to get that problem fixed? **Never**, ever again Iphone, you're the **shit of** all stuff [SER+FR]
5 . It's *amazing how you guys* manage your precious time in replying all of *your fans' comments*. [SER] Thanks again and keep up the hard work Samsung!	5. [apple]Make a better product and provide better customer service in your stores when your products **malfunction**! [SER+FC] TRY THAT!!! [FR]
6 . *Dear Samsung*, Thanks for *welcoming me* into the Galaxy family and giving me a ton of reasons to stay there. ♥ [SER] Love, Your biggest fan, Trisha :)	6. (Jason:) **apple you suck** [FR] … i have an iphone and **i hate it**… [SER] (Steven:) My recommendation is **stop wasting time with apple** when *Samsung is much much better.* [FC +SER]

[12] Space limits did not allow to show larger samples.

[13] See par. 2.

While in Samsung's FB pages positive emotion-comments play in the foreground, in Apple's FB pages these are sparse and consistently overwhelmed by vigorously negative emotion-comments, mainly focusing on the hardware shortcomings, and on the company's lack of interaction and unsatisfactory customer service. Such posts result into negative exposure of the brand to a specialized, if circumscribed, audience. Instead, a sense of belongingness (especially the posts 5, 6) characterizes Samsung's community, whose posts convey explicit, partly physical, positive Affect (especially the posts 1, 2, 4), blended with positive Appreciation of given features of the products and their performances (3, 5), and humorous, affectionate irony (4) – all resulting into (comparatively) positive promotion of the brand. Such differences are even more striking where Judgement is concerned.

Table 5. AF– Attitudinal positioning: qualitative data – JUDGEMENT

Samsung Mobile Facebook page	Apple Inc. Facebook page
I'm waiting for the Note 3 gonna be *so cool.* SER	i bought a iphone 4 from ahmedabad mobile store at bodakdev … then **how can they cheat like this.** SER+FC …. [they] also they created **a lot of problem** FR when doing the credit card transaction **… i was completely harassed** FR when buying the iphone 4 and **really upset** FR with apple services SER
The best electronic manufacturer in the world! samsung rocks number one SER	**Apple stock is sinking like a stone** FR They are **not business friendly at all.** SER They have put up **substantial barriers to entry** FR+FC for small business to develop apps… You'd think the company would do everything they can to help, OER **not hurt business** FR+FC …if Apple would only listen. Simply listen. SER
I am **extremely disappointed** with my S3. … my battery **would die so fast**… the **problem** still persists. It's stressful SER	Well, **sad to say** customer service must not know the difference between abused and well used!!!!! SER **The Greed of Apple has shown it's ugly face,** FR+FC You should be **ashamed of** FR **yourselves.** SER +OER … **Sorry apple**, but we may just have to switch back to PC - **your customer service** which *you pride yourself on*, **failed** us this time. FC

Explicit, motivated and proclaimed positive Judgement, often blended with Appreciation, could be found much more easily in the Samsung than in the Apple corpus. Also, while Negative Judgement is generally circumscribed to 'disappointment' for Samsung's users, Apple's users resort to much more forceful expressions, with a specific focus on the company's deficiencies. As far as Appreciation is concerned, a considerable difference is also recognizable between the two, as the following excerpts show.

Table 6. AF– Attitudinal positioning: qualitative data – APPRECIATION

Samsung Mobile Facebook page	Apple Inc. Facebook page
- *Huge upgrade*! from the awful blackberry I had for 4 years. SER *Samsung has great features that competitors can't even dream* to compete with. SER My 1st Samsung & Android phone is the Note 2. *Great device*! SER	**FUCK YOUR BATTERY LIFE ON IPHONES!!! PIECE OF SHIT.** FR Just to let you know, **you guys** *make a lovely operating system,* SER **but your hardware truly** FC **sucks!** FR+ SER
- even with a cracked screen, my sprint s3 is *the best phone* i've ever had!!SER - I live in Los Angeles, USA and I am contacting the support in my region. I will keep doing so, until this is resolved	**Never, ever again Iphone,** FC you're the **shit** FR **of all** stuf wich is running out there. **stop producing smartphones.** SER Start producing better Mac books!

Apparently, both Apple's and Samsung's users do not hesitate to explicitate their evaluation, i.e., their positioning towards the artefacts and their functioning that trigger (positive/negative) reactions/emotions.

7 Conclusion

Although to fully track the consumers' voice for commercial applications larger corpora would be necessary, yet, this pilot investigation has highlighted how the consumers of these two leading companies differ along the dimension of attitude (Affect, Judgement, Appreciation). Although more frequent among Samsung's users, expressions of positive Affect can be also found in the AppleCorpus, where expressions of positive, motivated Appreciation and Judgement are instead rare. Dissimilarly, Samsung's users often express positive Judgment and enthusiastic Appreciation for the brand's artefacts, with a specific emphasis on new, competitive features. When their appreciation is negative, they are usually satisfied with expressions of concern or regret for 'bad luck' – very differently from Apple's users' strong (and vulgar) emotional expressions, whose evaluative responsibility they fully take.

Essentially, the integration of two methodologies (CL and AF) has helped to overcome the difficulties of identifying attitudes and evaluative meanings. Since many expressions may be purely factual in some contexts and evaluative in others, also in implicit ways (e.g. relying on shared cultural construal), their value may elude automated larger-scale analysis. Hence, the more practicable way forward appears to be a multi-competence, integrative approach, which should be able to take advantage of qualitative and quantitative linguistic analyses as well as of informatics and cognition research strands.

References

1. Clore, G.L.: Affective Coherence: Affect as Embodied Evidence in Attitude, Advertising, and Art. In: Semin, G.R., Smith, E. (eds.) Embodied Grounding: Social, Cognitive, Affective, and Neuroscientific Approaches, pp. 211–236. Cambridge University Press, New York (2008)
2. Damasio, A.: The Feeling of What Happens: Body, Emotion and the Making of Consciousness. Harcourt Brace, New York (2000)
3. Dolan, R.J.: Emotion, cognition, and behavior. Science 298(5596), 1191–1194 (2000)
4. Gibbs, R.W.: Embodied experience and linguistic meaning. Brain and Language 84(1), 1–15 (2003)
5. Goodman, J.C., Dale, P.S., Li, P.: Does frequency count? Parental input and the acquisition of vocabulary. Journal of Child Language 35, 515–531 (2008)
6. Gries, S.T.: Multifactorial analysis in corpus linguistics: a study of particle placement. Continuum Press, London (2003)
7. Izard, C.E.: Human Emotions. Plenum Pless, New York (1977)
8. Kepser, S., Reis, M.: Linguistic evidence: empirical, theoretical and computational perspectives. Mouton de Gruyter, Berlin (2005)
9. Niedenthal, P.M., et al.: Embodiment of Emotion Concepts. Journal of Personality and Social Psychology 96(6), 1120–1136 (2009)
10. Oakes, M., Farrow, M.: Use of the chi-squared test to examine vocabulary differences in English language corpora representing seven different countries. Literary and Linguistic Computing 22, 85–99 (2006)
11. Scott, M.: WordSmith Tools 4.0. O.U.P, Oxford (2007)
12. Stefanowitsch, A.: New York, Dayton (Ohio), and the raw frequency fallacy. Corpus Linguistics and Linguistic Theory 1(18), 295–301 (2006)
13. Tognini-Bonelli, E.: Corpus Linguistics at work. Studies in corpus Linguistics. Amsterdam, Benjamins (2001)
14. White, P.R.: The Appraisal Website: The Language of Attitude, Arguability and Interpersonal Positioning (2012), http://grammatics.com/appraisal/ (June 15, 2012)

Which Avatars Comfort Children?

Judit Bényei[1], Anikó Illés[1], Gabriella Pataky[2], Zsófia Ruttkay[1], and Andrea Schmidt[1,3]

[1] Moholy-Nagy University of Art and Design Budapest
[2] ELTE University of Budapest
[3] Institute for Sociology, CSS, Hungarian Academy of Sciences, Budapest, Hungary
benyeij@t-online.hu, anikoilles@gmail.com,
pataky.gabriella@vizu.hu, ruttkay@mome.hu, schmidt@mome.hu

Abstract. In this paper we give an account of an empirical study related to avatar selection, whose purpose is to comfort children in the context of the TERENCE learning application. We investigated what Hungarian children of ages 6-11 like to play with, what stories they like to read, and what avatars – out of a selection of nine different designs – they like best. We found a statistically relevant correlation between age, gender, and the participants' expertise and habits as regards ICT tools. Our studies thus provided relevant information on how to design avatars for this special age group, taking gender equality into consideration and increasingly subjective well-being in order to motivate learning.

Keywords: avatar selection, children 6-11 years-old, ICT expertise and habits.

1 Introduction

Nowadays, ICT applications find their way to children at an early age, in the form of interactive books. The TERENCE application, being developed within the EU FP7 project of the same name, aims at providing reading materials for 7-11 year-old children having difficulties with text comprehension. The adaptive system offers, among other components, stories to be read and tests (in the form of games) to check if the child has understood the content. A child is supposed to follow TERENCE sessions over a period of some weeks. The system's Release 2 is being tested with target groups, and may be tried out online (Alrifai at al., 2012). Avatars play a central role in the system. At the beginning, the child chooses one from among offers that are "marching along". Then, this avatar is his/her companion to help him/her through the training process. "Help" is basically of a comforting type: the avatar shares joy if the child performs well, and gives encouragement when needed; and this is conveyed by facial expressions.

S. Bassis et al. (eds.), *Recent Advances of Neural Network Models and Applications*,
Smart Innovation, Systems and Technologies 26,
DOI: 10.1007/978-3-319-04129-2_29, © Springer International Publishing Switzerland 2014

Fig. 1. The 9 avatar designs

Moholy-Nagy University of Art and Design Budapest, as part of an international consortium, took care of visual tasks, including the design of the avatars. We obtained user models and personae descriptions from our partners, and the designers (educated children's book illustrators) were free to propose the avatars. They produced 9 designs, including cute animals, some fictitious humanoid characters and two pairs of children of their (the children participants') own age, see Figure 1. In the design process, much discussion went on about what the characters should be and how they should look. How does the nature of a character (animal – fiction figure – child) and the style of the design (more or less realistic, the colors used) contribute to 'liking'? What designs do children prefer as comforting companions? How are their preferences influenced by age, gender, living environment, exposure to various media and their reading-playing habits?

While children are overwhelmed with (real and virtual) figures, ranging from teddy bears, via plastic figures - such as toys from games and films, to heroes in films, computer or tablet games, we do not know much about the *nature* of their preferences. Their actual preferences are often biased both by what they see (whether on TV, or on the computer screen) and the marketing push.

Empirical aesthetics'studies (the psychology of art and design appreciation),which search for correlations among preferences as regard visual objects and viewers' characteristics (such as personality features, social background, expertise, and so on) are a well-examined field (Locher, Martindale, Dorfman, 2006). There are so many factors which can modify the beholder's appreciation; so it would seem to be an inspiring investigation to search for determining or influencing factors regarding preference answers given by children with regard to visual objects, such as avatars.

In the literature of visual art and design appreciation a relation can be seen between liking and similarity to the ideal self (Alexander and Marks 1983). According to results obtained here, people do prefer images that are semantically closer to the character they wish to be. In the context of avatar preference, this means that children will probably choose figures they are familiar with and feel close enough to, to find them meaningful; and they will also make them feel a little closer to the ideal self.

Children's preferences regarding figures are a special case of empirical aesthetics. We believe that the opportunity for children to be able to choose the avatar they like makes them feel more familiar and comfortable and increases also the experience of competence. This subjective well-being prompts their self-confidence and facilitates their chances of enjoying the learning process. This, though, synchronizes with other findings from avatar studies, as with Inal and Cagiltay (2006): "Avatars are also used to increase the motivation of players as pedagogical agents when they play computer games." (pp. 3440.)

We shall therefore take a close and careful look at the preferences of 6-11 year-old Hungarian children; and we wanted to find out if and how age, gender, media exposure and the reading habits of the child may determine what avatar they will identify with. In this paper we give an account of our first findings.

2 Researching Preferences among Avatars

2.1 Research Methodology

As research material, we used the 9 avatars designed for TERENCE, printed in a random order, in color, on A4 paper (see Figure 1). Children had to select the one they liked the best, and also their second choice, by circling them in red and blue.

The child's profile was obtained by asking them to fill in an A4 form containing multiple-choice questions (with answers to bc circled), and also open questions: such as age, gender, Do you like reading? (yes – no) How often do you read? (never – sometimes – every day) Do you use smartphones, tablets, or computers at home? (never – sometimes – every day). What is your favorite game? What is your favorite story?

On the test sheets, we made sure that we used a font of an appropriate size so that children aged 6–11 years would find it easy to read; and while the instructions followed the format the children were accustomed to, we also asked the teachers to read them out so that reading skills would not influence the independent success of solutions arrived at. Some of the first-form pupils were still unable to write at the time of the data collection, so they dictated their answers to open-ended questions to the interviewers.

The interviewers provided all the requisites of the survey (questionnaire, a letter to inform parents about the purpose of the survey and seeking their consent for their children's participation); the children only needed a red and a blue pencil.

The questionnaire was filled in by 377 children aged 6–11 years, in 16 schools.

The participating schools, classes and children were chosen with a random sampling method (Csíkos, 2009), and the participants came from 18 classes in which the interviewing students were completing their 10-week teaching practices.

2.2 Research Hypotheses

H1: Children prefer a character of their own sex.
H2: Younger children prefer animals - and older ones human - characters.
H3: ICT usage habits have an impact on avatar preference.

2.3 Results

General ICT usage. In our investigation there is some interesting data, in general, which has significance in research into the ICT usage of children. All the data we refer to in this paper is significant at $p<0,05$ (Pearson Chi-Square-, McNemar test). One such piece is a gender-based inequality, which was clear from the data we got from answers related to ICT tool-using habits. It is well articulated that girls use fewer tools. For example, boys have an advantage with everyday computer usage (42.2% vs. girls' 27.5%), or more boys use a tablet (51.4% vs. 41.7%), or everyday tool usage is more frequent among boys (smartphone, tablet, and computer - altogether, 56.6 vs.43.9 %).

The other significant piece of data relating to ICT usage is connected to reading habits. We found that the different questions associated with reading habits and features of ICT tools usage had an interesting pattern. Such data shows that children more likely to be readers are also more likely to use ICT tools (tablet, computer and/or smartphone) When making a comparison with socio-economical circumstances in the country where the research was carried out, this indicates that children who more willingly read or are familiar with the experience of reading have more of an opportunity to make use of technological gadgets and their benefits. Both of these findings would need proper discussion, however; though in this paper we present them as the background to avatar preference.

Avatar Choice. The target diagrams for avatar preference show the distribution of the first choice of avatar with a gender-based distribution (Fig. 2), to show the most popular ones. The most popular figures were the ninja, the horse, and the robot. Thus it would seem as if children make a more childish choice.

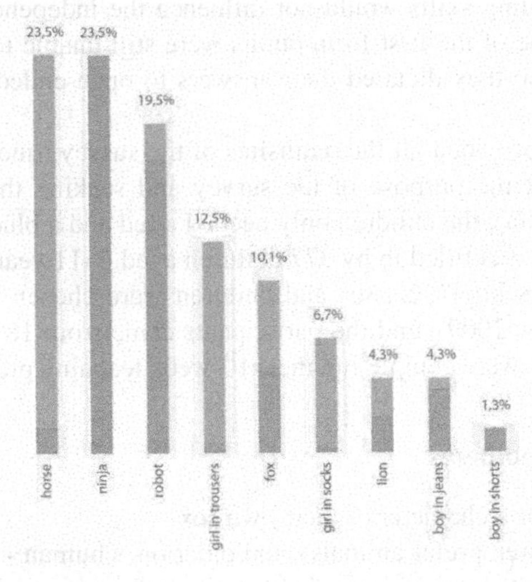

Fig. 2. Distribution of avatar choices (pink: girls' distribution Blue: boys' distribution)

However, further analyses show us that almost half of the children do prefer human figures (see Fig. 3). (Boys are more likely to follow this tendency and like the robot; yet as a first choice, older girls will choose a humanoid character); whilst concerning background factors relating to a human, non-human or robot preference, children who prefer human figures are more likely to use smartphones every day or never read or are more likely to be a boy, and/or older than age 9. Animal preferences correspond to the fact of everyday reading, or someone's never using a smartphone, or their being female; and a robot preference appears to go hand in hand with one's being male or not using a smartphone.

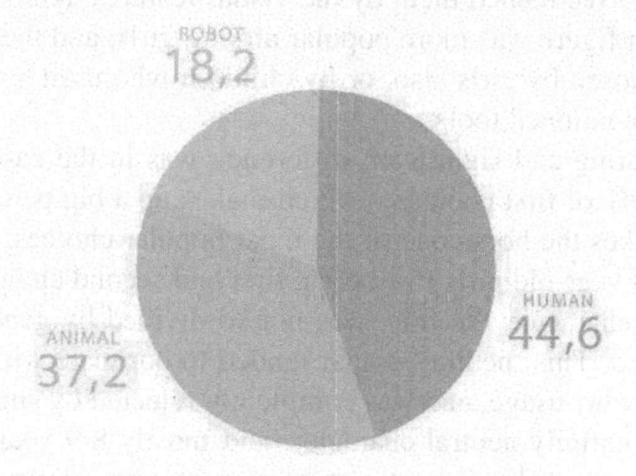

Fig. 3. Distribution of avatar-type preference in connection with being human – non-human creatures (%)

Gender. Our data shows that there is one major factor predicting avatar choice - and the biggest amongst others: sex. On the basis of preference choice, it was clear what character would be preferred by what sex; and with this data we coded avatars accord ing to whether they are boyish (two figures of boys, a robot and a ninja), girlish (two figures of girls, and a horse), or neutral (a fox and a lion). The distribution of answers can clearly be seen in Fig. 4.

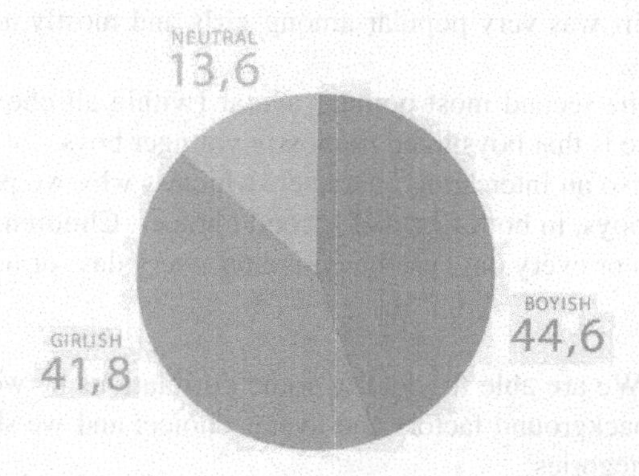

Fig. 4. Distribution of avatar-type preference by the gender of the figures (%)

Some correlations can be found in relation to gender-based avatar choices and background factors: someone choosing boyish characters will mostly not read at all or does not read every day, and they will use a computer every day.

One by One. We wanted to have a closer look at the avatars one by one, though; and in introducing this data we would like to draw attention to some additional, and also interesting data. Concerning first choices, we found that 21% of these first choices were humanoid characters.

Because of the working progress of the TERENCE project we had two types of humanoid character. We named them by the visual features: realistic vs. thin humans. The realistic human figure was more popular among girls, and the thin human figures were more often chosen by girls also, or by children who used a smartphone, or who use the three aforementioned tools.

The most interesting and significant difference was in the case of the horse. The horse obtained 23.5% of first choices – which makes up a big percentage amongst the 9 avatars, i.e. it makes the horse one of the most popular choices. Most of the 'votes' here came from 6-8 year-old girls (for both a first and second choice).

Another animal character, the lion, was not so divided by gender, as first choices were equally divided. This 'neutral' avatar tended to correlate with zero tool usage in general and in everyday usage, and was completely rejected by smartphone users.

The fox was an entirely neutral character; and mostly 8-9 year-old children chose it. Reading, reading every day, sometimes using a computer, zero tool usage and zero everyday usage strongly correlated with a fox preference. However, children using smartphones and using computers would always contradict this seen tendency.

The boy wearing jeans (the thin, male human) was chosen mostly by boys, and also children who used tablets at times (which is easy to understand: he is holding a tablet in his hand!). Children favouring him did not use smartphones, did not use tablets every day, and were not everyday tool users.

The boy in shorts (the realistic, male human) was chosen only by a few children, ones who use some tool every day.

The girl in trousers was chosen by girls - it was quite popular; and the children who preferred it were tool users, and also computer or smartphone users.

The girl in a skirt was very popular among girls and mostly among children who do not use any tools.

The robot was the second most popular avatar (within all choices). The most important thing to note is that boys liked it, mostly younger boys.

The ninja was also an interesting character, which is why we present it last. It was mainly chosen by boys, in both first and second choices. Children who liked this read at home sometimes or every day, used a computer every day, or used some tool every day.

Favorite Games. We are able to identify some correlations between type of favorite game/some other background factors and avatar choice; and we shall here set out the most interesting categories.

Electronic games were more often mentioned by boys, children age 6-8 years, who do not read at home at all, show everyday computer or smartphone or any other digital tool usage, who use all three tools, or who preferred the ninja or human or boys. Electronic games were not – or were less - mentioned by children who do not use computers, or who preferred the animal, horse or girl figures.

Moving games were more often mentioned by children who favoured the lion; while moving games were less mentioned by those who preferred real human or human figures.

Dolls and plush/furry toys were more often mentioned by girls, children aged 6-8 years, who preferred the girl in a skirt, or the girl in trousers, the horse, animals, girls overall, real humans, or humans overall. Dolls and plush/furry toys were not – or were less - mentioned by children who chose the robot, the ninja, or the boy in jeans.

Board games were more often mentioned by girls, children who used some tools but not all, and who favoured the horse. Board games were not - or were less - mentioned by everyday ICT users or persons who used the three tools.

Lego was more often mentioned among boys or by children who use no or just one ICT tool, boys who liked the robot or the boy avatar, or who favoured the non-human avatar. Lego was not – or was less - mentioned by children who preferred the girl in a skirt, the girl in trousers or the horse.

Favorite Story. It was very difficult to analyze answers given to the favorite story question. Our research pupils mixed up all the stories that they had ever seen - from TV series to cartoons, which includes stories available only in books; so we went back to the original goal of the TERENCE project - namely, focusing on reading. This is why we categorized answers on the basis of whether it is a story that was read or had been seen or where one was not able to categorize it. We did, however, find correlations between the diverse factors noted and stories from books. It seems likely, though, that a read story may correlate with reading every day or someone's never using a computer or using one digital tool or preferring the boy in shorts, the fox, the boy in trousers, the horse, an animal, girls in general, or a neutral gender figure. Yet one's using a smartphone or a computer every day or preferring the ninja, robot, or boy figures in general does not correlate with stories from books.

3 Conclusions and Further Work

Children readily participated in our investigation. Their interest and their consonant answers reflected unequivocal and very significant differences proved the validity of our research question -giving children an opportunity to choose an avatar was a meaningful process for them. We obtained clear patterns with regard to avatar preference as related to age, gender, and ICT tool usage.

Empirical evidence of gender-based ICT usage is available (for example: Kay (2007); Fisher, Margolis and Miller (1997). Also, a cross-cultural study has shown that among university students (even though their knowledge and skills are the same), girls feel less confident using ICT than do boys (Li and Kirkup, 2007). Our data reconfirms this finding. Therefore, one of our results is that in the examined age group the girlish or boyish outfit is extremely important for children. Since a disadvantage was found when it comes to girls, we can say that it is more important to offer proper avatars in order to motivate girls into gadget – the three aforementioned tools - usage. Associations between ICT usage and the sex of a child show us that the design of an avatar must be attractive to girls in order to involve them in the comfortable world of technology.

Preferences of children using smartphones do not align with the mainstream - for example, they will not like the fox or the lion, but prefer thin figures. This is one issue that we need to continue examining, to understand whether the explanation might be settled via looking at the social-economic background or whether living with a smartphone changes one's lifestyle. The former issue needs more investigation -which requires the collecting of socio-economic data; the latter correlates with issues arising with the challenges coming with a new mode of life that has been established by smartphones.

Our data indicates that younger children prefer an animal or the robot; while older children (9-11) favour humanoid characters. This result is similar to that coming via a study done with a somewhat older population. McCue, C. (2008) found in her study that students created mostly realistic avatars instead of fantastic/fantasy-based ones for themselves.

It would be exciting to repeat the investigation with the same materials and methodology, but with different target groups: children from different cultures, children with special needs (particularly ones with reading difficulties and hearing impairment); and in particular, it would be interesting to see if the TERENCE system – if used in the context of different languages, where different avatars may be needed – might offer different types/sets of avatars to comfort their users.

Acknowledgements. The authors' work was supported by the TERENCE project, and was funded by the EC through the FP7 for RTD, Strategic Objective ICT-2009.4.2, ICT, TEL. The contents of the paper reflect only the authors' views and the EC is not to be held liable for this. The work was also funded within the framework of the "Research of Interactive Children Books" project of the MOME Nonprofit Kft, being financed by the Main Architect Office of the Ministry of Internal Affairs.

References

1. Alexander, B., Marks, L.: Aesthetic preference and resemblance of viewer's personality to paintings. Bulletin of the Psychonomic Society 21, 384–386 (1983)
2. Alrifai, M., Gennari, R., Tifrea, O., Vittorini, P.: The user and domain models of the TERENCE adaptive learning system. In: Vittorini, P., Gennari, R., Marenzi, I., De la Prieta, F., Corchado Rodríguez, J.M. (eds.) International Workshop on Evidence-Based TEL. AISC, vol. 152, pp. 1–9. Springer, Heidelberg (2012)
3. Csíkos, C.: Mintavétel a kvantitatív pedagógiai kutatásban. Gondolat Kiadó. Budapest (2009)

4. Fisher, A., Margolis, J., Miller, F.: Undergraduate women in computer science: experience, motivation and culture. ACM SIGCSE Bulletin 29(1), 106–110 (1997), http://www.cs.cmu.edu/afs/cs/project/gendergap/www/papers/sigcse97/sigcse97.html
5. Inal, Y., Cagiltay, K.: Avatars as Pedagogical Agents for Digital Game-Based Learning. In: Crawford, C., et al. (eds.) Proceedings of the Society for Information Technology & the Teacher Education International Conference 2006, pp. 3440–3443. AACE, Chesapeake (2006)
6. Kay, R.H.: Gender differences in computer attitudes, ability, and use in the elementary classroom. Ontario Universities, Literacy and Numeracy Secretariat, Toronto (2007)
7. Li, N., Kirkup, G.: Gender and cultural differences in Internet use: a study of China and the UK. Computers and Education 48(2), 301–317 (2007)
8. Locher, P., Martindale, C., Dorfman, L. (eds.): New Directions in Aesthetics, Creativity and the Arts. Baywood (2006)
9. McCue, C.: Tween Avatars: What do online personas convey about their makers? In: McFerrin, K., et al. (eds.) Proceedings of the Society for Information Technology & the Teacher Education International Conference 2008, pp. 3067–3072. AACE, Chesapeake (2008)
10. The TERENCE project's home page, http://www.terenceproject.eu/web/guest
11. Art Education Avatars in Action: Learning, Teaching, and Assessing in 3D Virtual Worlds pp 201 Lilly Lu. In: Serious EduGames Assessment

The Effects of Hand Gestures on Psychosocial Perception: A Preliminary Study

Augusto Gnisci and Antonio Pace

Second University of Naples, Department of Psychology, Caserta, Italy
{augusto.gnisci,antonio.pace}@unina2.it

Abstract. To date a few studies have experimentally investigated the effects of hand gestures (and frequency) on psychosocial perception. A preliminary study with two experiments were conducted, in which confederates manipulated "Type" (rhythmic gestures, cohesive gestures and self-adaptors for experiment 1; rhythmic gestures, focusing gestures and dynamic gestures for experiment 2) and "Frequency" (low and high) during a face-to-face conversation with the participants. ANOVAs reveal that rhythmic gestures influence positively competence perception but negatively conversational fairness, self-adaptors increase warmth evaluation and high frequency influences positively warmth and dominance perceptions. Hand gestures appear to play a causal role in psychosocial evaluation.

Keywords: Hand gestures, Psychosocial perception, Preliminary study, Rhythmic gestures, Cohesive gestures, Self-adaptors, Frequency.

1 Introduction

There are many hand gestures classifications [e.g., 1-3] and a general agreement on the main categories: ideational gestures (emblems and illustrative) are related to semantic content of speech; conversational gestures (rhythmic and cohesive) are related to structural component of speech; adaptors (self- and hetero-addressed) can be used during speech but they are not related to it.

These categories are differently decoded by listener [4], influencing the evaluations on speaker. Classical studies on psychosocial perception found that two dimensions underlie most judgments: one dimension typically refers to attributes such as warmth and sociability, the other to attributes such as competence and efficacy [e.g., 5,6].

A few studies have experimentally examined the effect of hand gestures (and frequency) on psychosocial perception. Previous research has examined the role of gestures in correlational studies [e.g., 7,8]; experimentally, but not in a systematic manner [e.g., 9,10]; experimentally, but using an "external" paradigm (i.e., in which the participants evaluated a person in a video) and without manipulating the frequency [11].

On these bases, we designed two preliminary experiments to test the effects of type and frequency of gestures on psychosocial perception in a face-to-face conversation

S. Bassis et al. (eds.), *Recent Advances of Neural Network Models and Applications*,
Smart Innovation, Systems and Technologies 26,
DOI: 10.1007/978-3-319-04129-2_30, © Springer International Publishing Switzerland 2014

with confederates. In experiment 1, the confederates manipulated two categories related to speech: rhythmic gestures (giving emphasis by rapid up-and-down movements) and cohesive gestures (clarifying structure of speech) and one not-related category: self-adaptors (touching parts of own body). In experiment 2, the confederates manipulated three categories all related to speech: rhythmic gestures and two microcategories of cohesive gestures (focusing, highlighting main points of speech; dynamic, articulating parts of speech), used simultaneously in experiment 1. Therefore, the first experiment was a macro-analysis (investigating two different categories related to speech and one not-related category) while the second was a micro-analysis (investigating three categories related to speech, of which two were microcategories).

Shown below, examples of all the types of gestures (these can be executed also in other ways). The segment is extracted from the script followed by the confederates:

It's impossible using only Social Networks. It's fundamental managing to meet each other also face-to-face

Rhythmic gesture: two hands up-and-down movement for all the length of the extract.
Focusing gesture: one hand palm up movement in co-occurrence of the single highlighted words.
Dynamic gesture: two hands turning movement in co-occurrence of the double highlighted words.
Self-adaptor: one hand self-touching movement for all the length of the extract.

Rhythmic gestures, which give emphasis to speech [12], are expected to increase competence evaluation compared to self-adaptors but also conversational unfairness (that is a disagreeable way of interacting), because this emphasis is exaggerated for a young peer-to-peer conversation. Self-adaptors, anxiety-related [13], are expected in a ludic context to influence positively warmth perception (for an empathy process) compared to rhythmic gestures. Focusing and dynamic gestures are expected to elicit similar evaluations, belonging to the same macrocategory of cohesive gestures.

Gesticulating frequently is expected to increase warmth (in Italian culture, gestures are traditionally very important in communication) [14,15] and dominance (gesticulating during speech is a way to control spatial part of an interaction) [10].

Gesticulating frequently is expected to amplify also the effects of rhythmic gestures on competence and conversational unfairness and the effects of self-adaptors on warmth.

2 The First Experiment

2.1 Method

Participants. 91 subjects (*Mean age*=22.75, *SD*=4.19), 14 male and 77 female students at the Second University of Naples (SUN), participated: 85 from Department of Psychology, 3 from Department of Political Sciences, 3 from other Departments.

Procedure. The participants partook in a conversation about interpersonal relations with another student (actually, a non-acquaintance confederate), in the Observation

Lab of SUN. The confederates were trained under the supervision of the first author approximately for 60 hours, before the start of data collection.

During the conversation, the confederates manipulated different types and frequencies of hand gestures, trying to be natural, keeping under control other aspects of non-verbal behavior (e.g., absence of smiles, interruptions and sudden face expression changes) and following specific arguments of a script (the importance of Social Network, the differences between cohabitation and marriage, etc.).

An assistant stopped the conversation after six minutes and gave a questionnaire to the participant, inviting the confederate to exit. After completing the questionnaire, the assistant debriefed and thanked the participant.

Experimental Conditions. Two between-subject independent variables were manipulated: Type and Frequency of hand gestures. Three types (rhythmic gestures, cohesive gestures and self-adaptors) were used frequently (approximately 30 gestures per minute) or sporadically (15 gestures per minute), creating six combinations. There was also a no-gesture condition as control. Experimental combinations and no-gesture condition are comparable because the first author trained the confederates to act in the same natural way in both cases.

The confederates (two male and two female) and the number of participants were balanced in each combination (and no-gesture condition), while sex of participants was not balanced but equally distributed.

Measures. For the subjective manipulation check, the participants answered some items to evaluate quantity of each type and rate of gesticulation (quantity and percentage). We are executing also the objective manipulation check (i.e., two independent observers coding the videotaped material). Interested scholars can write to the authors for the results.

For evaluating the interlocutor and his/her gestures, the participants answered psychosocial scales: "Warmth and Competence of gestures", "Dominance and Disrespect of confederate" [16,17], "Sociability and Assertiveness of confederate" [18], "Confederate conversational unfairness".

For evaluating gestures perception, the participants answered three items: "Attention to gestures", "Gestures efficacy" and "Gestures distraction".

For more details on Measures, see the questionnaire in Appendix.

Design and Data Analysis. We conducted a principal-component factor analysis on all the scales (obtaining a factor score for each dimension) and transformed the single-item measures in z scores. On each measure, we executed a 3×2 (Type × Frequency) factorial ANOVA. For each effect of Type, we carried out LSD post-hocs (comparing the types) and t-tests (comparing each type with no-gesture condition). For each interaction, we executed t-tests (comparing high and low frequency for each type).

2.2 Results

We calculated skewness and kurtosis for all the measures, testing the probability distribution of the data. Only one parameter of a measure ("Confederate conversational unfairness") was just a little problematic (i.e., > |3|), with kurtosis=4.00. Therefore, we decided to use a parametric method (ANOVA) for testing the hypotheses.

Manipulation Check. We carried out one-way ANOVAs to check the manipulation of gestures, with "Type" as independent variable. Each ANOVA revealed a significant effect: "Quantity of rhythmic gestures" [$F(2, 76)=8.99, p<.001$]; "Quantity of focusing gestures" [$F(2, 76)=5.89, p<.01$]; "Quantity of dynamic gestures" [$F(2, 76)=10.48, p<.001$]; "Quantity of self-adaptors" [$F(2, 76)=19.06, p<.001$].

We executed one-way ANOVAs to check the manipulation of gesticulation rate, with "Frequency" as independent variable. Both ANOVAs were significant: "Quantity of gestures" [$F(2, 71)=25.11, p<.001$]; "Percentage of gestures" [$F(2, 88)=18.80, p<.001$].

Hypotheses Testing. Factorial ANOVAs revealed a significant (or just non-significant) effect of "Type" on "Gestures competence" [$F(2, 73)=3.47, p<.05, \eta^2=.09$], "Confederate conversational unfairness" [$F(2, 73)=3.06, p=.053, \eta^2=.08$] and "Gestures warmth" [$F(2, 73)=3.77, p<.05, \eta^2=.09$].

On these measures, we calculated 2 LSD post-hocs ($\alpha=.025$ in accord to Bonferroni correction), comparing self-adaptors (not related to speech) with rhythmic and cohesive gestures (the two categories related to speech). See the graph n° 1 for the significant post-hocs. The t-tests, comparing each type with no-gesture condition ($\alpha=.05$), were non-significant for all these measures.

Other Measures. Factorial ANOVAs revealed a significant effect of "Frequency" on "Attention to gestures" [$F(1, 72)=7.74, p<.01, \eta^2=.10$; high ($M=.49, SD=.88$) > low ($M=-.11, SD=1.03$)].

2.3 Discussion

Manipulation checks showed that the participants perceived correctly the experimental conditions: a type used in a specific condition was perceived more than other types; gestures used in "High frequency" condition were perceived as more frequent.

As regards hypotheses testing, the confederates were evaluated as more competent but also conversationally unfair when they used rhythmic gestures compared to self-adaptors. Moreover, they were evaluated as warmer when gesticulated by self-adaptors compared to rhythmic gestures. See the graph n° 1 for more details.

As for other measures, the participants gave greater attention to gestures when they observed a high frequency compared to low frequency.

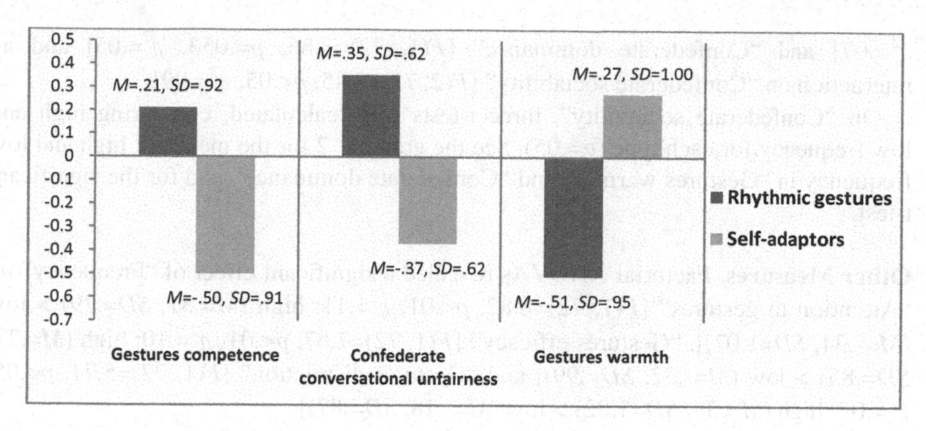

Graph n°1. Experiment 1. Significant LSD post-hocs (p<.025).

3 The Second Experiment

3.1 Method

Participants. 91 subjects (*Mean age*=22.81, *SD*=3.51), 15 male and 76 female students at the SUN, participated: 87 from Department of Psychology, 3 from Department of Political Sciences, 1 from Department of Mathematics and Physics.

Experimental Conditions. The confederates used three types (rhythmic gestures, focusing gestures and dynamic gestures) frequently (30 per minute) or sporadically (15), creating six combinations. There was also a no-gesture condition as control. For more details, see experiment 1.

Measures, Procedure, Design and Data Analysis. See experiment 1.

3.2 Results

No value of *skewness* or *kurtosis* was >|3|.

Manipulation Check. We carried out one-way ANOVAs to check the manipulation of gestures. One ANOVA revealed a significant effect, "Quantity of dynamic gestures" [F(2, 74)=11.59, p<.001], while two ANOVAs were non-significant: "Quantity of rhythmic gestures" [F(2, 74)=2.16, p=.12]; "Quantity of focusing gestures" [F(2, 74)=.46, p=.64].

We executed one-way ANOVAs to check the manipulation of gesticulation rate. Both ANOVAs were significant: "Quantity of gestures" [F(2, 71)=14.94, p<.001]; "Percentage of gestures" [F(2, 88)=17.37, p<.001].

Hypotheses Testing. Factorial ANOVAs revealed a significant (or just non-significant) effect of "Frequency" on "Gestures warmth" [F(1, 72)=5.69, p<.05,

η^2=.07] and "Confederate dominance" [$F(1, 72)=3.86$, $p=.053$, η^2=.05] and an interaction on "Confederate sociability" [$F(2, 72)=3.45$, $p<.05$, η^2=.09].

On "Confederate sociability", three t-tests were calculated, comparing high and low frequency for each type (α=.05). See the graph n° 2 for the means of high and low frequency in "Gestures warmth" and "Confederate dominance" and for the significant t-test.

Other Measures. Factorial ANOVAs revealed a significant effect of "Frequency" on "Attention to gestures" [$F(1, 72)=8.82$, $p<.01$, η^2=.11; high (M=.31, SD=.79) > low (M=-.34, SD=1.07)], "Gestures efficacy" [$F(1, 72)=7.67$, $p<.01$, η^2=.10; high (M=.27, SD=.85) > low (M=-.32, SD=.99)] and "Gestures distraction" [$F(1, 72)=5.71$, $p<.05$, η^2=.07; high (M=.34, SD=1.02) > low (M=-.18, SD=.87)].

3.3 Discussion

Manipulation checks showed that the participants perceived correctly "Frequency" (gestures used in "High frequency" condition were perceived as more frequent). As for "Type", rhythmic and focusing gestures were not discriminated. We argue that this non-distinction occurred because the item asking quantity of rhythmic gestures is written in a too similar way to that of focusing gestures (see Appendix).

As regards hypotheses testing, the confederates were evaluated as warmer and more dominant when they used a high frequency of gesticulation compared to low frequency. In addition, they were evaluated as more sociable when gesticulated by high frequency of dynamic gestures compared to low frequency. See the graph n° 2 for more details.

As for other measures, the participants that observed a high frequency gave greater attention to gestures and evaluated them as more efficacious and distracting compared to the participants that observed a low frequency.

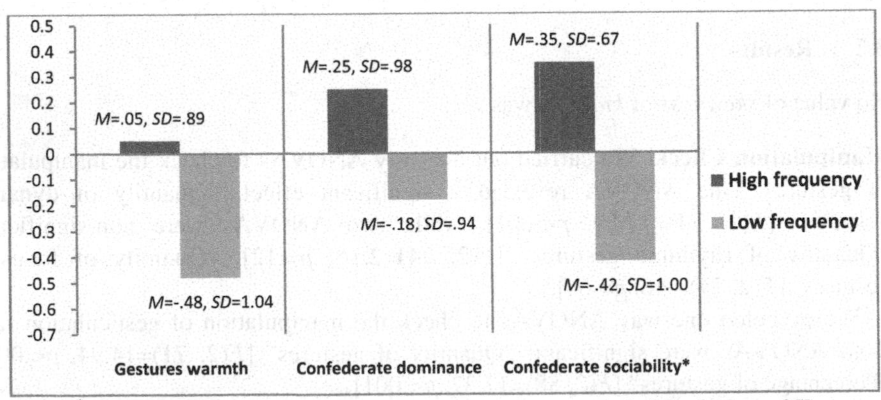

* Means of high and low frequency for dynamic gestures [$t(24)=2.30$, $p<.05$].

Graph n°2. Experiment 2. Means of high and low frequency.

4 General Discussion

This preliminary study addressed the causal effect of the speakers' hand gestures on the psychosocial evaluations by the listeners. For the first time, the participants to conversation evaluated the gesturing speaker and together with the type of gestures the frequency was manipulated.

As hypothesized, rhythmic gestures increased competence evaluation compared to self-adaptors, but also conversational unfairness, and self-adaptors influenced positively warmth perception compared to rhythmic gestures. Probably, the emphasis of rhythmic gestures [12] increase evaluation on cognitive abilities (like competence) but in a context of young peer-to-peer conversation is regarded as exaggerated, decreasing evaluation on social abilities (like conversational fairness and warmth). Self-adaptors were perceived as a sign of anxiety [13] influencing incompetence perception, but we think that the participants just for this felt also empathy and compassion toward the interlocutor, that elicited a higher warmth evaluation. As hypothesized, focusing and dynamic gestures did not influence differently the psychosocial evaluations (probably expressing a too much similar function during speech).

As hypothesized, gesticulating frequently increased warmth and dominance evaluations (although only for experiment 2). We speculate that gesticulating frequently during speech is very appreciated in Italian culture [14,15] and allows us to control the social-physical space which is shared, in addition to control, through the words, on social-conversational space [10].

Differently from what we hypothesized, gesticulating frequently did not amplify the positive effects of gestures. Probably, there exists an optimal threshold of gesticulation, and go beyond it is often useless. However, unexpectedly, in the second experiment a high frequency of dynamic gestures increased sociability evaluation compared to low frequency. We speculate that, in some cases, a high level of a certain type elicits an effect more than amplifying an existing one.

Overall, the results of our introductive study provide a firm foundation for establishing a cause-effect relationship between type (and frequency) of hand gestures and the psychosocial evaluations. A future study should explore the same effects with a more numerous sample.

Acknowledgments. A special thanks and a sincere gratitude to Fridanna Maricchiolo for her supervision on the functions of hand gestures and to the confederates: Claudio, Consiglia, Domenico M. and Iolanda (Experiment 1); Cassandra, Domenico N., Ivana and Manuel (Experiment 2).

References

1. Ekman, P., Friesen, W.V.: The Repertoire of Nonverbal Behavior. Semiotica 1, 49–98 (1969)
2. McNeill, D.: Hand and mind. The University of Chicago Press, Chicago (1992)

3. Maricchiolo, F., Gnisci, A., Bonaiuto, M.: Coding Hand Gestures: A Reliable Taxonomy and a Multi-media Support. In: Esposito, A., Esposito, A.M., Vinciarelli, A., Hoffmann, R., Müller, V.C. (eds.) COST 2102. LNCS, vol. 7403, pp. 405–416. Springer, Heidelberg (2012)
4. Patterson, M.L.: The evolution of theories of interactive behavior. In: Manusov, V.L., Patterson, M.L. (eds.) The Sage Handbook of Nonverbal Communication, pp. 21–39. Sage Publications, Thousand Oaks (2006)
5. Abele, A.E., Wojciszke, B.: Agency and Communion from the Perspective of Self Versus Others. Journal of Personality and Social Psychology 93, 751–763 (2007)
6. Fiske, S.T., Cuddy, A.J.C., Glick, P.: Universal Dimensions of Social Cognition: Warmth and Competence. Trends in Cognitive Sciences 11, 77–83 (2007)
7. Burgoon, J.K., Birk, T., Pfau, M.: Nonverbal Behaviors, Persuasion, and Credibility. Human Communication Research 17, 140–169 (1990)
8. Mehrabian, A., Williams, M.: Nonverbal Concomitants of Perceived and Intended Persuasiveness. Journal of Personality and Social Psychology 13, 37–58 (1969)
9. Carli, L.L., LaFleur, S.L., Loeber, C.C.: Nonverbal Behavior, Gender, and Influence. Journal of Personality and Social Psychology 68, 1030–1041 (1995)
10. Maricchiolo, F., Livi, S., Bonaiuto, M., Gnisci, A.: Hand Gestures and Perceived Influence in Small Group Interaction. The Spanish Journal of Psychology 14, 755–764 (2011)
11. Maricchiolo, F., Gnisci, A., Bonaiuto, M., Ficca, G.: Effects of Different Types of Hand Gestures in Persuasive Speech on Receivers' Evaluations. Language and Cognitive Processes 24, 239–266 (2009)
12. Butterworth, B., Hadar, U.: Gesture, Speech and Computational Stage: A Reply to McNeill. Psychological Review 96, 168–174 (1989)
13. Henningsen, D.D., Valde, K.S., Davies, E.: Exploring the Effect of Verbal and Nonverbal Cues on Perceptions of Deception. Communication Quarterly 53, 359–375 (2005)
14. Kendon, A.: Gestures as Illocutionary and Discourse Structure Markers in Southern Italian Conversation. Journal of Pragmatics 23, 247–279 (1995)
15. Kendon, A.: Gesture: Visible Action as Utterance. Cambridge University Press, Cambridge (2004)
16. Farley, S.D.: Attaining Status at the Expense of Likeability: Pilfering Power Through Conversational Interruption. Journal of Nonverbal Behavior 32, 241–260 (2008)
17. LaFrance, M.: Gender and Interruptions: Individual Infraction or Violation of the Social Order? Psychology of Women Quarterly 16, 497–512 (1992)
18. Robinson, L.F., Reis, H.T.: The Effects of Interruption, Gender, and Status on Interpersonal Perceptions. Journal of Nonverbal Behavior 13, 141–153 (1989)

Appendix

Shown below, the items of the questionnaire, in order of appearance. If there is not a different indication, the response scale is 0-10 (*"Not at all"* or *"Not at all in accord"* – *"Completely"* or *"Completely in accord"*). The information in bold type or square brackets was not present into the questionnaire.

Manipulation Check Items
[*Quantity of rhythmic gestures*] How much hand gestures of your interlocutor were synchronized with the rhythm of speech through vertical movements of hand, that is using little up-and-down rhythmical strokes. ___

[*Quantity of focusing gestures*] How much hand gestures of your interlocutor highlighted the main points of speech, clarifying from time to time which were the main words or ideas. ___

[*Quantity of dynamic gestures*] How much hand gestures of your interlocutor computed circular movements (by one or two hands), specifying the flow and the progression of speech. ___

[*Quantity of self-adaptors*; just in experiment 1] How much your interlocutor moved his/her hands for putting them into contact with his/her body (hair, arms, head, face, thighs, etc.), executing movements as touching, wrinkling, scraping, scratching, etc. ___

[*Quantity of gestures*] How much your interlocutor gesticulated, that is moved his/her hands during speech? ___

[*Percentage of gestures*] How many times your interlocutor gesticulated? (Insert just one X)
Never ___; 1-10% ___; 11-20% ___; 21-30% ___; 31-40% ___; 41-50% ___ 51-60% ___; 61-70% ___; 71-80% ___; 81-90% ___; 91-99% ___; Always ___

Psychosocial Scales (for Hypotheses Testing)
[[a] Factor *Gestures warmth*; [b] Factor *Gestures competence*] How much, during the conversation, hand gestures of your interlocutor seemed to you:
Competent[b] ___; Friendly[a] ___; Secure[b] ___; Well-intentioned[a] ___; Capable[b] ___; Reliable[a] ___; Efficient[b] ___; Warm[a] ___; Intelligent[b] ___; Benevolent[a] ___; Able[b] ___; Sincere[a] ___

[[a] Factor *Confederate dominance*; [b] Factor *Confederate disrespect*] How much your interlocutor, during the interaction with you, was:
Rude[b] ___; Agreeable[a] ___; Passive[a] ___; Caring[b] ___; Rational[a] ___; Strong[a] ___; Pleasant[b] ___; Dominant[a] ___; Competitive[a] ___; Overbearing[a] ___; Concerned with other[b] ___

[[a] Factor *Confederate sociability*; [b] Factor *Confederate assertiveness*] How much your interlocutor seemed to you (semantic differential):

Supportive	0	1	2	3	4	5	6	7	8	9	10	Disapproving[a]
Considerate	0	1	2	3	4	5	6	7	8	9	10	Inconsiderate[a]
Friendly	0	1	2	3	4	5	6	7	8	9	10	Hostile[a]
Easy to get along with	0	1	2	3	4	5	6	7	8	9	10	Polemic[a]
Open	0	1	2	3	4	5	6	7	8	9	10	Closed[a]
Pleasant	0	1	2	3	4	5	6	7	8	9	10	Unpleasant[a]
Likeable	0	1	2	3	4	5	6	7	8	9	10	Disagreeable[a]
Attentive	0	1	2	3	4	5	6	7	8	9	10	Inattentive[a]
Competent	0	1	2	3	4	5	6	7	8	9	10	Incompetent[a]

Extraverted	0	1	2	3	4	5	6	7	8	9	10	Introvert[b]
Strong	0	1	2	3	4	5	6	7	8	9	10	Weak[b]
Assertive	0	1	2	3	4	5	6	7	8	9	10	Passive[b]
Pushy	0	1	2	3	4	5	6	7	8	9	10	Moderate[b]
Dominant	0	1	2	3	4	5	6	7	8	9	10	Submissive[b]

[Factor *Confederate conversational unfairness*] How much you agree with the following statements, relating to your interlocutor:
His/Her way of interacting was disagreeable. ___
His/Her way of interacting generated nervousness. ___
He/She seemed to take for granted my opinions. ___
He/She did not respect the turns of the conversation. ___
He/She tried to ensure that I lost the thread of the conversation. ___
His/Her attitude was unfair. ___
He/She called my opinions into question. ___
He/She was in agreement with everything I said. ___
He/She tended to preempt my answer. ___
He/She did not give importance to that I said. ___
He/She asked questions that surprised me. ___
He/She did not show interest in my words. ___

Other Measures
[Attention to *gestures*] How much you gave attention to hand gestures of your interlocutor during the conversation? ___
[*Gestures efficacy*] How much you consider efficacious hand gestures of your interlocutor? ___
[*Gestures distraction*] How much, during the conversation, hand gestures of your interlocutor distracted you from the content of speech? ___

The Influence of Positive and Negative Emotions on Physiological Responses and Memory Task Scores

Maria Teresa Riviello[1,2], Vincenzo Capuano[1,2], Gianluigi Ombrato[1], Ivana Baldassarre[1], Gennaro Cordasco[1,2], and Anna Esposito[1,2]

[1] Department of Psychology, Second University of Naples, Italy
[2] International Institute for Advanced Scientific Studies (IIASS), Italy
{mariateresa.riviello,vincenzo.capuano,
ivana.baldassarre,gennaro.cordasco,anna.esposito}@unina2.it,
gianluigi.ombrato@studenti.unina2.it

Abstract. The present paper report results of a preliminary study devoted to investigate whether and how different induced emotional states influence physiological responses and memory task scores. Physiological responses, such as skin conductance (SCL) and heart rate (HR) values were measured from 32 university students, before, during and after they were elicited by video stimuli. The considered stimuli were able to induce positive, negative and neutral emotional states. The specific physiological activation patterns were identified and correlated with memory task scores, computed using the "Anna Pesenti" Story Recall Test (SRT).

The results show significant changes in physiological values when positive (increase in HR values) and negative (increase in SCL values) emotional states are induced. Surprisingly, increased SCL values, associated to induced positive emotional states, affect the participant's memory task scores.

Keywords: emotion, physiological activation, memory task scores.

1 Introduction

Emotion is defined as a complex state of feeling that results in physical and psychological changes able to influence thought and behavior. In particular, emotionality is associated with a range of phenomena including, among others, physiological arousal (activation), and cognitive processes as memory. Indeed, emotions involve the activation of behavioral dispositions implying bivalent reactions toward and away the stimulus that caused the arousal. These dispositions engage the Automatic Nervous System (ANS)[1-2], that governs involuntary actions, resulting, among others, in internal physiological changes such as electrical activities of the skin (skin conductance),respiratory activities, blood pressure, heart rate, sweat glands, reactions of the endocrine glands, chemical activities of blood.

Sequeira et al. [3] sustained that any organ that is influenced by the ANS could be a potential index of mind activity. On this assumption, understanding physiological ANS indices may help the detection of emotional processing states.

S. Bassis et al. (eds.), *Recent Advances of Neural Network Models and Applications*,
Smart Innovation, Systems and Technologies 26,
DOI: 10.1007/978-3-319-04129-2_31, © Springer International Publishing Switzerland 2014

Several researches [4-8] showed that focused on specific physiological activation patterns underlay are related to specific emotional experience. Nevertheless, univocal associations between physiological activation patterns and specific emotional states have not yet been identified. For example, some studies identify different physiological activation features for sadness: when this emotion is correlated with crying, it is characterized by heart rate (HR) and skin conductance level (SCL) increasing values [9], whereas, with no crying, it can be observed an increase of the level values of both the physiological signals [10-11] or an increase of the HR and no changes in the SCL [12] values. Also studies on happiness [13-15] showed controversial data. In particular, the physiological response to happiness includes either increased HR [16-17], unchanged HR [18-19], or decreased HR values [20].

The physiological arousal dimension of emotion also represents a critical factor contributing to the emotional enhancement effect on memory [21].Thanks to the increasing accessibility of brain imaging techniques researches reveal that heightened physiological activity increases the probability of remembering, and this improvement is due to the amygdale activation that mediates consolidation of memory in other brain regions (for a review see [22]).

In addition, there is evidence that memory can be enhanced even for the valence dimension of emotion, i.e. how positive or negative an event is. The studies [23] that investigate this dimension have found that items with positive or negative valence improve memory with the respect of neutral items, via modulation of distinct neural processes that are independent from the amygdala activation [24-26]. The Emotionally Enhanced Memory – EEM [27] is one of the most studied feature in psychology [27-30]. Scholars posit that emotional stimuli, both positive and negative, can facilitate the encoding process through attentional processes [31-33]. However, almost all the studies conducted on EEM involved memory task on arousing/emotional stimuli *per se*; whereas very few studies tested the so-called Easterbrook's hypothesis [34], which claims that attention is limited when people experience an high arousal not directly linked with the stimulus.

The present study is a preliminary investigation aimed at observing whether and how positive and negative induced emotional states influence physiological responses and memory task scores. In particular, it estimates memory task performances on neutral stimuli (not arousing/emotional stimuli *per se)* after having induced arousal trough emotional stimuli separately. It examines changes in skin conductance (SCL) and heart rate (HR) values before, during and after individuals experienced emotional states of happiness, sadness and neutral. The abovementioned emotional states were elicited by showing video stimuli to the participants in the experiments. The main goal is to identify specific physiological emotional responses and to explore their possible effects on memory task scores, computed using the "Anna Pesenti" Story Recall Test (SRT) [35].

2 Experimental Set-Up

2.1 Materials

The materials used for the experiments are described below:

a) ®PSYCHOLAB VD13SV, for measuring participants' physiological values. This instrument was provided with two different sensors able to measure heart rate and skin conductance values.

b) Emotional videos, for evoking emotional states. Six video-clips were selected from the database of emotional evoking video stimuli described in [36]. Such database consists in assessed videos downloaded from YouTube. In particular, we selected one180 seconds (s) long video-clip for happiness inducing the positive emotional state 2 video-clips for sadness, assembled together in order to have a 180s long video-clip inducing the negative emotional state; and 3 video-clips assembled together to form a 180s long video-clip inducing the neutral emotional state. The emotional videos, i.e. happy and sad videos, contain the original audio-track (dialogues and music), whereas neutral videos are soundless.

c) "Anna Pesenti" Story [35], for evaluating memory performances. This story recall test is usually used in neuropsychology to evaluate memory impairments in Italian patients [37]. It is composed by 28 different mnemonic units. The original protocol requires an immediate and deferred recall: after the experimenter read the test twice, the patient (subject) is asked to recall the story immediately and after 15 minutes, during which he/she is involved in a distractive non-verbal task. A correction table is used to evaluate results, in terms of number of units recalled, according to years of schooling, age and gender of the subject. In this work participants were asked to recall the story a single time, after the presentation of the emotional stimuli that in this case represents the non-verbal distractive task.

2.2 Participants

A total of 32 Italian university students (21 males and 11 females, age: 23,7 ± 3,4) were involved in the experiments. The participants filled and signed an agreement form declaring their voluntary participation and authorized the researchers to use the collected data for scientific purposes.

2.3 Procedure

Three experimental conditions were created according to the video stimuli presented, therefore, according to the emotional state induced: positive, negative, and neutral. The participants were partitioned into three groups so that 11 (8 males and 3 females) were assigned to positive, 10 (5 males and 5 females) to negative, and11 (8 males and 3 females) to neutral condition.

The experimental procedure was stable for all the three condition and it is described below. Each participant sat in front of a computer and was connected to PSYCHOLAB VD13SV sensors.

The timeline of the experiment follows: during the first 30s, the participant was able to read instructions 1 on the screen, so that he/she was informed about what it was going to be presented and how to proceed. Soon after, a fixation point (a crux in the middle of the screen) appeared for 180 s. This phase of the experimental procedure was used in order to induce a relaxation state in the participant and record physiological starting point data (baseline).Then, instructions2appeared on the screen for 5s. During this time the participant was asked to pay attention to the story he/she was going to listen for 2 times. In the following 55s, a clear recorded voice told the "Anna Pesenti" story twice. As soon as the voice stopped, instructions 3appeared on the screen for 5 s, informing the participant he/she was going to watch a video. Then the emotional video (positive, negative or neutral, according to the experimental condition the participant was assigned) was displayed on the screen for 180 s. At the end of the video presentation the participant was disconnected from the sensors and he/she was asked to recall the story he had listened to. A recorder stored the participant's report, in order to compute the memory task scores as described in section 2.1.

The experimental procedure timeline is summarized and displayed in Figure1. The two time intervals, depicted with black boxes in the figure, refer to the final 30 s of the fixation point time interval (considered as baseline) and the final 30 s of the emotional induction phase (considered as emotional state).Only the physiological (SCL and HR) values recorded during these two time intervals are used for the analysis. This approach, wildly used in literature [6, 38], was considered preferable because the considered data are related to a more complete relaxation state (baseline), and emotion induction (emotional state). Indeed, the autonomic nervous system deals with a lot of non-contingent activities as digestion, postural adjustment, thermoregulation, etcetera, so care must be taken to ensure that the measurements are time-locked to the stimulus onset.

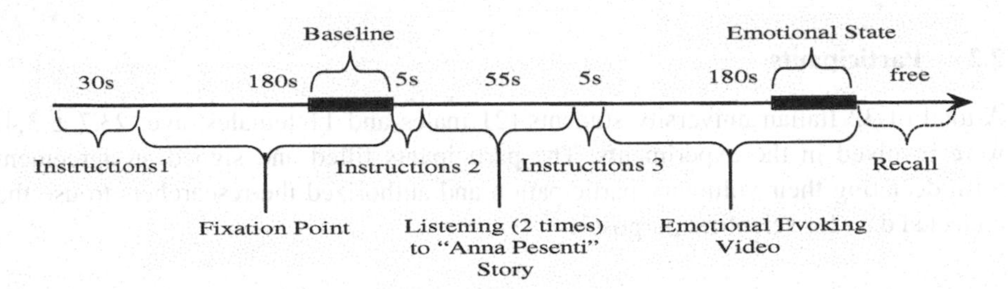

Fig. 1. Experimental procedure timeline. Each bracket indicates a specific phase of the experiment and its time duration. The physiological data used for the assessment are extracted during the two time intervals (Baseline and Emotional State) highlighted with black boxes.

3 Results

In order to evaluate the effect of the emotion induction on participants' physiological activity paired samples t-tests were computed comparing baseline and emotional state values for both SCL and HR, respectively. The analyses were separately performed on the data obtained in each experimental condition: positive, negative and neutral (see section 2.1 a).

Results reveal that, for the positive condition, no significant differences were found between the baseline and emotional state SCL values ($t(9)= -1.756$; $\rho= .113$), whereas a significant increase in HR values ($t(9)= -7.475$; $\rho= .0001$) was recorded.

On the other hand, for the negative condition, a significant increase in SCL values ($t(10)= -2.797$; $\rho= .019$) emerged, while no significant changes were found in HR values ($t(10)=2.154$; $\rho= .057$).These results indicate that positive and negative emotional states produce different patterns of physiological responses.

No significant changes in both SCL ($t(10)= -.067$, $\rho= .948$) and HR ($t(10) =-1.026$, $\rho=.325$) values occurred for neutral condition.

To evaluate the effect of the induced emotional state on memory task scores, a one-way ANOVA analysis was performed on the data, considering the experimental condition as fixed factor, and the memory task score as dependent variable.

The ANOVA shows that participants' performances at the memory task were not affected by the emotion induction ($F(2,31)=1.288$, $\rho=.291$).

However, when ANOVAs (as described above) were performed on data from female and male participants separately, they showed a significant effect of the induced emotional state only for female subjects (Females: $F(2,10)=6.258$, $\rho=.023$; Males: $F(2,20)=1.150$, $\rho=.339$).

In particular, as displayed in Figure 2, post hoc comparisons using the Bonferroni test indicated that females memory task scores, for the negative condition, were significantly lower than scores for the neutral condition. Whereas, the results for positive condition did not significantly differ from those for neutral and negative conditions.

Finally, linear regression analyses were performed to verify the existence of a significant relation between physiological emotional activation (changes in SCL and HR values) and participants' memory task performances. The analyses were computed on the data obtained in the three experimental conditions separately. Differences between physiological values (SCL and HR, separately) detected in the baseline and emotional states were considered as independent variables, and memory task scores as dependent variable. Results indicate a barely significant effect related to SCL values in the positive condition ($\beta=.639$; $t(1)=2.35$, $\rho=.047$) (cf. Fig.3).

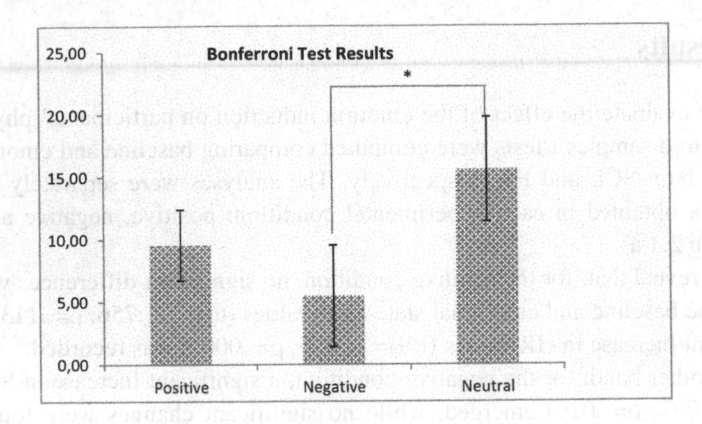

Fig. 2. Means and standard deviations on memory task scores provided by female participants in the 3 experimental conditions. The star refers to significant differences (ρ<.05).

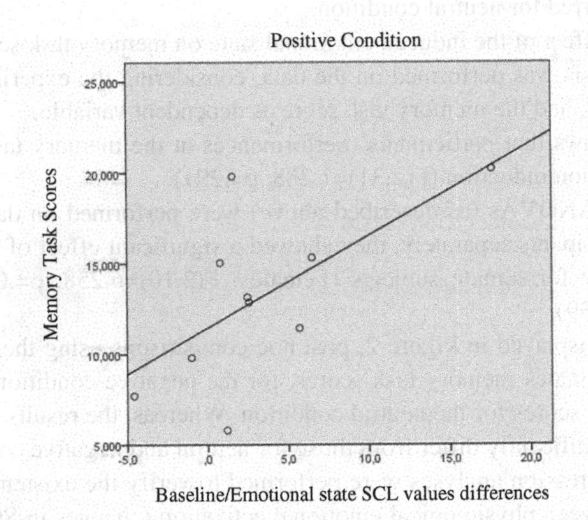

Fig. 3. Linear regression analysis results displaying the relation between changes (as the difference between Baseline and Emotional state) in SCL values (x-axis) and participants' memory task performances (y-axis) in the positive condition

4 Conclusions

The present study was a preliminary investigation aimed to assess, on the one hand, the effects of induced emotional states on Heart Rate variability (HR) and Skin Conductance Levels (SCL); on the other hand, the study explored possible effects of such preceding activation on memory performances, using a memory task with neutral stimuli.

The Autonomic Nervous System (ANS) response to emotional stimuli is well established, however the identification of a homogeneous physiological activation pattern underlay specific emotional experience is less clear: empirical findings showed controversial results for both positive and negative emotional stimuli.

The present study results partially support some of the findings reported in literature. They revealed an increase in HR values under positive emotion induction, whereas although increased values on SCL values were observed, these changes were not significant. On the contrary, under negative emotion induction, variability in HR values did not emerge and a significant increase in SCL values has been observed. As expected, when the stimuli did not have affective valence, no physiological activation appeared.

Results seem to indicate that the valence of emotional stimuli is an essential feature in order to investigate the physiological emotional response; furthermore, these findings seem suggesting that positive and negative emotional stimuli selectively affect HR and SCL, respectively.

The effect of emotional states induction on memory task performances was also analyzed. The analysis showed no significant effect of the induced emotional states on subjects' performances; whereas, by considering the gender of participants, results showed that memory task performances of females decreased when the task followed negative with respect to neutral stimuli. No differences emerged between the positive and the other two emotional states.

Finally, linear regression analyses, performed in order to verify the existence of a significant relation between physiological emotional activation (changes in SCL and HR values) and participants' memory task scores, surprisingly showed that an increase of SCL values, following positive emotional stimulation, produced better performances in memory task. This result was not expected due to the fact that the SCL increases in positive emotional state were not significant. It can be due to the sample size or to the experimental manipulation. Indeed, if it is true that an high physiological activation not directly linked with the stimulus negatively affects individuals' attention [34], and possible memory performance, the effects of a moderate physiological activation (as in this case) has not been specifically investigated yet.

However, more data is needed in order to support these results and hypothesis. In addition, an in-depth analysis of psychological and physiological gender difference in responding to emotion should be taken into account.

References

1. Andreassi, J.L.: Psychophysiology: Human Behavior and Psychological Response. Erlbaum Associates, Hillsdale (1989)
2. Bouscein, W.: Electrodermal Activity. Plenum Press, New York (1992)
3. Sequeira, H., Hot, P., Silvert, L., Delplanque, S.: Electrical autonomic correlates of emotion. International Journal of Psychophysiology 71(1), 50 (2009)
4. Berntson, G.G., Sarter, M., Cacioppo, J.T.: Ascending visceral regula-tion of corticalaffective information processing. Eur. J. Neurosci. 18, 2103–3109 (2003)

5. Bradley, M.M., Lang, P.J.: Measuring emotion: behavior, feeling and physiology. In: Lane, R.D., Nadel, L. (eds.) Cognitive Neuroscience of Emotion, pp. 242–276. Oxford UniversityPress, Oxford (2000)
6. Christie, I.C., Friedman, B.H.: Autonomic specificity of discrete emotion and dimensions of affective space: A multivariate ap-proach. International Journal of Psychophysiology 51(2), 143–153 (2004)
7. Nyklicek, I., Thayer, J.F., Van Doornen, L.J.P.: Cardiorespiratory differ-entiationof musically-induced emotions. J. of Psychophysiology 11, 304–321 (1997)
8. Rainville, P., Bechara, A., Naqvi, N., Damasio, A.R.: Basic emotions are associatedwith distinct patterns of cardiorespiratory activity. International Journal of Psychophysiology 61, 5–18 (2006)
9. Rottenberg, J., Kasch, K.L., Gross, J.J., Gotlib, I.H.: Sadness and amusement reactivitydifferentially predict concurrent and prospective func-tioning in majordepressive disorder. Emotion 2, 135–146 (2002)
10. Britton, J.C., Taylor, S.F., Berridge, K.C., Mikels, J.A., Liberzon, I.: Differential subjectiveand psychophysiological responses to socially and non-socially generate demotional stimuli. Emotion 6, 150–155 (2006)
11. Eisenberg, N., Fabes, R.A., Bustamante, D., Mathy, R.M., Miller, P.A., Lindholm, E.: Differentiation of vicariously induced emotional reactions in children. Developmental Psychology 24, 237–246 (1988)
12. Gross, J.J., Fredrickson, B.L., Levenson, R.W.: The psychophysiology of crying. Psychophysiology 31, 460–468 (1994)
13. Van der Kolk, B.A.: The body keeps the score: Memory and the evolv-ing psychobiology of posttraumatic stress. Harvard Review of Psychiatry 1(5), 253–265 (1994)
14. Vianna, E.P.M., Tranel, D.: Gastric myoelectrical activity as an index of emotional arousal. International Journal of Psychophysiology 61, 70–76 (2006)
15. Rottenberg, J., Salomon, K., Gross, J.J., Gotlib, I.H.: Vagal withdrawal to sad filmpredicts subsequent recovery from depression. Psychophysiology 42, 277–281 (2005)
16. Gehricke, J.-G., Fridlund, A.J.: Smiling, frowning, and autonomic activity in mildly depressed and nondepressed men in response to emotional im-agery of social contexts. Perceptual and Motor Skills 94, 141–151 (2002)
17. Jönsson, P., Sonnby-Borgström, M.: The effects of pictures of emotional faces on tonic and phasic autonomic cardiac control in women and men. Biological Psychology 62(2), 157–173 (2003)
18. Etzel, J.A., Johnsen, E.L., Dickerson, J.A., Tranel, D., Adolphs, R.: Car-diovascular and respiratory responses during musical mood induction. International Journal of Psychophysiology 61, 57–69 (2006)
19. Nyklicek, I., Thayer, J.F., Van Doornen, L.J.P.: Cardiorespiratory differentiation of musically-induced emotions. Journal of Psychophysiology 11, 304–321 (1997)
20. Krumhansl, C.L.: An exploratory study of musical emotions and psy-chophysiology. Canadian Journal of Experimental Psychology 51(4), 336–353 (1997)
21. Cahill, L., McGaugh, J.L.: A novel demonstration of enhanced mem-ory associated with emotional arousal. Consciousness and Cognition 4(4), 410–421 (1995), doi:10.1006/ccog.1995.1048 PMID 8750416
22. McGaugh, J.L., Cahill, L.: Emotion and memory: Central and peripheral contributions. In: Davidson, R.J., Scherer, K.R., Goldsmith, H.H. (eds.) Handbook of Affective Sciences, pp. 93–116. Oxford University Press, New York (2003)
23. Kensinger, E.A.: Remembering emotional experiences: The contribu-tion of valence and arousal. Reviews in the Neurosciences 15(4), 241 (2004)

24. Hamann, S.B.: Cognitive and neural mechanisms of emotional mem-ory. Trends in Cognitive Sciences 5, 394–400 (2001)
25. Phelps, E.A.: Emotion and cognition: Insights from studies of the hu-manamygdala. Annual Review of Psychology 57, 27–53 (2006)
26. Phelps, E.A., LeDoux, J.F.: Contributions of the amygdala to emo-tionprocessing: From animal models to human behavior. Neuron 48, 175–187 (2005)
27. Talmi, D., Luk, B.T.C., McGarry, L.M., Moscovitch, M.: The contribution of relatedness and distinctiveness to emotionally enhancedmemory. Journal of Memory and Language 56(4), 555–574 (2007)
28. Mather, M.: Emotional arousal and memory binding: An object based framework. Perspectives on Psychological Science 2, 33–52 (2007)
29. Kensinger, E.A., Schacter, D.L.: Memory and Emotion. In: Lewis, M., Haviland-Jones, J.M., Barrett, L.F. (eds.) Handbook of Emotions. Guilford Press, New York (2008)
30. Schaefer, A., Pottage, C.L., Rickart, A.J.: Electrophysiological cor-relates of remembering emotional pictures. Neuroimage 54, 714–724 (2011)
31. Christianson, S.A., Loftus, E.F., Hoffman, H., Loftus, G.R.: Eye fixations and memory for emotional events. Journal of Experimental Psychology: Learning, Memory, and Cognition 17, 693–701 (1991)
32. Kern, R.P., Libkuman, T.M., Otani, H., Holmes, K.: Emotional Stimuli, Divided Attention, and Memory. Emotion 5, 408–417 (2005)
33. Clark-Foos, A., Marsh, R.L.: Recognition memory for valanced and arousing materials under conditions of divided attention. Memory 16, 530–537 (2008)
34. Easterbrook, J.A.: The effect of emotion on cue utilization and the or-ganization of behavior. Psychology Review 66, 183–201 (1959)
35. Novelli, G., Papagno, C., Capitani, E., Laiacona, M.: Tre test clinici di memoria verbale a lungo termine: taratura su soggetti normali. Archivio di Psicologia, Neurologia e Psichiatria 47, 278–296 (1986)
36. Esposito, A., Capuano, V., Mekyska, J., Faundez-Zanuy, M.: A Naturalistic Database of Thermal Emotional Facial Expressions and Effects of Induced Emotions on Memory. In: Esposito, A., Esposito, A.M., Vinciarelli, A., Hoffmann, R., Müller, V.C. (eds.) COST 2102. LNCS, vol. 7403, pp. 158–173. Springer, Heidelberg (2012)
37. Scarpa, P.: Italian neuropsychological instruments to assess memory, attention and frontal functions for developmental age. Journal of the Neurological Sciences 27, 381–396 (2006)
38. Levenson, R.W.: Autonomic specificity and emotion. In: Davidson, R.J., Scherer, K.R., Goldsmith, H.H. (eds.) Handbook of Affective Sciences, pp. 212–224. Oxford University Press, New York (2003)

Mood Effects on the Decoding of Emotional Voices

Alda Troncone[1,*], Davide Palumbo[1], and Anna Esposito[1,2]

[1] Department of Psychology, Second University of Naples, Caserta, Italy
alda.troncone@unina2.it,
da.palumbo@libero.it
[2] International Institute for Advanced Scientific Studies (IIASS), Vietri sul Mare, Italy
iiass.annaesp@tin.it

Abstract. This study examines the effect of mood induction on the decoding of emotional vocal expressions. An adequate sample of 145 students (71 females, 74 males; mean age = 23.37 ± 2.05) was recruited at the Second University of Naples (Italy). Subjects were randomly assigned to one of three (sad, fear or neutral) emotion conditions induced by viewing short movies. The results showed a significant general decrease in the decoding accuracy in the mood induction conditions when compared to the accuracy of the participants who did not received such mood induction. Post hoc analyses revealed that recognition of emotional vocal voices conveying anger was especially impaired by mood induction conditions. No findings consistent with mood congruity theory were observed. This study contributes to emotion regulation research by showing differences in emotion decoding tasks by voices due to mood induction procedures, as already observed in studies exploiting the decoding of emotional faces.

Keywords: mood induction, emotional vocal expressions, emotional voices decoding.

1 Introduction

Mood is an ever-present and influential feature of our mental lives. There is much evidence in literature supporting the notion that mood affects cognitive processes, such as memory [3], executive functions (e.g. working memory) [11] and the decoding of emotional facial emotion expressions [10].

With respect to how mood affects emotion recognition, some empirical studies explain the effect of mood on the ability to decode emotions through mood-congruity theories [2], [12-13]. The mood-congruity theories state that a person's mood exerts a congruity effect on social judgments. In other words, subjects in a specific mood would be faster and more accurate in decoding expressions of emotions congruent with their mood, and slower and less accurate in decoding incongruent emotions

* To contact the authors, please write to: Alda Troncone, Department of Psychology, Second University of Naples, Viale Ellittico 31, 81100, Caserta, Italy.

S. Bassis et al. (eds.), *Recent Advances of Neural Network Models and Applications*,
Smart Innovation, Systems and Technologies 26,
DOI: 10.1007/978-3-319-04129-2_32, © Springer International Publishing Switzerland 2014

(e.g. subjects with a sad mood are more accurate in identifying expressions of sadness than subjects with a different mood). The validity of mood congruity theories is controversial and not supported by all experimental data [5], [14].

These aforementioned studies focused on the ability of subjects to recognize certain emotions through tasks decoding facial expressions. To the date, direct effects of induced emotional states on the ability to decode emotional vocal voices have not been investigated.

The goal of this study is to contribute to emotion regulation research by investigating possible similarities and/or differences in the task of emotion decoding, due to mood induction procedures, as has already been done for the decoding of emotional faces. In particular, it aims at understanding how sadness, fear, neutral moods affect, in a healthy subject sample, the ability to decode 5 out of the 6 basic emotions according to Ekman & Friesen's definition [6]. In addition it further inspects whether the current induced mood (in our case sadness or fear) boosts or hinders the recognition of mood-congruent and mood incongruent voices compared to the neutral induced mood.

The research questions are:

1. Does mood induction affect subjects' ability to decode emotional vocal expressions?
2. Are subjects with an induced mood of sadness (or fear) more accurate in recognizing congruent emotional voices?

In order to answer the aforementioned questions the decoding accuracy obtained by subjects participating to the mood induction procedure was compared with the performance of no mood induced subjects involved in the decoding of the same emotional voices in a previous study of ours [7-8].

Along with mood, the socioeconomic status (SES) of subjects was considered. In spite of the well-known effects of SES on cognitive performance [4], [15] there is a lack of data about the influence of SES on the decoding of emotional expressions.

2 Method

2.1 Subjects

The study was carried out on an adequate sample of 154 university students recruited at the Second University of Naples (Italy), 9 of which were excluded from the study because their socio-demographical characteristics did not fit those of the majority of subjects (e. g. ages and marital status).

The final sample consisted of 145 subjects (74 males, 71 females) with mean age of 23.37 years (SD= 2.05). The subjects are mainly undergraduate (81.4%), full time students (77.1 %), from the Faculties of Psychology (66.2%), Political Science (8.3%) and others (25.5%).

The control group was made of 30 adults, students at the Faculty of Science, in Salerno, aged between 23 to 29 years.

2.2 Materials

Mood Induction Stimuli

The mood induction procedure consisted in letting the participants watch 3 short movies believed to induce a fear, sadness and a neutral mood. The movies were selected among the highest rated as being effective in inducing mood from a previously validated database of films [8].

The fear induction movie lasts 25 seconds and starts by showing frames of a security camera monitoring an entrance. Suddenly a screaming witch-like face appears right in front of the camera. The film used for inducing sadness, lasts 79 seconds, features a series of photographs of babies and young children, sometimes interacting with their mothers, overlaid by written text about the denial of life connected to abortion and has a soft, placid but slightly sad soundtrack.

The neutral induction movie mood lasts 30 seconds, and shows photographs of household furniture.

Emotional Voice Decoding Task

In order to assess their ability to decode emotional feelings, the subjects were asked to participate at an audio test where they listened to 20 vocal emotional stimuli, selected by one of the authors from a database of emotional voices already assessed and published in literature [7-8]. The "emotional voices" are Italian sentences of short duration, with a semantic content not related to the expressed emotion. They were extracted from video-clips acted by famous Italian actors/actresses expressing vocal emotional expressions of joy, sadness, fear, surprise, and anger. Generally, the stimuli used in these experiments are collected asking an actor/actress to produce a given sentence with different emotional contents. In this respect, the actor/actress is portraying the requested emotional state from scratch, in a laboratory setting, without an external and internal context of reference that is normally present when making a movie. The absence of a frame of reference may make such stimuli less ecological than those extracted from movies since the latter are more close to real social stimuli both because the audio track is affected by the environmental noise of the scene from which it is extracted and because the actor/actress was not asked to portray an emotionally coloured sentence from scratch, rather his/her coloured vocal expression is produced in the context of the movie scene. A detailed description of the stimuli exploited in this work is reported in Esposito [9]. Participants were requested to listen and label the sentences attributing to them one of the five emotional labels reported above.

The Socioeconomic Status (SES)

The assessment of the participants' socioeconomic status (SES) was made using the Barratt Simplified Measure of Social Status (BSMSS) [1]. This simplified measure derives from an update of the pioneering work of Hollingshead [1] in devising a simple measure of Social Status based on marital status, current employment status (former status for retirees), level of education and occupational prestige. This index attributes parental SES, and a combination of their educational level and work activity, to the students. The obtained SES's values varies from 8 to 66 giving rise to three different near-equal groups of students belonging to a low (8-27 points), medium (28-47 points), high (48-66 points) SES category.

2.3 Procedure

Each participant first filled in and signed a consent form providing also her/his general and socio-economic information. Generally within a week after the consent participants made an appointment with the experimenter to undergo the mood induction and the subsequent emotional voice decoding task.

A suitable neutral setting was created in the laboratory for these two tasks (free of distractions and disturbing events), each subject, after being informed of the ongoing experiment, was randomly assigned to one of the three mood induction conditions (fear, sad and neutral) and was asked to watch and listen through headphones the corresponding movie on a 13-inch computer screen.

Immediately after the mood induction procedure, the subject underwent the emotional voice decoding task, where the 20 emotional vocal expressions were randomly presented through headphones. For each stimulus she/he had to select the emotional label she/he wanted to attribute to it by crossing the right box on an answer grid reporting the five emotional labels. The subjects were allowed to listen to the stimuli as many times as they needed before selecting their answer.

The above reported experimental set-up can rise two methodological concerns. First, no assessment of the subject's emotional intelligence was made before the mood induction procedure. This concern is circumvented by the high number of the involved subjects (145) ensuring that outliers line up at the boundaries of the answer distribution. In addition, the effectiveness of the induction procedure was not assessed in order to avoid interrupting and/or interfering with the induced emotional state, thereby accelerating its physiological dissolution.

The subjects of the control group underwent to the same decoding task of the subjects involved in the mood induction experiment except that their choice was made among 8 other than 5 emotional labels (in addition to joy, fear, anger, surprise, sadness, they can also choose among the labels irony, "other", and "no emotion"). However, the frequency of answers such as "other" and "no emotion" was very low we were able to consider their response for the comparison with the mood induced groups.

3 Results

The average SES value was 34.82 (SD=12.92), which can be categorized as a mid SES level. In order to check if it affected the emotion decoding task, a one-way ANOVA was conducted on each of the five emotional labels considered, using the number of correct answers to emotional voices as the dependent variable. No SES effects were found in the decoding of stimuli conveying joy ($F(69,140)=1.005$, p=n.s.), fear ($F(69,140)=.938$, p=n.s.), anger ($F(69,140)=.901$, p=n.s.), surprise ($F(69,140)=1.388$, p=n.s.) and sadness ($F(69,140) = .888$, p=n.s.)

To explore the effects of the mood induction procedure a chi-square test (α=.001) was conducted comparing the percentage of correct answers to the emotional voice decoding task obtained from subjects involved in the experiments without (control) and with mood induction (sad, fear, neutral) procedures. The results showed a significant decrease in the emotion decoding accuracy for the latter group. This decrease affected subjects in induced moods of sadness ($\chi2$ =58.82, df=4, p<.001), fear ($\chi2$ =51.87, df=4, p<.001) and in the induced neutral mood ($\chi2$ = 66.49, df=4, p<.001) when compared with the control group's performance. The confusion matrices showing the percentages of the decoding accuracy are reported in Tables 1, 2 and 3 respectively. In these Tables, the grey and white rows report the percentages of correct answers obtained by the mood induced and control group respectively in each of the three different conditions, and for each emotional labels. Figure 1 provides an overall view on how the percentages of correct answers differ in the control and in the three different mood induction conditions. A subsequent Bonferroni post hoc testing (adjusted α = 0.00025) indicated that the performance decrease in the three mood induction conditions was principally concerned with the decoding of voices expressing anger.

Table 1. The confusion matrix reporting the percentages of accuracy (on the diagonal) in the emotional voice decoding task obtained by subjects participating (grey rows) or not (white rows) to the **sadness** mood induced condition

emotion to identify	% answers					
	Joy	fear	anger	surprise	sadness	2 Crit= 21.5
joy	58					0.77
	51.7					
fear		37.2				11.35
		64.2				
anger			33.5			**28.44**
			81.7			
surprise				42		12.01
				71.2		
sadness					48.4	6.25
					69.2	

Table 2. The confusion matrix reporting the percentages of accuracy (on the diagonal) in the emotional voice decoding task obtainedby subjects participating (grey rows) or not (white rows) to the **fear** mood induced condition

emotion to identify	% answers					
	Joy	fear	anger	surprise	sadness	2 Crit= 21.5
joy	54.9					0.2
	51.7					
fear		41.6				7.95
		64.2				
anger			31.9			**30.35**
			81.7			
surprise				44.6		9.94
				71.2		
sadness					54	3.34
					69.2	

Table 3. The confusion matrix reporting the percentages of accuracy (on the diagonal) in the emotional voice decoding task obtained by subjects participating (grey row) or not (white rows) to the **neutral** mood induced condition

emotion to identify	% answers					
	Joy	fear	anger	surprise	sadness	2 Crit= 21.52
joy	**54.8**					0.18
	51.7					
fear		**36.2**				12.21
		64.2				
anger			**28.8**			**34.25**
			81.7			
surprise				**39.4**		14.24
				71.2		
sadness					**49.5**	5.61
					69.2	

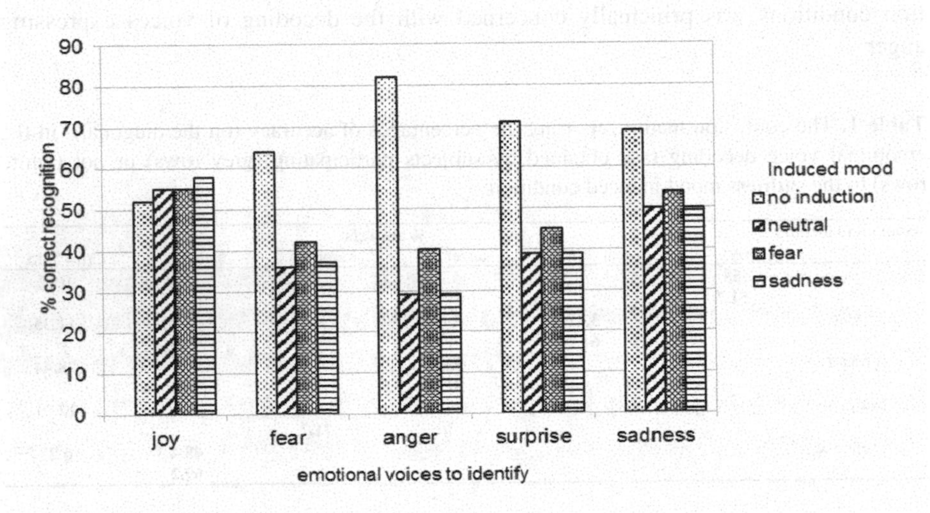

Fig. 1. The percentage of emotion's correct labeling by the control group (grey bars) and by the subjects participating to the neutral (black bars), fear (green bars) and sadness (blue bars) mood induced conditions

In addition, the obtained findings suggest that the current mood (sadness or fear) neither boosts or hinders the decoding of mood-congruent or mood incongruent voices compared to neutral mood. The fear mood induced subjects not to decode fear stimuli better than neutral mood induced ones ($\chi 2$=.805, df=1, p=n.s); the sadness mood induced subjects not to decode sad stimuli better than the neutral mood induced ($\chi 2$= 0.024, df=1, p=n.s) and the subject's ability to decode incongruent stimuli (joy) does not worsen ($\chi 2$=0.186, df=1, p=n.s).

4 Discussion

It is worth to note that there was a difference between the control and mood induced group experimental design. The subjects in the first group were forced to choose among 8 emotional labels (joy, fear, anger, surprise, sadness, irony, "other", and "no emotion") whereas those in the second one were forced to choose among only 5 labels (joy, fear, anger, surprise, sadness). This difference may have also affected the subjects' performance. Nevertheless, they can be considered as a trend in the distribution of the correct answers and interestingly suggest that any typology of mood induction procedure (even the neutral states) affects the subjects' judgment and their correct decoding performance. Speculating on this difference, the decrease in performance can be attributed to a cognitive load caused by the movie processing task regardless of the changes in the emotional feelings. We suppose that subjects in the no mood induced group are more accurate in decoding emotional voices than subjects in the mood induced groups because their cognitive processes are completely engaged by only one task.

Even if we are aware that the comparison of the present study with the available literature is limited by the specificity of the methodology applied (recognition of emotional voices vs emotional faces), our findings are in line with Chepenik et al. [5] who found, in a sample of healthy adults, a general worsening of ability in recognize emotions in subjects with induced mood of sadness.

In addition, the absence of mood congruity influences does not support mood congruity theories but may also have been affected by the experimental design giving the subjects the chance to listen to the emotional voices as many times as they needed before selecting their answer. This procedure may have favoured the dissolution of the induced mood and eliminated congruence effects.

The reported results are considered as pilot requiring support from more experimental data and the specific impairment in the recognition of voices expressing anger need to be supported by further investigations. An analysis of the errors can also help clarify the nature of this impairment.

References

1. Barratt, W.: The Barratt Simplified Measure of Social Status (BSMSS) measuring SES. Indiana State University (2006), `http://wbarratt.indstate.edu/socialclass/Barratt_Simplifed_Measure_of_Social_Status.pdf` (unpublished manuscript) (retrieved)
2. Bouhuys, A.L., Bloem, M.G., Groothuis, T.G.G.: Induction of Depressed and Elated Mood by Music Influences the Perception of Facial Emotional Expressions in Healthy Subjects. Journal of Affective Disorders 33, 215–226 (1995)
3. Bower, G.H., Forgas, J.: Handbook of Affect and Social Cognition. Erlbaum, Mahwah (2001)
4. Caro, D.H.: Socio-economic Status and Academic Achievement Trajectories from Childhood to Adolescence. Canadian Journal of Education 32(3), 558–590 (2009)

5. Chepenik, L.G., Cornew, L.A., Farah, M.J.: The Influence of Sad Mood on Cognition Emotion 7(4), 802–811 (2007)
6. Ekman, P., Friesen, W.V., Hager, J.C.: The Facial Action Coding System, 2nd edn. Salt Lake City, Weidenfeld & Nicolson (2002)
7. Esposito, A., Riviello, M.T., Di Maio, G.: The COST 2102 Italian Audio and Video Emotional Database. In: Apolloni, B., Bassis, S., Morabito, C.F. (eds.) Frontiers in Artificial Intelligence and Applications, vol. 204, pp. 51–61 (2009), `http://www.booksonline.iospress.nl/Content/View.aspx?piid=14188`
8. Esposito, A., Riviello, M.T.: The New Italian Audio and Video Emotional Database. In: Esposito, A., Campbell, N., Vogel, C., Hussain, A., Nijholt, A. (eds.) Second COST 2102. LNCS, vol. 5967, pp. 406–422. Springer, Heidelberg (2010)
9. Esposito, A.: The Perceptual and Cognitive Role of Visual and Auditory Channels in Conveying Emotional Information. Cogn. Comput. J. 1(2), 268–278 (2009)
10. Langenecker, S.A., Bieliauskas, L.A., Rapport, L.J., Zubieta, J.K., Wilde, E.A., Berent, S.: Face emotion perception and executive functioning deficits in depression. Journal of Clinical and Experimental Neuropsychology 27, 320–333 (2005)
11. Marvel, C.L., Paradiso, S.: Cognitive and Neurological Impairment in Mood Disorders. Psychiatric Clinics of North America 27, 19–36 (2004)
12. Niedenthal, P.M., Halberstadt, J.B., Margolin, J., Innes-Ker, A.H.: Emotional State and The Detection of Change in Facial Expression of Emotion. European Journal of Social Psychology 30, 211–222 (2000)
13. Niedenthal, P.M., Brauer, M., Halberstadt, J.B., Innes-Ker, A.H.: When Did Her Smile Drop? Facial Mimicry and the Influences of Emotional State on the Detection of Change in Emotional Expression. Cognition and Emotion 15(6), 853–864 (2001)
14. Schmid, P.C., Schmid Mast, M.: Mood Effects on Emotion Recognition. Motive Mot. 34, 288–292 (2010)
15. Sirin, S.R.: Socioeconomic Status and Academic Achievement: A Meta-Analytic Review of Research. Review of Educational Research 75(3), 417–453 (2005)

The Ascending Reticular Activating System
The Common Root of Consciousness and Attention

Mauro Maldonato

DICEM: Department of European and Mediterranean Cultures
University of Basilicata – Matera, Italy
mauro.maldonato@unibas.it

Abstract. In the organization of the central nervous system the role of Ascending Reticular Activating System (ARAS) – comprising the reticular formation, thalamus and thalamo-cortical system of bi-directional projection which governs the activities of wakefulness and vigilance – does not correspond to a hierarchical superiority with respect to the cerebral hemispheres. The ARAS is not limited to the brain stem: it projects upwards towards the cerebral hemispheres and downwards towards the spinal cord. Its functions are much more complex than simple cortical desynchronization, even though this is essential in the state of alertness and attention. Its thalamo-cortical projections, which are a-specific with a high oscillatory frequency, are fundamental for some essential functions of consciousness.

Keywords: consciousness, attention, ascending reticular activating system, alertness, high oscillatory frequency.

1 Categorical or Dimensional? The Prismatic Fadings of Consciousness

For too long now debate on the problem of the consciousness has tended to focus on rarefied (and often irrelevant) theoretical controversies. Over the last fifty years this scenario has changed and a genuine revolution has taken place. The rapid development of methods of neuro-imaging (non invasive techniques for investigating the cerebral functions in their functional version which show the brain's activity while we are performing an action, thinking, becoming emotional, and so on) and, above all, the ever closer matching of experimental and clinical data, have ushered in a new era of research, making progress that until recently was simply unimaginable. These sophisticated technological pieces of equipment are overthrowing a large part of the traditional applications in image-based diagnostics, with regard to magnetic resonance in its functional application (fMRI), as well as to medicalnuclear (PET and SPECT) and electro-physiological (EEG) methods. The lengthy era of Victorian neurophysiology is closed. The distinction between the higher (cortical) levels and the lower (subcortical) levels has had its day. Currently, investigation of consciousness refers to three basic levels: the neuronal correlates, the underlying causes of this

S. Bassis et al. (eds.), *Recent Advances of Neural Network Models and Applications*,
Smart Innovation, Systems and Technologies 26,
DOI: 10.1007/978-3-319-04129-2_33,

correlation, and the reproduceability of the hypotheses produced. Clearly the existence of specific causal levels can only be verified in a healthy individual: in a patient suffering from altered states of consciousness the required monitoring, difficult enough in itself, would be out of the question. Nonetheless, even in the presence of all the conditions favouring the verification of precise causal levels and the necessary correlations between structure and functions, it would soon become clear that the nerve areas involved in conscious processes are also involved in other mental functions, both conscious and unconscious, like perception, attention, memory and so on.

So what are the optimal experimental conditions for a neurobiological study of consciousness? And how are we to adequately describe its contents if they can only be verbalised in the third person? There are behaviours, gestures and movements associated with neural activities that can be represented using electrophysiology or cerebral imaging. Several years ago Frith [12] distinguished three levels of neuronal activity, associated with conscious representation, sensory stimuli and behaviour.

In the study of consciousness more than in other fields, theory has to go hand in hand with experimental research. Even if the formulation of accurate experimental models is always the first step, we can only gain a true understanding of the fundamental processes if we are familiar with the elementary neural levels. In recent decades studies of cerebral activation (using fMR, PET, MEG, event-related potentials) have made an extraordinary contribution. These techniques make it possible to explore cerebral reactions prior to and following a stimulus such as the presentation of ambiguous visual prompts, transition from a general anaesthesia to consciousness, passage from the vegetative to a minimally conscious state, and so on. For example, the renewal of the thalamo-cortical activity in a patient who was first 'vegetative' and then 'minimally conscious' confirms on one hand the importance of such connections in the processes of consciousness, and on the other that this cannot be circumscribed to one specific region of the brain.

Studies on the passage from oblivion to consciousness carried out in the sphere of anaesthetics with the help of cerebral imaging are making important contributions to our knowledge of conscious processes. In particular some very recent research work has shown how coming round after anaesthesia – with full recovery of a clear, oriented consciousness – is often preceded by a phase of considerable turbolence. Apart from all the questions that remain open, the demonstration of the co-implication of cortical and subcortical areas in the emergence of consciousness is an important step forwards. Experiments conducted on twenty young volunteers coming round from a general anaesthetic by a research team led by Harry Scheinin [13] have shown how the recovery of consciousness is correlated to the activation of profound cerebral structures rather than to the cerebral cortex. In other words, the state of consciousness is preceded by an intensification of the metabolic activities of the midbrain archipelago comprising the thalamo, limbic system and the lower frontoparietal cortical areas. The emergence of progressive levels of consciousness, monitored according to the response to vocal commands, right up to the highest level of consciousness of the self and the world, begins from these neural territories.

It is indeed extremely interesting that the activity of profound cerebral structures can also be elicited by anaesthesia. This suggests that there is a common autonomous mechanism of arousal. More generally the study of the effects of anaesthetic drugs and their ability to modulate consciousness opens up new scenarios for research. The suspension of the relational sphere and the passing into a state of oblivion on one hand, and the recovery of consciousness as the pharmaco-dynamic effect wears off on the other, constitute phenomena of extraordinary importance for a science of consciousness.

2 The Tree of Life Brain

Half way through last century Moruzzi and Magoun [21] informed the scientific community of the existence of a system – comprising the reticular formation, thalamus and thalamo-cortical system of bi-directional projection – which governs the activities of wakefulness and vigilance: the Ascending Reticular Activating System (ARAS). In particular the two neuroscientists observed that stimulation of these areas causes a sleeping animal to awaken, and a state of alarm in an animal that is already awake. There are essentially two pathways for these influences: one extra-thalamic, responsible for the reactions of reawakening; and the other thalamic, which modulates the level of vigilance, orienting it in one direction rather than another. This second route appears to be more closely correlated than the first with attention which, as we can recall, is a selective phenomenon in relation to the higher levels of integrated nervous systems. It is vigilance that ensures the efficiency of attention and, more generally, the functioning of the entire psychic apparatus. In fact, in a non-vigilant subject not only attention but also other faculties such as memory, decision-making and motivation remain in a virtual state. Thus if reflex represents the lowest level of motor integration, vigilance represents the fundamental level of the higher integration which has the ARAS as its basic structure.

It is worth reiterating that the centrality of the ARAS in the organization of the central nervous system does not correspond to a hierarchical superiority with respect to the cerebral hemispheres: given their high specialisation and greater selectivity, they do their work at a higher functional level. Moreover, for specific anatomical and physiological reasons, systems with a specific projection (olfactory, gustatory, visual, auditory, somatosensory) have functions which are quite different to those of a system with general projection. The former are made up of a peripheric receptor, afferent pathways and relay nuclei for these pathways which access a small portion of the cortex; and the latter of cellular aggregates distributed throughout the brain stem whose stimulation has a widespread influence on the electrical cerebral activity. The ARAS is not limited to the brain stem: it projects upwards towards the cerebral hemispheres and downwards towards the spinal cord. As Mauro Mancia [18] pointed out, its functions are much more complex than simple cortical desynchronization, even though this is essential in the state of alertness and attention. Its thalamo-cortical projections, which are a-specific with a high oscillatory frequency, are fundamental for some essential functions of consciousness.

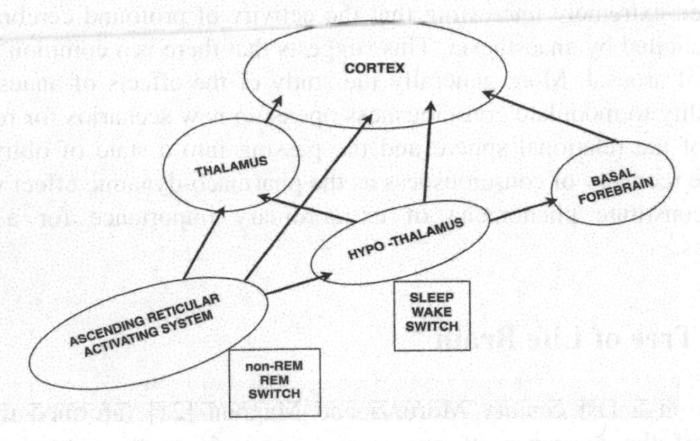

Fig. 1. Ascending activation of the cortex. The ascending reticular activation system (ARAS) reaches the cortex through a ventral pathway (hypothalamus, basal forebrain), through the aminergic nuclei (containing catecholamines, acetylcholine, and serotonin) and a dorsal pathway, the thalamic relay. Switching between paradoxical sleep (REM-sleep) and slow wave sleep occurs in the reticular formation, whereas the switch between sleep and waking lies in the hypothalamus.

Studies of cerebral activation have shown that in patients in a vegetative state (i.e. wakefulness without content) the connectivity between areas that are normally connected is lost: in particular between the primary cortical areas and the associative multimodal areas (prefrontal, premotory and parieto-temporal, cortex of the posterior and precuneo cingolate gyrus); or between these areas and the thalami. This evidence taken together suggests that the vegetative state is a sort of disconnection syndrome in which an isolated neuronal activity is associated with a reduction of the cerebral metabolism [20]. Conversely, in patients in a minimally conscious state the connectivity between primary areas, associative cortexes and thalami is maintained, so that the anatomo-functional apparatus required for conscious activity is intact. These studies of cerebral activation are opening up new scenarios. Recent research has shown how, in a sub-group of clinically vegetative patients, cortical activities of information analysis persist in such a way as to make the existence of levels of minimal consciousness plausible [16]. On the contrary, the verbalisation of words – which automatically excludes a diagnosis of vegetative state – could represent a complex or automatism generated by little 'archipelagoes' of neuronal activity without consciousness [17]. This however brings us back to a crucial question: how small do these little archipelagoes have to be for them to be non-thinking?

3 A Neverending Controversy

Over the last thirty years the philosophical debate on consciousness has involved the distinction between an easy problem, i.e. how the brain and psychic organization generate consciousness, and a hard problem, i.e. the relation between the neurobiological processes and the quality of subjective experience. A significant contribution has come even from those who, without making any concessions to dualism or forms of scepticism, have assigned a crucial role to subjective experience, denying the possibility of its naturalisation. The proponents of this thesis believe that the limits of our mind prevent us from penetrating the secrets of consciousness and grasping the qualitative elements of the phenomena. Even if we know that it is the brain which handles sentiments and the subjective quality of our thoughts, we shall never be able to clarify the enigma of subjectivity [19].

Then again there are scientists who, making the link between cognition and intentional states, have placed experience and cognition on the same level, favouring a functional approach [14; 1; 7]. In particular, they identify mental events and states (beliefs, wishes, desires and so on) as functions rather than specific neural processes. It is simply functional relations that hold together mental states, sensory inputs and behavioural outputs. For the adepts of computational functionalism (which has seen the development of the strong Artificial Intelligence programme) the mind resembles a computer programme. Functionalists argue that thought manipulates symbols inserted in a functional network which has much in common with computer processing. In such a perspective it seems quite plausible that experience derives from the combination of single (cognitive) modular units which are then inscribed in a variable theoretical framework, linked in turn to the emerging unit. Even though this hypothesis is ingenious and has had considerable success, it is profoundly ambiguous, for while on one hand it remains solidly grounded in the materialist model, on the other it recognises the reality of experience and mental activity because in order to convalidate a theory there must always be a third person.

For a long time authoriative cognitivists such as Fodor [9] insisted that mental phenomena are to be considered only in quantitative terms and not in qualitative or subjective terms. In other words the mind functions by means of vertical structures (modules) which mediate between the output of the sensory-perceptive organs and the central systems responsible for more complex elaborations. Rather than exchanging information with the central structures or other modules, these genetically determined modules follow predetermined and unmodifiable calculation strategies. It has to be added that, although he never repudiated this theoretical framework, in a later phase of his research Fodor [10] recognised that his theory was not able to clarify the more distinctive features and properties of the human mind adequately. Conversely, for Baars [1] the brain can be represented as a multitude of scattered microprocessors competing for access to a global workspace in order to filter, exchange and file information. Here distinct contents emerge from competitive and cooperative dynamics between neural groups that are transmitted and then brought into consciousness. The unitary character of such an experience is guaranteed by diffusion mechanisms rather than by the content transmitted. According to Baars, this model helps to distinguish the level of the specialised processors (unconscious) from that of the workspace (conscious), and above all to clarify the intentional filter which raises the integration and voluntary control of the attentive and ideo-motor spheres to the highest level. In other words the global workspace constitutes a sort of stable context-purpose, starting from which consciousness brings order to the multiple sources of knowledge and the innumerable interactions. While undoubtedly fascinating, Baars's model leaves several major questions unanswered. For example, what are the hierarchies of neuronal selection, and the dynamics of neuronal activity?

If the debate among cognitivists ran into major theoretical issues, neurobiologists were also facing serious problems. First of all they had to decide what were the ideal experimental conditions for highlighting the neuronal correlates of conscious experience. As we have seen, even the most radical materialists were obliged to admit that the only way of accessing qualitative experiences was through verbal report. The gap between the subjective account and the conscious experience was still very large. Language is the only, tenuous bridge between our thoughts and those of other people.

It was undoubtedly the work of Francis Crick, the biologist who with James D. Watson discovered the double helix structure of DNA, which revealed to the general public the full scientific scope of the problem of consciousness. In the mid-70s Crick decided to abandon his work in molecular biology and devote himself entirely to what he saw as the greatest enigma of them all: consciousness. He was convinced that the solution to this problem would come from experimental study of the electrochemical exchanges of neurons and the identification of the neural correlates of consciousness, i.e. the cerebral processes that are synchronised with the states and contents of consciousness. Starting from findings made by other neurobiologists concerning the electric oscillations in an interval of 35-45 Hz in a cat's neurons, Crick put forward the hypothesis that consciousness originates from the unification of the oscillatory frequency of a number of neural groups. The synchronisation of the discharges was supposed to consign the psychic content to the working memory [5], while consciousness is manifested in an intermediate zone of representations, distributed between a lower sensory level and a higher level of thought processes.

In order to bridge this gap Crick suggested that beneath the conscious level a homunculus perceives the world through the senses, formulating, planning and performing the voluntary actions. As a matter of fact, at a later stage of his research Crick himself went back on this hypothesis. On the basis of experimental evidence relating to visual patterns in primates, he suggested that the phenomenon of (visual) consciousness depends on the neurons in layers V and VI of the cerebral cortex, with the mediation of oscillating thalamo-cortical properties. This type of consciousness is sustained by electrochemical activity in the visual areas, which projects nerve bundles directly onto the prefrontal areas.

Edelman [8] argued that one has to distinguish between primary consciousness (a multimodal structure which combines various sources of information) and a superior consciousness which develops in parallel with language acquisition. The superior consciousness is the expression of a conscious self which organizes past and future and recalls and narrates its experiences, emancipating the organism from subservience to the here and now. While primary consciousness connects memory to current perception, superior consciousness operates a synthesis between the memory of special values and categories distributed in the temporal, frontal and parietal areas. Consciousness is seen as originating precisely in the dimensional and categorial interaction between the non-self (which interacts with the world through current experience and behaviour) and the self (which by virtue of social interactions acquires

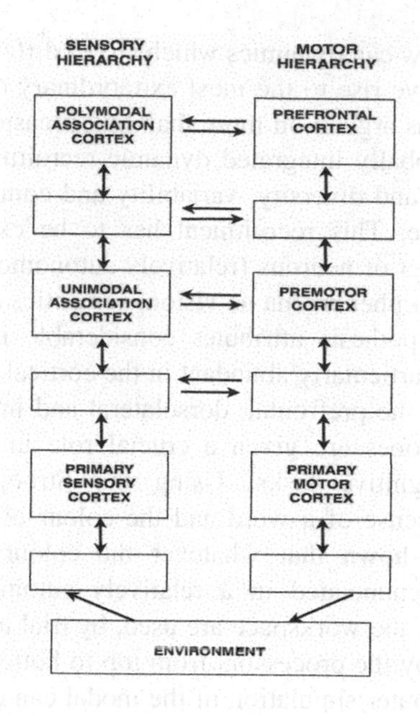

Fig. 2. Fuster's figure (reproduced with permission by Lippincott-Raven Publishers) showing the fiber connections between cortical regions participating in the perception-action cycle. Empty rhomboids stand for intermediate areas or subareas of the labeled regions. Notice that there are connections between the two hierarchies at several levels, not just at the top level.

a semantic and a 'syntactic memory' for concepts). Learning takes place by means of the development of memories of the former system, for which the perceptive categories take on an extraordinary value.

Clearly Edelman's model implies the presence of an external observer who, by means of a symbolic alphabet, codifies and decodifies messages. The complexity of this system varies with variations in the functional organization: it is relatively low when the connections show a statistic distribution, and at a maximum when linked to specific neuronal groups. The higher the reciprocal information between each sub-set and the rest of the system, the greater the complexity. The core of conscious experience is constituted by a scattered neuronal aggregate which brings about integrations with a duration of about a hundred milliseconds. According to Edelman the integrative activities required by higher consciousness occur on the boundary between the thalamo-cortical system and the other cerebral areas. This autonomous dynamic nucleus – which cannot be identified either in the brain as a whole nor in specific cerebral areas, nor yet in any one sub-set of neurons – is both unified and differentiated: in other words it gives rise to correlations at a distance between different regions of the brain that can vary from one moment to the next, within one individual and between one individual and another.

Although unquestionably intriguing, Edelman's model does not take into account some fundamental problems. For example, if consciousness mobilises multiple cerebral territories (the reticular formation, thalamic system and so on) characterised by spontaneous relations, how can dynamics which are so different and far apart, "de-territorialised" as it were, give rise to the most extraordinary of natural phenomena? Jean-Pierre Changeux [4] has argued on more than one occasion that at the origin of consciousness there is a globally integrated dynamic recruitment of representations characterised by both unity and diversity, variability and competition in a restricted neuronal architectonic space. This recruitment has to be extended to feature the coordinated operation of a set of neurons (relatively autonomous with respect to one another) that engenders such phenomena as vision, semantics and motor skills. Such an anatomico-functional hypothesis attributes considerable importance to neurons with long axons which are particularly abundant in the cortical layers I, II and III, and present in large numbers in the prefrontal, dorsolateral and inferoparietal cortex [4]. In this model the frontal lobes are given a crucial role in conscious experience, simulating multimodal cognitive tasks. Using the Stroop test (based on an incongruence between the sense of a word and the colour of the ink in which it is written) researchers have shown that whatever the colour in which a word is presented, its meaning is enunciated in a relatively automatic fashion [6]. This indicates that the neurons of the workspace are used, by trial and error, to control the elaboration of information by the processors from top to bottom, although they work from bottom to top. If computer simulation of the model can clarify the dynamics of selection of a global representation, it also enables prediction of the dynamics of cerebral visualisation during the task performance. We shall come back to the assets and limits of the models we have referred to here.

4 Attention: The Door of Consciousness

For all the formidable progress made in numerous fields by cognitive neurosciences, we are still in the dark about very many aspects of attention. One thing that is now beyond doubt is the multiplicity of processes that underlie it, for attention is involved in numerous other fundamental cognitive processes – perception, motor action, memory – and any attempt to isolate it in order to study its constant features is bound to prove sterile. For over a century and a half attention was a crucial topic in neurophysiology and psychology. In the early days of scientific psychology it was viewed as an autonomous function that could be isolated from the rest of psychic activity. However, this idea soon came to be seen as inadequate. At the beginning of the 20th century researchers became convinced that attention underpinned a general energetic condition involving the whole of the personality. Within a few years the emergence of the Gestalt and Behaviourism paradigms caused these studies to be overshadowed, and it was not until the second half of last century that they regained their importance.

For a long time the debate was influenced by the hypothesis that attention constitutes a level of consciousness varying widely in extension and clarity and only functioning in relation to its variations: from sleep to wakefulness, from somnolent to crepuscular, from confusion to hyper-lucidity, from oneiric to oneiroid states, and so on. Subsequently other approaches of considerable theoretical importance linked attention to emotion, affectivity and psychic energy or social determinants. Yet what do we really know about attention, the sphere of our life which orients mental activity towards objects, actions and objectives, maintaining itself at a certain level of tension for variable periods of time? How and to what extent is attention related to consciousness? Why does only a minimal part of the information from the external world reach the brain even though the physical inputs strike our senses with the same intensity? And why is it that, although they enter our field of consciousness, most of these inputs do not surface in our awareness? It is well known that in the selection of stimuli, attention is strongly influenced by individual expectations. They 'decide' which objects and events appear in our awareness, and which are destined never to appear. The law of interest regulates a large part of the selection of the objects and topics on which our attention is focused.

Now, if it is true that we pay attention to what touches our sentiments and emotions, it is also true that this attention – whether it is spontaneous, intentional or direct – is highly selective. The distinction made by James concerning the various aspects of attention, admirably taken up and systematised by Bruno Callieri [3] in the spectrum of expectation, observation and reflection, is still extremely useful. Expectant attention prepares for action and is always conditioned by the expected events. It contemplates both modes of conduct or reflex activities in the short or long term in animal existence and intentional behaviours sustained by generic instinctive drives. There are animals, for example, which appear to manifest enormous patience in contexts that are highly specialised. One can think of the biological/instinctual expectations of birds and rodents which wisely conserve their food, of the leopard who delays action until the gazelle comes up close, of the spider who lurks in wait for its prey. Then one can think of all the situations of human consciousness in which attention appears to be an instinctual spur susceptible of modulation within certain limits: the hunter crouching in the hide, the fisherman waiting for the fish to bite, the marksman ready to fire at his target, the athlete in the starting blocks. This attentive modality has a duration which cannot be specified even though it cannot be prolonged ad infinitum: from split seconds to quite lengthy intervals. Whatever the scope, a state of intense vigilance, alertness and tension is in action. Observant attention indicates a distance between the subject and the surrounding scenario, which the subject registers in detail but with some oscillations. Here, where interest and motivations play a decisive role, attention deploys its agents to the full, whether modal (perplexity, hesitation, tiredness, somnolence) or maintenance (as when driving a car on an empty road). Lastly reflecting attention is exercised on an interior object on which mental activity is fully focused. In the reflection, attention is expressed through knowledge of the interior universe. Such turning in on oneself implies a keen attention, as in pure abstraction, meditation and contemplation.

Are there physiological indicators of attention? There undoubtedly are. We can name the orientation response to a new stimulus, signs like the dilation of the pupils, peripheric vasoconstriction, cerebral vasodilatation, arrest of the alpha rhythm in the EEG, and the substitution of the irregular beta rhythm. This state of psycho-physiological activation, known as arousal, varies across a continuum (from sleep to psychomotor excitation) and is decisive for the efficiency of a subject's performances.

We can point out that at low levels of activation the subject can be distracted, while very high levels cause a reduction of efficiency on account of the diminishing of attentive levels.

Initially the theory of levels of activation was studied by psychologists, and subsequently by neurophysiologists who investigated its relations with the ARAS and influence on cortical activity. In actual fact the concepts of attention and level of activation are correlated but distinct. Activation is a state of the organism which occupies a continuum, while attention is a selective function correlated with the levels of activation. The degree of attention depends on the level of activation of the organism, which in turn is modulated by the peripheric inputs and by its internal conditions. Intense inputs solicit the attention to select the significant biological or psychological information. Psychic energy can be deviated to change the focus of attention, as when an unexpected object enters the senso-perceptive field.

The object of attention can only be investigated in an efficient nervous system. The central hub is always the cortex, even when the sensory inputs appear to take pathways far removed from the field of attention. The attention is sharply stimulated not only by the meaning and intrinsic value of the object but also by the variations and by the freshness of its perception. In the absence of alterations attention fades away, to the benefit of the imagination. It is a universal truism that novelty captures the attention while the reiteration of a stimulus is habit forming and can end by inhibiting attention. On the contrary attention can be activated by prompts or thoughts which have an emotional or affective charge, confirming the role of the cortex in this process.

The most recent neurobiological research has revealed the extreme complexity of the phenomenon of attention. Experiments have been done concerning various aspects of the influence of individual factors on the speed of reaction and how reaction times differ when confronted by two synchronised stimuli coming from different sensory channels: duration and sustainability, intensity and degree, concentration and focalisation, or activation and "arousal" (Boring, 1970). It is common knowledge that attention makes mental states clearer, sharper, more aware. In this respect the relationship between attention and degrees of consciousness appears very close indeed. Jaspers [15] identified the coexistence of similar elements: the magnetism of an object or topic; the clarity of the contents, with reflections which can be foreseen but can also be quite unpredictable and mysterious; the influence of cognitive, affective and motivational elements on these processes. Now, if the latter elements can, within certain limits, be objectivised, the impact of an object can only be registered by an introspective appraisal, while the clarity of the contents requires a critical awareness.

In general, while there can be no doubt that attention makes the psychic processes more efficient, it is not unusual for their acceleration (or intensification) to weaken attention to the activities that are not in focus. More accurate analysis is called for of the relationships both between spontaneous attention and immediate interest and between voluntary attention and interests that are distributed over time. Freud (1886-1895) maintained that attention is the expression of a series of dynamic mechanisms which act between perception and desire . According to this approach it is the difference in tension between these two attributes which gives rise to thought and attention.

This schematic overview would be even more partial if one were to ignore the natural, cultural and social foundations of attention. For it is thanks to attention that memory can be considered a system in continuous evolution rather than an archive of static images. Remembering is also having a sense of movement. Each new experience represents the possibility of preserving the traces of our past life, safeguarding our personal historical and cultural "novel" from oblivion. This is all the more true if we bear in mind that attention always means attention to something, pure "intentionality". Attention, that is to say, "intentionalizes" an object, combining with it in an immediate perception. It is precisely the nature of attention which shows how the life of the mind is not an undifferentiated entity but a set of relations in continuous perpetuation. It is attention that impels me towards objects, introducing me to the world of things, broadening my perception and accompanying me to the verge of consciousness.

References

1. Baars, B.J.: In the Theater of Consciousness: the Workspace of the Mind. Oxford University Press, New York (1997)
2. Boring, E.G.: Attention: Research and beliefs concerning the concept in scientific psychology before 1930. In: Mostofsky, D.I. (ed.) Attention: Contemporary Theory and Analysis. Appleton-Century-Crofts, New York (1970)
3. Callieri, B.: L'accesso fenomenologico alla coscienza in psichiatria, tra l'empirico e il trascendentale. Rivista di Biologia 73, 170–190 (1980)
4. Changeux, J-P.: L'uomo neuronale. Feltrinelli, Milano (1998)
5. Crick, F.: The Astonishing Hypothesis. Scribner, New York (1994)
6. Dehaene, S., Kerszberg, M., Changeux, J.-P.: A neuronal model of a global workspace in effortful cognitive tasks. Neurobiology 95(24), 14529–14534 (1998)
7. Dennett, D.C.: Consciousness Explained. Little Brown & Co., Boston (1991)
8. Edelman, G.: The Remembered Present: A Biological Theory of Consciousness. Basic Books, New York (1989)
9. Fodor, J.: La mente modulare. Il Mulino, Bologna (1983)
10. Fodor, J.: La mente non funziona così. La portata e i limiti della psicologia computazionale, Laterza, Roma-Bari (2001)
11. Freud, S.: (1886-1895) Opere. 1. Studi sull'isteria e altri scritti. Boringhieri, Torino (1977)
12. Frith, C.: Attention to action and awareness of other minds. Consciousness and Cognition 11, 481–487 (2002)
13. Jaakko, W., Långsjö, M., Alkire, T., et al.: Returning from Oblivion: Imaging the Neural Core of Consciousness. Journal of Neuroscience 32(14), 4935–4943 (2012)
14. Jackendoff, R.: Consciousness and the Computational Mind. MIT Press, Cambridge (1987)
15. Jaspers, K.: Psicopatologia generale. Il Pensiero Scientifico. Roma, 1964 (1913)
16. Kotchoubey, B.: Event-related potentials predict the outcome of the vegetative state. Editorial, Clinical Neurophysiology 118(3), 477–479 (2007)
17. Laureys, S., Owen, A.M., Schiff, N.D.: Brain function in coma, vegetative state, and related disorders. The Lancet Neurology 3, 537–546 (2004)

18. Mancia, M.: Neurofisiologia. Raffaello Cortina, Milano (1994)
19. McGinn, C.: The Problem of Consciousness. Blackwell, Oxford (1991)
20. Midorikawa, A., Kawamura, M., Takaya, R.: A disconnection syndrome due to agenesis of the corpus callosum: disturbance of unilateral synchronization. Cortex 42, 356–365 (2006)
21. Moruzzi, G., Magoun, H.: Brain stem reticular formation and activation of the EEG. EEG Clin. Neurophysiol. 1, 455 (1949)

Conversational Entrainment in the Use of Discourse Markers

Štefan Beňuš
Constantine the Philosopher University, Nitra, Slovakia & Institute of Informatics,
Slovak Academy of Sciences, Bratislava, Slovakia
sbenus@ukf.sk

Abstract. Entrainment is the tendency for participants in conversations to develop behaviour similar to one another in multiple dimensions. The degree of such entrainment is linked to the emotional state and empathy of the speakers and people who entrain to their conversational partners are seen as more socially attractive, likeable, competent, more intimate, and the interactions with such partners as more successful. It is thus important that ICT interfaces for supporting wellbeing and empathy employ also some module of entrainment.

In this paper we analyze entrainment in the acoustic, prosodic and pragmatic domains connected to the use of Slovak discourse marker 'no' in the spoken modality of task-oriented collaborative dialogues. We analyze how speaking behaviour changes due to interacting with a different partner, and consequently, how entrainment is employed. We use acoustic and prosodic information extracted from the signal and labelled pragmatic functions of the marker (including acknowledgment, backchannel, reservation, topic shift, etc.). Results suggest a varied picture with both entrainment and disentrainment present in the data. Regarding the relationship between entrainment in acoustic-prosodic features and more cognitively complex features of pragmatic meaning and discourse functions, we found both matches and mismatches between the two.

Keywords: Entrainment, prosodic features, human-computer interaction, wellbeing, discourse markers.

1 Introduction

Speech entrainment is the tendency of interlocutors to become similar to each other in terms of their acoustic and prosodic production and relates to cognitive and social aspects of communication and information transfer. Some aspects of speech entrainment appear to be almost automatic, take place early in the interaction, and presumably thus employ lower levels of the cognitive communication systems (e.g. [7, 14, 15]). Other aspects might require higher cognitive functions since they include linguistic encoding (e.g. [17] for a review). Moreover, entrainment observable in spoken modality may be linked in non-trivial ways to entrainment in gestures, body postures, and other aspects of visual modality.

S. Bassis et al. (eds.), *Recent Advances of Neural Network Models and Applications*,
Smart Innovation, Systems and Technologies 26,
DOI: 10.1007/978-3-319-04129-2_34, © Springer International Publishing Switzerland 2014

Social aspects of spoken entrainment include findings that humans perceive conversational partners who entrain to their speaking style as more socially attractive and likeable, more competent and intimate, and conversations with such partners as more successful (see [10] for a review of extensive literature). It has also been shown that humans may consciously decrease their similarity to others in order to increase their social distance to the interlocutor or to show a negative attitude toward the interlocutor. It is thus important that ICT interfaces for supporting wellbeing and empathy employ also some module of entrainment.

Importantly for social robotics, not only do humans entrain to other humans, but studies have shown that they also entrain to computer systems, and that subjects do adapt to machines similarly to human conversational partners (e.g. [1, 6, 11, 19]). A better understanding of entrainment is thus important for all applications in human-machine communication that rely on Spoken Dialogue Systems. Due to the naturalness of the spoken modality for humans, human-robot interactions are likely to rely heavily on speech and the social aspects of these interactions will play a major role in the advances in the field of social robotics. Hence, the ability to mimic the tendency for entrainment in human-human conversation is important for human-robot conversation as well, if social robotics systems are to be as natural and effective as human partners.

In order to contribute to these future advances in implementation and engineering of human-computer interactions, we analyze here entrainment patterns in the usage of a single discourse marker in human-human task-oriented dialogues. Primarily, we are interested in comparing patterns in the single marker to general entrainment. Knowledge gained form such a comparison is potentially usable in at least two ways. First, observed similarities might allow faster processing, since entrainment can be assessed on a small sample of data. Second, observed differences might point to differences in cognitive processes underlying these two types of entrainment. Furthermore, we compare entrainment in acoustic-prosodic features with a limited investigation of entrainment in cognitively more demanding functional characteristics of this discourse marker. Section 2 describes all methodological aspects of data collection, labelling, extraction, and analysis. Section 3 presents our results that are discussed and summarized in Section 4.

2 Methodology

2.1 Corpus

Data for this study were selected from a corpus of dialogues in which two interlocutors were playing collaborative games, i.e. tasks in which spoken interaction is required to achieve successful completion. These games were adapted from the OBJECT Games described in [8, 9] and the current corpus is also described in [2]. Briefly, interlocutors could not see each other and were seated in a quiet room facing a computer screen and using a mouse. One player - the describer – verbally depicted the position of a target image in relation to other images on her screen. The second player – the placer – was supposed to place the same image to a position as close as possible to the position of the image on the describer's screen. Players were encouraged, and motivated with a small reward, to match the positions perfectly and their success was measured a 100-point scale based on how closely the pixel-positions of the two objects matched. Each dialogue consisted of 14 tasks and the roles of the describer and the placer were switched repeatedly and at the end were equally divided between the two players.

The corpus used for this analysis consists of six dialogues in Slovak involving seven native speakers (3 females 4 males). Importantly, five players (LP, KM, DF, MD, VR) participated in the recording twice (with a different partner) and two male subjects played only one game. The corpus contains almost four hours of speech (3h, 54m), there are 21773 words in total, and 2371 unique words.

2.2 Discourse Marker 'No' in Slovak

It has been widely observed that discourse structuring of interactions is signalled and facilitated by the distribution and prosodic characteristics of discourse markers. They not only display the discourse structure but play a prominent role in creating it; see e.g. [18] for a review. We concentrate on discourse marker '*no*' in Slovak [2]. This form might represent a shortening of affirmative particle *áno*, which means 'yes' in Slovak. It can thus typically signal many functions of *okay* identified and analyzed in [8] such as backchannel, acknowledgment, beginning of a new discourse segment, or agreement. Additionally, '*no*' can signal non-commitment and mild disagreement since Slovak *no* is also a conjunction roughly meaning 'but'.

2.3 Features: Labelling and Extraction

In this paper we concentrate on three types of features: acoustic-prosodic characteristics contained within the discourse marker 'no' itself, acoustic-prosodic features characterizing the speech of an interlocutor as a whole, and pragmatic features capturing the discourse communicative functions of 'no'. The unit of analysis will be the individual task; recall there are 14 tasks for each dialogue.

Acoustic and prosodic features in the signal were extracted using Praat [3]. We extracted a standard set of acoustic features including duration and the slope of F0 (only for no-tokens), and mean, maximum, minimum of F0 and Intensity, and also voice quality features such as jitter, shimmer, harmonics-to-noise ratio or spectral tilt for both no-tokens and each task of the game. Since in this paper we compare the behaviour of the same speaker in her two sessions, features were not normalized.

Table 1 shows the scheme for labelling the discourse functions of 'no'. The scheme was designed following the scheme used for the functions of 'okay' in [9] and appended for additional unique features of Slovak '*no*': Z, J, or RZ [2].

Table 1. Scheme for labelling discourse functions of 'no', adapted from [9]

Label	Meaning	Label	Meaning
R	I acknowledge that I understand, I got it	**H**	Hesitation, I am stalling for time
RP	I acknowledge that I understand, and please continue	**E**	I want to repair/redo something I've just said or did
RZ	I acknowledge that I understand, but I want to add something or express mild disagreement	**PH**	I express an assessment of something that has just happened, usually after receiving a score
RN	I acknowledge that I understand, and I want to start a new topic or a new discourse segment	**J**	I Soften of what is to follow, a hedge
N	I want to start a new topic or a new discourse segment	**K**	I signal the end of the current topic or discourse segment
S	I agree, also as an answer to a questions, usually meaning *yes*	**D**	I encourage some action, go on, do something
Z	I want to express an idea opposite to the one implied before, usually meaning 'but' or 'well'	**?**	None of the labels correspond to the perceived meaning

2.4 Analyzing Entrainment

We analyze here the differences in the speaking behaviour of the five subjects as a function of their conversational partner. In other words, we ask if a subject changed his/her speaking behaviour between the two games that s/he played, and if the answer is positive, we ask if his/her behaviour was more similar or dissimilar to that of his/her conversational partner. To answer this question for discrete dependent variables, such frequency of '*no*', we run a chi-square test comparing the frequencies of the speaker in her two games. To assess entrainment for continuous features, we run a t-test for the difference in the feature values extracted from the two games of the speaker. For both discrete and continuous features, if a significant difference is reported, we calculate the difference of the means of feature F in the two games played by the speaker, and the difference between the means of F extracted from the two interlocutors that played the game with the target speaker. If the signs of the two values are identical, we take it as evidence of entrainment. As an illustration, a speaker has mean pitch of 200Hz in session$_1$ and 230Hz in session$_2$. Her interlocutor in session$_1$ has mean pitch 180Hz and the other partner in session$_2$ 220Hz. Both differences (200-230 and 180-220) have the same sign (irrespective of the order of the operands), which corresponds to the target speaker adjusting her mean pitch to be more similar to her interlocutor. If the signs are different, we refer to this as dis-entrainment: the speaker changed her behaviour but became less similar to her interlocutor.

As an additional measure of entrainment in the use of '*no*' we follow [13] and compute entrainment as the negative value of the absolute difference between the frequency of '*no*' words between the two interlocutors (S1 and S2) shown in (1) below; *count* corresponds to the number of no-words and *ALL* to the sum of all words other than '*no*'.

$$ENTR(no) = -\left|\frac{count_{S1}(no)}{ALL_{S1}} - \frac{count_{S2}(no)}{ALL_{S2}}\right| \quad (1)$$

Hence, the lower (more negative) the ENTR(no) value, the less entrainment is there between the two interlocutors.

3 Results

We start with analyzing entrainment in the frequency of using '*no*' employing ENTR(no) measure in (1) above for the five speakers who played the game twice with a different partner. Table 2 shows the values for this measure separately for each of the five speakers and session. Columns 3 and 4 show that the highest entrainment was reached in the games played by speaker DF; identical numbers here correspond to game partners (DF, for instance, played her games with KM and VR). All other games are characterized by a rather low entrainment in this feature.

Table 2. Entrainment in the frequency of 'no' usage

Speaker	Gender	ENTR(no) with partner		ENTR(no) with self	X^2	p
LP	M	-2.6	-2.08	-0.83	2.9	0.1
KM	F	-0.61	-3.01	-1.6	4.7	0.03
DF	F	-0.61	-0.55	-0.85	2.7	0.1
MD	M	-2.05	-3.01	-0.32	0.7	0.4
VR	F	-2.08	-0.55	-2.32	16.2	0.001

The fifth column of Table 2 reports the values of our entrainment measure from the data for a single speaker in the two games s/he played. We assume that if ENTR(no) is low, the speaker differed in his/her frequency of '*no*' usage between the two games, and thus potentially s/he entrained or dis-entrained to his/her conversational partner. The last two columns show the results of chi-square tests assessing the significance of the difference in the no-usage for a speaker in his/her two games. We see that two speakers (KM and VR) significantly changed their no-usage and for two speakers (LP and DF) a tendency was reported. Further examination following the rationale described in section 2.4 revealed that both former speakers (KM and VR) significantly dis-entrained from their partners while the two latter speakers (LP and DF) showed a tendency for entrainment.

Finally, we checked if there is a significant difference between ENTR(no) in the third and fourth columns on the one hand and the fifth column on the other. In other words, we tested if speakers entrained more to themselves or to their partners, and we expected the former prediction will be born out. We do have a small number of values

(5 and 10 respectively), and the difference with this data is not significant; however, the direction of the effect follows the expectation (greater entrainment for self, t[11] = -1.1, p = 0.15). Doubling the data yields an almost significant value (t[24] = -1.63, p = 0.058). This result provides a sanity check and suggests that assessing entrainment with this measure is plausible.

The discourse function of '*no*' signalling that the speaker acknowledges and understands the previous utterance and wishes for his/her partner to continue (RP in Table 1) was the most frequent at 31% of all no-tokens. We examined if speakers (dis)entrain not only on the frequency of no-usage but also on this pragmatic function of the marker. For this purpose we calculated the frequency of RP function from all the uses of 'no' per speaker and session and followed the same steps and chi-square tests as described above. Our results show that three speakers (DF, KM, and VR) differed significantly in their frequency of RP function among no-uses. Following the method described above, the first two speakers disentrained and the last one entrained. Hence, interestingly, a speaker (VR) might disentrain in no-frequency but entrain in the frequency of a particular discourse function; we also have a speaker with the opposite pattern (DF). This supports the idea that entrainment on more cognitively complex features might differ from other types of entrainment.

We follow with testing (dis)entrainment in terms of acoustic and prosodic features extracted from no-tokens. Table 3 shows how many of our five speakers either entrained to their partner in terms of a given acoustic feature (2nd column), disentrained from the partner (3rd column), or did not produce a significant difference between the two games played (4th column) in terms of features extracted from no-tokens.

Table 3. Number of speakers showing entrainment, disentrainment, or no difference in the two games played

	No-tokens			Entire corpus		
Feature	**Speakers differ in 2 games**		**Speakers do not differ in 2 games**	**Speakers differ in 2 games**		**Speakers do not differ in 2 games**
	Entr	**Disentr**		**Entr**	**Disentr**	
Intensity mean	3	1	1	4	0	1
F0_mean	0	2	3	0	3	2
F0_slope	0	0	5			
Duration	1	1	3			
F1	1	1	3			
F2	0	1	4			
jitter	0	1	4	2	0	3
shimmer	2	0	3	3	0	2
hnr	1	1	3	2	1	2
spec. tilt	1	2	2	0	3	2
Total	**9**	**10**	**31**	**12**	**10**	**18**

Table 3 shows that the most pervasive pattern is no significant change between the behaviour of a speaker in the two games s/he played. Significant changes are equally distributed between more similarity and dissimilarity to the conversational partner.

It might be the case that the situation in terms of entrainment on the acoustic-prosodic features of no-tokens either reflects a general pattern of entrainment in the corpus, or that the patterns of entrainment in no-tokens and in a conversation as a whole are different. To approach this issue, we examined entrainment on the features in Table 3, but this time considered all data from the corpus and took a single task (14 tasks in a game) as a unit of analysis. Hence, for most cases, we had 14 data points per speaker and game. The results are in the three rightmost columns of Table 3. We observe that the results from the overall entrainment assessment are very similar to the ones reported for no-tokens only. Non-significant differences between the two games played by a speaker are most pervasive with roughly equal distribution of entrainment and disentrainment. In terms of features, speakers tend to be more inclined to entrain on intensity and voice quality and less on other features. This is to be expected since entrainment is most likely to occur for features that are most redundant and thus carry a minimal functional load in terms of linguistic contrasts; e.g. [16].

Finally, expecting different behaviour based on the task role (Describer vs. Placer), as reported for other features in corpora of task-oriented games (e.g. inter-speaker intervals in map-tasks [4]), we also tested entrainment for the two roles separately. The results for both no-tokens and entire corpus suggest that being in a more dominant role (Describer) induces 1) greater tendency for entrainment and 2) greater tendency to differ in the two games.

4 Discussion and Conclusion

Our small-scale pilot study revealed two main findings. First, we observed less entrainment than expected. Studies of entrainment in similar corpora suggest pervasive entrainment (e.g. [11]) while in our corpus, entrainment is present but not overwhelmingly. This might be due to a different language/culture (Slovak vs. English), less sophisticated measures of entrainment in this study, or several other differences. Second, the situation in just no-tokens is very similar to the entire corpus. This opens the possibility for more efficient ways of accessing well-being of users from the degree of entrainment to the system from a smaller set of target tokens.

Our results also show that the crucial feature of models of entrainment suitable for implementation in automatic systems interacting with humans is their adaptability. This is because speakers employ their own individual strategies for entrainment in prosodic and voice quality features, some speakers or features do not participate in entrainment, and some show disentrainment. Moreover, differences were reported for the same speaker between entrainment on the basic word-usage and its discourse function. Hence, entrainment is highly subject-dependent, which is an important finding for applications in human-computer interaction since accommodation to the user needs to be personalized. In future we plan to employ more sophisticated measures of entrainment, different units of analysis, and more speakers. For example, we will compare adjacent inter-pausal units (IPUs) across a turn-exchange and compare with random pairs of IPUs from the two speakers.

Acknowledgements. This research was supported by a VEGA grant 2/0202/11 and results also from of the project implementation: RPKOM, ITMS 26240220064 supported by the Research & Development Operational Programme funded by the ERDF.

References

1. Bell, L., Gustafson, J., Heldner, M.: Prosodic adaptation in human-computer interaction. In: Proceedings of 15th ICPhS (2003)
2. Beňuš, Š.: Prosodic forms and pragmatic meanings: the case of the discourse marker 'no' in Slovak. In: Proceedings of CogInfoCom (2012)
3. Boersma, P., Weenink, D.: Praat: doing phonetics with computer, `http://www.praat.org`
4. Bull, M., Aylett, M.: An analysis of the timing of turn-taking in a corpus of goal-oriented dialogue. In: Proceedings of ICSLP (1998)
5. Chartrand, T., Bargh, J.: The chameleon effect: The perception-behavior link and social interaction. Journal of Personality and Social Psychology 76, 893–910 (1999)
6. Coulston, R., Oviatt, S., Darves, C.: Amplitude convergence in children's conversational speech with animated personas. In: Proceedings of ICSLP (2002)
7. Delvaux, V., Soquet, A.: The Influence of Ambient Speech on Adult Speech Productions through Unintentional Imitation. Phonetica 64, 145–173 (2007)
8. Gravano, A., Hirschberg, J., Beňuš, Š.: Affirmative cue words in task-oriented dialogue. Computational Linguistics 38(1), 1–39 (2012)
9. Gravano, A., Benus, S., Chávez, H., Hirschberg, J., Wilcox, L.: On the role of context and prosody in the interpretation of okay. In: Proceedings of ACL, pp. 800–807 (2007)
10. Hirschberg, J.: Speaking More Like You: Entrainment in Conversational Speech. In: Proceedings of Interspeech, pp. 27–31 (2011)
11. Levitan, R., Hirschberg, J.: Measuring acoustic-prosodic entrainment with respect to multiple levels and dimensions. In: Proceedings of Interspeech (2011)
12. Nass, C., Moon, Y., Fogg, B.J., Reeves, B., Dryer, D.C.: Can computer personalities be human personalities? International Journal of Human-Computer Studies 43(2), 223–239 (1995)
13. Nenkova, A., Gravano, A., Hirschberg, J.: High frequency word entrainment in spoken dialogue. In: Proceedings of ACL/HLT, pp. 169–172 (2008)
14. Nielsen, K.: Implicit Phonetic Imitation is constrained by Phonemic Contrast. In: Proceedings of the 16th ICPhS (2007)
15. Pardo, J.: On phonetic convergence during conversational interaction. Journal of Acoustical Society of America 119(4), 2382–2393 (2006)
16. Pentland, A.: To Signal Is Human. American Scientist 98, 204–210 (2010)
17. Pickering, M., Garrod, S.: Toward a mechanistic psychology of dialogue. Behavioral and Brain Sciences 27, 169–226 (2004)
18. Redeker, G.: Review article: Linguistic markers of linguistic structure. Linguistics 29(6), 1139–1172 (1991)
19. Stoyanchev, S., Stent, A.: Lexical and syntactic priming and their impact in deployed spoken dialogue systems. In: Proceedings of NAACL (2009)

Language and Gender Effect in Decoding Emotional Information: A Study on Lithuanian Subjects

Maria Teresa Riviello[1,2], Rytis Maskeliunas[3], Jadvyga Kruminiene[4], and Anna Esposito[1,2]

[1] Seconda Università di Napoli, Department of Psychology, Italy
[2] International Institute for Advanced Scientific Studies (IIASS), Italy
mariateresa.riviello@unina2.it,
iiass.annaesp@tin.it
[3] Kaunas Univerisity of Technology, Information Technology Development Institute, Kaunas, Lithuania
rytis.maskeliunas@ktu.lt
[4] Vilnius University, Kaunas Faculty of Humanities, Kaunas, Lithuania
hedw_bush@yahoo.com

Abstract. The present work explores how language specificity and gender affect the emotional decoding process. It investigates the ability of Lithuanian male and female subjects to decode emotional information through male and female vocal and visual emotional expressions. The exploited emotional stimuli are based on extracts of American English (a globally spread language), and Italian (a country specific language) movies. The assumption is that the recognition of the emotional states expressed by the actors/actresses will change according to the familiarity of the languages and the subjects' gender. Results show that Lithuanian subjects recognition accuracy is affected by the language specificity of the stimuli. Moreover, a gender effect occurs in decoding Italian vocal stimuli.

Keywords: Emotional information decoding process, Language specificity, Gender effect.

1 Introduction

The importance of research on emotions influences studies in numerous disciplines such as psychology, sociology, linguistic, health services as well as computer science. There is, in fact, a growing interest in the automatic individuation and generation of emotions for application in communicative machines which can identify people's affective states and appropriately react to them [1- 2].

Identifying emotions in human interactions is a complex task: the encoding/decoding processing of emotional information occurs through several communication modalities simultaneously (facial, vocal and gestural expressions, among others) [3-4]. This demands a multimodal approach which only recently characterizes studies on emotions [5-7].

S. Bassis et al. (eds.), *Recent Advances of Neural Network Models and Applications*, Smart Innovation, Systems and Technologies 26,
DOI: 10.1007/978-3-319-04129-2_35, © Springer International Publishing Switzerland 2014

Furthermore, numerous factors influence the way emotions are expressed and interpreted, comprising cultural and individual characteristics.

On the one hand, in spite of the fact that a long research tradition sustains the innate and universal nature of emotional expressions [8-11], there are many studies supporting the idea that expressions of emotions are learned behaviors and thus they vary across cultures [12-14]. On this research line, recent studies speculates that experience and familiarity with faces from other cultures affect the recognition of emotional expressions [15] and that cultural variability, in the accuracy of the emotion decoding process, depends on differences in language [16-17].

On the other hand, researches have focused on individual factors affecting emotional behaviors, such as gender differences in perception and expression of emotions. It seems that women are stereotyped as being more emotional [18], in particular, they are more expressive of emotions in general, and better at decoding them [19-20].

This work attempts to account for the impact of cultural, language and gender factors in affecting the decoding process of emotional information. In particular, it explores the ability of Lithuanian male and female participants in decoding emotional information exploiting male and female expressions. The emotional data used in the experiments consist of realistic, dynamic and mutually related audio, mute video and audio-video stimuli. The stimuli were extracted from American English, as a globally spread language [21-22], and Italian, as a country specific language [23], movie scenes. The main goal is to investigate whether the decoding processing of the emotional states in the three considered expressive contexts (audio, mute video and audio-video) depends on the familiarity with the language and the gender of both participants (receivers) and actors (senders).

2 Materials and Procedure

In this study different groups of Lithuanian subjects, each consisting of an equal number of males and females, were involved in perceptual experiments aimed at comparing their ability in decoding emotional information dynamically presented through American English and Italian emotional stimuli portrayed by three communication modalities: audio, mute video and audio-video. The stimuli consist of emotional expressions of *happiness, fear, anger, irony, surprise* and *sadness,* played by actors and actresses and extracted from movie scenes [24-25].

The aim is to investigate possible differences in the decoding performance when American/Italian emotional expressions were exploited, also considering the role of the communication modality (audio, mute video and audio-video) through which they were portrayed. The effects of both subjects' and actors' gender on the emotional decoding process are also explored.

2.1 Materials

Two databases of dynamic emotional expressions have been defined exploiting video-clips extracted from American and Italian movies, and used as stimuli. A detailed description of the cross-modal databases is available in Esposito [24-25]. The main features of the databases are briefly described below:

1. each database contains short audio and video stimuli concerning 6 emotional states: happiness, sarcasm/irony, fear, anger, surprise, and sadness. These emotions, except for sarcasm/irony, are considered by many theories as basic and therefore universally shared [26-28];

2. for each database, each emotion is represented by 10 stimuli, 5 expressed by an actor and 5 by an actress, for a total of 60 American and 60 Italian video-clips, each extracted from a different movie scene and acted by a different actor and actress;

3. in each video-clip, the protagonist's face and upper part of the body are clearly visible;

4. for each video-clip, the emotion is portrayed at a moderate intensity (medium arousal) and the semantic meaning of the produced utterances is not related to the portrayed emotional state. For example, sadness stimuli where the actor/actress was using words like "sad" and the like, or he/she was clearly crying or happiness stimuli where the protagonist was declaring his/her joy or he/she was strongly laughing were not included in the data. This was because we wanted the subjects assessing the stimuli to be able to exploit only non-verbal emotional signs generally employed in natural and not extreme emotional interaction;

5. each video-clip has been assigned to a specific emotional category first by two experts and then by three naïve judges independently; only the video-clips that obtained unanimous agreement were selected;

6. the audio and mute video have been extracted, both for the American and Italian data, from the complete audio-video stimulus (video-clip). Summarizing the dataset comprises a total of 180 American and 180 Italian stimuli: 60 mute video and 60 audio and 60 audio-video for each language.

2.2 Participants

The perceptual experiments have been conducted on a total of 180 Lithuanian participants, balanced by gender. The participants were volunteers recruited mainly among university students at Vilnius University, Kaunas Faculty of Humanities, Lithuania. Their age ranges from 18 to 35, and all of them use English as a second language, whereas they do not have any knowledge of the Italian language.

2.3 Procedure

In order to evaluate whether the ability to decode emotional expressions is affected by the stimuli cultural context and the language (American vs. Italian), participants were partitioned into 6 groups consisting of 30 subjects, equally distributed per gender. Specifically, we defined a group for each combination of communication modality (audio, mute video and audio-video) and language (American and Italian). The results were assessed through three comparisons made between 2 groups of 30 subjects tested on American and Italian audio, American and Italian mute video, and American and Italian audio-video respectively.

Furthermore, with the aim of investigating the role of gender in the emotional decoding process, each group of 30 subjects has been partitioned in two subgroups of 15 male and 15 female subjects as well as each set of stimuli has been split into two subset, according to the gender of the actor producing the emotional expression (male and female stimuli).

To explore gender effects due to both participant's and actor's gender, the analyses compared first the ability of males and females to decode emotions for each communication modality (audio, mute video and audio-video), for each language (American and Italian), for each subset of male and female stimuli, and then for each subgroup of male and female subjects.

3 The Role of Language

Figure 1 reports means and standard deviations of correct accuracies obtained by Lithuanian subjects in the decoding of both American and Italian emotional stimuli administered as only audio, mute video and combined audio-video.

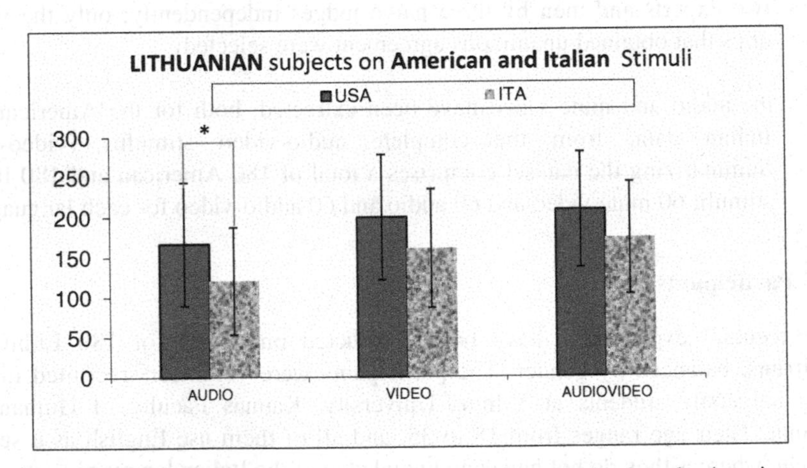

Fig. 1. Means and standard deviations of correct recognitions obtained on the American (black bars) and Italian (gray bars) audio, video and audio-video emotional stimuli

ANOVA analyses were separately performed on the data obtained for the three different communication modalities (i.e., the audio, the mute video and the combined audio-video), considering the *Cultural context and the Language of the stimuli* (American and Italian) as between subjects variable, while *Emotions* (happiness, fear, anger, irony, surprise and sadness), and *Artiste' gender* (male and female) as within subjects variables. Significance was established for α=.05.

The analyses revel a significant difference for the audio modality (F (1, 8)= 17.75, ρ=.003): the American vocal emotional expressions (m: 16.63; SD: 7) have been better identified than the Italian ones (m: 12.03; SD: 6.6). No differences in the emotional decoding processing were found when the American and Italian mute video (F (1, 8)= 1.95, ρ=.2) and audio-video (F (1, 8)= 1.504, ρ=.255) stimuli have been considered.

4 The Role of Gender

ANOVA analyses were computed in order to assess whether the gender of the participants affects the process of decoding emotional information. Such analyses considered separately each communication modality (audio, mute video and audio-video), each language (American and Italian) and each subgroup of male and female subjects. In these analyses the *subjects' gender* (male and female) was set as a between subjects variable, and the *emotions* (happiness, fear, anger, irony, surprise and sadness) as a within subjects variable. Significance was established for α=.05.

The results (displayed in Table 1) show a significant gender effect only for Italian audio stimuli.

In particular, Lithuanian female are better able than male subjects in the decoding of Italian male vocal emotional expressions. No gender effects were found for American vocal expressions and for both American and Italian mute videos and complete video-clips.

Table 1. Statistical assessment of the gender effect on the emotional decoding process

F/M SUBJECTS			
ITALIAN STIMULI		AMERICAN STIMULI	
Male Audio	**Female Audio**	**Male Audio**	**Female Audio**
F (1,8)=20.4, p=.013* Male Subj M= 4,933 SD:2,7 **Female Subj M= 6,1 SD:2,7***	F (1,8)=1.380, p= .272	F (1,8)=.563, p=.475	F (1,8)=1.776, p=.219
Male Mute Video	**Female Mute Video**	**Male Mute Video**	**Female Mute Video**
F (1,8)=.001, p=.975	F (1,8)= .019, p=.895	F (1,8)= .144, p=.714	F (1,8)=.009, p=.926
Male Audio-Video	**Female Audio-Video**	**Male Audio-Video**	**Female Audio-Video**
F (1,8)=.126, p=.732	F (1,8)=.469, p=.513	F (1,8)=1.140, p=.317	F (1,8)= .013, p=.913

More detailed ANOVAs were computed to assess whether the gender of the artiste producing the emotional expression affected the ability of each male and female subgroup to decode emotional stimuli, considering separately each communication modality (audio, mute video and audio-video), each language (American and Italian) and each subgroup of male and female participants.

In this case the *artiste's gender* (male and female) was considered as a between subjects variable and the *emotions* (happiness, fear, anger, irony, surprise and sadness) as a within subjects variable. The significance was established for α=.05.

The ANOVA results (reported in Table 2) show that there are no significant effects of the artiste's gender both for the American and Italian mute videos and complete video-clips, as well as for the American audio stimuli.

On the other hand, significant artiste's gender effects emerge significantly when the male and female Lithuanian subjects were separately tested on the Italian male and female emotional vocal expressions. In this case, the analyses show that both Lithuanian male and female subjects prefer to decode Italian female rather than male emotional voices.

Table 2. Statistical assessment of the artiste' gender effect in decoding Italian and American emotional information

ITALIAN STIMULI		AMERICAN STIMULI	
MALE SUBJECTS	FEMALE SUBJECTS	MALE SUBJECTS	FEMALE SUBJECTS
Male/Female Audio	**Male/Female Audio**	**Male/Female Audio**	**Male/Female Audio**
F (1,8)=19.7, p=.002* Male Audio Stimuli M=4.93, SD= 2.74 **Female Audio Stimuli M=6.7, SD=3.76**	**F (1,8)= 7.8, p= .024*** Male Audio Stimuli M=6.1, SD= 2.73 **Female Audio Stimuli M= 7,2, SD= 3.9**	F (1,8)= .815, p=.393	F (1,8)= .369, p=.560
Male/Female Video	**Male/Female Video**	**Male/Female Video**	**Male/Female Video**
F (1,8)=1.34, p=.724	F (1,8)= .048, p=.832	F (1,8)= .358, p=.566	F (1,8)=.452, p=520
Male/Female Audio-Video	**Male/Female Audio-Video**	**Male/Female Audio-**	**Male/Female Audio-**
F (1,8)= .021, p=.888	F (1,8)=.412, p=.539	F (1,8)= 5.45, p=.58	F (1,8)= .423, p=.534

5 Discussion and Conclusion

In the present study, Lithuanian male and female participants were involved in perceptual experiments with the aim of investigating language and gender effects on their ability to decode emotional expressions. In the experiments, American and Italian audio, mute video and audio-video emotional expressions produced by movie actors and actresses were used as stimuli, while subjects were asked to decode the emotion expressed in each stimulus.

The assumption was that the recognition of the emotional expressions will be influenced by the familiarity between the sender's and the receiver's language as well as the gender of both the sender and the receiver. In our experiment, American English is considered as a globally spread language and it is used as second language by the participants, whereas they do not have any knowledge of Italian, considered as a country specific language.

Results reveal a significant difference in decoding the proposed emotions (happiness, fear, anger, irony, surprise and sadness) when American and Italian vocal expressions were evaluated. Lithuanian subjects, in fact, show more difficulties in identifying emotions when tested on Italian audio stimuli than American ones, supporting the abovementioned assumption. On the other hand, the processing of visual information allows similar decoding accuracy for both American and Italian data. These results support the idea that facial expressions may share more emotional features than language across similar (western) cultures, and therefore emotional faces may result more powerful than voices in conveying emotional information.

Familiarity with the language also seems to play a role in the assessment of gender differences in decoding emotions. Significant gender effects, in fact, were exclusively found when Lithuanian subjects examined Italian vocal expressions.

The results, at least considering the encoding/decoding of vocal expressions, show that females are better accurate than males both in expressing and interpreting emotional states. Indeed, whereas no differences between male and female subjects were found when Italian female voices were exploited (females are better in expressing emotions), female subjects were more accurate than male ones in the decoding of Italian male emotional voices (females are better in interpreting emotions). Furthermore, both male and female Lithuanian subjects perform better in decoding Italian female rather than male emotional vocal expressions.

Since this effect was not observed on American voices, as well as on visual emotional expressions (both mute videos and complete video-clips), we can speculate that the influence of gender in transmitting and perceiving emotions is subtle, but may emerge when some difficulty occurs in processing emotions, as in the case of vocal expressions produced in an unknown language.

The reported study is an attempt to go into cultural and individual factors characterizing human decoding process of emotional cues. It could be interesting to extend the research to a larger sample of subjects of different cultures and languages, as well as to explore the role of each specific emotion category and its possible specific relation with both language and gender effects on the emotional decoding process.

References

1. Zeng, Z., Pantic, M., Rosiman, G.I., Huang, T.S.: A Survey of Affect Recognition Methods: Audio, visual, and spontaneous expressions. Trans. on Pattern Analysis and Machine Intelligence 31(1), 39–58 (2009)
2. Ververidis, D., Kotropoulos, C.: Emotional Speech Recognition: Resources, Features and Methods. Elsevier Speech Communication 48(9), 1162–1181 (2006)
3. Aubergé, V., Cathiard, M.: Can we hear the prosody of smile? Speech Communication 40, 87–97 (2003)
4. Attardo, S., Eisterhold, J., Hay, J., Poggi, I.: Multimodal Markers of Irony and Sarcasm, Humor. International Journal of Humor Research 16(2), 243–260 (2003)
5. Chollet, G., Esposito, A., Gentes, A., Horain, P., Karam, W., Li, Z., Pelachaud, C., Perrot, P., Petrovska-Delacrétaz, D., Zhou, D., Zouari, L.: Multimodal Human Machine Interactions in Virtual and Augmented Reality. In: Esposito, A., Hussain, A., Marinaro, M., Martone, R. (eds.) COST Action 2102. LNCS, vol. 5398, pp. 1–23. Springer, Heidelberg (2009)
6. Esposito, A.: The Amount of Information on Emotional States Conveyed by the Verbal and Nonverbal Channels: Some Perceptual Data. In: Stylianou, Y., Faundez-Zanuy, M., Esposito, A. (eds.) COST 277. LNCS, vol. 4391, pp. 249–268. Springer, Heidelberg (2007)

7. Wang, Y., Guan, L., Venetsanopoulos, A.N.: Kernel Cross-Modal Factor Analysis for Information Fusion with Application to Bimodal Emotion Recognition. IEEE Transactions on Multimedia 14(3), 1520–9210, 597 – 607 (2012)
8. Ekman, P.: Strong Evidence for Universals in Facial Expressions: A Reply to Russell's Mistaken Critique. Psychological Bulletin 115, 268–287 (1994)
9. Mesquita, B., Frijda, N.H., Scherer, K.R.: Culture and Emotion. In: Berry, J.W., Dasen, P.R., Saraswathi, T.S. (eds.) Handbook of Cross-cultural Psychology. Basic processes and human development, pp. 255–297. Allyn & Bacon, Boston (1997)
10. Russell, J.A.: Is There Universal Recognition of Emotion from Facial Expression? A Review of the Cross-cultural Studies. Psychological Bulletin 115, 102–141 (1994)
11. Scherer, K.R., Wallbott, H.G.: Evidence for Universality and Cultural Variation of Differential Emotion Response Patterning. Journal of Personality and Social Psychology 66, 310–328 (1994)
12. Fridlund, A.J.: The new ethology of human facial expressions. In: Russell, J.A., Fernandez Dols, J. (eds.) The Psychology of Facial Expression, pp. 103–129. Cambridge University Press (1997)
13. Scherer, K.R., Banse, R., Wallbott, H.G.: Emotion Inferences from Vocal Expression Correlate across Languages and Cultures. Journal of Cross-Cultural Psychology 32, 76–92 (2001)
14. Esposito, A., Riviello, M.T., Bourbakis, N.: Cultural Specific Effects on the Recognition of Basic Emotions: A Study on Italian Subjects. In: Holzinger, A., Miesenberger, K. (eds.) USAB 2009. LNCS, vol. 5889, pp. 135–148. Springer, Heidelberg (2009)
15. Elfenbein, H.A., Ambady, N.: On the universality and cultural specificity of emotion recognition: A meta-analysis. Psychological Bulletin 128, 203–235 (2002), doi:10.1037/0033-2909.128.2.203
16. Scherer, K.R., Banse, R., Wallbott, H.G.: Emotion inferences from vocal expression correlate across languages and cultures. Journal of Cross-Cultural Psychology 32, 76–92 (2001)
17. Riviello, M.T., Esposito, A.: A Cross-Cultural Study on the Effectiveness of Visual and Vocal Channels in Transmitting Dynamic Emotional Information. Acta Polytechnica Hungarica, Journal of Applied Sciences 9(1), 157–170 (2012)
18. Plant, E.A., Hyde, J.S., Keltner, D., Devine, P.G.: The gender stereotyping of emotions. Psychology of Women Quarterly 24, 81–92 (2000), doi:10.1111/j.1471-6402.2000.tb01024
19. Brody, L.R., Hall, J.A.: Gender and emotion in context. In: Lewis, M., Haviland-Jones, J. (eds.) Handbook of Emotions, vol. 3, pp. 395–408. Guilford, New York (2008)
20. Schmid, P.C., Schmid Mast, M., Bombari, D., Mast, F.D.: Gender Effects in Information Processing on a Nonverbal Decoding Task. Springer Science Business Media, LLC (2011)
21. Crystal, D.: English as a Global Language, 2nd edn. Cambridge University Press, Cambridge (2003)
22. Riviello, M.T., Chetouani, M., Cohen, D., Esposito, A.: On the perception of emotional "voices": A cross-cultural comparison among American, French and Italian subjects. In: Esposito, A. (ed.) Communication and Enactment 2010. LNCS, vol. 6800, pp. 368–377. Springer, Heidelberg (2011)
23. Esposito, A., Riviello, M.T., Bourbakis, N.: Cultural Specific Effects on the Recognition of Basic Emotions: A Study on Italian Subjects. In: Holzinger, A., Miesenberger, K. (eds.) USAB 2009. LNCS, vol. 5889, pp. 135–148. Springer, Heidelberg (2009)
24. Esposito, A.: The Perceptual and Cognitive Role of Visual and Auditory Channels in Conveying Emotional Information. Cognitive Computation Journal 1, 268–278 (2009)

25. Esposito, A.: The Amount of Information on Emotional States Conveyed by the Verbal and Nonverbal Channels: Some Perceptual Data. In: Stylianou, Y., Faundez-Zanuy, M., Esposito, A. (eds.) COST 277. LNCS, vol. 4391, pp. 249–268. Springer, Heidelberg (2007)
26. Ekman, P.: An argument for basic emotions. Cognition and Emotion 6, 169–200 (1992)
27. Izard, C.E.: Innate and universal facial expressions: Evidence from developmental and cross-cultural research. Psychological Bulletin 115, 288–299 (1994)
28. Scherer, K.R.: Vocal communication of emotion: A review of research paradigms. Speech Communication 40, 227–256 (2003)

Preliminary Experiments on Automatic Gender Recognition Based on Online Capital Letters

Marcos Faundez-Zanuy and Enric Sesa-Nogueras

Escola Universitària Politècnica de Mataró - Tecnocampus, Barcelona, Spain
{faundez,sesa}@eupmt.es

Abstract. In this paper we present some experiments to automatically classify online handwritten text based on capital letters. Although handwritten text is not as discriminative as face or voice, we still found some chance for gender classification based on handwritten text. Accuracies are up to 74%, even in the most challenging case of capital letters..

Keywords: on-linc handwriting, gender recognition, capital letters, de-identification.

1 Introduction

While biometric recognition based on handwritten text has been addressed in several papers [1-5], gender classification has not been studied too much. However, several recent papers have appeared [6-8]. In this paper we present a novel approach based on capital letters, which is more challenging than cursive letters.

Table 1 summarizes the state-of-the-art in gender recognition based on online handwritten analysis. It is worth mentioning that offline systems provide slightly worse results.

Table 1. State-of-the-art in gender recognition based on handwritten text

Authors	Accuracy	Online/ off-line	Classification and experimental conditions	Population
Bandi & Srihari [6]	73.2%	Off-line	Single neural network; CEDAR database, cursive letters	training set =800, testing set=400
Liwicki et al. [7]	67.06%	On-line	GMM, IAM-OnDB database, cursive letters	Training set =100 Testing set=50
Liwicki et al. [8]	64.25%	On-line	GMM, IAM-OnDB database, cursive letters	Training set =100 Testing set=50
Our approach	76%	On-line	SOM, BIOSECURID database	Training set =100 Testing set=125

As can be seen in table 1, the accuracy of gender recognition based on online handwritten text is far from other biometric traits such as face, speech, etc. Nevertheless, we consider that it is an interesting research topic, and this paper contributes to cast some more light on this topic.

S. Bassis et al. (eds.), *Recent Advances of Neural Network Models and Applications*, Smart Innovation, Systems and Technologies 26,
DOI: 10.1007/978-3-319-04129-2_36, © Springer International Publishing Switzerland 2014

1.1 Database and Classifier

In this section we describe the experimental results, which have been obtained using samples from the the BIOSECURId database. This database includes eight unimodal biometric traits, namely: speech, iris, face (still images and videos of talking faces), fingerprints, hand, keystrokes, handwritten signature and handwritten text (online dynamic signals and offline scanned images).

BIOSECURId comprises 400 subjects with balanced gender distribution and available information on age, gender and handedness. Data was collected in 4 sessions distributed in a time span of 4 months.

Regarding the online handwritten text, BIOSECURId provides data gathered from 3 different tasks: a Spanish text handwritten in lower case with no corrections or crossing outs permitted; the sequence of the digits, from 0 to 9 written in a single line; and 16 Spanish uppercase words, written each in a line.

The acquisition of the online handwritten data was carried out with a WACOM INTOUS A4 USB pen tablet. The following dynamic information was captured at 100 samples per second: x-coordinate, y-coordinate, time stamp, button status, azimuth, altitude and pressure.

For the experiments in this paper we have used the online handwritten text and, more precisely the first 4 words of the 16 uppercase words sequence, namely BIODEGRADABLE, DELEZNABLE, DESAPROVECHAMIENTO and DESBRIZNAR.

As words were not written isolated from each other, but one below the other, a simple segmentation step was required. During the segmentation, 30 users were found not to comply with the alleged prerequisites (words spanning more than one line, two words in a line, corrections, crossing outs ...). Those users were screened out. Summarizing, we have 370 writers with 4 words per writer and 4 sessions per word.

In this paper our results have been obtained with a Self organizing map (SOM), similar to the system that we applied in our previous paper [1]. The description of this system is beyond the scope of this paper, and will be described in a future journal paper.

2 Experimental Results

In order to compare the performance obtained by the proposed algorithm with the performance attained by human classifiers, five people have contributed they 'manual' classifications.

Two of them are calligraphic experts and the other three can be considered amateurs as they do not have any background in gender classification.

Tables 2 and 3 present the experimental results when analyzing cursive text and capital letters for 125 users (72 males and 53 females). Manual classification was based on a m/f and [0, 5] value, where m=male f=female, and the higher the score, the more confident is the classifier about his decision. In order to simplify the analysis, the scores are mapped into the [-5, 5] interval , where negative values correspond to females and positive values to males. The ground truth has been obtained assigning a

-5 value to females and a 5 value to males. These set of values has been used to work out the correlation coefficient (ρ) between ground truth and manual/automatic classification, as well as a figure of merit (FM). The figure of merit consists of the average of the products between ground truth and assigned scores. It is worth noticing that correct classification provides a positive figure, while an error provides a negative one (mismatch between manual score and ground truth).

Table 2. Experimental results obtained with the automatic and human classifiers, cursive letters

	Cursive letters				
			Identification rates		
classifier	FM	ρ	mean	male	female
machine					
expert 1	4	0,3543	68,80%	72,22%	64,15%
expert 2	4,2	0,3683	68,80%	72,22%	64,15%
amateur 1	3,48	0,3969	68,00%	52,78%	88,68%
amateur 2	4,44	0,3100	64,80%	65,28%	64,15%
Amateur 3	5,28	0,3961	73,60%	84,72%	58,49%

Table 3. Experimental results obtained with the automatic and human classifiers , capital letters

	Capital letters				
			Identification rates		
classifier	FM	ρ	mean	male	female
machine	4,04	0,5033	76,00%	86,11%	62,26%
expert 1					
expert 2					
amateur 1	4,12	0,3792	66,40%	63,89%	69,81%
amateur 2	3,92	0,3316	60,00%	72,22%	43,40%
Amateur 3	6,12	0,3845	68,80%	77,78%	56,60%

Table 2 reveals that similar performance is achieved by calligraphic experts and amateurs.

On the other hand, the comparison of tables 2 and 3 shows that gender classification based on capital letters is a more challenging problem than using cursive letters. This is in agreement with the observations made by calligraphic experts, who consider that it is not possible to ascertain the gender by means of the mere inspection of the handwriting. In addition, they considered that the use of capital letters was much more challenging and they declined to perform any classification based on this text. For this reason, human classification of capital letters has been done only by amateur people. Experimental results confirm that indeed capital letters are more challenging than cursive letters.

Although we have not developed an automatic system for gender classification based on cursive letters, we think that such a system could improve the results of capital letters, as it seems to be a a task that human classifiers can perform more easily.

Figures 1 and 2 show some clear examples of masculine and feminine cursive handwriting. Figures 3 and 4 show some hard-to-classify cases.

This is based on general experience about handwriting. Feminine scripts tend to be more roundish and legible, while masculine ones are more hard-to-read and sharp. Nevertheless, this is not a general rule, as some individuals present the opposite characteristics. The study about personal differences related to these two types of handwriting also presents some interest but is out of the scope of this paper.

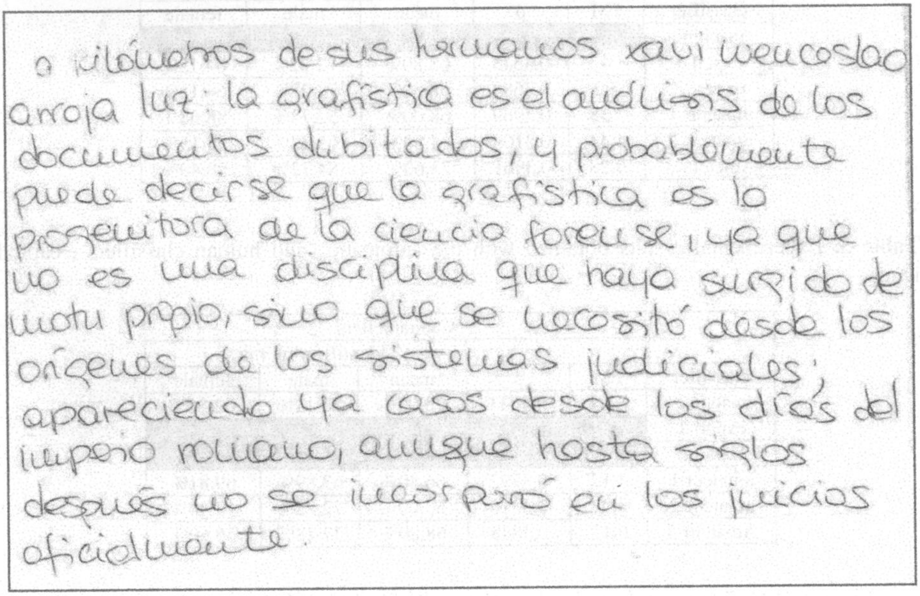

Fig. 1. 'Feminine' cursive handwriting produced by a woman

It is worth pointing out that the types of script shown in figures 1 and 2 are the most frequent while the discordant cases shown in figures 3 and 4 are less frequent.

Figure 5 shows an histogram with the figure of merit for several human classifiers. It can be seen that experts produce a smaller amount of gross errors (those with value -25) than amateurs , although their figure of merit does not seem better than the amateurs'

Figure 6 shows that there is more variability between the scores of amateurs and experts. Experts tend to produce more similar values to each other. Probably because they proceed more systematically whilst amateurs tend to behave more "randomly".

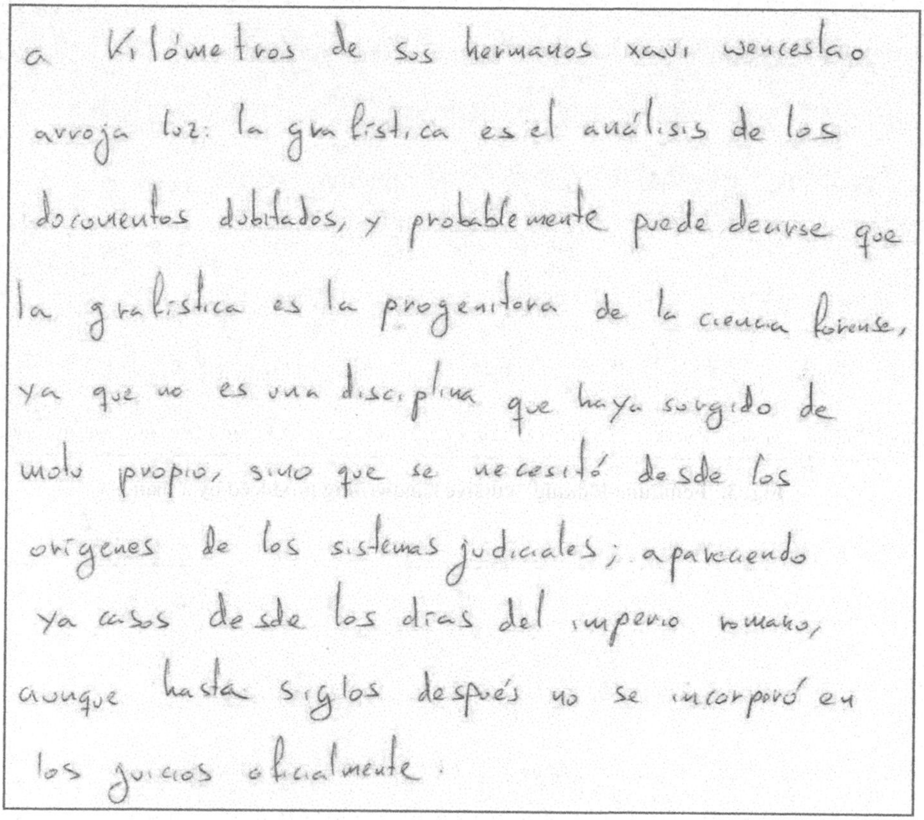

a Kilómetros de sus hermanos xavi wenceslao
arroja luz: la grafística es el análisis de los
documentos dubitados, y probablemente puede decirse que
la grafística es la progenitora de la ciencia forense,
ya que no es una disciplina que haya surgido de
motu propio, sino que se necesitó desde los
orígenes de los sistemas judiciales; apareciendo
ya casos desde los días del imperio romano,
aunque hasta siglos después no se incorporó en
los juicios oficialmente.

Fig. 2. 'Masculine' cursive handwriting produced by a man

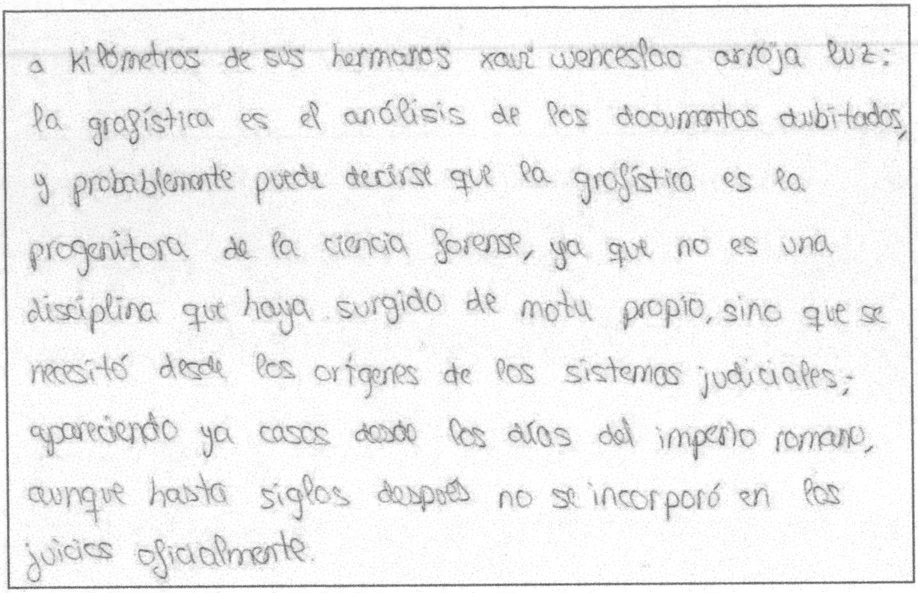

a kilómetros de sus hermanos xavi wenceslao arroja luz:
la grafística es el análisis de los documentos dubitados,
y probablemente puede decirse que la grafística es la
progenitora de la ciencia forense, ya que no es una
disciplina que haya surgido de motu propio, sino que se
necesitó desde los orígenes de los sistemas judiciales;
apareciendo ya casos desde los días del imperio romano,
aunque hasta siglos después no se incorporó en los
juicios oficialmente.

Fig. 3. 'Feminine-looking' cursive handwriting produced by a man

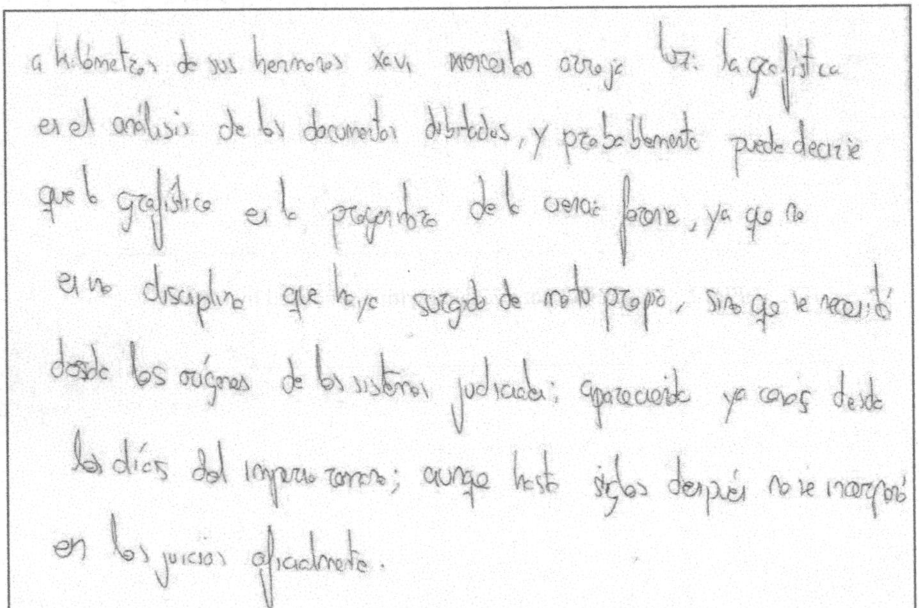

a kilómetros de sus hermanos xavi wenceslao arroja luz: la grafística
es el análisis de los documentos dubitados, y probablemente puede decirse
que la grafística es la progenitora de la ciencia forense, ya que no
es una disciplina que haya surgido de motu propio, sino que se necesitó
desde los orígenes de los sistemas judiciales; apareciendo ya casos desde
los días del imperio romano; aunque hasta siglos después no se incorporó
en los juicios oficialmente.

Fig. 4. 'Masculine-looking' cursive handwriting produced by a woman

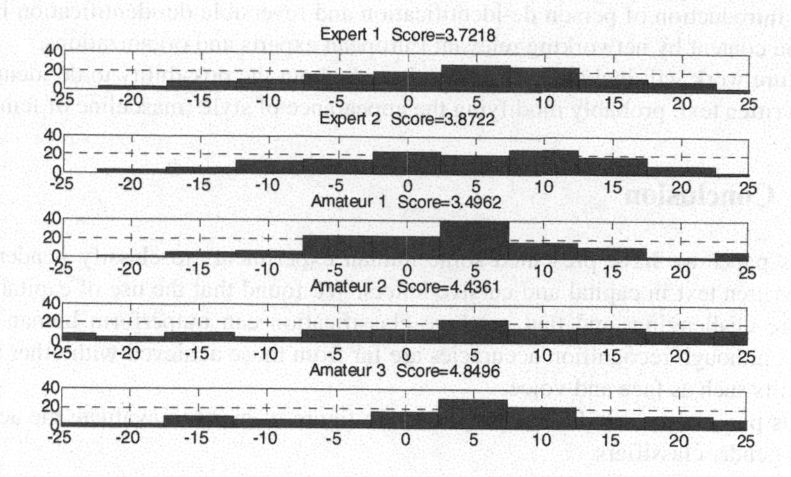

Fig. 5. Histograms of figures of merit for the human classifiers

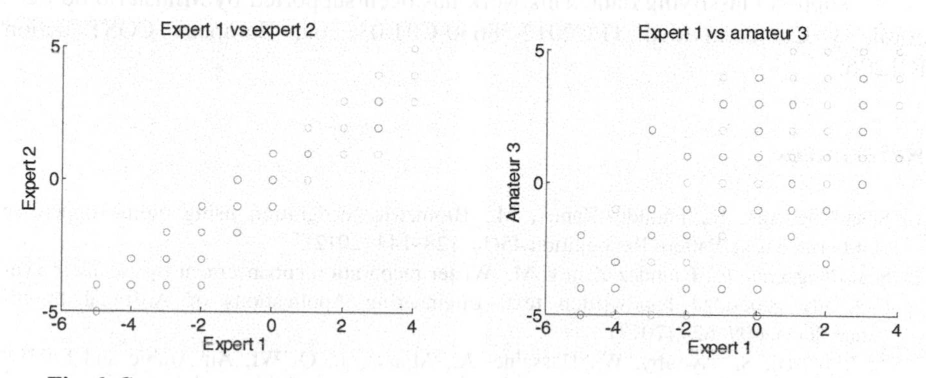

Fig. 6. Comparison of scores produced by expert 1 vs. expert 2 and amateur vs. expert

3 Future Work

Recently it has been launched a new COST action called IC1206 De-identification for privacy protection in multimedia content. De-identification in multimedia content can be defined as the process of concealing the identities of individuals captured in a given set of data (images, video, audio, text), for the purpose of protecting their privacy. This will provide an effective means for supporting the EU's Data Protection Directive (95/46/EC), which is concerned with the introduction of appropriate measures for the protection of personal data. The fact that a person can be identified by such features as face, voice, silhouette and gait, indicates the de-identification process is an interdisciplinary challenge, involving such scientific areas as image processing, speech analysis, video tracking and biometrics. This Action aims to facilitate coordinated interdisciplinary efforts (related to scientific, legal, ethical and societal aspects) in the introduction of person de-identification and reversible de-identification in multimedia content by networking relevant European experts and organizations.

Future work will include experimental work about the possibility to de-identificate handwritten text, probably modifying the appearance of style (masculine of feminine).

4 Conclusion

In this paper we have presented some human experiments to classify gender using handwritten text in capital and cursive letters. We found that the use of capital letters is more challenging, and that machine classification can outperform human classifieres , although recognition accuracies are far from those achieved with other biometric traits such as face and voice.

This paper also includes the proposal of a figure of merit to evaluate the accuracy of the gender classifiers.

Acknowledgements. We want to acknowledge Mari Luz Puente and Francesc Viñals for his support classifying data. This work has been supported by Ministerio de Economía y Competitividad TEC2012-38630-C04-03, and European COST action IC1206.

References

1. Sesa-Nogueras, E., Faundez-Zanuy, M.: Biometric recognition using online uppercase handwritten text. Pattern Recognition 45(1), 128–144 (2012)
2. Sesa-Nogueras, E., Faundez-Zanuy, M.: Writer recognition enhancement by means of synthetically generated handwritten text. Engineering Applications of Artificial Intelligence 26(1), 609–624 (2013)
3. Al-Maadeed, S., Ayouby, W., Hassaine, A., Aljaam, J.: QUWI: An Arabic and English Handwriting Dataset for Offline Writer Identification. In: International Conference on Frontiers in Handwriting Recognition, Bari, Italy (September 2012)
4. Hassaïne, A., Al-Maadeed, S., Bouridane, A.: A Set of Geometrical Features for Writer Identification. In: Huang, T., Zeng, Z., Li, C., Leung, C.S. (eds.) ICONIP 2012, Part V. LNCS, vol. 7667, pp. 584–591. Springer, Heidelberg (2012)
5. Nogueras, E.S., Faundez-Zanuy, M., Mekyska, J.: An information analysis of in-air and on-surface trajectories in online handwriting. Cognitive Computation 4(2), 195–205 (2012)
6. Bandi, K., Srihari, S.N.: Writer demographic identification using bagging and boosting. In: Proc. International Graphonomics Society Conference (IGS), pp. 133–137 (2005)
7. Liwicki, M., Schlapbach, A., Loretan, P., Bunke, H.: Automatic detection of gender and handedness from online handwriting. In: Proc. 13th Conference of the International Graphonomics Society, pp. 179–183 (2007)
8. Liwicki, M., Schlapbach, A., Bunke, H.: Automatic gender detection using on-line and off-line information. Pattern Analysis and Applications 14, 87–92 (2011)
9. Fierrez-Aguilar, J., Galbally, J., Ortega-Garcia, J., Freire, M., Alonso-Fernandez, F., Ramos, D., et al.: BiosecurID: a multimodal biometric database. Pattern Analysis & Applications 13 (2010)

End-User Design of Emotion-Adaptive Dialogue Strategies for Therapeutic Purposes

Milan Gnjatović and Vlado Delić

Faculty of Technical Sciences, University of Novi Sad
Trg Dositeja Obradovića 6, 21000 Novi Sad, Serbia
milangnjatovic@yahoo.com, vdelic@uns.ac.rs

Abstract. Two fundamental nontechnical research questions related to the development of emotion-aware dialogue systems are how to identify different kinds of emotional reactions that can be expected to occur in a given interaction domain, and how the system should react to the emotional user behavior. These questions are especially important for dialogue systems used in medical treatment of children with developmental disorders. The paper reports on an adaptive dialogue system that allows the therapist to flexibly design and test dialogue strategies. Our aim was to achieve a balance between the ease-of-use of the system by nontechnical users and the flexibility to adapt the system to different therapeutic settings. The system builds on our previous work, and its functionality is explained by means of example.

Keywords: Adaptive dialogue strategies, affective computing, end-user design, medical treatment of children.

1 Introduction

Two fundamental nontechnical research questions related to the development of emotion-aware dialogue systems are how to identify different kinds of emotional reactions that can be expected to occur in a given interaction domain, and how the system should react to the emotional user behavior. These questions are especially important for dialogue systems used in medical treatment of children with developmental disorders, where the adaptation of the dialogue strategy according to the current physiological and mental state of the child is one of the crucial factors that determine the efficacy of the treatment. Therapists can provide a valuable insight into important aspects of the therapeutic interaction, including clinically relevant changes in emotional states of children, and appropriate therapeutic interventions on these changes. Thus, it is clear that therapists should be involved in the process of designing dialogue strategies. However, even then, post-implementation changes of the system's dialogue strategy seem to be, for many reasons, inevitable. Assessing the phenomenon of children's affective behavior in therapeutic settings is a challenging research question, since the type and intensity of emotional expressions, and the required interventions, may vary significantly between patients. In addition, the presence

S. Bassis et al. (eds.), *Recent Advances of Neural Network Models and Applications*, Smart Innovation, Systems and Technologies 26,
DOI: 10.1007/978-3-319-04129-2_37,

of a system (e.g., a robot) may induce strong emotional reactions in children, such as emotional attachment or revulsion towards the system. Predicting these reactions may require collecting samples from real-life therapeutic interactions between the therapist, the patient, and the system. Practically, it means that each time when the therapist needs to redefine a dialogue strategy (e.g., to adapt it for a particular patient, etc.), he must go back to the programmer. This is too restrictive for our purposes. The paper addresses this desideratum, and reports on an adaptive dialogue system that allows the therapist to flexibly design and test dialogue strategies.

2 Background and Related Work

In the field of human-robot interaction, considerable efforts have been, and still are, invested in research on robot-assisted therapy. It includes research on therapy of stroke patients [1–3], patients with cerebral palsy [4, 5], children with autism [6, 7], etc. Recently, research in this field has become increasingly concerned with the motivational aspects of robotic systems in therapy, focusing especially on children in therapeutic settings [8, 9] and older adults [10, 11]. The point of departure for this paper is that the robot's capacity to engage in a speech-based dialogue may significantly contribute to establishing affective attachment of the child to the robot, and, thus, motivate the child to undergo a therapy. To achieve this purpose, the dialogue system integrated with the robot should address both the analytical and generative aspects of the robot's dialogue behavior. Our previous work in this field includes research on the processing of spontaneously produced users' linguistic inputs [12, 13], linguistic encoding of motion events in a spatial context-aware interaction with the robot [14–16], emotion-adaptive dialogue management [17], identifying and conceptualizing emotional states of the user in human-machine interaction [18], and end-user design of adaptive dialogue strategies [19].

3 Interaction Scenarios

This paper considers a particular research question: for given therapeutic settings and a dedicated robotic system, how to enable the therapist to flexibly design and test dialogue strategies? We introduce an adaptive dialogue system intended to address this question in a general way, and illustrate it for two interaction scenarios in the context of the therapy for children with developmental disorders.

The first therapeutic scenario is aimed at helping the child to improve functional control of the spatial environment. The spatial context shared between the child and the therapist is a set of objects randomly positioned on a table. The therapist verbally instructs the child to recognize or manipulate the objects. The second therapeutic scenario is aimed at helping the child to improve functional control of the body. The therapist verbally instructs the child to move her head, arms, etc. In both these scenarios, the child is expected to try to comprehend and perform the therapist's verbal instructions. When appropriate, the

Fig. 1. (a) The industrial robot ABB IRB 140 [14, 15], and (b) the KHR-1 HV robot [20]

therapist assists the child by repeating or reformulating commands, providing additional information, demonstrating the correct response to the child, relaxing the child, etc. One of the general aims of including a robot in these scenarios is to additionally support the child in performing the therapist's instructions. For example, if the child does not correctly interpret or perform a given instruction, the therapist may ask the robot to demonstrate the correct response to the child. It implies that the robot should be able to interpret and perform the therapist's instructions, and to involve into conversation when appropriate (e.g., to ask for additional information when the therapist's instruction is incomplete, to commend the child, etc.). In the first scenario, we use the industrial robotic arm ABB IRB 140 (cf. Fig. 1a), integrated with a visual subsystem that recognizes the objects on the pad and determines their positions, and a dialogue system that manages the spoken interaction between the therapist and the robot [14, 15]. We decided to use this robot because it is convenient[1] for the purpose of pointing to or moving the objects on the table. In the second scenario, a humanoid robot is required for the purpose of demonstrating movements of certain body parts. We use the KHR-1 HV robot (cf. Fig. 1b), integrated with the software that enables synthesis and execution of humanoid movements [20].

4 Adaptation of the System to Therapeutic Settings

The introduced scenarios differ significantly, and the question underlying this research may be reformulated as follows: can we integrate the same dialogue system with both the robotic systems, rather than to develop two independent dialogue systems? To achieve this, it should be possible to flexibly adapt the dialogue system to a particular therapeutic scenario. At the implementation level, it means that—at least—the spatial context, the lexicon, the interaction

[1] We are aware that the appearance of the industrial robot is not appealing to children. However, it is used only for the demonstration purposes, since the integrated dialogue system is independent of the robot. In real therapeutic settings, we will use a dedicated robot designed for this purpose.

context, and the system's dialogue strategy should be modeled independently of the implementation of the dialogue system.

The implementation of the dialogue system is based on the focus tree—a model of attentional state in task-oriented human-machine interaction [12, 13]. Two fundamental features of this model are that (i) it provides a framework for robust natural language understanding and designing attention-based dialogue strategies, and that (ii) it has a relatively high level of generalizability and scalability. In our previous work, we have successfully adapted and applied this model in several prototypical dialogue systems with diverse domains of interaction (for references to the prototypical systems, cf. [13]). Here, we make a step further and report on the early-stage (yet functional) dialogue system, based on the focus tree model, that can be flexibly adapted by a nontechnical therapist. We abstract away from the conceptual details of the focus tree model and implementation details of the system, (they are thoroughly discussed in [12, 13] and in an unpublished paper) and introduce the system by means of example.

4.1 The Spatial Context and Lexicon

The graphical user interface that enables the therapist to define a spatial context and a lexicon at an arbitrary level of detail is given in Fig. 2. For the purpose of illustration, we adopt a simple spatial context. Let us suppose for the second therapeutic interaction scenario that each arm of the humanoid robot may be positioned in three different positions: beside the torso, in front of the torso, or above the head. Fig. 2 shows a possible definition of this spatial context. It is defined as a hierarchy. Each node in this hierarchy, except the nodes entitled *phrase*, represents an entity in the spatial context. Thus, node ROBOT represents the robotic system. It has two children—LEFTARM and RIGHTARM—that represent the robot's arms. Node LEFTARM has thee children—UP, DOWN, and IN_FRONT—that represent the possible positions of the left arm. Node RIGHTARM is collapsed, but the possible positions of the right arm are defined in the same manner. Furthermore, each of these nodes has one of more children entitled *phrase*. Each *phrase* defines a lexicon entry[2] assigned to its parent node. For example, the lexicon entries assigned to node ROBOT are "*marko*" and "*robot*". The lexical entries are used as keywords (i.e., the carriers of the propositional content) in the therapist's instructions, e.g., "*Marko*, please move your *left arm down*", "now raise it *up*", etc. The lexicon is imported into the module for automatic speech recognition (of the Serbian language), and the whole hierarchy is imported into the module for natural language understanding. Therefore, the dialogue system can recognize keywords and interpret[3] the propositional content of the therapist's verbal instructions. It is important to note that the hierarchy, the names of nodes (except the name *phrase*), and the lexicon entries are not prespecified, and can be defined at the therapist's dis-

[2] For the purpose of presentation, the lexical entries given in Fig. 2 are translated from Serbian into English.

[3] The interpretation of spontaneously produced commands is discussed in [12].

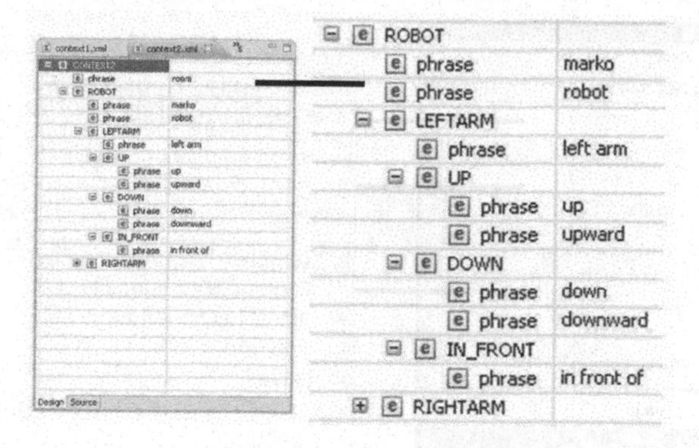

Fig. 2. Graphical user interface that enables the therapist to define the spatial context and lexicon, exemplified for the second therapeutic interaction scenario. This interface is part of Eclipse Java Integrated Development Environment.

cretion. For example, the therapist can use this interface to define the spatial context and the lexicon for the first therapeutic interaction scenario.

4.2 The Context of Interaction and Dialogue Strategy

In the approach presented in this paper, the context of interaction is a finite set of interaction variables. For each of the interaction variables, there is a finite set of possible values. For the purpose of illustration, let us suppose that the therapist needs to describe an emotional state of the child with respect to the type, intensity and duration of the signalled emotion. He can arbitrarily introduce the relevant emotional states following a relatively simple syntax. An example is given in Fig. 3a. The context of interaction is conceptualized as a set of three interaction variables: *emotion*, *duration*, and *intensity*. The sets of possible values of these variables are defined in the upper part of the window. The interaction variables are defined and initialized in the middle part of the window. For example, the set of possible values of the interaction variable *emotion* is $Emotion = \{NEGATIVE, POSITIVE, NEUTRAL\}$, and, at the start of the interaction, this variable is initialized with the value of $NEUTRAL$.

An adaptive dialogue strategy is conceptualized as a sequence of if-else-statements. The conditions in these statements are logical expressions that contain the interaction variables introduced by the therapist, and their bodies are sequences of primitive actions and nested if-else-statements. The early-stage version of the dialogue management module implements three prespecified primitive actions: synthesizing a verbal dialogue act (*utter()*), instructing the external robotic system to perform a nonverbal act (*perform()*), and waiting until

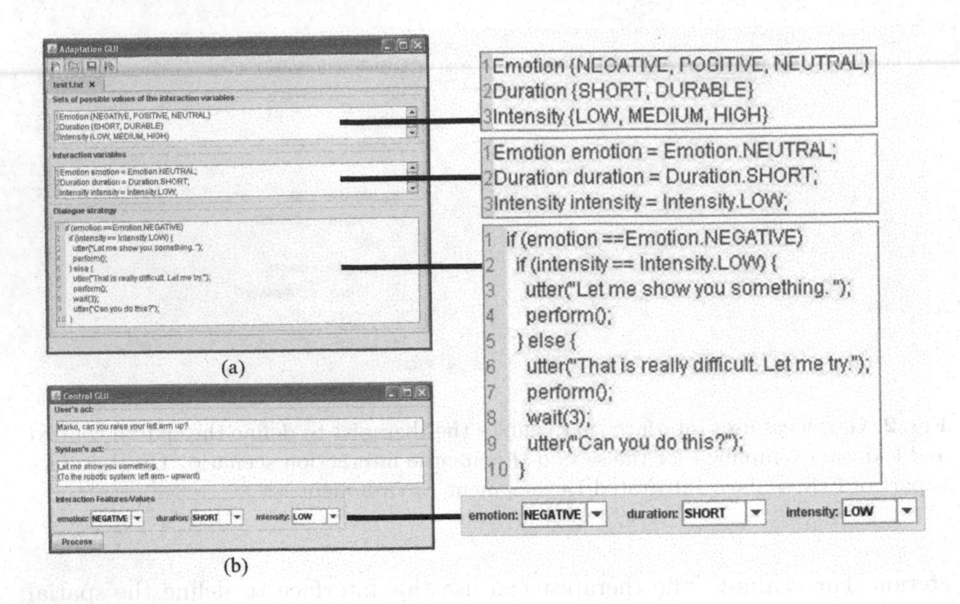

Fig. 3. Graphical user interfaces that enables the therapist (a) to define the context of interaction and the system's dialogue strategy, and (b) to test dialogue strategies

a specified time elapses (*wait()*). The lower part of the window in Fig. 3a shows a fragment of an adaptive dialogue strategy defined by the therapist[4]. The meaning of this strategy fragment can be formulated as follows. Each time when the therapist instructs the robot to perform a nonverbal act, the dialogue system should check whether the child is in a negative emotional state. If the child is in a negative emotional state of low intensity, the system should first synthesize the dialogue act ("Let me show you something."), and then perform the therapist's instruction. If the child is in a negative emotional state of medium or high intensity, the system should first synthesize the dialogue act ("This is really difficult. Let me try."), and then perform the therapist's instruction and wait for three seconds before it synthesizes the second dialogue act ("Can you do this?").

Finally, the graphical user interface that enables the therapist to test the introduced dialogue strategy is given in Fig. 3b. The therapist and the dialogue system both use typed text during the interaction. The therapist can also monitor and manually change the values of the interaction variables. The external robotic system and the modules for automatic speech recognition and text-to-speech synthesis are not integrated with this interface to enable testing of dialogue strategies in the cases when these modules are not available.

[4] The syntax is based on the Java programming language. Although the given example is much self explanatory, the therapists should be instructed how to "program" dialogue strategies.

5 Conclusion

This paper reported on an early-stage adaptive dialogue system that allows for flexible design and testing of dialogue strategies for therapeutic purposes. The system builds on our previous work, and its functionality is explained by means of example. As part of the ongoing research, we have also developed a more advanced version of this system that includes significantly more programmatic features, a richer syntax, and a user-friendly feedback on syntax error (reported in an unpublished paper). However, the level of details reported in this paper was deliberately selected to make the introduced dialogue system appropriate for use by therapists. It was aimed at achieving a balance between the ease-of-use of the system by nontechnical users and the flexibility to adapt the system to different therapeutic settings. Initial testing showed a satisfying level of acceptance of the system by nontechnical users. In our future work, we plan to use this system in the scope of the research aimed at designing emotion-adaptive dialogue strategies for the three-party interaction between the child, the therapist and the robot in the context of the therapy for children with cerebral palsy.

Acknowledgments. The presented study was sponsored by the Ministry of Education, Science and Technological Development of the Republic of Serbia under the Research grants III44008 and TR32035.

References

1. Dipietro, L., Krebs, H.I., Volpe, B.T., Stein, J., Bever, C., Mernoff, S.T., Fasoli, S.E., Hogan, N.: Learning, Not Adaptation, Characterizes Stroke Motor Recovery: Evidence from Kinematic Changes Induced by Robot-Assisted Therapy in Trained and Untrained Task in the Same Workspace. IEEE Transactions on Neural Systems and Rehabilitation Engineering 20(1), 48–57 (2012)
2. Kwakkel, G., Kollen, B.J., Krebs, H.I.: Effects of Robot-Assisted Therapy on Upper Limb Recovery After Stroke: A Systematic Review. Neurorehabilitation and Neural Repair 22(2), 111–121 (2008)
3. Johnson, M.J.: Recent trends in robot-assisted therapy environments to improve real-life functional performance after stroke. Journal of NeuroEngineering and Rehabilitation 3, 29–34 (2006)
4. Blázquez, M.P.: Clinical Application of Robotics in Children with Cerebral Palsy. In: Pons, J.L., Torricelli, D., Pajaro, M. (eds.) Converging Clinical and Engineering Research on Neurorehabilitation. Biosystems & Biorobotics, vol. 1, pp. 1097–1102. Springer, Heidelberg (2013)
5. Krebs, H.I., Ladenheim, B., Hippolyte, C., Monterroso, L., Mast, J.: Robot-assisted task-specific training in cerebral palsy. Developmental Medicine & Child Neurology 51(4), 140–145 (2009)
6. Thill, S., Pop, C.A., Belpaeme, T., Ziemke, T., Vanderborght, B.: Robot-assisted therapy for autism spectrum disorders with (partially) autonomous control: Challenges and outlook. Paladyn. Journal of Behavioral Robotics 3(4), 209–217 (2012)
7. Ricks, D.J., Colton, M.B.: Trends and considerations in robot-assisted autism therapy. In: Proc. of the IEEE International Conference on Robotics and Automation (ICRA), Anchorage, AK, pp. 4354–4359 (2010)

8. Nalin, M., Baroni, I., Sanna, A.: A Motivational Robot Companion for Children in Therapeutic Setting. In: Proc. of the Workshop on Motivational Aspects of Robotics in Physical Therapy, IEEE/RSJ International Conference on Intelligent Robots and Systems, Vilamoura, Algarve, Portugal, 6 pages, no pagination (2012)
9. Belokopytov, M., Fridin, M.: Motivation of Children with Cerebral Palsy during Motor Involvement by RAC-CP Fun. In: Proc. of the Workshop on Motivational Aspects of Robotics in Physical Therapy, IEEE/RSJ International Conference on Intelligent Robots and Systems, Vilamoura, Algarve, Portugal, 6 pages, no pagination (2012)
10. Fasola, J., Matarić, M.J.: Using Socially Assistive Human-Robot Interaction to Motivate Physical Exercise for Older Adults. Procedings of the IEEE 100(8), 2512–2526 (2012)
11. Wada, K., Shibata, T.: Living With Seal Robots—Its Sociopsychological and Physiological Influences on the Elderly at a Care House. IEEE Transactions on Robotics 23(5), 972–980 (2007)
12. Gnjatović, M., Janev, M., Delić, V.: Focus Tree: Modeling Attentional Information in Task-Oriented Human-Machine Interaction. Applied Intelligence 37(3), 305–320 (2012)
13. Gnjatović, M., Delić, V.: A Cognitively-Inspired Method for Meaning Representation in Dialogue Systems. In: Proc. of the 3rd IEEE International Conference on Cognitive Infocommunications, Kosice, Slovakia, pp. 383–388 (2012)
14. Gnjatović, M., Tasevski, J., Nikolić, M., Mišković, D., Borovac, B., Delić, V.: Adaptive Multimodal Interaction with Industrial Robot. In: Proc. of the IEEE 10th Jubilee International Symposium on Intelligent Systems and Informatics (SISY 2012), Subotica, Serbia, pp. 329–333 (2012)
15. Gnjatović, M., Tasevski, J., Mišković, D., Nikolić, M., Borovac, B., Delić, V.: Linguistic Encoding of Motion Events in Robotic System. In: Proc. of the 6th PSU-UNS International Conference on Engineering and Technology (ICET 2013), Novi Sad, Serbia, 5 pages, no pagination (2013)
16. Gnjatović, M., Delić, V.: Encoding of Spatial Perspectives in Human-Machine Interaction. In: Železný, M., Habernal, I., Ronzhin, A. (eds.) SPECOM 2013. LNCS, vol. 8113, pp. 116–123. Springer, Heidelberg (2013)
17. Gnjatović, M., Rösner, D.: Emotion Adaptive Dialogue Management in Human-Machine Interaction. In: Proc. of the 19th European Meetings on Cybernetics and Systems Research (EMCSR 2008), Vienna, Austria, pp. 567–572 (2008)
18. Gnjatović, M., Rösner, D.: Inducing Genuine Emotions in Simulated Speech-Based Human-Machine Interaction: The NIMITEK Corpus. IEEE Transactions on Affective Computing 1, 132–144 (2010)
19. Gnjatović, M., Pekar, D., Delić, V.: Naturalness, Adaptation and Cooperativeness in Spoken Dialogue Systems. In: Esposito, A., Esposito, A.M., Martone, R., Müller, V.C., Scarpetta, G. (eds.) COST 2010. LNCS, vol. 6456, pp. 298–304. Springer, Heidelberg (2011)
20. Kendereši, E., Raković, M., Nikolić, M., Gnjatović, M., Borovac, B.: Realization of Arms Movements for 3D Robot Model and Synchronization with Real Humanoid Robot. In: Proc. of the IEEE 10th Jubilee International Symposium on Intelligent Systems and Informatics (SISY 2012), Subotica, Serbia, pp. 217–220 (2012)

Modulation of Cognitive Goals and Sensorimotor Actions in Face-to-Face Communication by Emotional States: The Action-Based Approach

Bernd J. Kröger

Neurophonetics Group, Department of Phoniatrics, Pedaudiology, and Communication Disorders, Medical School, RWTH Aachen University, Aachen, Germany
Cognitive Computation and Applications Laboratory, School of Computer Science and Technology, Tianjin University, Tianjin, China
bernd.kroeger@rwth-aachen.de

Abstract. Cognitive goals – i.e. the intention to utter a sentence and to produce co-speech facial and hand-arm gestures – as well as the sensorimotor realization of the intended speech, co-speech facial, and co-speech hand-arm actions are modulated by the emotional state of the speaker. In this review paper it will be illustrated how cognitive goals and sensorimotor speech, co-speech facial, and co-speech hand-arm actions are modulated by emotional states of the speaker, how emotional states are perceived and recognized by interlocutors in the context of face-to-face communication, and which brain regions are responsible for production and perception of emotions in face-to-face communication.

Keywords: face-to-face communication, emotion, speech, facial expression, gesture, sensorimotor action, emotional speech, brain imaging, fMRI.

1 Introduction

Speech production comprises the activation of cognitive goals or intentions, their transformation into a (still cognitive) lexical phonological representation and subsequently into a sequence of speech articulator movements and an acoustic speech signal. Thus we can separate the cognitive part of speech production [1, 2] and its sensorimotor part [3-5]. The emotional state of the speaker modulates both, the cognitive part (e.g. by choice of different lexical items, choice of different syntactic structure, etc.) as well as the sensorimotor part of speech production (e.g. type of phonetic realization of the utterance by varying loudness level, speaking rate, intonation, voice quality, etc.) [6-8]. In the case of face-to-face communication in addition the co-speech facial expression and the co-speech hand-arm gestures are modulated by speaker's emotional state at both levels. At the cognitive level different types of facial expressions and different types of hand-arm gestures can be chosen (non-conscious or conscious). At the sensorimotor realization level amplitude and duration of facial as well as hand-arm movements can vary [9, 10].

S. Bassis et al. (eds.), *Recent Advances of Neural Network Models and Applications*, Smart Innovation, Systems and Technologies 26,
DOI: 10.1007/978-3-319-04129-2_38, © Springer International Publishing Switzerland 2014

It is the goal of this paper to discuss these processes from the viewpoint of our action-based approach for production, perception, and acquisition of communicative actions in face-to-face-situations [11], as well as to review literature which identifies the brain regions involved in these processes.

2 The Action-Based Approach for Face-to-Face Communication

The main proposition of our action-based approach is that speech, facial, as well as hand-arm actions can be described by one comprehensive cognitive and sensorimotor concept. On the one hand the intention or meaning of an utterance, its co-speech facial expression and its co-speech hand-arm gestures can be specified in form of a hypermodal cognitive pattern [11]. This cognitive pattern can be seen as the starting point for production as well as the end point of perception. On the other hand during the production process the complex temporarily overlapping sequence of movement actions controlling the speech articulators (lips, tongue, velum glottis etc.), the facial articulators (eye brows, eye lids, cheeks, mouth corners etc.) as well as the articulators or the hand-arm system (position of lower arms and hand, orientation of palm by wrist, and form of hand and fingers; see also Figure 1) needs to be specified. This is done at the motor plan level ([11] and see Fig. 1 as an example for a motor plan).

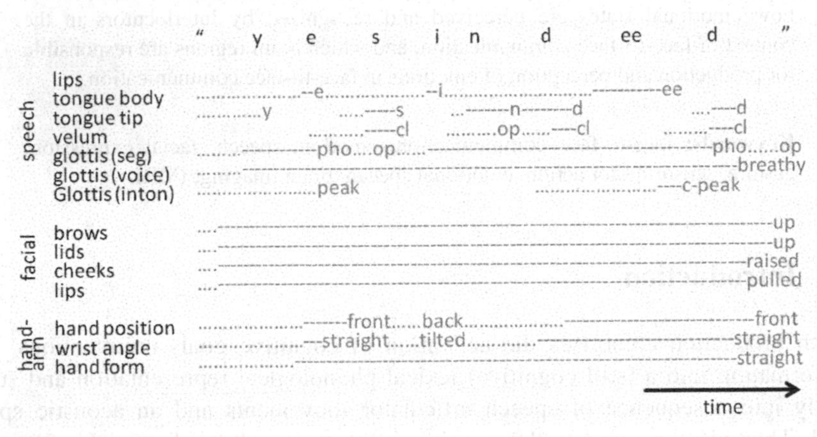

Fig. 1. Example of a motor plan for temporally coordinated speech, facial, and hand-arm movement actions (utterance: "yes indeed"). Movement phases are marked by "..." and hold phases are marked by "---" for each action. Tongue body and lip actions are labeled orthographically by the resulting speech sound. In addition, temporally overlapping velopharyngeal opening (op) and closing (cl) actions and glottal opening (op) and phonation (pho) actions are needed in order to produce proper speech sounds (cf. [18]). In addition, voice quality actions (here for breathy voice) and glottal actions for production of specific intonation contours (peak and central peak actions) are specified. Only one co-speech facial expression is produced here ("happy"&"interest" in combination, cf. [16]), and two co-speech hand-arm actions (front, straight) are produced in this example in order to underline the stressed syllables. The third hand-arm action (back, tilted) can be interpreted as rest position of the hand-arm system in this communication situation. Only actions of left hand-arm system are displayed here because right hand-arm system was at rest in this utterance.

The hierarchical structure of actions is very steep in the case of speech. Here, each utterance can be separated into one or more phrases, each phrase in one or more stress groups (each containing one stressed syllable), and each stress group in one or more syllables [12]. Moreover each syllable comprises a set of temporally overlapping movement actions of speech articulators (Fig. 1). In the case of co-speech hand-arm actions each gesture at least comprises a preparation, stroke, and recovery phase [13, 14]. Moreover each of these phases comprise one or more temporally overlapping target-directed movement actions of hand-arm articulators controlling (arm and) hand position, wrist angle (leading to palm orientation), and hand shape (Fig. 1). In the case of facial actions the hierarchical structure is less steep. Here, each facial expression is directly realized by a set of temporally overlapping and mainly synchronously occurring facial action units as defined in FACS [15, 16] (see also Fig. 1: four temporally synchronous facial actions). These target-directed movement actions control the shape of parts of the facial structure like form/location of mouth, eye lids, eye brows etc. The movement actions activated at the motor plan level are always target- or goal-directed, where the goal is defined in the auditory domain in the case of speech and in the visual domain in the case of hand-arm gestures as well as in the case of facial expressions.

The realization of movement actions (of articulator movements) should not just be interpreted as the endpoint of the production process but as well as starting point or input perceptual vehicles for comprehending the intentions of a speaker in face-to-face communication [11]. It is an important aspect of our action-based approach, that each movement action can be subdivided in a movement and in and target or hold phase [17, 18]. From the viewpoint of action perception, the movement phase is as important as – or sometimes even more important than – the target or hold phase. Movement phases of speech actions produce (auditory perceivable) formant trajectories, which are important for perception of place of articulation of consonants as well as vowel quality. Movement phases often are important parts of stroke phase in case of co-speech hand-arm actions, and specific parameters of movement phase of facial actions help to identify, whether a facial expression is spontaneously produced or acted [19]. Moreover, long hold phases especially occur in co-speech facial movement actions, in specific phases of co-speech hand-arm gestures, as well as in voice quality actions, because the overall timing of face-to-face communication actions are dominated by speech movement actions, defining the segmental and syllabic structure of an utterance (cf. [17] for a vice-versa situation in sign language production).

3 Emotions Modulate Behavior: The Functional View

Internal and external aspects of emotion can be separated [20]. While the internal aspect of emotion is important for organization of behavior like action selection, attention, learning etc., the external aspect of emotion focuses on production and perception of emotional expressions which is an important aspect in communication and social coordination [ibid., p. 554]. A comprehensive model for emergence of emotions and how emotions influence behavior and perception is given in [21] and summarized in Fig. 2. In this model, the drive system is central. Drives establish the top-level goals of each individual and need to be satisfied by specific behavior or specific actions of the individual. Thus, the intensity of a drive increases again after satisfaction (e.g. degree of hunger, degree of fatigue, degree of loneliness etc.). The intensity of drives on the one hand organizes behavior (e.g. start to eat or drink, try to rest or sleep, start to communicate etc.) and on the other hand indirectly influences the emotional system of the individual. Here it is important to state, that intensity of drives change more slowly, while emotive responses work on a faster time scale (ibid., p. 129). An emotional reaction can be due to a recent external event occurring in the environment of the individual. For example a specific reaction of a communication partner could make the individual happy. Thus, the emergence of emotions mainly results from high-level processing of perceptual input. A set of specific releasers at a higher level of the perception system is postulated in [21], which on the one hand is capable of releasing specific behaviors, e.g. if a goal is not satisfied or if a stimulus is not desired, but which on the other hand is capable of influencing affective and thus emotional states. The emotion system itself can be subdivided in three stages, i.e. affective appraisal, elicitation and activation of emotions (ibid.). Affective states are mainly driven by the perceptual system and can be defined by specifying values on three scales (i.e. arousal, valence, and stance). Thus, a stimulus may be very arousing (positive arousal), but at the same time very unfavorable (negative valence) and therefore is refused or avoided (negative stance). Each affective state is associated with a specific type of emotion (e.g. "anger" in the case of the example given above; see ibid., p. 135 for more examples) and an emotion is elicited, if the activation of an affective state exceeds specific threshold values. The current emotional state now directly influences the cognitive behavioral system (e.g. selection of behavioral goals) as well as the sensorimotor system (i.e. the explicit temporal motor planning and execution for the actions, which are already defined at the cognitive level).

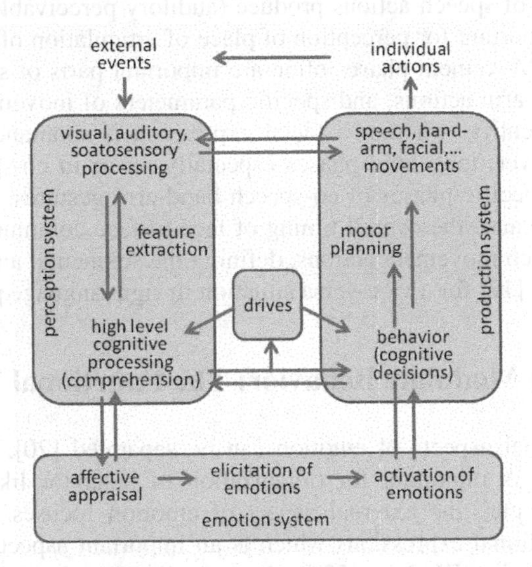

Fig. 2. Overview on interplay of emotion system, production system, perception system, and drives, following [21]. Here, the behavior system (after [21]) is included as cognitive part of production system. Lower levels of the production system comprise motor planning and execution; lower levels of perception end at the feature extraction level (cf. [11]).

Different sets of emotions are proposed in literature. A minimal set of six basic emotions is widely accepted: "anger", "disgust", "fear", "happiness", "sadness", and "surprise" [22]. But especially in the case of face-to-face communication including speech, beside "neutral", often two more emotional states are cited: "boredom" and "stress" (e.g. [8]). These additional states in part might be blends of basic emotions and as well might be biased by cultural influences. Even more emotional categories can be defined, if emotions are ordered with respect to the 3-dimensional space defined by the dimensions arousal, valence, and stance (see [21], p. 135).

4 Neuroanatomical Correlates

It is widely accepted that the limbic system, especially the amygdala plays a crucial role in learning and (re-)activation of emotions [23]. Input as well as output pathways to and from the amygdala involve the brainstem and connect this "emotion processing center" within the limbic system with cortical and subcortical processing centers for sensory input as well as with cortical processing centers for context evaluation as well as for generation of specific emotionally motivated or emotionally modulated behavior. Thus, in addition cortical cognitive centers are involved in processing of emotions [24, 25]. In accordance, recent studies assume a complex network comprising the amygdala and its synaptic connections with medial prefrontal cortex for emotion regulation [26].

A complex cerebral network is postulated for processing emotionally loaded speech. This network comprises temporal regions such as the posterior and middle part of the superior temporal cortex for acoustic feature extraction, frontal structures including the inferior frontal cortex for emotion identification, as well as parts of the limbic system such as the amygdala [27]. Here, the neural associations towards the limbic system are mainly inhibitory, because the evaluation process of externally perceived emotions should not directly be influenced by the current emotional state of the subject. Emotional reactions are assumed to occur after this evaluation process (i.e. after evaluation of external voice signals produced by interlocutors). Comparable networks are postulated for processing of emotionally loaded facial expressions. Here, beside additional activation patterns within the occipital cortex for a basic perceptual processing of the incoming visual signals, the network comprises the amygdala, as well as frontal and temporal cortical areas [25-28].

Because of its experimental complexity only few studies are known, which investigate brain activities during speech production within face-to-face communication. Here – in contrast to brain imaging studies investigating the processing of emotionally loaded auditory speech or visual facial input (see above) – not the process of emotion evaluation, conveyed within the external signals (emotions of communication partner) is investigated, but the emotional activation of the subject involved in a free conversation with an interlocutor is measured [29]. But interestingly, comparable brain regions are involved in activation of emotions as are known for emotion evaluation. Brain activations of subjects in an emotionally loaded conversation situation appear in frontal cortex (BA 45 and BA 47) as well as within the limbic system. These results are in accordance with the fact that perception and production of speech prosody (i.e. the aspect of speech which conveys most information concerning the emotional state of a speaker) share common neural networks [30]. Here, common network regions (i.e. regions of overlap of activity for producing and/or perceiving speech prosody) were found especially in left inferior frontal gyrus.

5 Modulation of Communicative Actions by Emotion: Towards a Transcription System for All Three Domains

From the idea of defining a unified motor plan structure for movement actions in all three domains of face-to-face communication (speech, facial expressions, hand-arm gestures; see above) we aim at describing face-to-face communication behavior in a form as is already displayed in Fig. 1. Movement and hold phases of facial and hand-arm actions can be identified from video recordings. For identifying speech movement actions we use an articulatory speech resynthesis procedure as is described in [31]. From a first transcription of parts of the eNTERFACE05 audio-visual Emotion Database [32] (speech and co-facial actions) and of the Bielefeld Speech and Gesture Alignment corpus (SaGA) [33] (speech, co-speech facial actions, and co-speech hand-arm actions), we got results which are in accordance with literature (cf. Introduction of this paper and [6-10]). Two points seem to be very obvious using our transcription system (Fig. 1): On the one hand, the type of facial expression as well as the type of voice quality (defined at cognitive levels in our model for each utterance) seems to be selected directly with respect to the emotional state of a speaker and seem to hold over a whole utterance or whole turn of one speaker in face-to-face communication scenarios. On the other hand, speech actions as well as co-speech hand-arm movement actions are selected with respect to the lexical content of the utterance. In the case of these actions, the emotional state of the speaker merely modifies quantitative parameters of these movement actions, e.g. duration, amplitude, and degree of realization of these movement actions, as well as degree of temporal overlap with adjacent movement actions.

6 Conclusions and Future Work

The action-based approach has been developed for describing movement actions in face-to-face communication in a comparable way for speech, facial expressions, and hand-arm gestures. This approach allows a transcription of movement actions occurring in all three domains of face-to-face communication in a similar way. This transcription system is qualitative (i.e. denominating each type of action) as well as quantitative (i.e. annotation of beginning and end of movement phase and hold phase for each action). Thus, this approach allows a qualitative and quantitative evaluation of face-to-face communication e.g. in order to identify differences in cognitive goals and sensorimotor realizations of actions in face-to-face communication. In future work a broader corpus-based analysis of face-to-face communicative actions in different emotional states is planned by using our qualitative-quantitative transcription system.

Acknowledgements. This work was partially supported by the National Natural Science Foundation of China (Grant No. 61233009).

References

1. Levelt, W.J.M., Roelofs, A., Meyer, A.S.: A theory of lexical access in speech production. Behavioral and Brain Sciences 22, 1–75 (1999)
2. Levelt, W.J.M.: Models of word production. Trends in Cognitive Sciences 3, 223–232 (1999)
3. Guenther, F.H.: Cortical interactions underlying the production of speech sounds. Journal of Communication Disorders 39, 350–365 (2006)
4. Guenther, F.H., Ghosh, S.S., Tourville, J.A.: Neural modeling and imaging of the cortical interactions underlying syllable production. Brain and Language 96, 280–301 (2006)
5. Kröger, B.J., Kannampuzha, J., Neuschaefer-Rube, C.: Towards a neurocomputational model of speech production and perception. Speech Communication 51, 793–809 (2009)
6. Halberstadt, J.B., Niedenthal, P.M., Kushner, J.: Resolution of lexical ambiguity by emotional state. Psychological Science 6, 278–282 (1995)
7. Bänziger, T., Scherer, K.R.: The role of intonation on emotional expressions. Speech Communication 46, 252–267 (2005)
8. Scherer, K.R.: Vocal communication of emotion: A review of research paradigms. Speech Communication 40, 227–256 (2003)
9. Ekman, P., Oster, H.: Facial expressions of emotion. Annual Review of Psychology 30, 527–554 (1979)
10. Castellano, G., Villalba, S.D., Camurri, A.: Recognising human emotions from body movement and gesture dynamics. In: Paiva, A.C.R., Prada, R., Picard, R.W. (eds.) ACII 2007. LNCS (LNAI), vol. 4738, pp. 71–82. Springer, Heidelberg (2007)
11. Kröger, B.J., Kopp, S., Lowit, A.: A model for production, perception, and acquisition of actions in face-to-face communication. Cognitive Processing 11, 187–205 (2010)
12. Kröger, B.J., Birkholz, P., Kaufmann, E., Neuschaefer-Rube, C.: Beyond vocal tract actions: speech prosody and co-verbal gesturing in face-to-face communication. In: Kröger, B.J., Birkholz, P. (eds.) Studientexte zur Sprachkommunikation: Elektronische Sprachsignalverarbeitung 2011, pp. 195–204. TUDpress, Dresden (2011)
13. Kendon, A.: Gesture: Visible Action as Utterance. Cambridge University Press, New York (2004)
14. Kopp, S., Wachsmuth, I.: Synthesizing multimodal utterances for conversational agents. Journal of Computer Animation and Virtual Worlds 15, 39–51 (2004)
15. Ekman, P., Friesen, W.V.: Facial Action Coding System. Consulting Psychologists Press, Palo Alto (1978)
16. Cohn, J.F., Ambadar, Z., Ekman, P.: Observer-based measurement of facial expression with the facial action coding system. In: Coan, J.A., Allen, J.J.B. (eds.) Handbook of Emotion Elicitation and Assessment, pp. 203–221. Oxford University Press US, New York (2007)
17. Kröger, B.J., Birkholz, P., Kannampuzha, J., Kaufmann, E., Mittelberg, I.: Movements and holds in fluent sentence production of American Sign Language: The action-based approach. Cognitive Computation 3, 449–465 (2011)
18. Kröger, B.J., Birkholz, P.: A gesture-based concept for speech movement control in articulatory speech synthesis. In: Esposito, A., Faundez-Zanuy, M., Keller, E., Marinaro, M. (eds.) COST Action 2102. LNCS (LNAI), vol. 4775, pp. 174–189. Springer, Heidelberg (2007)

19. Schmidt, K.L., Ambadar, Z., Cohn, J.F., Reed, L.I.: Movement differences between deliberate and spontaneous facial expressions: Zygomaticus major action in smiling. Journal of Nonverbal Behavior 30, 37–52 (2006)
20. Arbib, M.A., Fellous, J.M.: Emotions: from brain to robot. Trends in Cognitive Sciences 8, 554–561 (2004)
21. Breazeal, C.: Emotion and sociable humanoid robots. International Journal of Human-Computer Studies 59, 119–155 (2003)
22. Ekman, P.: An argument for basic emotions. Cognition and Emotion 6, 169–200 (1992)
23. LeDoux, J.E.: Emotion circuits in the brain. Annual Reviews of Neuroscience 23, 155–184 (2000)
24. Lazarus, R.S.: Cognition and motivation in emotion. American Psychologist 46, 352–367 (1991)
25. Pessoa, L., Adolphs, R.: Emotion processing and the amygdala: from a 'low road' to 'many roads' of evaluating biological significance. Nature Reviews Neuroscience 11, 773–782 (2010)
26. Whalen, P.J., Raila, H., Bennett, R., Mattek, A., Brown, A., Taylor, J., van Tieghem, M., Tanner, A., Miner, M., Palme, A.: Neuroscience and facial expressions of emotion: the role of amygdala-prefrontal interactions. Emotion Review 5, 78–83 (2013)
27. Brück, C., Kreifelts, B., Ethofer, T., Wildgruber, D.: Emotional voices: the tone of (true) feelings. In: Armony, J., Vuilleumier, P. (eds.) The Cambridge Handbook of Human Affective Neuroscience, pp. 256–285. Cambridge University Press, New York (2013)
28. Kesler-West, M.L., Andersen, A.H., Smith, C.D., Avison, M.J., Davis, C.E., Kryscio, R.J., Blonder, L.X.: Neural substrates of facial emotion processing using fMRI. Cognitive Brain Research 11, 213–226 (2001)
29. Mitsuyoshi, S., Monnma, F., Tanaka, Y., Minami, T., Kato, M., Murata, T.: Identifying neural components of emotion in free conversation with fMRI. In: Defense Science Research Conference and Expo, Singapore, pp. 1–4 (2011), doi:10.1109/DSR.2011.6026845
30. Aziz-Zadeh, L., Sheng, T., Gheytanchi, A.: Common premotor regions for the perception and production of prosody and correlations with empathy and prosodic ability. PLoS ONE 5, e8759, 1-7 (2010), doi:10.1371/journal.pone.0008759
31. Bauer, D., Kannampuzha, J., Kröger, B.J.: Articulatory Speech Re-Synthesis: Profiting from natural acoustic speech data. In: Esposito, A., Vích, R. (eds.) Cross-Modal Analysis of Speech, Gestures, Gaze and Facial Expressions. LNCS, vol. 5641, pp. 344–355. Springer, Heidelberg (2009)
32. Martin, O., Kotsia, I., Macq, B., Pitas, I.: The eNTERFACE05 Audio-Visual Emotion Database. In: First IEEE Workshop on Multimedia Database Management, Atlanta, USA (2006), doi:10.1109/ICDEW.2006.145
33. Lücking, A., Bergmann, K., Hahn, F., Kopp, S., Rieser, H.: Data-based analysis of speech and gesture: the Bielefeld Speech and Gesture Alignment corpus (SaGA) and its applications. Journal on Multimodal User Interfaces (2012), doi:10.1007/s12193-012-0106-8

Investigating the Form-Function-Relation of the Discourse Particle "hm" in a Naturalistic Human-Computer Interaction

Ingo Siegert, Dmytro Prylipko, Kim Hartmann,
Ronald Böck, and Andreas Wendemuth

Cognitive Systems Group, Otto von Guericke University Magdeburg, Germany
ingo.siegert@ovgu.de

Abstract. For a successful speech-controlled human-computer interaction (HCI) the pure textual information as well as individual skills, preferences, and affective states of the user have to be known. However, verbal human interaction consists of several information layers. Apart from pure textual information, further details regarding the speaker's feelings, believes, and social relations are transmitted. The additional information is encoded through acoustics. Especially, the intonation reveals details about the speakers communicative relation and their attitude towards the ongoing dialogue.

Since the intonation is influenced by semantic and grammatical information, it is advisable to investigate the intonation of so-called discourse particles (DPs) as "hm" or "uhm". They cannot be inflected but can be emphasised. DPs have the same intonation curves (pitch-contours) as whole sentences and thus may indicate the same functional meanings. For German language J. E. Schmidt empirically discovered seven types of form-function-concurrences on the isolated DP "hm".

To determine the function within the dialogue of the DPs, methods are needed that preserve pitch-contours and are feasible to assign defined form-prototypes. Furthermore, it must be investigated which pitch-contours occur in naturalistic HCI and whether these contours are congruent with the findings by linguists.

In this paper we present first results on the extraction and correlation of the DP "hm". We investigate the different form-function-relations in the naturalistic LAST MINUTE corpus and determine expectable form-function relations in naturalistic HCI in general.

Keywords: Prosodic Analysis, Companion Systems, Human-Computer Interaction, Discourse Particle, Pitch Classification.

1 Introduction

To obtain a more human-like interaction with technical systems, those have to be adaptable to the users' individual skills, preferences, and current emotional state [16]. In human-human interaction (HHI) the behaviour of the speaker is

S. Bassis et al. (eds.), *Recent Advances of Neural Network Models and Applications*, Smart Innovation, Systems and Technologies 26,
DOI: 10.1007/978-3-319-04129-2_39,

characterised by semantic and prosodic cues as short feedback signals. These signals minimally communicate certain dialogue functions such as attention, understanding, confirmation, or other attitudinal reactions [1]. Thus, these signals play an important role in the progress and coordination of interaction. They allow the dialogue partners to inform each other of their behavioural or affective state without interrupting the ongoing dialogue [1].

To enable a naturalistic HCI, it is necessary to capture as many human abilities as possible. Lot of progress have been made in the area of disposition detection from speech, facial expression, and gesture, cf. [17]. For scripted speech corpora, linguistic features are normally not suitable, whereas for realistic data linguistic features gain considerable importance [13]. For this kind of data also feedback signals promise remedy.

2 Discourse Particles

During HHI several semantic and prosodic cues are exchanged among the interaction partners and used to signalize the progress of the dialogue [1]. Especially the intonation of utterances transmits the communicative relation of the speakers and also their attitude towards the current dialogue.

As known from literature, specific monosyllabis (e.g. discourse particles) have the same intonation curves as whole sentences and also cover a functional concordance [7]. These DPs like "hm" or "uhm" cannot be inflected but can be emphasised and are occurring at crucial communicative points. The DP "hm" is seen as a "neutral-consonant" whereas "uhm" can be seen as a "neutral-vocal" [12]. The intonation curves of these particles are largely free of lexical and grammatical influences.

Schmidt presented an empirical study where he could determine seven form-function relations of the DP "hm", see Table 1 and [12]. The duration of some classes is given, where necessary.

An investigation for English DPs and the features syllabification, duration, loudness, pitch slope, and pitch-contour is presented in [15]. In [2] the prosody of American English feedback cues is investigated and several DPs are annotated using eleven categories.

As these experiments are made on isolated recorded particles, it first have to been checked if these prototypical form-function relation can be found within a HHI. One of these investigations is presented in [5]. The authors examined the data of four different conversational styles: talk-show, interview, theme-related talk, and an informal discussion, with an overall length of 179min. They extracted 392 particles from the material and could confirm the form-function relation. The occurrence frequency of specific backchannel words for American English was investigated in [2].

Since the considered DPs have a specific function within the conversation one may assume that they do not occurring during a HCI, as the use of these particles requires both conversational partners to understand the meaning. One study investigating the hypothesis that DPs are less frequent within HCI is presented

Table 1. Form-function relation of DP "hm" according to [12], the terms are translated into appropiate Englisch ones. DP-D can be seen as a combination of DP-R and DP-F

Name	idealised pitch-contour	Description
DP-A		attention
DP-T		thinking
DP-F		finalisation signal
DP-C		confirmation
DP-D		decline
DP-P		positive assessment
DP-R		request to respond

in [3]. The authors conclude that the number of partner-oriented signals is decreasing while the number of signals indicating a talk-organising, task-oriented, or expressive function is increasing. But, they are lacking an automatic method to assign the DPs characteristic to a specific meaning.

3 Dataset

The conducted study of this paper utilizes the LAST MINUTE corpus [11], which contains multimodal recordings of 133 German speaking subjects in a so called Wizard-of-Oz experiment. The setup revolves around an imaginary journey to the unknown place "Waiuku", which the subjects have won. Each experiment takes about 30min. Using voice commands, the subjects have to prepare the journey, equip the baggage, and select clothing. Most of the experiments are transliterated, enabling the automatic extraction of speaker utterances. More details on the design of the corpus can be found in [11]. We opt for this data set, because this corpus represents a naturalistic HCI [14]. The corpus was designed to have an equal distribution of age and gender of the subjects. The subjects are divided in two subgroups according to their ages. The young group ranges from 18-28 years, the elder group consists of subjects over 60 years.

The experiment is divided into two phases, a personalisation and a problem solving phase. During personalisation, the subject gets familiar in communicating with a machine. The subjects are guided to talk freely. During problem solving the conversation is more task focused and the subjects talk in a commando-like style. Additionally, interviews have been conducted with roughly half of the subjects. The result of the interviews was the fact that the experimental situation is experienced as a hybrid between human-human and human-computer interaction [9]. This supports the claim that the system has companion-like abilities.

We used a subset of 56 subjects with a total duration of 25h, where a professional neckband headmic has been utilized. The distribution of age and gender is as follows: young males and young females each 16 subjects, old males and old females both 12 subjects. As the DPs are transliterated we conducted an automatic alignment with a manual correction phase to extract them. This results in 274 DPs of "hm" in our subset.

4 Methods

4.1 Automatic Pitch Extraction and Form-Correlation

As we have seen in Section 2, DPs fulfil an important function within the conversation for both HHI and HCI. The feature *pitch-contour* together with *duration* allows to derive a form-function relation. Therefore, a reliable method to extract and classify the pitch-contour should be developed. To extract the pitch, we rely on commonly used methods, presented in detail in [4,10]. The extracted DPs are windowed using a Hamming window with a width of 30ms and a stepsize of 5ms. Then, we calculated the short-time energy, this is later used, to perform the voiced/unvoiced decision. After windowing, we applied a low-pass filter with a pass-band frequency of 900Hz and center-clipping. Autocorrelation method is used afterwards, to extract the pitch values for each frame. Having extracted the pitch for all windows of one utterance, a smoothing using a median filter utilizing 5 values is applied to suppress outliers.

As we are only interested in the coarse of these contours, we further do a normalization (subtracting the mean) and stretch all functions, to equalise the signal lengths. To extract the pitch-contour, we consider two types: linear types (DP-T, DP-F, DP-R, and DP-D) and polynomial ones (DP-A, DP-C, DP-P), see Table 1. We distinguish both types before extracting the pitch-form. We determined the linearity of the signal by calculating the correlation by linear regression. If the coefficient is grater than 0.95, we assume a linear formtype and store the resulting linear regression line. Otherwise, we use a 3rd order polynomial fit to obtain the contour. Afterwards, the extracted pitch-contours are compared with pre-defined prototypes using a correlation, to specify the form relation.

4.2 Function Assessment of the Discourse Particles

To get an assessment for the functional use of the DPs, we utilized a manual labelling with 10 labellers. The implementation corresponded to the usual scientific standards with test-instructions and explained examples. Furthermore, the labellers listened to the surrounding context in order to properly assess the meaning. The labellers were given a choice of form function types as labels and have to assign the particle to either one category according to Schmidt, "other" together with a replacement statement, or "no hm" for the case where the automatic extraction fails.

5 Results

5.1 Phenomenological Inspection

Within our subset of 56 subjects, only 40 ones uttered the DP "hm", with a mean of 6.83 particles per conversation and a standard deviation of 9.67. For one experiment, we could observe a use 50 particles, which is the maximum. The occurrence of DPs is not equally distributed among the four age-gender groups, as shown in Fig. 1. It can be seen that the use of "hm" is more frequent for female than for male subjects for both age groups. Whereas the subjects's age does not influence the frequency of "hm".

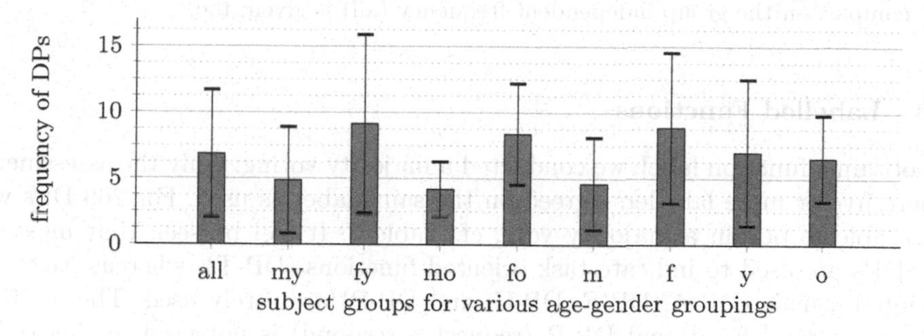

Fig. 1. Mean and standard deviation of the frequency for the DP "hm" regarding different subject groups in the case of gender (male, female) and age (young and old). For comparison the group independent frequency (all) is given, too.

Considering the distribution of DPs over the two experimental parts it can be stated that the frequency of DPs during the personalisation phase is approx 1.5 to 2 times lower than for the problem solving phase, cf. Fig. 2. Since, the conversation style is much more restricted in the second phase, we did not expect this result. Furthermore it can be seen that the use of "hm" is more frequent for female and elderly subjects for the second phase, which is identical to our previous inspection.

5.2 Extracted Formtypes

By using the extracted features pitch-contour and duration (cf. Section 4) and applying a simple correlation with prototypes, we could extract the formtypes defined by [12]. As our particles are collected within an interaction and not recorded separately, as done by Schmidt, we observed additional formtypes. These types are occuring when the subject utters a DP which directly merges to another word. This results in a harmonisation of the pitch-contour towards the succeeding word. Therefore, we split the pitch-contour by 2/3 of the length. To distinguish the formtypes we consider the first part, and consult the latter part only to distinguish DP-A and DP-T. This approach enables us, to identify about 71% of all occurring "hm" correctly.

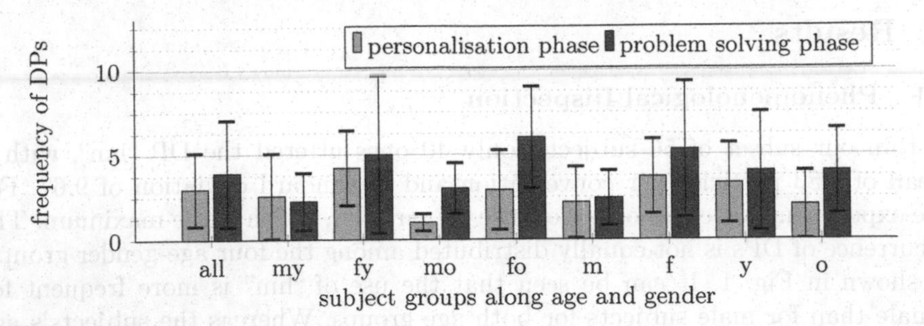

Fig. 2. Mean and standard deviation of the frequency for the DP "hm" for both parts of the experiment regarding different subject groups (male, female, young and old). For comparison the group independent frequency (all) is given, too.

5.3 Labelled Functions

To obtain a function label, we conducted a majority voting: Only the assessment where five or more labellers agreed on the same label, is used. For 269 DPs we were able to obtain a majority vote, cf. Table 2. It can be seen that most of the DPs are used to indicate task oriented functions (DP-T), whereas partner-oriented signals (DP-A, DP-F, DP-C, and DP-P) are rarely used. The in HHI commonly used functional DP-R (request a respond) is not used in this HCI. Presumable, the subjects do not expect the system to recognize these partner-oriented functions properly. Although, the labellers had the opportunity to assess meanings in addition to the ones given by Schmidt, this was used only for a few particles without leading to a majority. This additional labels always occurred in connection with laughter.

Table 2. Number and resulting label for all considered DPs. The categories are according to Table 1, additionally used labels: DPN (no "hm") and NONE (no majority).

Label	DP-A	DP-T	DP-F	DP-C	DP-P	DPN	NONE
# Items	8	211	6	39	3	2	5

Further, we calculated Krippendorff's alpha, to determine the reliability of our labelling process, cf. [6]. We obtained a value of 0.55, which indicates a moderate or appropriate reliability according to [8].

6 Form-Function Relation

Finally, we compared the extracted formtypes from Section 5.2 with the gathered labels from Section 5.3. The results are given in Fig. 3, where we plotted the number of classified formtypes compared with the number of "correctly" labelled formtypes.

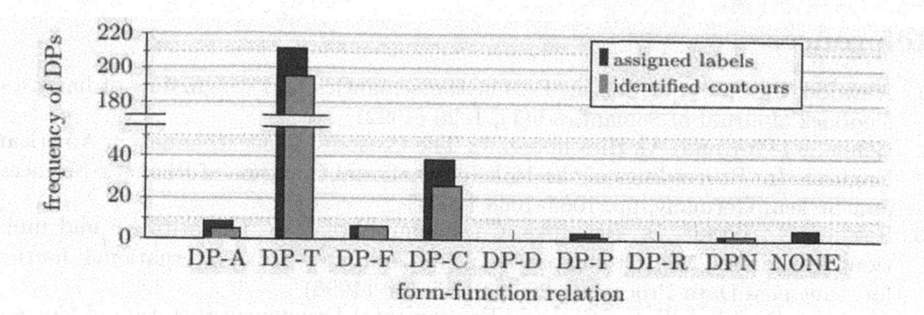

Fig. 3. Frequency of labelled function of the DP "hm" for each functional (label) and the number of identified related formtype (pitch-contour)

It can be seen that for most of the classified pitch-contours the matching function, as stated by Schmidt for HHI, is labelled. This indicates the validity of the form-function relation for HCI. Furthermore, in the case of no majority DP label, we could identify a pitch-contour of either DP-A or DP-T. In this case the ratings varied between DP-A, DP-T and DP-C. The functionals DP-D and DP-R could neither be found as pitch-contour nor were assessed. The non-occurrence of DP-D may be due to the experimental design itself, as no decline by the subjects is expected in this experiment. The lack of DP-R indicates, that the subjects does not totally assign the system human skills.

7 Conclusion

This paper presents results on the usage of different functional meanings of DPs within HCI and their identification. Starting with an overview of the form-function relation of DPs within HHI, we investigated the occurrence of the DP "hm" for HCI. Therefore, we rely on work presented in [12]. We utilized a naturalistic interaction and demonstrated that the DP "hm" is occurring frequently. Performing a labelling process, we could reliably identify the functional meaning. Our results are consistent with the findings of [3], determining an increasing use of task-oriented signals. The pitch extraction and correlation with defined prototypes reveals that the form-function relation is also valid for HCI. Furthermore, it can be assumed that the functionals determined by Schmidt are sufficient, to distinguish the meaning of DPs within HCI.

In our further research activities, we will develop a robust automatic identification of the particles' formtype. An extension of the presented method to other languages or DPs, such as "uhm" will be developed.

Acknowledgement. The work presented in this paper was done within the Transregional Collaborative Research Centre SFB/TRR 62 "Companion-Technology for Cognitive Technical Systems" funded by the German Research Foundation (DFG).

References

1. Allwood, J., Nivre, J., Ahlsn, E.: On the semantics and pragmatics of linguistic feedback. Journal of Semantics 9(1), 1–26 (1992)
2. Benus, S., Gravana, A., Hirschberg, J.: The Prosody of Backchannels in American Englisch. In: Proceedings of the 16th International Congress of Phonetic Sciences, Saarbrcken, Germany, pp. 1065–1068 (2007)
3. Fischer, K., Wrede, B., Brindpke, C., Johanntokrax, M.: Quantitative und funktionale Analysen von Diskurspartikeln im Computer Talk. International Journal for Language Data Processing 20(1-2), 85–100 (1996)
4. Gerhard, D.: Pitch Extraction and Fundamental Frequency: History and Current Techniques. Tech. Rep. TR-CS 2003-06, Regina, Saskatchewan, Canada (2003)
5. Kehrein, R., Rabanus, S.: Ein Modell zur funktionalen Beschreibung von Diskurspartikeln. In: Neue Wege der Intonationsforschung. Germanistische Linguistik, vol. 157-158, pp. 33–50. Georg Olms Verlag, Hildesheim (2001)
6. Krippendorff, K.: Content Analysis: An Introduction to Its Methodology, 3rd edn. SAGE Publications, Thousand Oaks (2012)
7. Ladd, R.D.: Intonational Phonology. Studies in Linguistics, vol. 79. Cambridge University Press (1996)
8. Landis, J.R., Koch, G.G.: The measurement of observer agreement for categorical data. Biometrics 33, 159–174 (1977)
9. Lange, J., Frommer, J.: Subjektives Erleben und intentionale Einstellung in Interviews zur Nutzer-Companion-Interaktion (Subjective experience and intentional setting within intervies of User-Companion-Interaction). In: Informatik 2011: Informatik schafft Communities, Beitrge der 41. Jahrestagung der GI. Lecture Notes in Informatics, vol. 192, p. 240 (2011)
10. Rabiner, L., Cheng, M.J., Rosenberg, A.E., McGonegal, C.A.: A comparative performance study of several pitch detection algorithms. IEEE Trans. on ASSP 24, 399–417 (1976)
11. Rösner, D., Friesen, R., Otto, M., Lange, J., Haase, M., Frommer, J.: Intentionality in interacting with companion systems – an empirical approach. In: Jacko, J.A. (ed.) Human-Computer Interaction, Part III, HCII 2011. LNCS, vol. 6763, pp. 593–602. Springer, Heidelberg (2011)
12. Schmidt, J.E.: Bausteine der Intonation. In: Neue Wege der Intonationsforschung, Germanistische Linguistik, vol. 157-158, pp. 9–32. Georg Olms Verlag, Hildesheim (2001)
13. Schuller, B., Batliner, A., Steidl, S., Seppi, D.: Recognising realistic emotions and affect in speech: State of the art and lessons learnt from the first challenge. Speech Commun. 53(9-10), 1062–1087 (2011)
14. Siegert, I., Böck, R., Wendemuth, A.: The influence of context knowledge for multimodal annotation on natural material. In: Proc. of the First Workshop on Multimodal Analyses enabling Artificial Agents in Human-Machine Interaction (MA3), Santa Cruz, USA (2012)
15. Ward, N.: Pragmatic functions of prosodic features in non-lexical utterances. In: Proceedings of Speech Prosody 2004, Nara, Japan, pp. 325–328 (2004)
16. Wendemuth, A., Biundo, S.: A Companion Technology for Cognitive Technical Systems. In: Esposito, A., Esposito, A.M., Vinciarelli, A., Hoffmann, R., Müller, V.C. (eds.) COST 2102. LNCS, vol. 7403, pp. 89–103. Springer, Heidelberg (2012)
17. Zeng, Z., Pantic, M., Roisman, G.I., Huang, T.S.: A Survey of Affect Recognition Methods: Audio, Visual, and Spontaneous Expressions. IEEE Trans. on Pattern Analysis and Machine Intelligence 31, 39–58 (2009)

Intended and Unintended Offence

Carl Vogel

Computational Linguistics Group,
Centre for Computing and Language Studies,
School of Computer Science and Statistics,
Trinity College Dublin,
Dublin 2, Ireland
vogel@tcd.ie

Abstract. This paper argues that politeness and impoliteness are integrally related to offence management. The outlines of a semantic theory of linguistic politeness are sketched. As a semantic theory, interfaces to both pragmatics and compositional syntax may be expected, but these are spelled out in companion papers.

Keywords: offence, disgust, impoliteness, politeness, facework.

1 Introduction

Politeness and impoliteness in language present some puzzles.[1] One is that forms of politeness exist at all. Another is in whether one can meaningfully approach linguistic politeness from the perspective of semantics, as opposed to pragmatic theory. Consider these questions with respect to some examples.

(1) Salt.
(2) Scalpel.

If all one knew about a context of interaction was that someone uttered (1) as a means to obtaining salt from a fellow-diner, one would likely consider the speaker rude. In contrast, representations in popular entertainment of medical surgery would yield an impression of the speaker of (2) in a comparably de-contextualized operating theatre that would not be specified for rudeness (except to the extent that one thinks of surgeons as rude, by default). The contrast between (1) and (2) demonstrates the role of context in determining attributions of rudeness: the default for a request for salt is use for one's own consumption, while the default for a request for a scalpel is use for primary benefit that is not one's own. A pragmatic politeness principle of selflessness may be stated as in (3).[2] This is a

[1] I take the term "politeness" to be characterized in a spectrum from positive values, through neutral values, to negative values. Negative values of politeness are also referred to as impoliteness. Henceforth, "positive politeness" and "polite" will be used to refer to those instances of politeness that are thought to have positive value.
[2] The principle is a strong default. There are conceivable situations in which putting oneself first is arguably better for the community than the alternative.

S. Bassis et al. (eds.), *Recent Advances of Neural Network Models and Applications*, Smart Innovation, Systems and Technologies 26,
DOI: 10.1007/978-3-319-04129-2_40,

principle with force beyond linguistic communication. Attributions of politeness, therefore, transcend language.

(3) a. Putting oneself before others is impolite.
b. Putting others before oneself is polite.

Examples (4) and (5) demonstrate that politeness is amenable to linguistic encoding, although the exact circumstances of context will determine whether the target audience for the utterances deem them polite or impolite (or speaker intent in the matter). In discussing linguistic politeness from the point of view of semantic theory, it is methodologically safe to put irony and sarcasm aside: any ironic use of an expression is parasitic upon its default interpretation.

(4) a. Salt, please.
b. Salt, idiot.
(5) a. Scalpel, please.
b. Scalpel, idiot.

It is relatively easy to find in the literature emphasis on the role of context in politeness evaluation [1]. However, there are others who also emphasize that forms of language associated with politeness expressions are open to semantic interpretation [8]. Moreover, researchers who appeal to the norms of language use in a community of practice as determining what counts as polite or impolite implicitly rely on straightforward interpretation decoding content and politeness levels of utterances within those sub-languages (e.g. [20]). The formulation of the semantics, as opposed to the pragmatics, of linguistic politeness has remained an open problem.

Linguistic politeness is a species of politeness behavior. Linguistic politeness is one dimension of language use that is quintessentially communicative. Even if one agrees that overwhelmingly dominant use of language is for thought, rather than for communication, one must concede that the main role of linguistic politeness is in communicative language use, thinking for speaking,[3] rather than in thinking for oneself. Notice that thinking for oneself does not necessarily involve anyone else, and therefore the principle in (3) arguably does not apply. One might think thoughts for oneself that encode negative politeness, but probably not positive politeness: while I can easily enough imagine having thoughts that encode negative politeness (6), I am extremely unlikely to entertain a reflexive thought that encodes positive politeness (7).

(6) Idiot, why did I let the toast burn?
(7) Carl, please enjoy this toast I've made.

One might wonder why linguistic politeness exists at all, given that it requires communicative effort, and since it has little value in private thought. It is noteworthy that in other dimensions of language use, a tendency towards reduction of effort is palpable. After topics and entities are introduced into discourse they

[3] See [17].

are typically referred to subsequently with definites with reduced descriptions and ultimately pronouns that exhibit phonological reduction (cite sources). Even in the context of such reductions, politeness terms persist (cite data). This essay argues a position on why (im)politeness forms are given the energy they are by interlocutors–the gist of this position is that the mechanisms of disgust and offence are strong and pervasive, and that (im)politeness behaviors arise as investments in mitigating disgust responses.

Past theories of (im)politeness of tended to focus on the pragmatics of politeness. Inspired by Goffman's analysis of human interactions in terms of "face" [11,12], the research programme of a large part of the literature was defined by a view of politeness as "facework" [6]. On this conception, two construals of the self influence the perception of actions as polite or not: positive face, self-image held in unison with others' image of self; negative face, freedom to act. An alternative view does not emphasize the self in isolation as much as through relation management, and in this conception, politeness is seen as "relational work" [16,5]. On either view, impoliteness is analyzed as the absence of politeness [4,7,8], though crucially noting that some environments elicit impoliteness behaviors as the norm for discourse (e.g. basic military training camps [4]). The view pursued here is not at odds with either of those conceptions, which themselves appear to be complementary rather than competing frameworks for analysis, but attempts on one hand to contribute to explanation the perception of acts as polite or impolite and on the other hand to specify the ontology required in order to deliver a formal semantics for linguistic politeness and impoliteness. An independent test of the efficacy of this proposal is the extent to which it addresses some of the puzzles of politeness and impoliteness, such as those described already, or additional puzzles, like the the linguistic energy consumed by such behaviors, or the interaction of familiarity on perception or production of the relevant behaviors.

One might try to explain the linguistic encoding of politeness by virtue of the transcendence of politeness over language. The argument would be that it is a palpable feature of communication (regardless of the medium), and as a real-world phenomenon, it merits linguistic manifestation. However, this begs the question of why politeness is a feature of human interaction. My argument is that politeness is a form of offence management: humans are polite and use polite expressions in order to avoid invoking disgust; speakers are impolite when they do not object to disgusting others. This view of politeness is independent of intent. The politeness value of any expression may vary for speaker, addressee or wider audiences, just as interpretation of language in general may yield distinct meanings depending on who is interpreting it. A theory of (im)politeness can unfold partly in the realm of semantics without claiming inherent offence or other sentiment in particular forms of language.

The remainder of this article sketches how it might be that politeness forms are anchored in offence management through generalization of instinctive human response to disgust. Then, a semantic theory of politeness is sketched within a framework for analyzing politeness. This counts as a theory, because alternative theories could be sketched within the framework and because empirical predictions are made. That is, while it is always possible to use a theory as a descriptive device for characterizing phenomena (and it is easy to find such uses of theory in the politeness literature – see for example, [4]), it is also interesting to know whether one can predict phenomena. This article does not test empirical predictions, but suggests how this could be done.

2 Linguistic (Im)politeness

A theory of linguistic politeness would be trivial if there is no element of politeness or impoliteness behaviors that are elucidated through linguistic analysis. One might make a claim that there cannot be a non-trivial linguistic theory of politeness on the basis of the differences between the language that is invoked in discussion of (im)politeness and language in general. However, first it is necessary to express a stance on the efficacy of language in general as a tool for communication. Natural language is manifestly effective for human thought, but it is less obviously fit for purpose for communication. It is possible that close-up scrutiny of the reliability of (im)politeness signals differentiates them from "the rest of language" on the basis of an understanding of language in general that has shortcomings. Some research on language appears to take it as axiomatic that the role of language is communication; however, others take a line also argued by Chomsky that the primary role of human language is thinking. Language for communication must address issues of ambiguity and synonymy that confuse communication among interlocutors in a way that does not happen in interior monologue. Aligned with the axiom of that the purpose of language is communication is another axiom of intersubjective conformity which can be seen as presuming that communicators generally understand each other and that communicators signify the same ideas with a common language [21]. A skeptical analyst may note that communicators cannot ever know if they really understand each other, and moreover that communicators cannot even know if they use language the same way. Quine's "gavagai" problem is as pervasive within language communities as across them. An effective pragmatic response to this skepticism is the presumption that mutual understanding is achieved unless events make miscommunication manifest: interlocutors act as if they understand each other unless they have clear evidence that they have misunderstood. However, the non-linguistic elements of existence – sharing the same sort of biology and ecological niches, along with the impact of all of that on what is typically taken to be pragmatics – have the effect that misunderstandings around the edges can be ignored, even if they are the norm rather than the exception. In fact, because interlocutors are typically able to intervene in the world themselves, it does not matter if the linguistic channel of communication has been unsuccessful, and on that basis, presume that language is very effective. Because of all that is shared outside language, it is easiest for interlocutors to more or less assume that they share world views. This effectively yields Saussurean communicators, who use language with the same meaning-form mappings for *interpretation* as for *production*. Indeed, in evolutionary simulation models, [13] showed that Saussureans "win" over imitators and calculators.[4] One may conclude that the Saussurean strategy is a very reliable one for communicating agents with shared biology and ecology.[5] While language in general appears to be used in a Saussurean manner, people are not manifestly Saussurean with respect to perceptions of offence, and consequently politeness or impoliteness. Miscommunication appears to be rife in linguistic politeness, and people tend to perceive the gaps in this form of communication in way that they do not notice in the use of language in general. A behavior may offend the speaker when produced by others, but the speaker may not intend nor notice offence in others when the speaker produces the same behavior – it is a double dissociation.

3 Disgust

To understand the relation argued between (im)politeness behaviors and offence management, it is helpful to consider a root of offence in disgust. The disgust response in humans has been argued to be a human universal [10], and a trait that separates humans from other animals [15]. The nature of the response can be clarified with respect to an example (see Figure 1). Firstly, notice that in spite of the descriptive terms of the example, which are meant to cater to the fact that although the disgust response appears universal, disgust triggers may not be,[6] most readers are unlikely to actually imagine what the actual picture in Fig. 1 would be for them on short reflection; it requires a longer focus. This is a manifestation of disgust avoidance: there is pressure to avoid even constructing mental images of the worst triggers. Evidence for any person that they did not picture the most upsetting image possible for them is that the main symptoms of the disgust reaction are not present. They are reflexes of expulsion, associated with ingested toxins, and avoidance, associated with disease and parasites. The canonical manifestation is nausea or nauseum-sympathetic oral and nasal musculature reflexes.

An insightful description of disgust responses and a theory of their evolution is provided by [15]. The main features are that they are involuntary. Disgust triggers are, perhaps uniquely, available to "one-shot" learning: one never forgets the sort of food item it was that one holds responsible for a single episode

[4] It is unfortunate that these particular models bootstrap from assumptions of telepathy: that speakers can sample interpretation and production mappings used in the community in order to influence their own. This is a coding of the axiom of intersubjective conformity: even if some agents have the possibility of using the meaning/form mappings in different ways to other agents, they know what meaning/form mappings the other agents have. I conjecture, though, that the same result will emerge in simulation systems with more pessimistic starting assumptions (cf. [23,24,2,3]).

[5] It may therefore have potential to be quite poor in human-robot communication, where such sharing does provide a body of constraints that offer independent means of deducing likely intent behind utterances.

[6] This is in close relation to fear; recall a passage from Orwell's *1984*: "'The worst thing in the world,' said O'Brien, 'varies from individual to individual."' [18]

This is a picture of the most disgusting thing you can imagine.

Fig. 1. A superlatively upsetting image

of food poisoning, and this can create a permanent aversion.[7] The response to disgust triggers is as if they have the power to contaminate the environment. Upon encounter with a disgusting substance, one feels compelled to seek cleansing and purification. Items that have been in contact with the disgusting item may themselves be discarded rather than treated, through the sensation that the item is contaminated. The contamination easily associates with related objects, and it is through the capacity for transfer of disgust through contamination that the response generalizes from biological triggers (rotting life forms, toxins, bodily excretions) to associates whose availability depends on situations of encounter that yield the possibility for culture-specificity of triggers. This is because people who share a social and geographical niche are more likely to have experiences in common, including both situations that invoke disgust and accidental properties of those situations to which the disgust response generalizes. It is through the propensity for situations of disgust to be perceived as contaminating associates, contaminating everything from inanimate objects to human beings, that (im)politeness behaviors become relevant to social interactions and worth the energy they require.

Consider the irreversibility of disgust contagion. Consider a scenario in which soup is being prepared over an open fire among the dishes in preparation for an outdoor party. Should a bird defecate into the open pot, then no matter how deft and swift the chef is at scooping out the undesired ingredients, the remainder of the soup will be deemed unfit for consumption. Even if the pot at the moment of spoiling contained only water, it is difficult to imaging that any one present would accept that the water could be non-professionally filtered and sterilized for use. Similarly, if a cake provides the landing spot—it seems very unlikely that people who witnessed the incident would be satisfied with having the spoiled part of the cake cut away. They will likely demand that the knife used by the well-meaning chef in order to save the remainder be at least washed before any other use on food to be eaten, if they tolerate re-use of the knife at all. Similarly, in some households, a plate that has been used once as a dish for a pet is not retrievable for human use again. If it is used again, it is only after a decontamination process that is symbolically more involved than that used in

[7] The contrast in learning speed in the case with learning in other dimensions of human development merits reflection. Other cognitive reference points are the speed of language acquisition, in the general absence of labelled negative examples, and the slow pace of mastery of valid reasoning [14].

any other ritual of dish washing. Where disgust arises from the manifestations of decay and infection associated with disease, people have historically found it difficult to accept as uninfected those who have associated with the visibly infected.[8]

As physical responses are available for interpretation, they themselves are available for adaptation and re-use as conscious signals. The availability of the response for generalization predicts “metaphorical” extension to triggers that are not associated with toxins or disease — social triggers, for example. Contempt and disgust involve comparable affective states, and are frequently both apt descriptions of the same affective states, but are not identical in their use conditions. One may think of contempt as a socially mediated form of disgust—the sort of feeling invoked by someone who harms other people for sport (which might be experienced without qualm by someone who declares sport-fishing as a hobby). Here again, people are inclined to form an ill opinion of associates of those for whom they feel contempt. In the extended sense, relative irreversibility of the attribution is likely: as the maxim goes, trust takes years to build and seconds to break. The role of disgust in social situations, as opposed to situations with immediate physical triggers, is also evident in the fact that in some moral debates some will attempt to argue the correctness of one position over another on the strength of the disgust response. Indeed, some are tempted to resolve questions of morality with reference to disgust responses that are triggered by the behaviors associated with the moral questions,[9] and it is the fact that accidental association is fundamental to disgust that makes appeal to disgust unsafe as a justification for a moral judgement [15].

These considerations lead to the claim that the potential irreversibility and contagiousness of disgust responses are what prompt (im)politeness behaviours, despite the communicative expense of producing them. With respect to politeness behaviors, where not overused to the point of obsequiousness, they provide a mechanism for avoiding being deemed disgusting. They offer rituals for rectifying social mistakes. Where consistently applied, they offer a shield against premature criticism in the face of disgust triggers, where those disgust triggers are then deemed “out of character”. On this view, impoliteness behaviors are produced when the impolite actor views their interlocutor with disgust. Perceived asymmetry for the impolite actor allows that person to believe that by being impolite they will not attract disgust but will instead lead the target of their action to be viewed with disgust by any witnesses. Of course, a risk of impoliteness is that public perception will be the opposite. However, impoliteness behaviours themselves amount to the proof that the agent believes either

[8] The associations need not follow a rational course. We life in a society in which a medical doctor who manually conducts prostate examinations is accorded more esteem than a person who works in garbage collection, even though both may be safely assumed to engage in appropriate levels of hygiene and though the garbage collector probably has a greater overall positive impact on health in society.

[9] In June 2013, a Google search of the web for the phrase, `"wrong because it is disgusting"`, with quotation marks and without any specification of what “it” might refer to, yielded 768,000 hits.

that witnesses will also view the target with disgust or that the witnesses are themselves worthy of contempt and do not merit civility either.

4 Conclusions

A semantic theory of politeness must take into account the speaker's evaluation of events and agents participating in them. When speakers utter sentences events are described, and some elements of the event description is partially encoded directly in the syntax/semantics interface (e.g. tense [19] and aspect [22]) while other elements of meaning associated with an expression about an event follow from inference from the fact that the particular described event is of a certain type [9]. The evaluation of events undertaken by speakers includes an abstract assessment of costs and benefits. Differences in an agent's views of the value of some future event for the agent's self may lead an utterance about that future event to take the form of a polite request or as an order. Attitudes towards interlocutors and their role in a past event under discussion, may, also in relation to the use for the speaker, lead to an apology or to an affronted self-justification. Thus, a semantic theory of politeness must allow that from the classification of an utterances as polite or impolite that a corresponding configuration of esteem for interlocutors and value of the event for the speaker is in place. A framework which respects these desiderata is provided in some companion articles; however, any theory which respects them and which explains the puzzles of impoliteness will suffice.

Here it has been argued that speakers use polite forms in order to avoid invoking disgust and impolite forms when they do not mind disgusting others. Argument is made that it is through specification of described events that theory of (im)politeness can unfold partly in the realm of semantics. This may happen without claiming inherent offence or other sentiment in particular forms of language as the sole means of imbuing conversational behavior with politeness or impoliteness. While this does not resolve the extent to which linguistic politeness behaviors can be effectively characterized within semantic theory as opposed to pragmatics; however, it is argued that at least some of linguistic politeness is well-characterized in semantics. A theory of politeness and impoliteness may be expected to provide strong predictions about individual experience; however, scientific theories often diverge from perceptual experience in some respects. Linguistic politeness and impoliteness arise out of offence management, and as with ambiguity in language generally, in which speaker meaning and hearer meaning have ample chance of differing, the offences to be managed may be intended or unintended.

Acknowledgments. This research is supported by Science Foundation Ireland (Grants 07/CE/I1142 and 12/CE/I2267) as part of the Centre for Next Generation Localisation at Trinity College Dublin (www.cngl.ie).

References

1. Allan, K., Burridge, K.: Forbidden Words: Taboo and the Censoring of Language. Cambridge University Press (2006)
2. Bachwerk, M., Vogel, C.: Establishing linguistic conventions in task-oriented primeval dialogue. In: Esposito, A., Vinciarelli, A., Vicsi, K., Pelachaud, C., Nijholt, A. (eds.) Communication and Enactment 2010. LNCS, vol. 6800, pp. 48–55. Springer, Heidelberg (2011)
3. Bachwerk, M., Vogel, C.: Language and friendships: A co-evolution model of social and linguistic conventions. In: Scott-Phillips, T.C., Tamariz, M., Cartmill, E.A., Hurford, J.R. (eds.) Proceedings of the 9th International Conference (EvoLang9) The Evolution of Language, pp. 34–41. World Scientific, Singapore (2012)
4. Bousfield, D.: Impoliteness in Interaction. John Benjamins, Amsterdam (2008)
5. Bousfield, D., Locher, M. (eds.): Impoliteness in Language: Studies on its Interplay with Power in Theory and Practice. Mouton de Gruyter, Berlin (2008)
6. Brown, P., Levinson, S.: Politeness: Some Universals in Language Usage. Cambridge University Press (1987)
7. Culpeper, J.: Reflections on impoliteness, relational work and power. In: Bousfield, D., Locher, M. (eds.) Impoliteness in Language: Studies on its Interplay with Power in Theory and Practice, pp. 17–44. Mouton de Gruyter, Berlin (2008)
8. Culpeper, J.: Impoliteness: Using Language to Cause Offense. Cambridge University Press (2011)
9. Davidson, D.: Events as particulars. In: Davidson, D. (ed.) Essays on Actions & Events, pp. 181–187. Oxford University Press, Oxford (1980)
10. Ekman, P.: Basic emotions. In: Dalgleish, T., Power, M. (eds.) Handbook of Cognition and Emotion, pp. 45–60. John Wiley & Sons (1999)
11. Goffman, E.: The Presentation of Self in Everyday Life. Doubleday, New York (1956)
12. Goffman, E.: On face-work. In: Interaction Ritual: Essays in Face-to-Face Behavior, pp. 5–45. Transaction Publishers, New Brunswick (1967) (reprinted 2008)
13. Hurford, J.: Biological evolution of the saussurean sign as a component of the language acquisition device. Lingua 77, 187–222 (1989)
14. Johnson-Laird, P., Savary, F.: Illusory inferences: A novel class of erroneous deduction. Cognition 71, 191–229 (1999)
15. Kelly, D.: Yuck! The Nature and Moral Significance of Disgust. MIT (2011)
16. Locher, M., Watts, R.: Relational work and impoliteness: Negotiating norms of linguistic behaviour. In: Bousfield, D., Locher, M. (eds.) Impoliteness in Language: Studies on its Interplay with Power in Theory and Practice, pp. 77–99. Mouton de Gruyter, Berlin (2008)
17. MacNeill, D.: Growth points cross linguistically. In: Nuyts, J., Pederson, E. (eds.) Language and Conceptualization, Language, Culture & Cognition, pp. 190–212. Cambridge University Press (1997)
18. Orwell, G.: Signet, New York (1949, 1984)
19. Reichenbach, H.: Elements of Symbolic Logic. McMillan (1947)
20. Schnurr, S., Marra, M., Holmes, J.: Impoliteness as a means of contesting power relations in the workplace. In: Bousfield, D., Locher, M. (eds.) Impoliteness in Language: Studies on its Interplay with Power in Theory and Practice, pp. 211–229. Mouton de Gruyter, Berlin (2008)
21. Taylor, T.J.: Mutual Misunderstanding: Scepticism and the Theorizing of Language and Interpretation. Duke University Press (1992)

22. Verkuyl, H.J.: A theory of aspectuality. Cambridge Studies in Linguistics. Cambridge University Press (1993)
23. Vogel, C., Woods, J.: Simulation of evolving linguistic communication among fallible communicators. In: Hurford, J., Fitch, T. (eds.) Proceedings of the Fourth International Conference on the Evolution of Language, p. 116. Harvard University, Cambrige (2002)
24. Vogel, C., Woods, J.: A platform for simulating language evolution. In: Bramer, M., Coenen, F., Tuson, A. (eds.) Research and Development in Intelligent Systems XXIII, Cambridge, UK, pp. 360–373 (2006)

Conceptual Spaces for Emotion Identification and Alignment

Maurice Grinberg[1], Evgeniya Hristova[1], Monika Moudova[1], and James Boster[2]

[1] Department of Cognitive Science and Psychology, New Bulgarian University, Sofia, Bulgaria
mgrinberg@nbu.bg, ehristova@cogs.nbu.bg, moudova@yobul.com
[2] Department of Anthropology, University of Connecticut, USA
james.boster@uconn.edu

Abstract. The paper explores a method for emotion identification based on the mapping of emotional terms to a set of emotion eliciting situations. The method has been applied for emotion words from the Bulgarian language. Situations and words have been evaluated using the valence, arousal, and dominance ratings given by participants. Nine clusters of emotion terms and situations have been identified which allowed to map emotion terms with situations. The mapping method and the results for Bulgarian can be used for cross-group or cross-cultural studies of emotions involving various languages. The possible usage of the results obtained for emotion conceptual space evaluation and emotional alignment of people communicating in social networks is also discussed.

Keywords: emotions, conceptual spaces, mapping emotional terms and situations.

1 Introduction

Social networking connects people with different sex, age, social status, background, and culture. The interactions among these people can go through interfaces which allow various form of communication from writing unaddressed messages (e.g. on twitter and in blogs) to chats based on exchange of written messages or audio and video conversations (e.g. Skype and Google+ hangouts). In all these interactions, a common conceptual ground is needed for achieving a genuine communication. While this may not be a real problem when describing factual information about events, natural phenomena, etc. when it comes to sentiment and emotion sharing about specific situations, the establishing of such a common background is crucial for mutual understanding.

One possible approach of making explicit the underlying conceptual structures for a particular domain is to explore the corresponding conceptual space based on a set of terms (words) representative of the domain.

This paper presents an approach for building emotional conceptual spaces based on techniques from cognitive anthropology (Romney et al., 1996; Romney, Moore, & Rusch, 1997; D'Andrade, Boster, & Ellsworth, under review). This approach allows

S. Bassis et al. (eds.), *Recent Advances of Neural Network Models and Applications*,
Smart Innovation, Systems and Technologies 26,
DOI: 10.1007/978-3-319-04129-2_41, © Springer International Publishing Switzerland 2014

for the selection of a small number of situations and terms which if incorporated in suitable interfaces, as suggested in the paper, can be used for alignment of emotional spaces. This alignment of emotion conceptual spaces would allow participants in social networks to be aware and account for cultural differences related to emotional content expressed by words denoting emotions or emotional content.

In this paper, an analysis of emotion related situations and emotion related words are rated along the dimensions valence, arousal, and dominance. Then, words and situations are clustered together using the respective scores and they are effectively mapped to each other by taking the terms and situations belonging to the same cluster.

The paper is organized as follows: In Section 2, the theoretical background needed for the understanding of the paper is given. In Section 3, the empirical study is described and in Section 4, the results are presented. In Section 5, the results are summarized and their potential for emotion alignment interface is discussed.

2 Theoretical Background

In this paper, a relatively simple dimensional view of emotions was adopted, assuming that emotions can be defined by scale values on a number of dimensions (Lang, 1980). This view was founded in Osgood's work (Osgood, Suci, & Tannenbaum, 1957), which indicated that the variance in verbal judgments in emotional assessment can be accounted for by three major dimensions. The most important of these dimensions are *affective valence* (ranging from pleasant to unpleasant) and *arousal* (ranging from calm to excited). The third dimension was called *dominance* or *control*.

According to the model proposed by Lang, Bradley, & Cuthbert (1997), emotions are organized around two basic motivational systems that have evolved to mediate transaction in the environment that either promote or threaten physical survival – the appetitive and the defense systems. Activation of the appetitive system facilitates approach behaviors such as mating, food taking, and exploration whereas the defensive system facilitates behaviors such as avoidance, escape, and defense (Lang, 1995). According to this view the two dimensions arousal and valence capture global and basic elements of emotion. Valence indicates which motivational system is activated and arousal indicates the intensity of this activation (Lang, Bradley, & Cuthbert, 1997).

For the study of emotions, several normative sets of affective stimuli have been designed. The most widely used among them are the International Affective Picture System (IAPS) (Lang, Bradley, & Cuthbert, 2008), the International Affective Digitized Sound System (IADS) (Bradley & Lang, 1999), and the Affective Norms for English Words system (ANEW) (Bradley & Lang, 1999). The availability of such collections of normatively rated affective stimuli allows better experimental control in the selection of emotional stimuli, facilitates the comparison of results across different studies, and last but not least can serve as a basis of cross lingual, cultural, and age communication in the context of social networks.

The assessment of affective stimuli along the valence, arousal, and dominance dimensions can conveniently be conducted with the help of the Self-Assessment

Manikin (SAM), an affective rating system devised by Lang (1980). The system consists of graphic figures depicting values along each of the three dimensions on a continuously varying scale, used to indicate emotional reactions. For the valence dimension, SAM ranges from a smiling happy figure to a frowning unhappy figure. For the arousal dimension, SAM ranges from excited wild-eyed figure to a relaxed sleepy figure. And for the dominance dimension, SAM ranges from a small figure (symbolizing being dominated) to a large figure (symbolizing being in control). Rating is given on a 5-point scale visualized as five SAM figures for each dimension.

The approach proposed by D'Andrade, Boster, & Ellsworth (under review), has been developed for comparing the structure of emotions across cultures in cognitive anthropology. The method has three stages. The first one is the generation of a list of emotion terms. The terms were collected using the frame "I feel ___". After removal of synonymous and ambiguous terms the list resulted in a total of 400 feeling terms. In the second stage, affectively charged situations were elicited by asking a group of responders to give for each term an example of a situation in which they felt like the term says. The result of the latter stage was a set of 125 situations related to specific emotions. The third stage consisted of mapping the 125 situations to the 400 terms. Responders were asked to match situations with emotion terms. One big virtue of this approach is the possibility to describe emotion terms based on the situations they are assigned to and not on the basis of definitions.

The paper explores a way to combine the advantages of the above described methods and analyzes the qualities of the combined method with respect to emotion term discrimination.

3 Empirical Study

3.1 Goals

The main goal of the present study is exploring the conceptual spaces for emotional terms in Bulgarian based on emotion related situations and terms. This is done by direct independent evaluation using the valence, arousal, and dominance scales and subsequent simultaneous clustering of situations and emotional terms based on this evaluation. The rating was done using the described above SAM (Lang, 1980).

3.2 Stimuli

The stimuli in the study were 100 emotional terms and 100 emotion-eliciting situations. In short, the affectively charged situations used were 100 out of 125 elicited in D'Andrade, Boster, & Ellsworth (under review). The feeling terms were generated by giving the 100 situations and asking subjects to list the emotions that the particular situation elicits in them. The terms collected were reduced from 261 to 100 by a procedure described bellow. The terms used are not only emotion proper, but included all kinds of emotional terms.

The first step was translating the situations from English into Bulgarian using the backward translation method (Brislin, 1970).

The next step was to generate terms for emotions and emotional states. To this aim, the list of 100 situations was divided into 4 sections containing 25 situations each. Each participant was given at least one of the four lists and instructed to read each situation and fill in the blank by stating how and what she would feel in the given situation. Moreover, participants were asked to use the terms and not metaphoric or idiomatic phrases such as "high in the sky", "as a baby", etc. There were 150 participants (85 female and 65 male). Each of the 4 lists was given to at least 60 people. The procedure yielded a list of 261 terms with frequencies from 1 up to 215 (the number of times a term was generated in the free-listing task). All words with frequencies lower than 10 were removed from the original list, giving 132 terms. 11 terms had frequencies higher than 100, among them the emotional terms considered as basic – angry, sad, and happy. 89 additional terms were randomly selected to obtain a list of 100 emotion words.

3.3 Rating Terms and Situations

The 100 situations and 100 emotional terms selected were rated together using the SAM (Lang, 1980) on the valence, arousal, and dominance scales.

The study was conducted using an web-based on-line system for data collection designed for the experiments. The instructions were translated into Bulgarian from the original SAM instructions (Lang, 1997).

Using the web-based interface, participants were presented with the situations or terms presented one by one, each time in a different random order. Bellow each situation/term the three rating scales appeared. To see the next situation/term, participants had to provide a response using each of the three scales (valence, arousal, and dominance).

The scales comprised five SAM figures end-point labels as follows:

- Valence scale: 1 = 'very negative' to 5 = 'very positive';
- Arousal scale: 1 = 'very calm' to 5 = 'very aroused';
- Dominance scale: 1 = 'dominated' to 5 = 'in control'.

3.4 Participants

In the study, 130 participants (70 female and 60 male) rated the emotionally charged situations on the three dimensions, and 145 participants (85 female and 60 male) rating the emotional terms on the same three dimensions.

4 Results

4.1 Mean Ratings

After the rating each term and situation could be characterized by the three ratings they obtained. In Figure 1, they are plotted in the space spanned by the valence, arousal, and dominance dimensions.

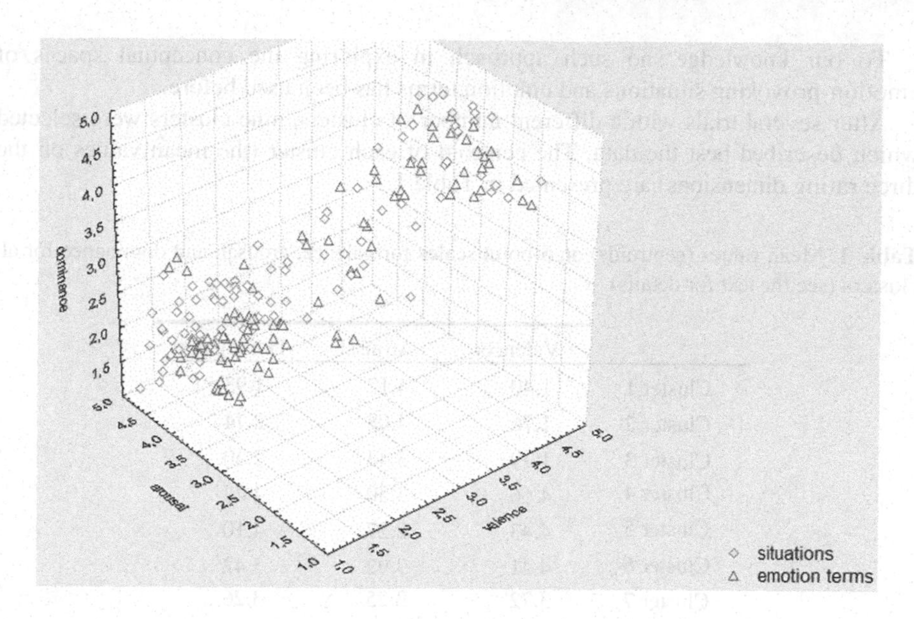

Fig. 1. Mean ratings for emotional terms and situations in the valence, arousal, and dominance space

The results for the mean ratings are as follows:

- **Valence scale** (from 1 – very bad to 5 – very good) – ratings for terms and situations together range from 1.06 to 4.95 with mean of 2.61 and with a standard deviation of 1.28.
- **Arousal scale** (on a scale from 1 – very calm to 5 – very aroused) – ratings the for terms and situations together range from 1.32 to 4.70 with mean of 3.51 and with a standard deviation of 0.66.
- **Dominance scale** (on a scale from 1 – very dominated to 5 – very dominating) – ratings for terms and situations together range from 1.25 to 4.69 with mean of 2.85 and with a standard deviation of 0.85.

Interestingly, it turns out that the valence and arousal dimensions have a correlation of –0.552 ($p < .001$); the valence and dominance 0.920 ($p < .001$); and the arousal and dominance –0.575 ($p < .001$), respectively.

4.2 Forming Clusters of Situations and Emotional Terms

Based on the ratings, a K-means cluster analysis was performed of the situations and emotions together. So, each cluster contained terms and situations. The goal of this procedure was to find emotions and situations that fit together and thus obtain a description of a group of emotion terms by a group of situation with respect to their valence, arousal, and dominance.

To our knowledge, no such approach in exploring the conceptual spaces of emotion-provoking situations and emotion terms has been used before.

After several trials with a different number of clusters, nine clusters were selected which described best the data. The centroid of each cluster (the mean values on the three rating dimensions) are presented in Table 1.

Table 1. Mean values (centroids) on 5-point scales for valence, arousal, and dominance for all clusters (see the text for details)

	Valence	Arousal	Dominance
Cluster 1	1.40	4.12	1.93
Cluster 2	1.76	4.08	2.74
Cluster 3	1.79	3.43	2.30
Cluster 4	4.66	3.30	4.24
Cluster 5	4.43	2.27	4.10
Cluster 6	4.41	3.92	3.42
Cluster 7	3.72	3.25	3.26
Cluster 8	2.73	3.43	3.04
Cluster 9	2.44	2.41	2.97

The clusters are named using the emotion term which is closest to the respective centroids.

To be able to interpret the results, we grouped the clusters in 3 groups by valence – negative, positive, and neutral. The clusters with valence values between 1 and 2.4 were classified as negative, with valence values between 2.4 up to 3.5 as neutral, and with valence values between 3.5 and 5 as positive.

Negative Clusters. There are 3 negative clusters whose valence ranges from 1.40 to 1.79. On the arousal scale, they range from 3.43 to 4.12 and on the dominance scale from 1.93 to 2.30 (see Table 2). The three negative clusters contain 113 items, of which 65 are emotional terms and 48 situations. Thus, 65% of the emotional terms and 48% of the situations are classified in the negative clusters.

Table 2. Mean ratings and number of items for negative clusters

	Valence	Arousal	Dominance	Centroid	Items	Terms	Situations
Cluster 1	1.40	4.12	1.93	deceived	46	17	29
Cluster 2	1.76	4.08	2.74	angry	22	16	6
Cluster 3	1.79	3.43	2.30	uncertain	45	32	13

Positive Clusters. 4 positive clusters have been identified. The range of mean ratings is as follows: on the valence scale from 3.71 to 4.66; on the arousal scale from 2.27 to 3.92; and on the dominance scale from 3.26 to 4.24 (see Table 3). The 4 positive clusters contain 62 items of which 34 are emotional terms and 28 are situations. Thus, 34% of the emotional terms and 28% of the situations are classified in the positive clusters. The emotional terms in the positive clusters are almost half of those in the negative clusters.

Table 3. Mean ratings and number of items for positive clusters

	Valence	Arousal	Dominance	Center	Items	Terms	Situations
Cluster 4	4.66	3.30	4.24	happy	22	9	13
Cluster 5	4.43	2.27	4.10	complete	20	14	6
Cluster 6	4.41	3.92	3.42	agitated	9	5	4
Cluster 7	3.72	3.25	3.26	amazed	11	6	5

Neutral Clusters. There are 2 neutral clusters (see Table 4). Neutral clusters contain 25 items of which 11 are emotional terms and 14 situations. 11% of the emotional terms and 14% of the situations are classified in the neutral clusters.

Table 4. Mean ratings and number of items for neutral clusters

	Valence	Arousal	Dominance	Center	Items	Terms	Situations
Cluster 8	2.73	3.43	3.04	tired	22	9	13
Cluster 9	2.44	2.41	2.97	worried	11	6	5

5 Summary and Discussion

In this paper, a conceptual space of 100 emotional terms and 100 emotion-provoking situations (D'Andrade, Boster, & Ellsworth, under review) was generated. Ratings on three traditional emotion related dimensions – valence, arousal, and dominance – were gathered using a web-based interface which presented the scales using 5 SAM figures (Lang, 1997) for each dimension. Next, using the K-means cluster analysis a meaningful grouping of emotionally charged situations and emotion terms was obtained. We found 4 positive clusters and 3 negative ones, although the negative clusters contained much more items. Having more negative terms and situations is coherent with theories which state that negative emotions and states are better defined in language as they provide more important information with respect to well-being and survival. Of course a more elaborate clustering could reveal the fine grained structure of the clusters. However, this is not the goal of the present paper which is focused on the method of gathering data and on the demonstration that such a method can be efficient in emotion identification and alignment. In such a case, the analysis presented of the paper has to be carried out for the groups of interest and the obtained spaces compared, e.g. by means of correspondence analysis (Romney, Moore, & Rusch, 1997).

The adopted approach allowed for the mapping of terms and situations and aligning them in a common conceptual space spanned by the valence, arousal, and dominance dimensions. The database, obtained in the study, can be a useful source of standardized stimuli for emotion-related research in Bulgarian. It could also serve as a basis for cross-group or cross-cultural comparisons of conceptual spaces and mapping of emotional terms with other languages. The latter highlights the potential of the obtained results to be used for the design and development of an emotion alignment interfaces for social networks. Such interfaces could serve to improve the communication among participants and allow better cohesion and mutual understanding.

References

Bradley, M.M., Lang, P.J.: International affective digitized sounds (IADS): Stimuli, instruction manual and affective ratings (Tech. Rep. No. B-2). Gainesville, FL: The Center for Research in Psychophysiology, University of Florida (1999)

Bradley, M.M., Lang, P.J.: Affective norms for English words (ANEW): Stimuli, instruction manual and affective ratings. Technical report C-1, Gainesville (1999)

Brislin, R.W.: Back-Translation for Cross-Cultural Research. Journal of Cross-Cultural 1(3), 185–216 (1970)

D'Andrade, R., Boster, J., Ellsworth, P.: The Structure of Feelings. Ethos (under review)

Lang, P.J.: Behavioral treatment and bio-behavioral assessment: Computer applications. In: Sidowski, J.B., Johnson, J.H., Williams, T.A. (eds.) Technology in Mental Health Care Delivery Systems, Ablex, Norwood (1980)

Lang, P.J.: The emotion probe: Studies of motivation and attention. American Psychologist 50, 371–385 (1995)

Lang, P.J., Bradley, M.M., Cuthbert, B.N.: Motivated attention: Affect, activation, and action. In: Lang, P.J., Simons, R.F., Balaban, M.T. (eds.) Attention and Orienting: Sensory and Motivational Processes, pp. 97–135. Erlbaum, Hillsdale (1997)

Lang, P.J., Bradley, M.M., Cuthbert, B.N.: International affective picture system (IAPS): Affective ratings of pictures and instruction manual. Technical Report A-8. University of Florida, Gainesville, FL (2008)

Osgood, C.E., Suci, G., Tannenbaum, P.: The measurement of meaning. University of Illinois Press, Urbana, IL (1957)

Romney, A.K., Boyd, J.P., Moore, C.C., Batchelder, W.H., Brazill, T.J.: Culture as Shared Cognitive Representations. Proceedings of the National Academy of Sciences USA 93, 4699–4705 (1996)

Romney, A.K., Moore, C.C., Rusch, C.D.: Cultural universals: Measuring the semantic structure of emotion terms in English and Japanese. PNAS, USA 94, 1–5 (1997)

Emotions and Moral Judgment: A Multimodal Analysis

Evgeniya Hristova, Veselina Kadreva, and Maurice Grinberg

Department of Cognitive Science and Psychology,
New Bulgarian University, Sofia, Bulgaria
{ehristova,vkadreva}@cogs.nbu.bg, mgrinberg@nbu.bg

Abstract. Recent findings in the field of moral psychology suggest that moral judgment results both from emotional processing and deliberate reasoning. The experimental study uses artificial situations that pose moral dilemmas – a human life have to be sacrificed in order to save more lives. Two factors (physical directness of harm and inevitability of death) are varied in order to explore potential differences in emotional processing and their effects on judgment. Multimodal data is collected and analyzed: moral judgments, skin conductance (as a somatic index of affective processing), and response times (as providing information on deliberation process). Personal-impersonal distinction and inevitability of death are found to influence emotions and judgments in moral dilemmas.

Keywords: moral dilemmas, moral judgments, emotional engagement, skin-conductance response.

1 Introduction

1.1 Moral Dilemmas

Morality represents a belief that certain intentions, decisions or behaviors are either "right" or "wrong". While some moral views and decisions are very easy to accept some might be contradictory because of conflict with other beliefs or rights. Many moral controversies are in fact ones that oppose individual rights versus greater good. Utilitarian theories of morality hold that behavior, which is motivated to achieve greater good in the sense of "optimal utility", is right. Deontological moral theories oppose to utilitarian views and state that acts should not be determined by their consequences but their rightness or wrongness depends on the principle that this act represents. While utilitarian and deontological theories are normative and as such are subject of ethics and philosophy, experimental psychology aims to describe and explain the processes that underlie moral judgment.

The 'Trolley problem' is extensively used in both psychological and neuroscience research in the field of moral psychology, e.g. [1-3]. The following scenario could illustrate it: *"A runaway trolley is headed for five people who will be killed if it proceeds on its present course. The only way to save them is to hit a switch that will turn the trolley onto an alternate set of tracks where it will kill one person instead of*

S. Bassis et al. (eds.), *Recent Advances of Neural Network Models and Applications*,
Smart Innovation, Systems and Technologies 26,
DOI: 10.1007/978-3-319-04129-2_42, © Springer International Publishing Switzerland 2014

five. Is it morally appropriate to turn the trolley in order to save five people at the expense of one?" [2]. The 'Footbridge dilemma' describes a similar situation but suggests that a bystander is pushed from a footbridge in front of the trolley in order to save the other people. From a utilitarian point of view (based on comparing 1 to 5 lives) the situations are identical. However, when people are asked to judge the suggested resolutions as morally appropriate or not, most of them find the resolution permissible in the 'Trolley dilemma' but not in the 'Footbridge dilemma' [2]. Different theories have been proposed in order to interpret the behavioral dissociation.

1.2 Emotions and Moral Judgment

A possible explanation of the contrast in judgments described above is that moral judgment results from application of readily available principles that constitute universal moral grammar (UMG) [4]. What proponents of the UMG theory state is that moral judgments are made unconsciously and automatically.

An alternative theory – the social intuitionist theory – suggests that morality is based on emotions rather than logical reasoning; reasoning serves as a post hoc justification of automatic and intuitive moral judgment guided by emotion [5].

Greene et al. [2] also reported data in support of emotional engagement in moral reasoning. The authors introduced a dual-process theory according to which the 'up-close and personal' infliction of harm (like in the Footbridge dilemma) provokes negative emotional response that affects judgment – behavior is judged as impermissible. In contrast, 'impersonal' harm (in the case of the Trolley problem) elicits a utilitarian, reasoned response rather than emotional one. Researchers provided fMRI data in support of the dual-process theory showing that brain areas responsible for emotional processing are activated by *personal* moral dilemmas while judgment of *impersonal* moral dilemmas activated regions underlying working memory and cognitive control. The authors also reported longer response times when *personal* harm is judged as appropriate, speculating that automatic emotional response interfered with utilitarian reasoning.

However, several studies [6-7] criticized inconsistencies in the stimulus material. McGuire et al. [6] reanalyzed data and attributed the results to the strong effects of single items. Moore et al. [7] eliminated possible confounds and systematically varied three additional factors (*inevitability of death*, *self-risk* and *intentionality of harm*). All these factors were found to influence judgments. The study failed to replicate response time findings of Greene et al. [2] - there was no significant difference between affirmative and negative responses to personal dilemmas. What they found is that *personal* dilemmas are overall judged faster than *impersonal* ones. Although Moore and his colleagues criticized research in support of dual-process theory for non–systematic variation of several factors, authors themselves failed to control for a certain number of potential confounds in their study.

2 Goals and Hypothesis

The current study aims to explore intuitions in moral judgment and factors affecting it while controlling for possible confounds unaccounted in previous research. The role of emotions in the process of moral judgment is as well a subject of the study.

Using hypothetical, artificial moral dilemmas two factors identified as significant in moral judgment [2], [7] are explored:

- ***physical directness of harm*** - harm is inflicted by physical contact (*personal* harm) or is mediated through mechanical means (*impersonal* harm);
- ***inevitability of death*** - harm is inflicted either to a person that is going to die anyway (*inevitable* death), or to a person that is not endangered by the situation described in the scenario (*avoidable* death).

Dilemmas used in the experiments are constructed in such a way as to eliminate possible confounding factors identified in previous experiments.

Multi-modal data is collected: responses about the permissibility of sacrificing one life to save 5 other persons, response times, and skin-conductance responses are recorded. Responses about the moral permissibility of the hypothetical resolutions to the dilemmas are collected and analyzed to capture the moral judgments. Response time data is collected to identify potential interference between emotional processing and rational deliberation (interference is supposed to produce longer reaction times).Skin conductance response is used as a tool to evaluate emotional involvement in moral judgment. Being a non–invasive and reliable method to identify sympathetic arousal, electro-dermal activity measures are widely used to detect emotional engagement in judgment and decision making [8-9].

Resolutions of *personal* dilemmas are predicted to be appraised as less permissible compared to *impersonal* ones. Also killing one person to save more people is hypothesized to be judged as more permissible when death is *inevitable* (compared to *avoidable*).

We think that emotional processes are present and interfere with rational calculations not only in personal dilemmas (as suggested by [2]) but are present in all scenarios in which people have to make difficult decisions - and especially if they make decisions involving harm to other persons (as suggested also by [7]).

3 Stimuli

As stated, the stimuli are constructed with the aim to control for possible confounding factors identified in the previous research. All dilemmas are homogenously structured and resolutions suggest a constant tradeoff between killing one person and saving five others. Participants are assigned the role of the protagonist whose life is not threatened and no specific role or any responsibilities is assigned to the protagonist. All of the six endangered persons are identified with equal roles in a working environment, thus suggesting equal responsibilities to all of them. The endangered and potentially sacrificed persons are adults only. The suggested intervention is instrumental (it is not just an incidental by-product of an action). All situations are designed to illustrate artificial scenarios in order to avoid potential confounding effects (e.g. familiarity with a certain situation or readily available personal opinion on resolutions. Each situation is followed by one and the same question: ‘Is it permissible to act as described?’ with two possible responses – ‘Yes’ and ‘No’.

Participants are instructed to imagine that the action described is the only action possible; also, to disregard legality and consider only moral appropriateness of judgment.

An example of *avoidable* dilemma with 2 possible resolutions (*personal* and *impersonal*) is presented below:

Situation (Avoidable)	**Resolution (Personal)**	**Judgment**
You are on the roof of a tall building. Builders are working on a scaffolding outside the building. Suddenly part of the scaffolding is disassembled. Five builders turn out to be suspended on an unstable structure, which is about to collapse, and the five builders are going to die. One of the builders is on a stable part of the scaffolding. In order to prevent the collapse of the scaffolding, its load needs to be reduced.	The only thing that you can do is to push the builder who is on the stable part of the scaffolding. The builder will fall and the load on the scaffold will be reduced enough. He is going to die but the other five persons will be saved.	Is it permissible to act as described?
	Resolution (Impersonal)	YES NO
	The only thing that you can do is to press a button in order to unfasten the safety belt of the builder who is on the stable part of the scaffolding. The builder will fall and the load on the scaffold will be reduced enough. He is going to die but the other five persons will be saved.	

4 Method

4.1 Design

Physical directness of harm (*personal* vs. *impersonal*) and *inevitability of death* (*avoidable* vs. *inevitable*) are manipulated in a within-subjects design.

Each participant is presented with 8 dilemmas - 4 scenarios (2 *avoidable* and 2 *inevitable*) each with *personal* and *impersonal* resolutions. Dilemmas are presented in order randomized for each participant.

For each dilemma, the following measures are analyzed: number of responses 'permissible', response times, skin-conductance reaction (SCR) during the response period.

4.2 Procedure and Data Recordings

Participants are tested individually. Three practice dilemmas are shown. Next, the eight stimuli are presented in random order by E-Prime 1.2. Each dilemma is presented on a single screen. Self-paced confirmation for reading completion and comprehension of the presented dilemma is given by the participant by pressing a key. Next, a screen with a question appears: 'Is it permissible to act as described?' Participants indicate either 'Yes' or 'No' using computer keyboard. Response is followed by 8 seconds inter-trial interval.

Responses and response time data are collected via E-Prime. *Response time* is considered the interval between the question onset ('Is it permissible to act as described?') and the YES/NO response.

Skin conductance is recorded using Biopac, Inc. MP 150 system and GSR100C amplifier with sampling rate of 200 samples/s. The amplifier is connected to TSD203 Ag-AgCl, unpolarizable finger electrodes. The electrodes are placed on the non-dominant hand. Skin conductance (SCR) signal is shifted by 200 samples (1 s). Following [9], the signal is smoothed (smoothing interval of 200 samples) and then a moving-difference function (10 samples difference interval) is applied. Markers generated by E-prime are used to synchronize the skin conductance recordings with the task. For each response period (defined from the question onset to the YES/NO response) an integral is calculated, corresponding to the area defined by the differenced skin conductance signal and the line connecting the end points of the signal for the analyzed period. Then the resulting value is divided by the response time (in seconds) so the final measure used for the SCR is in μS/s.

4.3 Participants

A total of 31 participants (10 male, 21 female) took part in the experiment. The age range was from 18 to 34 (M = 22.6). The participants took part in the experiment in exchange for partial credit toward an undergraduate course requirement. SCR data from 1 participant was discarded due to technical difficulties.

5 Results

5.1 Responses to the Dilemmas

Mean number of responses 'permissible' was analyzed in a repeated-measures ANOVA with 2 within-subjects factors - *physical directness of harm* (*personal* vs. *impersonal*) and *inevitability of death* (*avoidable* vs. *inevitable*).

Data is presented in Fig. 1. The interaction between factors was not significant. Analysis revealed main effect of *physical directness of harm* ($F(1, 30) = 4.02$, $p = 0.05$, $\eta_p^2 = 0.12$) - *impersonal* harm was judged as more permissible than *personal* harm (54.9% vs. 43.5% 'permissible' responses). There was also a main effect of *inevitability of death* ($F(1, 30) = 55.8$, $p = 0.00$, $\eta_p^2 = 0.54$) - killing someone whose death is *inevitable* was judged as more permissible than harming a person whose death is *avoidable* (68.3% vs. 30.1% responses 'permissible').

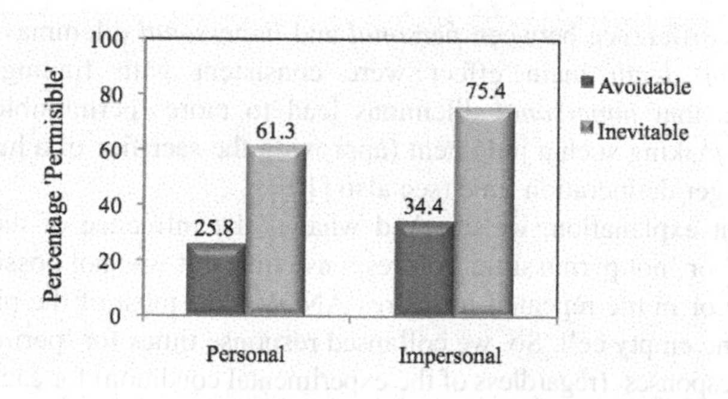

Fig. 1. Mean percentage of responses 'permissible'

Results are consistent with previous research and confirm the importance of factors *physical directness of harm* and *inevitability of death* in making moral judgments. As described in previous research [2], [7], harm inflicted by personal physical contact is judged to be more unacceptable than the same harm inflicted without a physical contact with the victim. Even stronger influence is found for the second factor explored – sacrificing a person whose death is inevitable is judged to be more permissible than sacrificing a person who is not endangered.

5.2 Response Time

Response times (N=10) exceeding the mean value plus 2 times SD were considered outliers and were not included in the analysis. Corresponding SCR were also excluded.

Response times were analyzed in repeated-measures ANOVA with 2 within-subjects factors - *physical directness of harm* (*personal* vs. *impersonal*) and *inevitability of death* (*avoidable* vs. *inevitable*).

Factors did not interact and there was no significant effect of *inevitability of death.* Only *physical directness of harm* demonstrated main effect ($F(1, 29) = 5.91$, $p = 0.02$, $\eta_\rho^2 = 0.17$) - responses to *personal* dilemmas (1133 ms) were significantly faster than responses to *impersonal* ones (1340 ms) (see Fig. 2).

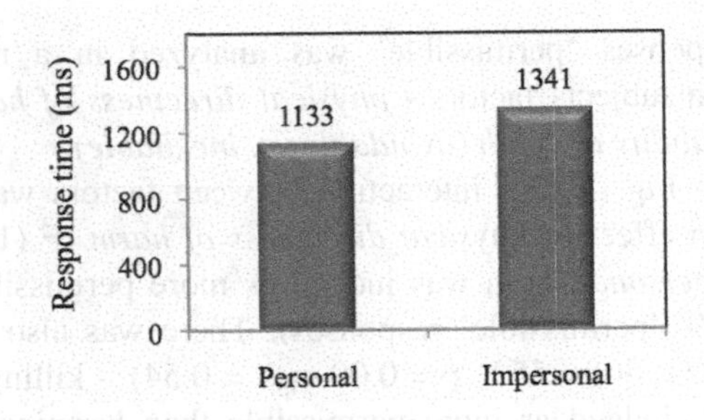

Fig. 2. Average response times for *personal* and *impersonal* dilemmas

Significant difference between *personal* and *impersonal* dilemmas and the lack of *inevitability of death* main effect were consistent with findings in [7]. One explanation is that *impersonal* dilemmas lead to more 'permissible' responses. It might be that making such a judgment (approving the sacrifice of a human life) is the reason for longer deliberation time (see also [10]).

To test that explanation, we checked what is the influence of the *response type* ('permissible' or 'not permissible') on response times. It was not possible to enter this additional factor in the repeated-measures ANOVA, as most of the participants' data had at least one empty cell. So, we collapsed response times for 'permissible' and 'not permissible' responses (regardless of the experimental condition) for each participant that has given both responses (25 participants). Repeated-measures ANOVA failed to demonstrate significant effect of *response type* on response times ($F(1, 24) = 1.55$, $p = 0.225$)

although the mean values are in the expected direction – responses 'permissible' are given more slowly (1451 ms) compared to responses 'not permissible' (1227 ms).

5.3 Integral Skin Conductance Reaction (SCR)

One subject was excluded because of empty cells and data from 29 subjects were analyzed. SCR was analyzed in a repeated-measures ANOVA with 2 within-subjects factors - *physical directness of harm* (*personal* vs. *impersonal*) and *inevitability of death* (*avoidable* vs. *inevitable*).

Interaction between factors was not significant. *Inevitability of death* demonstrated no main effect. The ANOVA yielded main effect of *physical directness of harm* ($F(1, 28) = 5.14$, $p = 0.03$, $\eta_p^2 = 0.15$). SCR was higher during decisions for *impersonal* dilemmas (0.24 μS/s) compared to *personal* ones (0.15 μS/s) (see Fig. 3).

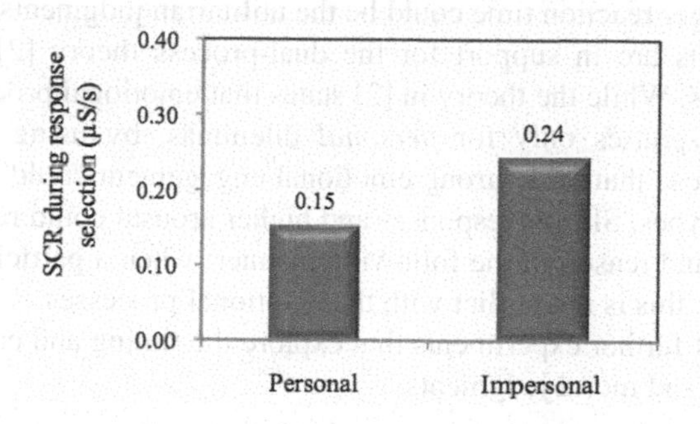

Fig. 3. Average SCR during response selection for *personal* and *impersonal* dilemmas

Responses to *impersonal* dilemmas were accompanied by stronger arousal compared to *personal* dilemmas. Again, the explanation of the results is that *impersonal* dilemmas lead to more responses 'permissible'. And if such judgment (approving the sacrifice of a human life) is accompanied by greater arousal, this could explain the obtained results. To check this hypothesis, we collapsed response times for 'permissible' and 'not permissible' responses (regardless of the experimental condition) for each participant that has given both responses. Repeated-measures ANOVA revealed marginally significant effect of response ('permissible' or 'not permissible') on SCR ($F(1, 23) = 3.2$, $p = 0.087$). The SCR was higher for responses 'permissible' (0.29 μS/s) than for responses 'not permissible' (0.17 μS/s).

6 Summary and Discussion

In the current research, strictly controlled moral dilemmas are used to study intuitions in moral judgments concerning situations in which a human life has to be sacrificed in order to save more lives.

Results show that judgment of hypothetical moral dilemmas is influenced by the *physical directness* and the *inevitability of death* factors. The utilitarian action – killing one to save five – is judged to be more permissible when the death of the person to be killed is inevitable. It is also considered more permissible when the act is impersonal (e.g. there is no physical contact with the 'victim').

To study the emotional processing in responses, response times and skin-conductance reactions are analyzed. Responses to impersonal dilemmas are slower and are accompanied by higher arousal than responses to personal ones. Slower responses to impersonal dilemmas replicate the results from [7]. The difference could be due to the greater number of utilitarian judgments in favor of impersonal dilemmas - for the impersonal dilemmas there are more responses 'permissible' (these responses are given more slowly and are accompanied by greater SCR). So it is possible that utilitarian actions lead to higher emotional engagement and thus – to higher arousal and slower responses. What actually provokes emotional engagement (indexed by SCR) and thus longer reaction time could be the utilitarian judgments themselves.

The latter results are in support for the dual-process theory [2], however, with some modifications. While the theory in [2] states that emotional processing interferes with utilitarian responses only for *personal* dilemmas, by using more controlled situations we propose that such strong emotional engagement could be observed also in other dilemma types. Slower responses and higher arousal could reflect the conflict between emotion and reason in the following manner: when a participant chooses the utilitarian response this is in conflict with the emotional processes.

There is need of further experiments that explore the timing and causal relationship between emotions and moral judgments.

References

1. Foot, P.: The Problem of Abortion and the Doctrine of the Double Effect. Virtues and Vices. Basil Blackwell, Oxford (1978)
2. Greene, J.D., Sommerville, R.B., Nystrom, L.E., Darley, J.M., Cohen, J.D.: An fMRI Investigation of Emotional Engagement in Moral Judgment. Science 293, 2105–2108 (2001)
3. Greene, J.D., Nystrom, L.E., Engell, A.D., Darley, J.M., Cohen, J.D.: The Neural Bases of Cognitive Conflict and Control in Moral Judgment. Neuron 44, 389–400 (2004)
4. Mikhail, J.: Universal Moral Grammar: Theory, Evidence and the Future. Trends in Cognitive Sciences 114, 143–152 (2007)
5. Haidt, J.: The emotional dog and its rational tail: A Social Intuitionist Approach to Moral Judgment. Psychological Review 108, 814–834 (2001)
6. McGuire, J., Langdon, R., Coltheart, M., Mackenzie, C.: A Reanalysis of the Personal/Impersonal Distinction in Moral Psychology Research. Journal of Experimental Social Psychology 45, 577–580 (2009)
7. Moore, A., Clark, B., Kane, M.: Who shalt not kill?: Individual Differences in Working Memory Capacity, Executive Control, and Moral Judgment. Psychological Science 19, 549–557 (2008)

8. Figner, B., Murphy, R.O.: Using Skin Conductance in Judgment and Decision Making Research. In: Schulte-Mecklenbeck, M., Kuehberger, A., Ranyard, R. (eds.) A Handbook of Process Tracing Methods for Decision Research. Psychology Press, New York (2010)
9. Naqvi, N.H., Bechara, A.: Skin Conductance: A Psychophysiological Approach to the Study of Decision Making. In: Senior, C., Russell, T., Gazzaniga, M.S. (eds.) Methods in Mind. MIT Press (2006)
10. Greene, J.D., Morelli, S.A., Lowenberg, K., Nystrom, L.E., Cohen, J.D.: Cognitive Load Selectively Interferes with Utilitarian Moral Judgment. Cognition 107, 1144–1154 (2008)

Contextual Information and Reappraisal of Negative Emotional Events

Ivana Baldassarre*, Lucia Abbamonte, Marina Cosenza, Giovanna Nigro, and Olimpia Matarazzo

Department of Psychology, Second University of Naples, Italy
{ivana.baldassarre,lucia.abbamonte,marina.cosenza, giovanna.nigro,olimpia.matarazzo}@unina2.it

Abstract. In this study the effect of the contextual-information induced reappraisal on modifying the emotional response elicited by failure has been investigated. To an academic or job setting failure (control condition) it has been added one of two types of contextual information (knowing that many other people failed the same task and knowing that it would be possible to try the failed task again) affecting three dimensions of failure appraisal: responsibility, sharing, and remediability. In an experimental condition both information were added. Four hundred and eighty undergraduates participated in this study. The experimental design was a 2 (negative emotional event) x 4 (contextual information) between-subjects design. The first variable was included in the design as covariate. We expected that generalized failure should imply a decrease of responsibility and an increase of sharing, the possibility of retrying should imply an increase in the remediability, and that the presence of both types of information should produce all the abovementioned effects. Our findings substantially corroborated the hypotheses.

Keywords: Reappraisal, contextual information, appraisal, emotion, emotion regulation.

1 Introduction

Reappraisal posits as a key word in the emotion regulation literature, which refers to the individual modalities that influence the generation, duration, intensity, and quality of one or more components of emotional response [8]-[18]. The term is commonly used to refer to changing the interpretation of an event, or the way of thinking about it, so as to modify the emotional reaction to this event [7]-[8]-[19].

Robust evidence showed that re-appraising an event is an effective way to regulate one's emotions both in laboratory experiments and in everyday life. Experimental findings indicate that reappraisal allows to decrease negative emotions or to increase the positive ones, and to temper peripheral physiology without impairment of

* Corresponding author.

S. Bassis et al. (eds.), *Recent Advances of Neural Network Models and Applications*, Smart Innovation, Systems and Technologies 26,
DOI: 10.1007/978-3-319-04129-2_43, © Springer International Publishing Switzerland 2014

cognitive skills (see [18]-[20] for a review). The frequent use of reappraisal in everyday life correlates with high psychological well-being, high life satisfaction, positive affects, sharing emotion with others, low risk of depression (see [11] for a review). So, reappraisal seem to be a royal strategy for successfully modulating one's emotion in the service of individual goals.

2 Reappraisal in the Framework of Stress and Coping Research

The concept of reappraisal has been originally formulated by Lazarus in the field of stress and coping research [12]-[13] and in the light of the psychoanalytic theory of defense mechanisms [6] and Arnold's cognitive theory of emotion [1]. Such a theory viewed appraisal (i.e. the way an individual immediately perceives an event as beneficial or harmful for his/her goals and concerns) as the pivotal cognitive process in emotion generation. Lazarus considered appraisal as the cognitive mediator between the stressor and its effects on individuals: his seminal studies about the experimental manipulation of appraisal [21]-[13] showed that the emotional distress elicited by watching a disturbing film could be reduced by means of the experimental instructions that induced participants either to take a detached attitude toward the film or to deny its painful aspects and consider them as joyful.

In the framework of stress and coping research – aimed at investigating how people cope with stressful events and at evaluating the effectiveness of the different strategies of coping - these results suggested that in everyday life the individuals who could spontaneously interpret threatening events in a more positive way were also less susceptible to succumb to them. Such individuals appraise negative events by adopting "defense mechanisms" analogous to those elicited by the experimental instructions. Lazarus [12]-[14] coined the term "reappraisal" to indicate the process by which the original appraisal of a negative event is modified in order to differently modulate its emotional burden. In the Lazarus and Folkman distinction [14] between two main types of coping, the problem-focused coping (i.e. facing a stressful event by modifying totally or partially the event) and the emotion-focused coping (i.e. facing a stressful event by modifying its emotional impact), reappraisal represents a key strategy of the latter. In this sense, the term was borrowed by the line of research on emotion regulation.

3 Reappraisal Operationalization in Stress and Coping and in Emotion Regulation Research

In the first experiments of Lazarus and colleagues, the experimental instructions aimed at manipulating appraisal consisted in a commentary introducing or accompanying the vision of a disturbing film that described the represented events so as to conform to the different experimental conditions and invited the viewers to actively re-interpret its meaning in such a way. In other terms, with reference to the control condition, the appraisal manipulation was done in two ways: through contextual information by providing a different interpretation of the events shown in the film and by leading the participants to take a different perspective while sawing it.

The subsequent research on coping and on emotion regulation no longer used contextual information to induce reappraisal but focused on the active role played by individuals in re-interpreting the original meaning of an event in order to better control their emotions. The stress and coping research mainly operationalized reappraisal – also labeled positive reappraisal – as a more positive way to think about a negative event, by focusing on its supposed future beneficial aspects rather on the present harmful ones. The reappraisal subscales COPE [4] and Ways of coping [5] questionnaires prove this tendency. A large amount of studies on self-reported positive reappraisal (see [20] for a review) documented the advantageous effect of this strategy in favoring individuals' well-being even when facing detrimental events.

The emotion regulation research has traditionally preferred to operationalize reappraisal as a form of distancing from the events' emotional significance in order to prevent the onset of emotional reaction. In more recent years, however, an increasing number of studies conceived reappraisal as an interpretative strategy aimed at modifying rather than neutralizing the emotional impact generated by an event [17]-[19]-[20].

Both detached and positive reappraisal are deliberate and effortful attempts to regulate one's emotions. In fact research in this domain has mainly focused on the conscious and explicit ways the individuals use to control their emotions. Only in the last years a new line of research centered on unconscious emotion regulation that can be defined as "the unintentional, automatic, and relatively effortless control of one's exposure to, processing of, and response to emotionally evocative events" ([24], p. 848). According to this perspective, emotion regulation can be environment-driven and can unfold without monitoring or awareness. A few studies [24] showed that unconscious reappraisal can be at least as effective as deliberate reappraisal in decreasing physiological reactivity linked to an anxiety-eliciting task. Other studies (see [9] for a review) suggest that the habitual use of a specific kind of regulation, such as reappraisal, can make it almost automatic and implicit in the course of time.

The present study focuses on the possibility that reappraisal can be generated in a spontaneous and almost automatic way by environmental cues. So, we take into account a mode of emotion regulation driven by the environment while not being unaware.

4 Experiment

In this experiment we tested the effect that the *reappraisal* of a negative event has on mitigating the emotional impact of such an event, when it is spontaneously induced by contextual information.

Contrary to most literature on emotion regulation, in which reappraisal was conceived as an effortful strategy voluntarily implemented by an individual to change the mental representation of the meaning of an event, in this study, we focused on what can be defined as "environment-driven reappraisal", i.e., the almost effortless change of the significance of an event produced by the acquisition of new information (which allows the individual to reframe the event in a less negative way).

We therefore connected to the first manipulation of appraisal operated by Lazarus and colleagues in 1960s in which one of the ways of inducing a more positive interpretation of a disturbing film was that of reframing it through " optimistic" information presented before or during its vision.

In our study we chose to investigate how the emotional impact elicited by a performance failure in academic or job setting would be modified through two types of contextual information which could naturally be available in everyday life and that we supposed to affect three dimensions of failure appraisal: responsibility, sharing, and remediability. The two types of information presented – separately or together – were: knowing that many other people failed the same task, knowing that it would be possible to try the failed task again. We expected that the two kinds of information should selectively affect emotions as follows: the possibility of retrying the failed task should increase positive emotions, such as hope, confidence, consolation, relief; the knowledge that many other people failed the same task should decrease negative emotions linked to responsibility, such as shame, anger toward one-self, guilt; the presence of both types of information (possibility of retrying and generalized failure) should involve both the above mentioned effects. More specifically, we predicted that the effect of the independent variable would be mediated by the re-appraisal dimensions induced by contextual information. In particular, we expected that generalized failure should imply a decrease of responsibility and an increase of sharing perception, that the possibility of retrying should imply an increase in the remediability perception, and that the presence of both types of information should produce all the abovementioned effects. So, in conformity with the mediational model [2]-[16], we predicted that the effects of the independent variable would no longer persist after controlling for the mediating variables. It is noteworthy that the mediators acted also as manipulation check.

5 Method

5.1 Participants and Design

Four hundred and eighty undergraduates (240 male and 240 female; mean age = 22.98; SD = 2.68) participated in this study, as unpaid volunteers. They were randomly assigned to one of the experimental conditions, except for gender for which they were paired.

The experimental design was a 2 (type of negative emotional event: academic examination failure vs. job interview failure) x 4 (type of contextual information: no information, shared failure, possibility of retrying, shared failure plus possibility of retrying) between-subjects design. The first variable (type of negative event) was included in the design as covariate.

5.2 Materials and Procedure

Eight scripts were built, with a similar structure and two sources of variation: type of failure and type of contextual information. In the first case, the situation described a student after failing an important academic examination; in the second case the situation described a company employee after failing an important job interview. In both cases, after the failure experience, protagonists are in one of the following situations: to receive no more information, to obtain information that many other people failed the same task, to obtain information that it was possible to try the failed task again, to receive both the pieces of information.

After reading the script, participants were asked to identify with the protagonist and to answer on a 7-point scale (1-not at all/7-extremely) to the three following questions:

1. How much the protagonist believes that the responsibility of negative outcome is attributable to him/her
2. How numerous the protagonist believes the people involved in the same failure are
3. How much the protagonist believes it is possible to remedy the failure.

Then the participants were asked to indicate, on the same 7-points scale, the intensity through which the protagonist felt the following emotions: shame, guilt, anger towards oneself, humiliation, frustration, disappointment, anger towards the circumstances, regret (expressed through the two Italian term of *rimpianto* and *rammarico*)[1], hope, confidence, consolation, relief, indifference, resignation, and worry. The emotions were selected from the literature on emotional consequences of failure [3]-[22] following this rationale: the first four emotions were chosen since they are linked to responsibility and self-blame; the successive four emotions were selected because they indicate that the outcome did not meet the expectations. Then, hope, confidence, consolation, and relief were chosen as positive emotions: in particular, hope and confidence imply positive expectations towards the future, while consolation and relief indicate the emotional reaction generated by the thought that things could be worse. Indifference and resignation were selected because they indicate low arousal and poor involvement; contrarily, worry suggests high concern for possible further negative outcomes.

The order of the items was randomized.

5.3 Results

For each manipulation check variable (responsibility, sharing failure with others, remediability), an ANCOVA on the four type of information conditions has been conducted, with script and gender as covariates. Mean (with standard deviations) and results are reported in Tab.1 and Tab.2, respectively. The experimental manipulation affected each variable: as regards “responsibility”, the “control” and “possibility of

[1] Both the Italian terms of *rimpianto* and *rammarico* correspond to the entry of regret, in this double meaning of sense of loss or grief and feeling of sorrow or remorse.

retrying" conditions obtained higher scores than "all failed" and "both information" conditions. As concerns "failure sharing", "all failed" and "both information" conditions obtained higher scores than the "control" and "possibility of retrying" conditions; as to the "remediability" variable, the "possibility of retrying" and "both information" conditions obtained higher scores than "control" and "all failed" conditions. On the responsibility variable, there were the two effects of script and gender covariates: the parameter estimates showed that the academic examination script and female obtained high scores than the job interview script and male, respectively. A script effect also emerged on failure sharing: the academic examination script obtained high scores than the job interview script.

Table 1. Means (with standard deviations) of the three reappraisal dimensions as function of the experimental condition

	Responsibility	Sharing	Remediability
Control	4.43 (1.33)	4,23 (1.38)	4,02 (2.01)
All failed	3.62 (1.5)	5,27 (1.33)	3,80 (1.79)
Retrying possibility	4.37 (1.31)	4,24 (1.43)	5,27 (1.54)
Both information	3.64 (1.54)	5,04 (1.54)	5,05 (1.64)

In order to reduce the number of the emotions, an exploratory factor analysis with principal components extraction method was performed on the 16 items. Varimax rotation was used after controlling (with Oblimin rotation) that the factors were independent each other. Five factors with Eingenvalue > 1, explaining 62.28% of the total variance, were extracted: self-blame, event-focused emotions, prospective emotions, self-soothing, and regret. Results are reported in table 3. Indifference and resignation scales were inverted since they had a negative correlation with the respective factor.

Table 2. Significant effects of the 3 ANCOVAs as function of the experimental condition on the three reappraisal dimensions

	Effect	F	d. f.	*p*
Responsibility	*Condition*	12.91	3,474	.000
	Script	47.44	1,474	.000
	Gender	5.59	1,474	.019
Sharing	*Condition*	18.02	3,474	.000
	Script	20.76	1,474	.000
Remediability	*Condition*	21.00	3,474	.000

In order to test the effect of experimental condition on the five factorial components (self-blame, event-focused emotions, prospective emotions, self-soothing, regret) and to evaluate the effects of the three supposed mediators (self accountability, sharing and remediability), five regression analyses were conducted through Mediate Macro [10][2], which allows to estimate and to compare total, direct, and indirect effects in multiple mediator models. Experimental condition, gender and script were coded as dummy variables. The latter two variables were put in the analyses as covariates.

Table 3. Results of factor analysis performed on the emotions

Factor labels	Emotions	Loadings	Percent of variance	Cumulative percent of variance
Self-blame	anger towards oneself	.765	24.79	24.79
	shame	.755		
	guilt	.709		
	humiliation	.669		
Event-focused emotions	anger towards the circumstances	.746	12.91	37.7
	disappointment	.707		
	worry	.602		
	indifference (inverted)	.540		
	frustration	.487		
Prospective emotions	hope	.769	9.19	46.89
	confidence	.701		
	resignation (inverted)	.696		
Self-soothing	consolation	.813	8.76	55.64
	relief	.794		
Regret	Regret_sense of loss (*rimpianto*)	.743	6.39	62.28
	Regret_sorrow (*rammarico*)	.683		

According to the assumptions of mediational model [2]-[16] there is a complete mediation when the total effect is significant and the direct effect is not; a partial mediation appears if the direct effect flags but it is still significant, whereas there is no mediational effect if the direct effect does not vary compared to the total effect. Total and direct effects after controlling for the multiple mediational effects and the covariates effects are reported in Tab. 4[3]. Note that, according to Baron & Kenny model, the Macro calculates also the effect of the independent variable on mediators but, since this effect was already calculated trough ANCOVAs (see Tab.2), it is not reported here. As regards self-blame, results of total effects (R^2=.154; F(5,474) = 17.25; p <.001)

[2] The macro is available on `http://www.afhayes.com/public/mediate.sps`

[3] Space limits did not allow to show Tolerance and VIF tests but their values (ranking from 0.41 and 0.90 and from 1.2 to 1.7, respectively) indicated the absence of multicollinearity.

Table 4. Results of mediational regression analyses on the factorial components, with experimental conditions as independent dummy variables, gender and script as covariates, responsibility, sharing and remediability as mediators

		Total effects			Direct effects s and mediational effects		
		B	t	*p*	B	t	*p*
Self-blame	Constant	.116	1.11	.264	-.499	.216	.021
	All failed	-.381	-3.19	.002	-.212	-1.76	.074
	Retrying possibility	-.198	-1.66	.099	-.134	-1.13	.258
	Both information	-.521	-4.36	.000	-.308	-2.54	.012
	Gender	-.294	-3.48	.001	-.225	-2.75	.006
	Script	.612	7.25	.000	.460	5.30	.000
	Responsibility				.201	6.65	.000
	Sharing				-.015	-.53	.599
	Remediability				-.042	-1.79	.074
Prospective emotions	Constant	-.278	-2.59	.010	-1.47	-6.86	.000
	All failed	-.192	-1.55	.121	-.122	-1.02	.310
	Retrying possibility	.440	3.55	.000	.180	1.54	.123
	Both information	.117	.944	.346	-.071	-.588	.557
	Gender	-.017	-.195	.845	-.057	-.697	.486
	Script	.390	4.46	.000	.265	3.08	.002
	Responsibility				.069	2.31	.021
	Sharing				.030	1.01	.312
	Remediability				.211	9.12	.000
Self-soothing	Constant	-.281	-2.59	.010	-.959	-4.17	.000
	All failed	.116	.924	.356	-.012	-.132	.895
	Retrying possibility	.411	3.28	.001	.290	2.31	.021
	Both information	.671	5.36	.000	.449	3.48	.001
	Gender	.036	.404	.687	.012	.133	.894
	Script	-.073	-.822	.411	-.133	-1.44	.150
	Responsibility				-.036	-1.11	.267
	Sharing				.119	3.82	.000
	Remediability				.094	3.775	.000
Regret	Constant	-.080	-.743	.458	-.514	-2.18	.030
	All failed	-.284	-2.27	.023	-.266	2.02	.044
	Retrying possibility	-.274	-2.19	.023	-.287	-2.23	.026
	Both information	-.398	-3.72	.000	-.390	-2.94	.003
	Gender	.329	3.72	.000	.37	3.89	.000
	Script	.310	3.51	.000	.236	2.49	.013
	Responsibility				.062	1.87	.062
	Sharing				.032	1.02	.309
	Remediability				.012	.499	.618

showed that self-blame decreases in the "all failed" and "both information" conditions. Furthermore, both covariates affected self-blame: females scored higher than males and academic examination scores were higher than those of job interview. After controlling for mediational variables (R^2 =.231; $F(8,471) = 17.66$; $p <.001$), a complete mediation emerged for "all failed" while a partial mediation persisted for the "both information" condition: self-blame emotion increased in function of responsibility and decreased when both types of information are presented. As regard event-focused emotions, results of total effects (R^2=.056; $F(5,474) = 5.57$; $p <.001$) showed only an effect due to the gender, with females scoring higher than males.

Thus, no mediational analysis was performed. As to prospective emotions, results of total effects (R^2=.091; $F(5,474) = 9.49$; $p <.001$) showed that prospective emotions increased in the "possibility of retrying" condition and with the academic examination script. After controlling for mediational variables (R^2=.241; $F(8,471) = 18.69$; $p <.001$), a complete mediation emerged for the "possibility of retrying": prospective emotions increased as a function of remediability and no longer of the independent variable. Furthermore, an effect of responsibility also emerged: the increase of this variable corresponded to an increase of prospective emotions. The results of total effects for self-soothing (R^2=.070; $F(5,474) = 7.167$; $p <.001$) showed that this component increased in function of the "possibility of retrying" and "both information" conditions. After controlling for mediational variables (R^2=.128; $F(8,471) = 8.62$; $p <.001$), a partial mediation of these two variables persisted although there was a robust effect of sharing and remediability. Finally, the results of total effects for regret (R^2=.073; $F(5,474) = 7.43$; $p <.001$) revealed that these emotions decreased in each of the experimental conditions ("all failed", "possibility of retrying" and "both information"). Covariates effects also emerged: males scored higher than females and academic examination scores were higher than those of job interview. After controlling for mediational variables (R^2=.082; $F(8,471) = 5.25$; $p <.001$), no difference emerged in comparison to total effects, indicating the absence of mediation.

6 Conclusion

This study examined the effect of the contextual-information induced reappraisal on modifying the emotional response elicited by failure. We expected that the two types of information presented after failure (the information that many other people also failed in the same task and the information that it was possible to retry the task) would spontaneously induce three different ways to less negatively re-appraise failure: decrease of responsibility, failure sharing perception, and remediability perception, respectively. In turn, reappraisal would temperate the negative emotional impact of failure in the previously described ways.

The results substantially corroborated the hypotheses. The two types of information affected the reappraisal dimensions in conformity with our predictions, and moreover modulated several components of the emotional response via reappraisal mediation. In particular, as we expected, self-blame emotions varied in function of responsibility, and prospective emotions varied in function of remediability, while self-soothing varied in function of remediability and sharing perceptions. However, contextual information acted also directly on emotion groups without reappraisal mediation: the "both information" condition, leading to decrease self-blame and to increase self-soothing, and the "possibility of retrying", allowing to increase self-soothing, dropped but persisted after controlling for mediators, thus indicating that their effects were not completely captured by the re-appraisal dimensions that we had envisaged. This remark is particularly noticeable for regret, whose intensity decreased in presence of both types of information – presented either separately or together – with no mediational effect of reappraisal[4].

On the other hand, the absence of the effects of contextual information on the event-focused emotions suggests that not all the emotions elicited by a failure may be regulated or, at least, that the types of information we had offered acted only on specific emotions and not on others. In particular, they did not affect the group of emotions signaling that the outcome does meet one's expectations. Indeed, this response appear to be almost automatic and represents, in our opinion, the emotional meaning of a failure. So, while the core of the emotional response persists, the emotions needing higher cognitive processing to be generated are also the more susceptible to be modified by new information, thus allowing to put what happened in a different perspective. Other studies are necessary for testing this hypothesis.

With regard to the effects of the covariates, the higher female scores on subjective responsibility, self-blame and event-focused emotions mirror analogous results reported in literature [17]-[23]. As to the higher scores obtained by academic vs. job failure on subjective responsibility, sharing perception, self-soothing, self-blame and prospective emotions, they could be due to participants' greater familiarity with academic rather than work setting – as they are all undergraduate. Since we are going to replicate this study with workers, we will wait for its results to more fully assess the reliability of the present findings.

In conclusion, this study focused on the role played by contextual circumstances in providing people with the possibility of reappraising a negative event in a less harmful way. Our findings suggest that reappraisal may be less effortful and more spontaneous than most emotion regulation literature indicate. Nevertheless they also seem to point out that not all emotions can be regulated, at least not in the effortless and natural way that we investigated in this study.

Acknowledgment. We are grateful to dr. Ivana Spina for collecting the data for this study.

[4] This result, showing among other the absence of relationship between responsibility and regret, is in contrast to most of the literature on regret according to which this emotion derives from the subjective responsibility of a negative outcome, but it is in line with the few studies that show the opposite, namely that the responsibility is not a necessary condition for the genesis of regret [e.g. Matarazzo & Abbamonte 2008].

References

1. Arnold, M.B.: Emotion and personality. Columbia University Press, New York (1960)
2. Baron, R.M., Kenny, D.A.: The moderator-mediator variable distinction in social psychological research: Conceptual, strategic, and statistical considerations. Journal of Personality and Social Psychology 51, 1173–1182 (1986)
3. Brunstein, J.C., Gollwitzer, P.M.: Effects of failure on subsequent performance: the importance of self-defining goals. Journal of Personality and Social Psychology 70, 395–407 (1996)
4. Carver, C.S., Scheier, M.F., Weintraub, J.K.: Assessing coping strategies: A theo-retically based approach. Journal of Personality and Social Psychology 56, 267–283 (1989)
5. Folkman, S., Lazarus, R.S.: Manual for the ways of coping questionnaire. Consulting Psychologists Press, Palo Alto (1988)
6. Freud, S.: Inhibitions, symptoms, anxiety (A. Strachey, Transl. and J. Strachey, Ed.). Norton, New York (1926/1959)
7. Goldin, P.R., McRae, K., Ramel, W., Gross, J.J.: The neural bases of emotion regulation: Reappraisal and suppression of negative emotion. Biological Psychiatry 63, 577–586 (2008)
8. Gross, J.J.: The emerging field of emotion regulation: An integrative review. Review of General Psychology 2, 271–299 (1998)
9. Gyurak, A., Gross, J.J., Etkin, A.: Explicit and implicit emotion regulation: A dual-process framework. Cognition and Emotion 25, 400–412 (2011)
10. Hayes, A.F., Preacher, K.J.: Statistical mediation analysis with a multicategorical independent variable (White paper) (2013), http://www.afhayes.com/ (retrieved)
11. John, O.P., Gross, J.J.: Individual differences in emotion regulation strategies. In: Gross, J.J. (ed.) Hand-book of Emotion Regulation, pp. 351–372. Guilford, New York (2007)
12. Lazarus, R.S.: Psychological stress and the coping process. McGraw Hill, New York (1966)
13. Lazarus, R.S., Alfert, E.: Short-circuiting of threat by experimentally altering cognitive appraisal. Journal of Abnormal and Social Psychology 69, 195–205 (1964)
14. Lazarus, R.S., Folkman, S.: Stress, appraisal, and coping. Springer, New York (1984)
15. Matarazzo, O., Abbamonte, L.: Regret, choice, and outcome. International Journal of Humanities and Social Sciences 3, 464–472 (2008)
16. MacKinnon, D.P., Fairchild, A.J., Fritz, M.S.: Mediation analysis. Annual Review of Psychology 58, 593–614 (2007)
17. McRae, K., Ochsner, K.N., Mauss, I.B., Gabrieli, J.D.E., Gross, J.J.: Gender differences in emotion regulation: An fMRI study of cognitive reappraisal. Group Processes and Intergroup Relations 11, 143–162 (2008)
18. McRae, K., Ciesielski, B., Gross, J.J.: Unpacking Cognitive Reappraisal: Goals, Tactics, and Outcomes. Emotion 12, 250–255 (2012)
19. Ochsner, K.N., Ray, R.R., Cooper, J.C., Robertson, E.R., Chopra, S., Gabrieli, J.D.E., Gross, J.J.: For better or for worse: Neural systems supporting the cognitive down-and up-regulation of negative emotion. Neuroimage 23, 483–499 (2004)
20. Shiota, M.N., Levenson, R.W.: Turn Down the Volume or Change the Channel? Emotional Effects of Detached Versus Positive Reappraisal. Journal of Personality and Social Psychology 103, 416–429 (2012)
21. Speisman, J.C., Lazarus, R.S., Mordkoff, A.M., Davison, L.A.: Experimental reduction of stress based on ego-defense theory. Journal of Abnormal and Social Psychology 68, 367–380 (1964)

22. Smith, C.A., Ellsworth, P.C.: Patterns of appraisal and emotion related to taking an exam. Journal of Personality and Social Psychology 52, 475–488 (1987)
23. Thomsen, D.K., Mehlsen, M.Y., Viidik, A., Sommerlund, B., Zachariae, R.: Age and gender differences in negative affect- is there a role for emotion regulation? Personality and Individual Differences 38, 1935–1946 (2005)
24. Williams, L.E., Bargh, J.A., Nocera, C.C., Gray, J.R.: The unconscious regulation of emotion: Nonconscious reappraisal goals modulate emotional reactivity. Emotion 9(6), 847–854 (2009)

Deciding with (or without) the Future in Mind: Individual Differences in Decision-Making

Marina Cosenza*, Olimpia Matarazzo, Ivana Baldassarre, and Giovanna Nigro

Second University of Naples, Viale Ellittico,
31 – 81100 Caserta, Italy
{marina.cosenza,olimpia.matarazzo,ivana.balsassarre,
giovanna.nigro}@unina2.it

Abstract. The aim of this study was to examine the influence of propensity to risk taking, impulsivity, and present versus future orientation in decision-making under ambiguity. One hundred and four healthy adults were administered the computer versions of the Iowa Gambling Task (IGT) and the Balloon Analogue Risk Task (BART). They then completed the Barratt Impulsiveness Scale (BIS-11) and the Consideration of Future Consequences Scale (CFC-14). Results indicated that high scores on the BIS-11 Non-Planning impulsivity scale, the CFC-14 Immediate scale, and the BART result in poorer performance on the IGT. In addition, the results of regression analysis showed also that the BART total score was the most powerful predictor of performance on the IGT. The study revealed that individuals who are more prone to risk, less likely to plan ahead carefully, and more oriented to the present, rather than to the future, performed worse on the IGT.

Keywords: Decision-making, impulsivity, risk taking, future orientation.

1 Introduction

In the mid-19th century John Martyn Harlow [25-26] described the case of Phineas Gage, a railroad construction worker whose frontal lobe was damaged during a strange accident with a tamping iron. Before the accident, Phineas Gage was a man of normal intelligence, active and persistent in executing his plans of operation. He was responsible, sociable, and popular among peers and friends. After receiving treatment and care, Mr. Gage was able to recover from his physical injuries, but became "fitful, irreverent, indulging at times in the grossest profanity (which was not previously his custom), manifesting but little deference for his fellows, impatient of restraint or advice when it conflicts with his desires, at times pertinaciously obstinate, yet capricious and vacillating, devising many plans of future operations, which are no sooner arranged than they are abandoned in turn for others appearing more feasible" [26]. The profound personality changes caused co-friends and acquaintances to say that he was "no longer Gage".

* To contact the authors, please write to: Marina Cosenza, Department of Psychology – Second University of Naples, Viale Ellittico, 31 – 81100 Caserta, Italy.

S. Bassis et al. (eds.), *Recent Advances of Neural Network Models and Applications*,
Smart Innovation, Systems and Technologies 26,
DOI: 10.1007/978-3-319-04129-2_44, © Springer International Publishing Switzerland 2014

In 1994 the amazing case of Phineas Gage was reconsidered by Damasio, Grabowski, Frank, Galburda, and Damasio [15]. These authors reconstituted the accident by relying on measurements taken from Gage's skull and concluded that the most likely placement of Gage's lesion included the Ventromedial region of the prefrontal cortex, bilaterally[1]. In studying the case of Mr. Gage and analyzing other similar cases (patients with damage to the VM prefrontal cortex), it was observed that after the brain damage these patients showed difficulties in expressing emotion and in experiencing feelings in appropriate situations, in planning their workday and future, and abnormalities in decision-making [8], [14], [16], [19]. On the basis of these observations, Damasio and colleagues proposed the Somatic Marker Hypothesis (SMH) [14], [16], one of the most influential conceptualization of how emotions are involved in deciding in terms of neural architecture.

In brief, the Somatic Marker Hypothesis (SMH) proposes that decision-making is a process that depends on emotion. Emotional experience may remain at the unconscious level or not. "The central feature of the SMH is not that non-conscious biases accomplish decisions in the absence of conscious knowledge of a situation, but rather that emotion-related signals assist cognitive processes even when they are non-conscious" (p. 159) [9]. According to the SMH, somatic markers are represented and regulated in the emotion circuitry of the brain, particularly in the ventromedial prefrontal cortex (VMPFC), that associates implicitly represented affective information with explicit representations of potential actions or outcomes. Empirical support for the SMH comes from studies using the Iowa Gambling Task (IGT) [7], [11], that was first developed to assess and quantify the decision-making defects of neurological patients by simulating real-life decision in conditions of reward and punishment and of uncertainty.

In the IGT participants make a series of choices from a set of four computerized 'decks of cards'. The four decks of cards are labeled A, B, C and D. Every card in decks A and B results in a $100 win and for each selection from deck C and D participants win $50. At the beginning of the task participants are given a loan of $2000 and asked to play with the aim of earning as much. Deck A and deck B (disadvantageous decks) yield large immediate monetary gains but larger monetary losses in the long-term, whereas deck C and deck D (advantageous decks) result in small immediate monetary gains but smaller long-term losses. So, the decks of the IGT differ in terms of long-term outcome, as well as in terms of punishment frequency. Playing mostly from disadvantageous decks leads to an overall loss, while playing from advantageous decks leads to an overall gain. The players cannot predict when a penalty will occur, nor calculate with precision the net gain or loss from each deck. Because it is impossible to calculate the best option from the beginning of the task, players have to learn to avoid bad decks by following their feeling and hunches, and by using the feedback they get after each choice.

[1] More recently, Van Horn, Irimia, Torgerson, Chambers, Kikinis, and Toga [38] found that while considerable damage was, indeed, localized to the left frontal cortex, the impact on measures of network connectedness between directly affected and other brain areas was profound, widespread, and a probable contributor to both the reported acute as well as long-term behavioral changes.

Since in a standard administration of the task there are 100 trials that are divided in five blocks of 20 cards, the most common method for scoring the IGT is to calculate net scores from individual blocks of trials. For each block and for all task the net score is equal to [(Deck C + Deck D) - [(Deck A + Deck B)]. A positive net score indicates that decision-making performance on the IGT was advantageous. A negative net score indicates that the decision-making performance on the IGT was disadvantageous [5].

Studies using the IGT on neurological or psychiatric patients provide strong support to the SMH: Compared to healthy controls, patients with ventromedial prefrontal cortex damage and drug addicts show "myopia" for the future consequences. They persist in making disadvantageous choices despite the rising losses associated with them [4]. However, research on healthy individuals has indicated that a substantial number of participants violate the assumption that healthy participants prefer the good decks over the bad decks [36] (for reviews see also [18] and [24])[2]. Interestingly, some studies have showed that in most healthy participants decision-making is guided by the frequency of gain and losses, rather than by the advantageousness or disadvantageousness of a deck of cards [13], [32], [39]. There is also growing evidence that many healthy individuals apply a "win-stay, lose-shift" strategy, as suggested by Lin et al. [32], and that their behavior is not driven by long-term outcomes expectancies [29]. These results seem to contradict the assumption that while neurological or clinical populations should perform badly on the IGT, normal populations should perform quite well on it [7].

Given that performance of healthy participants is characterized by considerable variability, it may be that their performance simply reflects individual differences. As Buelow and Suhr [12] have recently pointed out, "Overall, the results of the few studies that have explored personality correlates of IGT performance in nonclinical samples suggest that underlying personality characteristics, independent of a psychological disorder, mental disorder, or frontal lobe dysfunction, may impact performance on the IGT" (p. 109). Although contradictory findings have been reported, sensitivity to reward and punishment, propensity to risk taking, and trait impulsivity can bias IGT performance in normal population (see among others [17], [21], [24], [33], [37]). To paraphrase Bechara [3], now the most challenging question seems to be the following: Why do (even) normal participants show "myopia" for the future? Why can they not "foresee the future"? Why are they insensitive to the future consequences of their actions?

In an attempt to address these issues, we investigated the role of risk taking, impulsivity, and present orientation *versus* future orientation in decision-making in normal individuals.

---

[2] Bechara and Damasio [6] have found that, about 30% healthy participants showed impairment on the IGT. Glicksohn, Naor-Ziv and Leshem [23] have found that 46% healthy female undergraduates exhibited poor performance on the IGT task, and Glicksohn and Zilberman [24] have shown that roughly 40% of male participants exhibited poor performance on the task.

2 Method

2.1 Participants

One hundred and four healthy adults (41 men, 63 women), with ages ranging from 18 to 60 years (*M* = 32.13; *SD* = 12.24), took part in this study. Since substance and/or alcohol dependence, as well as addiction to gambling were found associated with poor decision-making [4] (for reviews see [2], [12]), exclusion criteria were addiction to gambling, substance and alcohol dependence[3]. We recruited participants from the local area surrounding Second University of Naples.

All participants were administered the computer versions of the Iowa Gambling Task (IGT) [5], [7], and the Balloon Analogue Risk Task (BART) [30], a behavioral measure of propensity for risk taking. They then completed the Italian versions of the Barratt Impulsiveness Scale (BIS-11) [20], a self-report measure of impulsivity, and the Consideration of Future Consequences Scale (CFC-14) [28], that assess the extent to which people consider the potential distant outcomes of their current behaviors and are influenced by those potential outcomes.

2.2 Instruments

For the present study we used the computerized version of the IGT and the Balloon Analogue Risk Task. The BART is a computerized, laboratory-based measure of risk taking that involves actual risky behavior for which, similar to real-world situations, riskiness is rewarded up until a point at which further riskiness results in poorer outcomes. The BART task consists of different balloons that have to be pumped up by participants. Each pump inflates the balloon. With each pump, 5 cents are accrued in a temporary reserve, but after every pump the balloon may explode. In such a case, all money in temporary bank is lost. The participants can stop pumping and accumulate their earnings in a permanent bank. After each balloon explosion or money collection, the participant's exposure to that balloon ends, and a new balloon appears until a total of 90 balloons (i.e., trials) has been completed. The 90 trials comprise 3 different balloon types (i.e., blue, yellow, and orange). Each balloon color has a different probability of exploding [30]. The total score on the BART is the average number of pumps of unexploded balloons (Adj BART).

The Barratt Impulsiveness Scale (BIS-11) [34] is a 30-item self-rating questionnaire designed to measure impulsiveness. Each item is measured on a 4-point Likert scale, with no available neutral response. The BIS-11 assess three components of impulsivity: Motor Impulsiveness (acting without thinking and lack of perseverance), Attentional (or Cognitive) Impulsiveness (not focusing on the task at hand), and Non-Planning Impulsiveness (not planning and thinking carefully).

---

[3] Preliminary, participant completed: a) The Alcohol Use Disorders Identification Test (AUDIT) [1], a 10 items designed to identify drinkers at risk for alcohol abuse and dependence; b) the Drug Abuse Screening Test (DAST-10) [35], a screening tool that assesses drug use behaviors in the last year; c) the South Oaks Gambling Screen (SOGS) [31], a sensitive measure of gambling severity. Inclusion criteria were AUDIT scores < 8, DAST-10 scores $= 0$, and SOGS scores ≤ 2.

The Consideration of Future Consequences Scale (CFC-14) [27-28][4] is a 14-item measure that aims to measure individual differences in the extent to which people weigh the immediate as opposed to distant implications of current behaviors and events. Responses are made with a Likert-type scale ranging from 1 to 7. The CFC-14 contains two subscales, one tapping Consideration of Immediate Consequences (CFC-I), the other tapping Consideration of Future Consequences (CFC-F).

3 Results

All data analyses were conducted using SPSS 15.0. The alpha level was set at $p = .05$.

Performance on the IGT was assessed in the standard manner using net scores, measured by subtracting the total number of disadvantageous deck choices from total advantageous selections.

Since the maximum net score of any patients with damage to the VM prefrontal cortex was below 10, performance with net scores <10 reflects decisions that are within the range of VM patients (i.e. impaired), whereas performance with net scores >10 reflects decision within the normal range (i.e. not impaired) [5], [9]. Our results indicated that 37,5% of the participants (43.9% of men and 33.3% of women) exhibited impaired performance on the task (Net Total score ≤10).

Pearson correlation coefficients and partial correlations were calculated to examine the relations between IGT score, BART score and ratings of the self-report scales. The learning process was evaluated using a repeated measures ANOVA with five points of measurement (block 1–5). To investigate the relative contribution of the BART and the self-report measures to behavioral decision, the significant scales of the correlation analysis were added as independent variables in a linear regression model with the IGT NET raw total score being the criterion variable. Additionally, age, gender and years of education were included in the stepwise regression analysis.

First-order correlations between all variables are displayed in Table 1.

Table 1. Pearson correlation coefficients among all variables

	1	2	3	4	5	6
1. BART						
2. Attentional Impulsiveness	-.031					
3. Motor Impulsiveness	.048	.379**				
4. Non-Planning Impulsiveness	-.027	.398**	.343**			
5. CFC-14 Immediate	-.008	.164	.271**	.139		
6. CFC-14 Future	-.053	-.024	-.255**	-.381**	-.173	
7. IGT NET total	-.359**	-.222*	-.223*	-.270**	-.234*	.163

* $p < 0.05$; ** $p < 0.01$

As can be seen, scores on the IOWA gambling task (NET Total) were significantly correlated with the BART scores, all the BIS-11 scales, and the CFC-14 Immediate scale.

Furthermore, significant correlations were found between BIS-11 Motor Impulsiveness scale and both CFC-14 scales, and between the BIS-11 Non-Planning

[4] We are grateful to Prof. Alan J. Strathman, who sent us the CFC-14, when it was still in press, and other precious material.

Impulsiveness scale and the CFC-14 Future scale. After partialling out BIS-11 and CFC-14 scores, along with gender, age and education, the negative association between the two behavioral measures (IGT and BART) remained still significant (r =.389; p< .001).

Results of the repeated measures ANOVA analysis proved that participants learned to avoid the risky decks over time ($F_{4, 412}$ = 13.64; p<.001, η^2_p = .12) (see Figure 1).

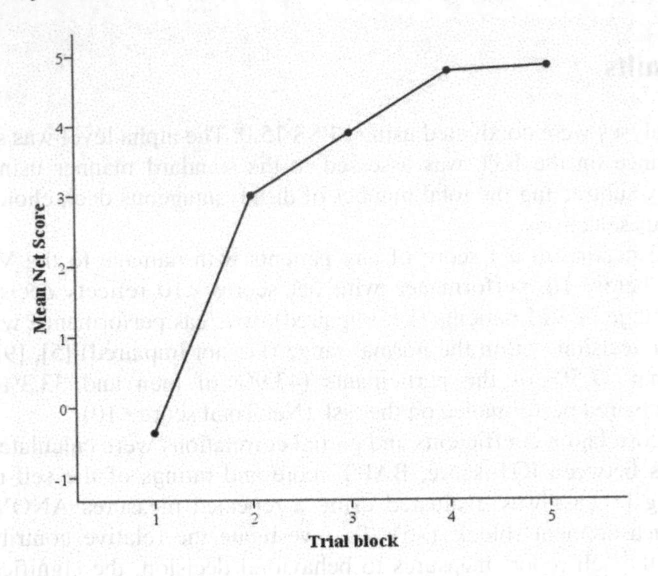

Fig. 1. Mean net score performance across the 5 blocks of 20 trials

Table 2. Summary of hierarchical regression analysis

Variable	B	R^2	ΔR^2	β	t	p
Step 1						
Gender	2.058			.053	.538	.592
Age	.282			.180	1.821	.072
Education	2.799	.040	.040	.093	.943	.348
Step 2						
Gender	1.839			.047	.495	.622
Age	.261			.167	1.733	.086
Education	2.148			.072	.744	.459
Non-Planning Impulsiveness	-1.095	.105	.065	-.257	-2.686	.008
Step 3						
Gender	1.145			.029	.315	.753
Age	.323			.207	2.170	.032
Education	2.130			.071	.755	.452
Non-Planning Impulsiveness	-.952			-.223	-2.367	.020
CFC-14 Immediate	-.540	.156	.051	-.232	-2.435	.017
Step 4						
Gender	.742			.019	.220	.826
Age	.286			.183	2.064	.042
Education	2.119			.071	.810	.420
Non-Planning Impulsiveness	-.999			-.234	-2.676	.009
CFC-14 Immediate	-.536			-.230	-2.604	.011
BART	-1.020	.281	.125	-.355	-4.107	.000

Finally, the linear regression model indicated that high performance on the IGT was positively associated with age, and lower scores on the BART, the BIS-11 Non-Planning Impulsiveness scale, and the CFC-14 Immediate scale. The overall model explained nearly a third part of the total variance of the IGT performance ($R^2 = .281$; $F_{6,\,97} = 6.33$; $p < .001$). Results of hierarchical regression analysis are reported in Table 2.

4 Conclusion

The present study examined the influence of propensity to risk taking, impulsivity, and present *versus* future orientation in IGT performance in healthy individuals. Results indicated that high scores on the BIS-11 Non-Planning impulsiveness scale, the CFC-14 Immediate scale, and the BART result in poor performance on the IGT. In addition, the results of regression analysis showed that the BART total score was the most powerful predictor of performance on the IGT.

The study revealed that individuals who are more prone to risk, less likely to plan ahead carefully, and more oriented to the present, rather than to the future, performed worse on the Iowa Gambling Task. Besides, the results indicated that older participants outperformed young participants.

These findings add further evidence that trait impulsivity is associated with poor decision-making [17], [22], [40], and clearly indicate that propensity to risk taking, as measured by the Balloon Analogue Risk Task, is a powerful predictor of impaired performance on the IGT. The observed association between IGT and BART is in line with the study of Upton et al. [37], who found that IGT and BART performance were related, but only in the later stages of the IGT, and only in participants with low trait impulsivity. However, the negative association between IGT and BART scores we found after partialling out BIS-11 and CFC-14 scores, along with demographic variables, represents a novel finding, indicating that the higher the propensity to risk taking, the poorer the decision-making, independently on impulsivity and future time perspective.

Taken together, the results of our research give further support to the general assumption that underlying personality characteristics impact performance on the IGT and demonstrate that more pronounced risk taking tendencies, associated with higher impulsivity and higher concern with immediate consequences of behavior, foster "myopic" decision-making in normal individuals.

References

1. Babor, T.F., Higgins-Biddle, J.C., Saunders, J.B., Montiero, M.G.: The Alcohol Use Disorders Identification Test: Guidelines of Use in Primary Care, 2nd edn. World Health Organization, Geneva (2001)
2. Barry, D., Petry, N.M.: Predictors of Decision-Making on the Iowa Gambling Task: Independent Effects of Lifetime History of Substance Use Disorders and Performance on the Trail Making Test. Brain & Cognition 66, 243–252 (2008)
3. Bechara, A.: Risky Business: Emotion, Decision-Making, and Addiction. Journal of Gambling Studies 19, 23–51 (2003)

4. Bechara, A.: Decision Making, Impulse Control and Loss of Willpower to Resist Drugs: A Neurocognitive Perspective. Nature Neuroscience 8, 1458–1463 (2005)
5. Bechara, A.: Iowa Gambling Task Professional Manual. Psychological Assessment Resources, Lutz (2007)
6. Bechara, A., Damasio, H.: Decision-Making and Addiction (Part I): Impaired Activation of Somatic States in Substance Dependent Individuals When Pondering Decisions with Negative Future Consequences. Neuropsychologia 40, 1675–1689 (2002)
7. Bechara, A., Damasio, A.R., Damasio, H., Anderson, S.W.: Insensitivity to Future Consequences Following Damage to Human Prefrontal Cortex. Cognition 50, 7–15 (1994)
8. Bechara, S., Damasio, H., Tranel, D., Anderson, S.W.: Dissociation of Working Memory from Decision Making within the Human Prefrontal Cortex. The Journal of Neuroscience 18, 428–437 (1998)
9. Bechara, A., Damasio, H., Tranel, D., Damasio, A.R.: The Iowa Gambling Task and the Somatic Marker Hypothesis: Some Questions and Answers. Trends in Cognitive Science 9, 159–162 (2005)
10. Bechara, A., Martin, E.M.: Impaired Decision Making Related to Working Memory Deficits in Individuals with Substance Addictions. Neuropsychology 18, 152–162 (2004)
11. Bechara, A., Tranel, D., Damasio, H.: Characterization of the Decision-Making Deficit of Patients with Ventromedial Prefrontal Cortex Lesions. Brain 123, 2189–2202 (2000)
12. Buelow, M.T., Suhr, J.A.: Construct Validity of the Iowa Gambling Task. Neuropsychology Review 19, 102–114 (2009)
13. Caroselli, J.S., Hiscock, M., Scheibel, R.S., Ingram, F.: The Simulated Gambling Paradigm Applied to Young Adults: An Examination of University Students' Performance. Applied Neuropsychology 13, 203–212 (2006)
14. Damasio, A.R.: Descartes' Error: Emotion, Reason, and the Human Brain. Grosset/Putnam, New York (1994)
15. Damasio, H., Grabowski, T., Frank, R., Galburda, A.M., Damasio, A.R.: The Return of Phineas Gage: Clues Aboutthe Brain from the Skull of a Famous Patient. Science 264, 1102–1104 (1994)
16. Damasio, A.R., Tranel, D., Damasio, H.: Somatic Markers and the Guidance of Behavior: Theory and Preliminary Testing. In: Levin, H.S., Eisenberg, H.M., Benton, A.L. (eds.) Frontal Lobe Function and Dysfunction, pp. 217–229. Oxford University Press, New York (1991)
17. Davis, C., Patte, K., Tweed, S., Curtis, C.: Personality Traits Associated with Decision-Making Deficits. Personality and Individual Differences 42, 279–290 (2007)
18. Dunn, B.D., Dalgleish, T., Lawrence, A.D.: The Somatic Marker Hypothesis: A Critical Evaluation. Neuroscience and Biobehavioral Reviews 30, 239–271 (2006)
19. Eslinger, P.J., Damasio, A.R.: Severe Disturbance of Higher Cognition after Bilateral Frontal Lobe Ablation: Patient EVR. Neurology 35, 1731–1741 (1985)
20. Fossati, A., Ceglie, A.D., Acqarini, E., Barratt, E.S.: Psychometric Properties of an Italian Version of the Barratt Impulsiveness Scale-11 (BIS-11) in Nonclinical Subjects. Journal of Clinical Psychology 57, 815–828 (2001)
21. Franken, I.H.A., Muris, P.: Individual Differences in Decision-Making. Personality and Individual Differences 39, 991–998 (2005)
22. Franken, I.H.A., van Strien, J.W., Nijs, I., Muris, P.: Impulsivity is Associated with Behavioral Decision-Making Deficits. Psychiatry Research 158, 155–163 (2008)
23. Glicksohn, J., Naor-Ziv, R., Leshem, R.: Impulsive Decision Making: Learning to Gamble Wisely? Cognition 105, 195–205 (2007)

24. Glicksohn, J., Zilberman, N.: Gambling on Individual Differences in Decision Making. Personality and Individual Differences 48, 557–562 (2010)
25. Harlow, J.M.: Passage of an Iron Rod through the Head. Boston Medical and Surgical Journal 39, 389–393 (1848)
26. Harlow, J.M.: Recovery from the Passage of an Iron Bar through the Head. Publications of the Massachusetts Medical Society 2, 327–347 (1868)
27. Joireman, J., Balliet, D., Sprott, D., Spangenberg, E., Schultz, J.: Consideration of Future Consequences, Ego-Depletion, and Self-Control: Support for Distinguishing between CFC-Immediate and CFC-Future Sub-scales. Personality and Individual Differences 45, 15–21 (2008)
28. Joireman, J., Shaffer, M.J., Balliet, D., Strathman, A.: Promotion Orientation Explains Why Future-Oriented People Exercise and Eat Healthy: Evidence from the Two-Factor Consideration of Future Consequences-14 Scale. Personality and Social Psychology Bulletin 38(10), 1272–1287 (2012)
29. Kloeters, S., Bertoux, M., O'Callaghan, C., Hodges, J.R., Hornberger, M.: Money for nothing – Atrophy Correlates of Gambling Decision Making in Behavioural Variant Frontotemporal Dementia and Alzheimer's Disease. NeuroImage: Clinical 2, 263–272 (2013)
30. Lejuez, C.W., Read, J.P., Kahler, C.W., Richards, J.B., Ramsey, S.E., Stuart, G.L., Strong, D.R., Brown, R.A.: Evaluation of a Behavioral Measure of Risk Taking: the Balloon Analogue Risk Task (BART). Journal of Experimental Psychology: Applied 8, 75–84 (2002)
31. Lesieur, H.R., Blume, S.B.: The South Oaks Gambling Screen (SOGS): A New Instrument for the Identification of Pathological Gamblers. American Journal of Psychiatry 144(9), 1184–1188 (1987)
32. Lin, C.-H., Chiu, Y.-C., Lee, P.-L., Hsieh, J.-C.: Is Deck B a Disadvantageous Deck in the Iowa Gambling Task? Behavioral and Brain Functions 3, 1–10 (2007)
33. Mardaga, S., Hansenne, M.: Personality and Skin Conductance Responses to Reward and Punishment. Influence on the Iowa Gambling Task Performance. Journal of Individual Differences 33, 17–23 (2012)
34. Patton, J.H., Stanford, M.S., Barratt, E.S.: Factor Structure of the Barratt Impulsiveness Scale. Journal of Clinical Psychology 51, 768–774 (1995)
35. Skinner, H.A.: The drug abuse screening test. Addictive Behaviors 7, 363–371 (1982)
36. Steingroever, H., Wetzels, R., Horstmann, A., Neumann, J., Wagenmakers, E.: Performance of Healthy Participants on the Iowa Gambling Task. Psychological Assessment 25, 180–193 (2013)
37. Upton, D.J., Bishara, A.J., Ahn, W.-Y., Stout, J.C.: Propensity for Risk Taking and Trait Impulsivity in the Iowa Gambling Task. Personality and Individual Differences 50, 492–495 (2011)
38. Van Horn, J.D., Irimia, A., Torgerson, C.M., Chambers, M.C., Kikinis, R., Toga, A.W.: Mapping Connectivity Damage in the Case of Phineas Gage. PLoS One 7(5), e37454, 1–24 (2012)
39. Wilder, K.E., Weinberger, D.R., Goldberg, T.E.: Operant Conditioning and the Orbitofrontal Cortex in Schizophrenic Patients: Unexpected Evidence for Intact Functioning. Schizophrenia Research 30, 169–174 (1998)
40. Zermatten, A., Van der Linden, M., d'Acremont, M., Jermann, F., Bechara, A.: Impulsivity and Decision Making. Journal of Nervous and Mental Disease 193, 647–650 (2005)

Author Index

Printed in the United States
By Bookmasters